CYTOCHROMES P-450 AND b$_5$

Structure, Function, and Interaction

ADVANCES IN EXPERIMENTAL MEDICINE AND BIOLOGY

Recent Volumes in this Series

Volume 50
ION-SELECTIVE MICROELECTRODES
Edited by Herbert J. Berman and Normand C. Hebert • 1974

Volume 51
THE CELL SURFACE: Immunological and Chemical Approaches
Edited by Barry D. Kahan and Ralph A. Reisfeld • 1974

Volume 52
HEPARIN: Structure, Function, and Clinical Implications
Edited by Ralph A. Bradshaw and Stanford Wessler • 1975

Volume 53
CELL IMPAIRMENT IN AGING AND DEVELOPMENT
Edited by Vincent J. Cristofalo and Emma Holečková • 1975

Volume 54
BIOLOGICAL RHYTHMS AND ENDOCRINE FUNCTION
Edited by Laurence W. Hedlund, John M. Franz, and
Alexander D. Kenny • 1975

Volume 55
CONCANAVALIN A
Edited by Tushar K. Chowdhury and A. Kurt Weiss • 1975

Volume 56
BIOCHEMICAL PHARMACOLOGY OF ETHANOL
Edited by Edward Majchrowicz • 1975

Volume 57
THE SMOOTH MUSCLE OF THE ARTERY
Edited by Stewart Wolf and Nicholas T. Werthessen • 1975

Volume 58
CYTOCHROMES P-450 and b_5: Structure, Function, and Interaction
Edited by David Y. Cooper, Otto Rosenthal, Robert Snyder,
and Charlotte Witmer • 1975

Volume 59
ALCOHOL INTOXICATION AND WITHDRAWAL: Experimental Studies II
Edited by Milton M. Gross • 1975

Volume 60
DIET AND ATHEROSCLEROSIS
Edited by Cesare Sirtori, Giorgio Ricci, and Sergio Gorini • 1975

CYTOCHROMES P-450 AND b₅
Structure, Function, and Interaction

Edited by

David Y. Cooper and Otto Rosenthal

Harrison Department of Surgical Research
University of Pennsylvania School of Medicine
Philadelphia, Pennsylvania

and

Robert Snyder and Charlotte Witmer

Department of Pharmacology
Thomas Jefferson University
Philadelphia, Pennsylvania

SPRINGER SCIENCE+BUSINESS MEDIA, LLC

Library of Congress Cataloging in Publication Data

Philadelphia Conference on Heme Protein P-450, 2d, 1974.
  Cytochromes P-450 and b$_5$.

  (Advances in experimental medicine and biology; v. 58)
  Includes bibliographical references and index.
  1. Cytochrome P-450—Congresses. 2. Cytochrome b5—Congresses. I. Cooper,
David Y. II. Title. III. Series. [DNLM: 1. Cytochrome P-450—Congresses. 2.
Cytochromes—Congresses. W1 AD559 v. 58 1974 / QU135 P544 1974]
QP671.C83P48 1974   574.8'74   75-15625
ISBN 978-1-4615-9028-6   ISBN 978-1-4615-9026-2 (eBook)
DOI 10.1007/978-1-4615-9026-2

Proceedings of the Second Philadelphia Conference on Heme Protein P-450,
held at the Thomas Jefferson University, Philadelphia, Pennsylvania,
April 5-6, 1974

© 1975 Springer Science+Business Media New York
Originally published by Plenum Press, New York 1975
Softcover reprint of the hardcover 1st edition 1975

Preface

P-450 has in common with our nation that it can call Philadelphia its hometown. Yet there are differences, too. The
U.S.A. was born and named in this city. P-450 was first recognized here -- an odd CO-combining pigment without family or function. Japanese workers identified it as an unusual member of the
cytochrome family. Finally, in Philadelphia, its biological function was established and its growth to a global power initiated.

Since discovery of its function as terminal oxidase of the
21-steroid hydroxylase system of adrenocortical microsomes, P-450
has proved to play the same role in a wide variety of other mixed
function oxidase systems involved in biosynthesis and catabolism
of specific cell or body components as well as in the metabolism
of foreign substances entering organisms. P-450-like oxygenating
enzymes appear to be fundamental cellular constituents in most
forms of aerobic organisms. Activation of molecular oxygen and
incorporation of one of its atoms into organic compounds as catalyzed by P-450 enzymes are reactions of vital importance not only
for biosynthesis and degradation of steroid hormones necessary for
sustaining life, but also for metabolic activation or inactivation
of foreign agents such as drugs, food preservatives and additives,
insecticides and carcinogens. Moreover, P-450 linked enzyme systems can either be induced or suppressed by these agents with
significant biological consequences.

Both the wide variety of reactions catalyzed by P-450 and
the versatility of these enzyme systems have attracted increasing
numbers of investigators to this area. Environmentalists, ento-
mologists, nutritional biochemists have joined endocrinologists,
biochemists, biophysicists, biochemical pharmacologists and toxicologists who originally explored this field. The number of investigators has proliferated so rapidly that the earlier students
of P-450 are little acquainted with the new entries in the field.

This is the second time we have taken advantage of the geo-
graphical location of Philadelphia to get together the people
interested in P-450. Both meetings have been successful, not
only scientifically, but socially, also. Therefore, if there
are any profits from this current publication, we plan to set up
a fund to form a society to perpetuate these Philadelphia Con-
ferences, so that the people working with P-450 can meet, exchange
ideas, stimulate new research and enjoy themselves. Such con-
ferences could be held in Philadelphia -- or anywhere else in
the world.

The organizers of this symposium wish to express their
thanks to the following: Dean Robert C. Baldridge and the
College of Graduate Studies for both financial support and use of
the facilities of Thomas Jefferson University; Mr. G. Frederick
Roll whose kind donation, along with that of the Smith Kline and
French Laboratory, helped defray expenses. We also wish to
specially acknowledge the invaluable contribution of Dr. Heinz
Schleyer as Consulting Editor and the editorial work of Mrs.
Beatrice G. Novack, without which publication of this volume
would not have been possible.

David Y. Cooper
Otto Rosenthal
Robert Snyder
Charlotte Witmer

Contents

Official Greetings. xiii
    Robert C. Baldridge

            Session I - Chairman, John Schenkman

Part 1

Partial Purification and Separation of Multiple Forms
    of Cytochrome P-450 and Cytochrome P-448 from
    Rat Liver Microsomes.   1
    Levin, W., Lu, A. Y. H., Ryan, D., West, S.
    Kuntzman, R. and Conney, A. H.

Biochemical Characterization of Highly Purified Cytochrome
    P-450 and Other Components of the Mixed Function
    Oxidase System of Liver Microsomal Membranes. . .   25
    Coon, M. J., van der Hoeven, T. A., Haugen, D. A.,
    Guengerich, F. P., Vermilion, J. L. and Ballou, D. P.

Immunochemical and Compositional Comparison of Cytochromes
    P-450$_{cam}$ of Pseudomonas putida and P-450$_{LM}$ of
    Phenobarbital-Induced Rabbit Liver Microsomes. . .   47
    Dus, K., Litchfield, W. J., Miguel, A. G., van der
    Hoeven, T. A., Haugen, D. A., Dean, W. L. and Coon,
    W. L.

Immunochemical and Functional Similarities and Differences
    among Iron-Sulfur Proteins Involved in Mammalian
    Steroidogenesis.   55
    Baron, J.

General Discussion.   73
    Peisach, Coon, Jerina, Schenkman, Levin, Peterson,
    Ballou, Kamin, Baron, Masters.

Part 2

Comparison of the Induction Course, Biophysical Chemical
    Interactions and Photochemical Action Spectra of
    Phenobarbital- and 3-Methylcholanthrene-Induced
    Hepatic Microsomal P-450.   81
    Cooper, D. Y., Schleyer, H., Thomas, J. H., Vars,
    H. M. and Rosenthal, O.

Cytochrome P-450 in the Activation and Inactivation of
    Carcinogens.   103
    Greim, H., Czygan, P., Garro, A. J., Hutterer, F.,
    Schaffner, F. and Popper, H.

Effect of Cyclic AMP on the Phenobarbital Induced Increase
    in Cytochrome P-450 and Hypertrophy of the Endo-
    plasmic Reticulum of the Rat Liver.   117
    Hutterer, F., Dressler, K., Greim, H., Czygan, P.,
    Schaffner, F. and Popper, H.

Evidence for the Activation of 3-Methylcholanthrene as a
    Carcinogen In Vivo and as a Mutagen In Vitro
    by P$_1$-450 from Inbred Strains of Mice.   127
    Nebert, D. W. and Felton, J. S.

Increased Translation as a Result of Elevated Initiation
    Factor Activity after Administration of 3-
    Methylcholanthrene.   151
    Bresnick, E., Hopkinson, J. and Prichard, P. M.

General Discussion.   165
    Holtzman, Hutterer, Coon, Cooper, Schenkman, Jerina,
    Peisach, Remmer, Schleyer, Cinti, Nebert, Bresnick,
    Narasimhulu, Orrenius.

Session II: Chairman, Daniel Nebert

Part 1

Optical and EPR Studies of Partially Purified Rabbit
    Liver Cytochrome P-450.   175
    Witmer, C., Nehls, P., Krauss, P., Remmer, H. and
    Snyder, R.

Studies on the Spin State of 3-Methylcholanthrene-Induced
    Cytochrome P-450 from Rat Liver.   189
    Stern, J. O., Peisach, E., Peisach, J., Blumberg,
    W. E., Lu, A. Y. H., West, S., Ryan, D. and Levin, W.

An Analysis of the Optical Titrations of the 430 and
        455 nm Chromophores of Ethyl Isocyanide
        Complexes of Mammalian Hepatic Cytochrome P-450 .   203
    Peisach, J.

Implications of Ligand Modified Spectra of Cytochrome
        P-450 Associated with Pregnenolone Synthesis
        in Mitochondria from Corpus Luteum.   213
    Uzgiris, V. I., McIntosh, E. N., Graves, T. and
    Salhanick, H. A.

General Discussion.   229
    Mannering, Witmer, Lu, Snyder, Levin, Nebert,
    Holtzman, Remmer, Vore, Narasimhulu, Stern,
    Kamin, Peisach, Schenkman, Hildebrandt,Coon

Part 2

Studies on the Interaction of Water with Microsomal
        Cytochrome P-450.   239
    Holtzman, J. L.

Drug Metabolism in Isolated Rat Liver Cells   251
    Grundin, R., Moldéus, P., Vadi, H., Orrenius, S.,
    von Bahr, C., Bäckström, D. and Ehrenberg, A.

Role of Phospholipids in Adrenocortical Microsomal
        Hydroxylation Reactions: Activation of Lipid-
        Depleted Microsomal Preparations by Non-Ionic
        Detergents.   271
    Narasimhulu, S.

On the Structure of Putidaredoxin and Cytochrome P-450$_{cam}$,
        and Their Mode of Interaction   287
    Dus, K.

Metabolic Control of Cytochrome P-450$_{cam}$ 311
    Peterson, J. A. and Mock, D. M.

General Discussion.   325
    Stern, Peterson, Dus, Peisach, Bresnick, Orrenius,
    Holtzman, Remmer, Cinti, Thurman, Nebert,
    Narasimhulu, Coon, Lu, Jerina

Session III - Chairman, Charlotte Witmer

Relationship between Microsomal Hydroxylase and Glucuronyl-
    transferase. 335
    Remmer, H., Bock, K. W. and Rexer, B.

A Possible Role of Copper in the Regulation of Heme
    Biosynthesis through Ferrochelatase. 343
    Wagner, G. S. and Tephly, T. R.

Mixed Function Oxidation and Intermediary Metabolism:
    Metabolic Interdependence in the Liver. 355
    Thurman, R. G., Marazzo, D. P. and Scholz, R.

Dissociation of Microsomal Ethanol Oxidation from
    Cytochrome P-450 Catalyzed Drug Metabolism. . . . 369
    Vatsis, K. P. and Schulman, M. P.

General Discussion. 383
    van der Hoeven, Remmer, Mannering, Thurman, Coon,
    Vatsis, Orrenius, Schenkman, Kamin, Bresnick, Wagner,
    Hildebrandt, Thurman.

Session IV : Chairman, Otto Rosenthal

Interaction between Microsomal Electron Transfer
    Pathways. 387
    Schenkman, J. B. and Jansson, I.

Role of Cytochrome b_5 in the NADH Synergism of NADPH-
    Dependent Reactions of the Cytochrome P-450. . . . 405
    Mannering, G. J.

The Role of Cytochrome b_5 in Cytochrome P-450 Enzymes. . . . 435
    Sasame, H. A., Thorgeirsson, S. S., Mitchell, J. R.
    and Gillette, J. R.

Role of Cytochrome b_5 in NADPH- and NADH-Dependent
    Hydroxylation by the Reconstituted Cytochrome
    P-450- or P-448-Containing System. 447
    Lu, A. Y. H., Levin, W., West, S. B., Vore, M.,
    Ryan, D., Kuntzman, R. and Conney, A. H.

The Role of Cytochrome b_5 in Mixed Function Oxidations:
    Effect on Microsomal Binding of the Hemoprotein
    on Hepatic N-Demethylations. 467
    Cinti, D. L. and Ozols, J.

Comparison of Methods to Study Enzyme Induction in Man. . .   485
        Roots, I., Saalfrank, K. and Hildebrandt, A. G.

General Discussion.   503
        Holtzman, Lu, Mannering, Kamin, Lichtenberger,
        Rosenthal, Hildebrandt, Schleyer, Narasimhulu,
        Orrenius, Schenkman, Levin, Sasame, Cinti,
        Lotlikar.

Contributors   511

Author Index   519

Subject Index   537

OFFICIAL GREETINGS

I should like to welcome you on behalf of Thomas Jefferson University. Our institution is celebrating its 150th year, having begun as Jefferson Medical College in 1824. We have been engaged in graduate education for the past quarter century, although the College of Graduate Studies was not established formally until 1969, the year following the adoption of our present name, Thomas Jefferson University. The medical school, which retains the name Jefferson Medical College, is the largest private medical school in the nation; last year 223 incoming students were admitted. In addition we have a College of Allied Health Sciences and, of course, Thomas Jefferson University Hospital.

It is a pleasure to acknowledge the efforts of Drs. David Y. Cooper and Otto Rosenthal, colleagues from our sister institution, the University of Pennsylvania who, along with Drs. Robert Snyder and Charlotte Witmer of Jefferson, have arranged this splendid symposium.

Robert C. Baldrige
Dean of Graduate Studies
Thomas Jefferson University
Philadelphia, Pennsylvania

PARTIAL PURIFICATION AND SEPARATION OF MULTIPLE FORMS OF CYTOCHROME

P-450 AND CYTOCHROME P-448 FROM RAT LIVER MICROSOMES

W. Levin, A. Y. H. Lu, D. Ryan, S. West, R. Kuntzman,
and A. H. Conney
Department of Biochemistry and Drug Metabolism
Hoffman-La Roche, Inc.
Nutley, New Jersey 07110

INTRODUCTION

The liver microsomal mixed-function oxidase system is one of
the most versatile enzyme systems found in mammalian systems.
This membrane-bound multicomponent system consisting of cytochrome
P-450, NADPH-cytochrome c reductase, and a lipid component (Lu and
Coon, 1968; Strobel et al., 1970), is responsible for the oxidative metabolism of a variety of foreign compounds and certain endogenous compounds such as steroids and fatty acids (Conney, 1967).
Despite the growing interest in the elucidation of the molceular
mechanisms of carcinogen and drug metabolism by this enzyme system,
mammalian cytochrome P-450 has not yet been obtained in a homogeneous
state primarily because of its instability to various treatments
and its firm associations with the microsomal membrane. Several
methods have been described for the partial purification of an enzymatically active cytochrome P-450 (Lu and Coon, 1968; Lu and
Levin, 1972: Fujita and Mannering, 1973), but most of the preparations have been contaminated with lipid, NADPH-cytochrome c reductase, cytochrome b_5 or epoxide hydrase. A recent report from
this laboratory (Levin et al., 1974) described the preparation of
a partially purified cytochrome P-450 (from phenobarbital-pretreated
rats) and cytochrome P-448 (from 3-methylcholanthrene-pretreated
rats) which were virtually free of lipid and these contaminating
enzymes. The partially purified hemeproteins were enzymatically
active when recombined with lipid and NADPH-cytochrome c reductase.
A similar preparation of cytochrome P-450 from phenobarbital-treated
rabbits has been described by Van der Hoeven and Coon (1974).

It has become increasingly apparent that the substrate

specificity of this enzyme system resides in the cytochrome P-450
or P-448 fraction (Lu et al., 1972; Lu et al., 1973) and that there
are different forms of this hemeprotein. Although multiple forms
of cytochrome P-450 apparently exist in animals treated with different inducers, a conclusion based on spectral (Sladek and
Mannering, 1966: Alvares et al., 1967) and catalytic (Lu et al.,
1972; Lu et al., 1973) properties of the cytochrome from animals
pretreated with phenobarbital or 3-methylcholanthrene, it has not
been definitively established whether multiple forms of this hemeprotein are present in liver microsomes obtained from the same
animal. Recently, Comai and Gaylor (1973) have separated three
forms of cytochrome P-450 from rat liver by a technique based on
the differential affinities of the hemeproteins for cyanide. The
separation and characterization of different forms of cytochrome
P-450 from the same animal should provide information essential for
an understanding of the mechanisms of microsomal hydroxylation. In
this paper, the partial purification of cytochrome P-450 and P-448
from rat liver is described, and spectral evidence is presented for
the existence of multiple forms of these hemeproteins.

METHODS

Male Long-Evans rats (Blue Spruce Farms, Altamont, New York)
weighing 50-60 g were treated intraperitoneally with phenobarbital
(75 mg/kg/day) or 3-methylcholanthrene (25 mg/kg/day) for 3 days.
Liver microsomes were prepared as previously described (Lu and
Levin, 1972).

Partial Purification of Cytochromes P-450 and P-448. Cytochrome P-450 was first partially purified by the method of Lu and
Levin (1972) using the ionic detergent sodium cholate followed by
ammonium sulfate fractionation and adsorption on calcium phosphate
gel. This preparation of cytochrome P-450, termed Step III, was
purified approximately three-fold (specific content, 5-6 nmoles/mg
protein) as compared to microsomes and was still contaminated with
NADPH-cytochrome c reductase, cytochrome b_5, epoxide hydrase, and
a small amount of lipid (Levin et al., 1972). When this preparation was treated with the nonionic detergent Emulgen 911 (1 mg/
mg protein) and chromatographed on DEAE-cellulose in the presence
of Emulgen, cytochrome P-450 was separated into two fractions.
The first peak, referred to as fraction A, eluted in the column
volume; and the second peak, fraction B, eluted at 100 mM KCl (or
50 mM potassium phosphate buffer, pH 7.7). Both hemeprotein fractions were free of cytochrome b_5 and NADPH-cytochrome c reductase,
which eluted at higher ionic strength. To minimize conversion of
the hemeprotein to cytochrome P-420, the A and B fractions were
routinely prepared by a stepwise elution.

The excess detergent was immediately removed from both

fractions either by passage through a Sephadex LH-20 column
(4 x 15 cm) that was equilibrated and eluted with 10 mM potassium
phosphate buffer (pH 7.7) containing 50% glycerol as described by
Gaylor and Delwiche (1969) or by treatment with Bio-Rad SM-2 beads
for 2 hours at 4°. By either method, approximately 50% of the detergent was removed. No additional Emulgen could be removed by
subjecting the cytochrome fractions to both methods. The cytochrome fractions were dialyzed, concentrated, and centrifuged.
The final samples were stable for at least several months when
stored at -15° under N_2.

The final A and B fractions from either phenobarbital or 3-
methylcholanthrene treated rats contained 1-2% and 3.5-5.0% of the
total cytochrome present in the starting microsomes, respectively.
These values represent the recoveries of the hemeproteins through
the isolation procedure and may not represent the actual amounts
of these fractions present in intact microsomes.

Assay Methods. The concentration of cytochrome P-450 was
determined by the method of Omura and Sato (1964a) using an absorption coefficient of 91 mM^{-1} cm^{-1}. The ethyl isocyanide spectra
were performed using a final concentration of 9 mM ethyl isocyanide.
Cytochrome b_5 was determined from the difference spectrum between
NADH-reduced and oxidized samples using a molar absorption coefficient between 424 and 410 nm of 185 mM^{-1} cm^{-1} (Omura and Sato,
1964b). Since the purified preparations of cytochrome P-450 contained only small amounts of NADH-cytochrome b_5 reductase, cytochrome b_5 was determined after the addition of a crude cytochrome
b_5 reductase preparation (Levin et al., 1972).

Protein was analyzed by the method of Lowry et al., (1951)
using bovine serum albumin as a standard. NADPH-cytochrome c reductase activity was determined by the method of Phillips and
Langdon (1962).

Benzo[a]pyrene hydroxylation and benzphetamine N-demethylation
were determined as previously described (Lu et al., 1972). Napthalene metabolism was assayed as described by Dansette et al., (1974).
Epoxide hydrase was assayed as described by Oesch et al., (1971)
using H^3-styrene oxide as substrate.

RESULTS AND DISCUSSION

Separation of Multiple Forms of Cytochrome P-450 and P-448.
Figure 1 shows the elution profile of cytochrome P-450 from a DEAE-
cellulose column after the microsomes were partially purified
through Step III (Lu and Levin, 1972), treated with Emulgen 911,
and chromatographed in the presence of Emulgen (Levin et al., 1974).

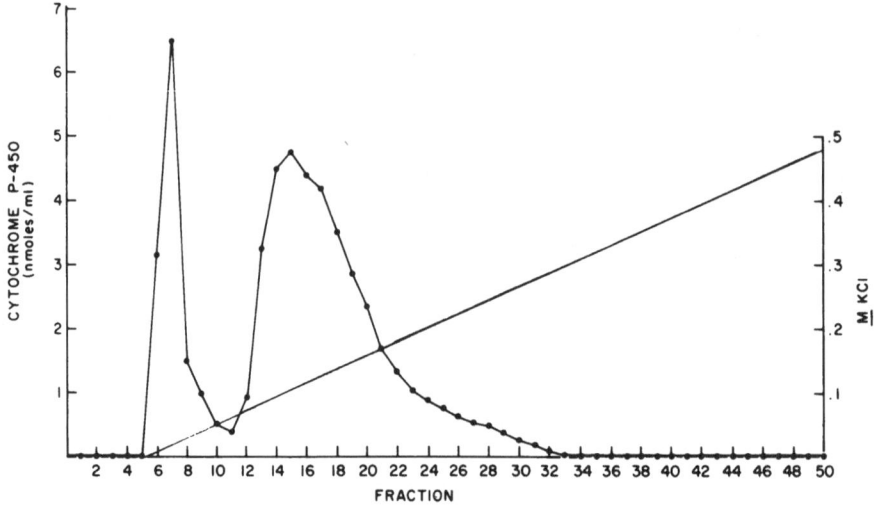

Fig. 1. Elution profile of cytochrome P-450 from a DEAE-
cellulose column. A partially purified cytochrome P-450 (Step
III) preparation was treated with 1 mg Emulgen per mg protein
(100 mg protein, 550 nmoles cytochrome P-450) and applied to a
DEAE-cellulose column equilibrated with 5 mM potassium phosphate
buffer (pH 7.7) containing 20% glycerol and 0.1% Emulgen. The
cytochrome P-450 was eluted with a linear gradient of KCl in the
same buffer mixture.

Cytochrome P-450 eluted as two peaks, one which eluted in the
column volume (fraction A) and one which eluted with 100 mM KCl
(fraction B). Cytochrome b_5 and NADPH-cytochrome c reductase
eluted at higher ionic strength (200-300 mM KCl). The residual
phosphatidylcholine present in the Step III preparation eluted
with the cytochrome P-450 in fraction A as did the epoxide hydrase.
A similar elution pattern was obtained with the Step III preparation of cytochrome P-448 except that the ratio of fraction A to
fraction B was lower than that obtained with cytochrome P-450.

Rechromatography of the two cytochrome fractions on separate
DEAE-cellulose columns revealed that these two hemeprotein fractions
were not interconvertible since each fraction rechromatographed as
a single fraction from a second column. A similar elution profile
of cytochrome P-450 or P-448 was also obtained if the DEAE-cellulose
column was replaced with a CM cellulose column, and the hemeprotein
eluted with potassium phosphate buffer, pH 6.5. The resolution of
the cytochrome P-450 or P-448 into two fractions required the
presence of the nonionic detergent Emulgen 911 since chromatography

of the Step III cholate preparation on DEAE cellulose in the presence or absence of cholate resulted in the elution of the hemeprotein as a single broad peak at higher ionic strength (200-300 mM KCl) with no further purification of the hemeprotein.

In Table I, the specific content of cytochrome P-450, phosphatidylcholine, and other enzymes in the cytochrome fractions is compared to that of microsomes. The cytochrome P-450 or P-448 obtained in the B fractions was purified approximately 5-7 fold as compared to microsomes, while the A fractions had a specific content of cytochrome P-450 or P-448 only 1.5-2 fold higher than the corresponding microsomal values. Fraction A was enriched in epoxide hydrase while the B fractions had negligible levels of this enzyme. The final A and B cytochrome P-450 or P-448 fractions were free of NADPH-cytochrome c reductase and cytochrome b_r. The B fractions obtained from either phenobarbital or 3-methylcholanthrene-treated rats were essentially free of phosphatidylcholine as judged by the absence of ^{14}C-phosphatidylcholine in these fractions when they were obtained from animals injected with ^{14}C-choline. Further proof for the purification of cytochrome P-450 and P-448 with respect to total lipid was obtained by extracting the lipids from cytochrome fractions by the method of Bligh and Dyer (1959) and chromatographing the extracts. These studies revealed that the cytochrome P-450 and P-448 in the B fractions were virtually free of all phospholipids, cholesterol, and triglycerides. A small but detectable amount of fatty acids was still present (less than 3% of that in microsomes, per nmole of hemeprotein). Thus, the more highly purified cytochrome P-450 and P-448 (B fractions) were greater than 99% free of virtually all lipids, per nmole of hemeprotein. In contrast, the Peak A hemeprotein fractions still contained a considerable amount of phospholipid as measured by the presence of ^{14}C-phosphatidylcholine (Table I).

Spectral Properties of Partially Purified Cytochrome P-450 and P-448. A. Absolute Spectra. Figure 2 shows the absolute spectra of the more highly purified cytochrome P-450 and P-448 (Step IV, fraction B) at pH 7.4. The oxidized form of the cytochrome showed a Soret peak at 417-418 nm and absorption bands at 535 and 568 nm--spectral properties which are characteristic of low-spin ferric hemeproteins. EPR studies by Stern et al., (1974) have shown that the fraction B cytochrome P-450 and P-448 are low-spin ferric hemeproteins containing essentially no high-spin species. When the oxidized form was reduced with dithionite, the Soret peak shifted downward to 411-414 nm, and a single broad peak was seen at 542 nm. The CO complex of the reduced cytochrome exhibited absorption maxima at 447-450 nm and 550 nm with a small peak at 423 nm, attributable to small amounts of cytochrome P-420 in the preparations (Levin et al., 1974). The spectra obtained with cytochrome P-448 differed only slightly from those seen with cytochrome P-450 (Table II).

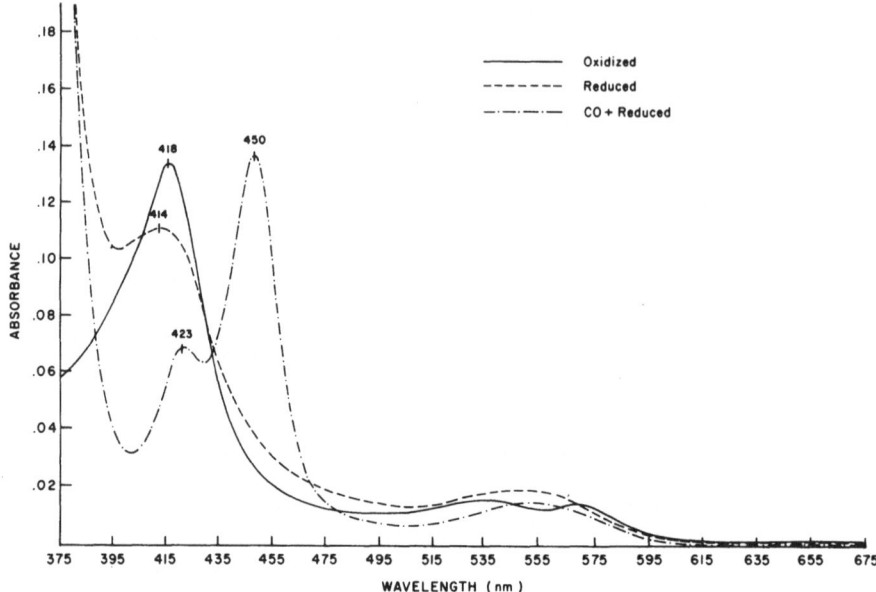

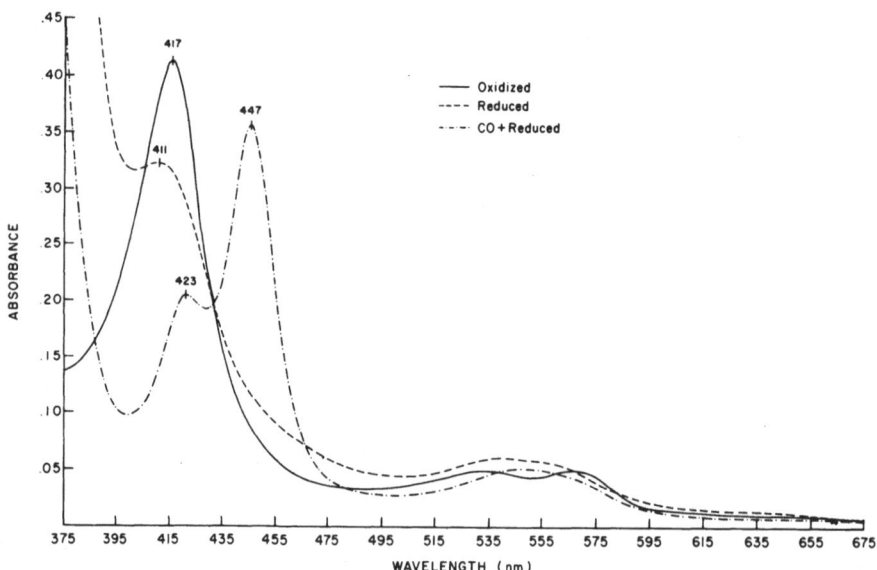

Fig. 2. Absolute spectra of Step IV (fraction B) cytochrome P-450 (top) and cytochrome P-448 (bottom) in 0.1 M potassium phosphate buffer. The hemeprotein concentration of cytochrome P-450 was 1.1 nmoles/ml, and the concentration of cytochrome P-448 was 2.7 nmoles/ml.

TABLE I

Analysis of Enzymes and Phosphatidylcholine in Microsomes and Partially Purified Cytochrome P-450 and P-448

Preparation	Cytochrome P-450 or P-448 (nmoles/mg protein)	Cytochrome b5 (nmoles/mg protein)	NADPH-cytochrome c reductase (units/mg protein)	Epoxide hydrase (units/mg protein)	C^{14}-Phosphatidylcholine (cpm/nmole hemeprotein)
Phenobarbital treated rats					
Microsomes	2.0	0.59	250	18	2250
Step IV					
Fraction A	2.8	0	0	200	955
Fraction B	10.6	0	0	2	30
3-Methylcholanthrene treated rats					
Microsomes	1.4	0.55	135	13	1810
Step IV					
Fraction A	2.5	0	0	165	850
Fraction B	9.7	0	0	2	10

One unit of reductase is defined as nanomoles of cytochrome c reduced per minute.

One unit of epoxide hydrase is defined as nanomoles of styrene dihydrodiol formed from styrene oxide per minute.

The incorporation of C^{14}-choline into phosphatidylcholine was performed as previously described (Levin et al., 1972).

The less purified cytochrome P-450 and P-448 (Step IV, fraction A) had spectral properties similar to the more highly purified fractions with the following exceptions. The fraction A derived from 3-methylcholanthrene-treated rats had additional peaks at 390 and 642 nm in the absolute oxidized spectrum, absorption maxima characteristic of high-spin ferric hemeprotein. In addition, the Soret maximum of the CO-reduced complex of the fraction A from 3-methylcholanthrene-treated rats was at 449 nm, a shift upward of approximately 2 nm compared to fraction B from the same animals (Table II).

    B. CO Difference Spectra. Figure 3 shows the CO-reduced difference spectra of the A and B hemeprotein fractions derived from 3-methylcholanthrene pretreated rats. The absorption peak of the reduced CO derivative of cytochrome P-448 in liver microsomes and in the Step III preparation is at 448 nm. This shift in the CO maximum of the hemeprotein induced by polycyclic hydrocarbon treatment has been shown to be dependent on RNA and protein synthesis (Alvares et al., 1968). In repeated experiments, the absorption peak of the CO-reduced hemeprotein complex of the A fraction derived from 3-methylcholanthrene treated rats was at 448.5-449.0 nm and the B fraction was at 446.8-447.0 nm. The CO-maximum of the Step III preparation was the average of these two hemeprotein fractions indicating that the presence of Emulgen in these fractions could not account for the different CO maxima of the reduced hemeproteins. This is further substantiated by the fact that the addition of increasing concentrations of Emulgen to the Step III preparation does not result in any peak shift in the CO spectrum. In addition, no shifts in the CO absorption maxima are observed in the A and B fractions derived from phenobarbital-treated rats even though both fractions contain the same amount of Emulgen per nmole as the fractions obtained from 3-methylcholanthrene-treated animals.

    C. Ethyl Isocyanide Difference Spectra. Reduced cytochrome P-450 combines with ethyl isocyanide to give spectral absorption maxima at 430 and 455 nm which exist in a pH dependent equilibrium (Imai and Sato, 1966). Sladek and Mannering (1966) showed that the ratio of the absorbance at 455 nm to the absorbance at 430 nm is significantly elevated at any given pH in animals pretreated with 3-methylcholanthrene. Alvares et al., (1968) also showed that the 455 nm peak was shifted to 453 nm in 3-methylcholanthrene treated rats and that this peak shift and the change in the absorbance ratio of the 455 to 430 nm peaks were both dependent on RNA and protein synthesis. Figure 4 shows the ethyl isocyanide spectra of the fractions A and B cytochrome P-448 compared to the starting preparation (Step III) at pH 7.4. The "455 nm" peak of the A fraction was shifted to 453 nm and the B fraction peak was shifted to 451 nm as compared to the Step III preparation which had an

TABLE II

Absorption maxima of partially purified cytochrome P-450 and P-448

	Absorption maxima		
Hemeprotein	Oxidized	Reduced	CO reduced
		(nm)	
Cytochrome P-450			
Fraction A	418,535,568	414,542	450,550
Fraction B	418,535,568	414,542	450,550
Cytochrome P-448			
Fraction A	390,414,530,564,642	411,542	449,550
Fraction B	417,535,568	411,542	447,550

Spectra were measured in 0.1 M potassium phosphate buffer (pH 7.4). A small

peak at 423 nm in the CO reduced absolute spectrum is attributable to

cytochrome P-420 (Levin et al.,1974).

absorption maximum of the "455 peak" at an intermediate wavelength, 452 nm. The location of the peak in the 455 nm region is independent of the quantity of the 430 nm absorbing species of cytochrome P-448. As with the CO spectra shown earlier, addition of increasing amounts of Emulgen to either Step III or to the A and B fractions did not result in any peak shifts in the 455 nm region.

Figure 5 shows the ethyl isocyanide spectra of the Step III and A and B fractions derived from phenobarbital-pretreated rats. In contrast to the results obtained with the hemeproteins isolated from 3-methylcholanthrene-treated animals, the A and B fractions from phenobarbital-treated rats showed no peak shifts in the 455 nm region but did have significantly different ratios (455/430) of the two peaks. This is the first reported instance of a hemeprotein (fraction A) derived from phenobarbital-treated rats that has an ethyl isocyanide ratio of greater than 1 at pH 7.4. The Step III preparation had a 455/430 ratio intermediate to those of the A and B fractions.

Figure 6 shows plots of the 430 and 455 peak heights at different pHs with the Step III preparation and the A and B fractions from 3-methylcholanthrene-treated rats. The pH intercepts observed were pH 6.90, 7.10, and 6.85, respectively. Although the pH intercepts of the three fractions differed only slightly, the Step III preparation exhibited a pH intercept which was nevertheless the average of the values of the A and B fractions.

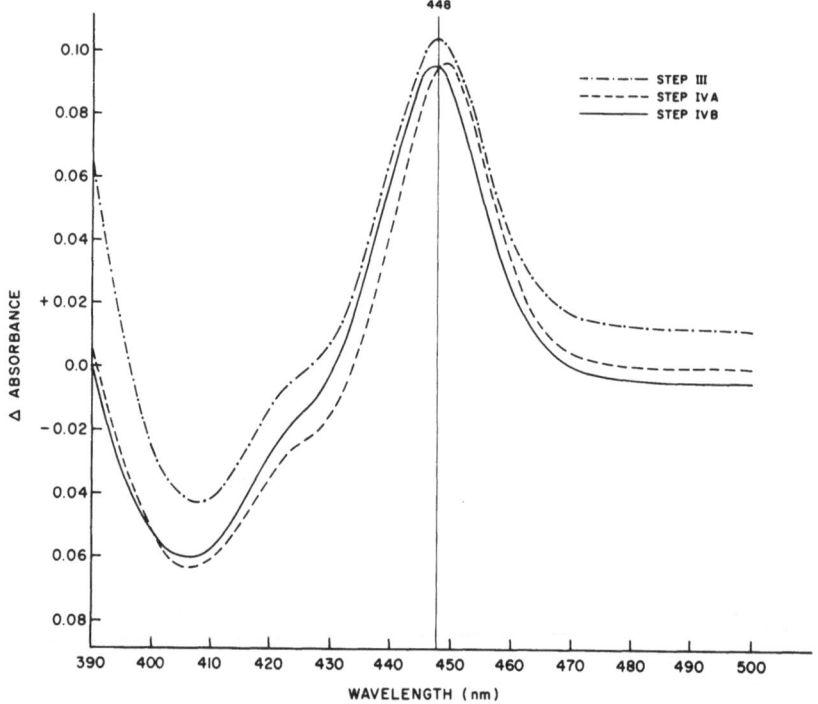

Fig. 3. CO-reduced difference spectra of Step III and Step IV (fractions A and B) cytochrome P-448 from 3-methylcholanthrene pretreated rats. The spectra were recorded at a hemeprotein concentration of 1 nmole/ml in the 0.1 M potassium phosphate buffer (pH 7.4).

Plots of the 430 and 455 peak heights at different pHs with the Step III preparation, A fraction, and B fraction from phenobarbital-treated rats are illustrated in Figure 6. The pH intercepts observed were pH 7.70, 6.90, and 7.85, respectively. The Step III preparation exhibited a pH intercept intermediate to the A and B fractions.

The A fraction from phenobarbital-treated rats was characterized by a 455/430 ratio of 1.8-2.0 at pH 7.4 and a pH intercept at 6.90. Although both the pH intercept and 455/430 ratio of this A fraction were similar to that of cytochrome P-448, no shifts in either the CO-reduced difference spectral maximum or ethyl isocyanide 455 nm peak were observed. It was observed in all cases that the relative size of the 430 peak did not influence the

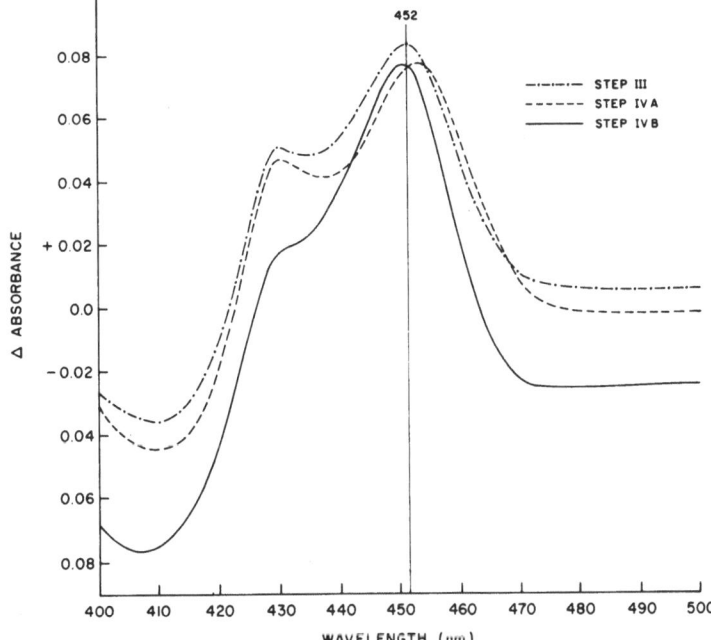

Fig. 4. Ethyl isocyanide difference spectra of Step III and Step IV (fraction A and B) cytochrome P-448 from 3-methylcholanthrene pretreated rats. The spectra were recorded at a hemeprotein concentration of 0.9-1.2 nmole/ml in 0.1 M potassium phosphate buffer containing 20% glycerol (pH 7.4).

location of the "455 region" peaks described. Therefore, this fraction differs not only from the B fraction from phenobarbital-treated rats, but also from the cytochrome P-448.

D.   Substrate Binding Spectra. Several studies in recent years have demonstrated the requirement for lipid in the metabolism of drugs by the mixed-function oxidase system of liver microsomes. Lu and Coon (1968) reported that one of the three components of the reconstituted system which was required for maximal hydroxylation activity was a lipid factor, later identified as phosphatidylcholine (Strobel et al., 1970).

Treatment of microsomes with phospholipase c or isooctane removed 50-70% of the total phospholipids, descreased the metabolism of Type I compounds, and virtually eliminated the binding of Type I

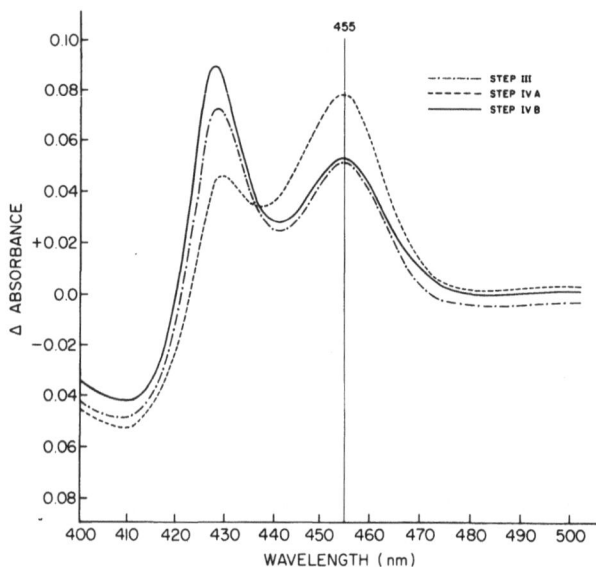

Fig. 5. Ethyl isocyanide difference spectra of Step III and
Step IV (fraction A and B) cytochrome P-450 from phenobarbital
pretreated rats. The spectra were recorded at a hemeprotein concentration of 1.0-1.3 nmole/ml in 0.1 M potassium phosphate buffer
containing 20% glycerol (pH 7.4).

compounds to cytochrome P-450 (Chaplin and Mannering, 1970; Eling
and DiAugustine, 1971; Leibman and Estabrook, 1971). These studies
with phospholipase c and isooctane suggested a role of phospholipid
in the binding of substrates to cytochrome P-450. Since the more
purified cytochrome P-450 and P-448 (Step IV, fraction B) are
essentially free of lipid, studies on the binding of substrates to
the hemeprotein were initiated. Figure 7 shows the binding spectra
of 3-methylcholanthrene, benzphetamine, and aniline to cytochrome
P-450 and P-448. Benzphetamine gave a typical Type I binding
spectrum when added to cytochrome P-450 or cytochrome P-448. The
addition of 3-methylcholanthrene to cytochrome P-450, however,
gave no detectable Type I binding spectrum. In contrast, a Type I
binding spectrum was seen on the addition of 3-methylcholanthrene
to cytochrome P-448. Aniline addition to cytochrome P-450 or P-448
resulted in a typical Type II binding spectrum. Since the purification procedure for the hemeprotein removes 20-40 fold more phospholipid per nmole cytochrome P-450 than does phospholipase c treatment or isooctane extraction, it is unlikely that phospholipid is
required for the binding of substrates to the cytochrome. Vore et

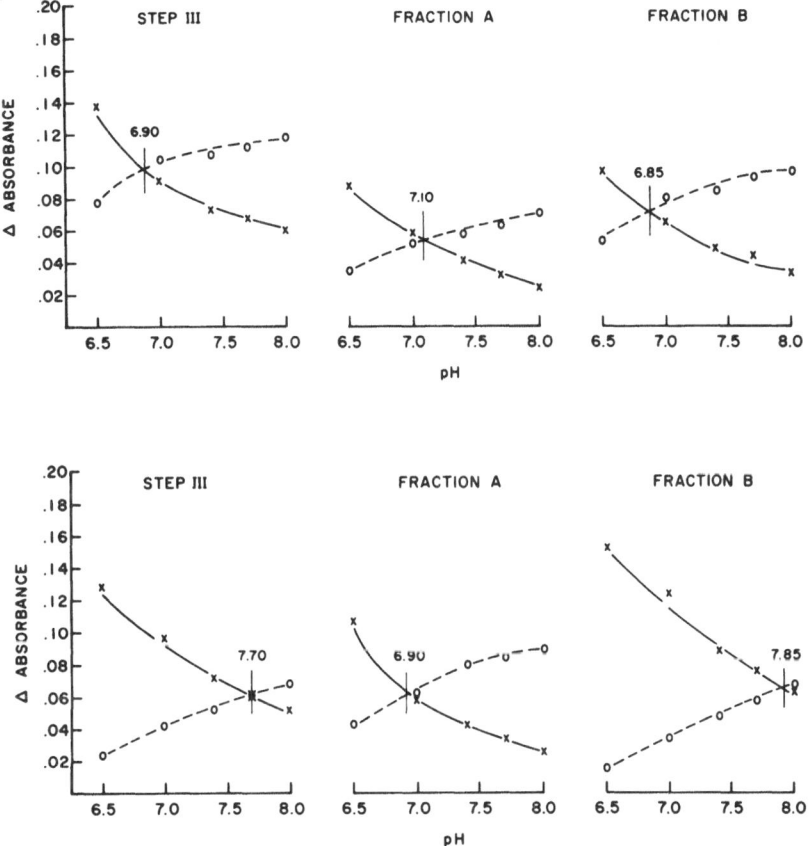

Fig. 6. Dependence of ethyl isocyanide difference spectra on pH using the Step III preparation, A fraction and B fraction from 3-methylcholanthrene treated rats (upper) and phenobarbital treated rats (lower). The spectra were recorded at a cytochrome concentration of 0.8-1.2 nmole/ml in 1.0 M potassium phosphate at the indicated pH values. The 455 nm peak height (0-0) and 430 nm peak height (x-x) are shown.

al., (1974a and b) have recently obtained similar results with organic solvent extracted lyophilized microsomes. The catalytic activity of these extracted microsomes was markedly decreased and could be restored upon addition of phospholipid. However, the binding spectra of several substrates were unaffected (or even

increased) after removal of 80-90% of the phospholipids. Further
studies will be required to determine the exact mechanism of lipid
function in drug metabolism, but it is clear that the site of lipid
function is located at the step of cytochrome P-450 reduction by
NADPH and NADPH-cytochrome c reductase as shown by Strobel et al.,
(1970), and not at the site of substrate binding to the hemeprotein.

Enzymatic Activity of the Reconstituted System. A. Absolute
Requirement for Lipid and NADPH-Cytochrome c Reductase. The requirement for lipid, reductase, and cytochrome P-450 in the reconstituted system for the N-demethylation of benzphetamine is shown
in Table III. Lipid and NADPH-cytochrome c reductase were absolutely required for activity when the Step IV B cytochrome P-450 was
used. The Step III cytochrome P-450 was still contaminated with a
significant amount of reductase and a small amount of phospholipid
(Levin et al., 1972) which accounted for the small but significant
metabolism of benzphetamine observed when either lipid or reductase
was omitted from the incubation system. This is further illustrated when a high concentration of cytochrome P-450 (1 nmole) is
used in the incubation. At this concentration of cytochrome P-450,
the lipid and reductase become rate-limiting in the system and
thus any small contamination of the hemeprotein with either of these
components would be more easily detected. Under these incubation
conditions, a small but detectable amount of metabolism is obtained
in the absence of lipid with Step IV B cytochrome P-450. Since
both the reductase and cytochrome P-450 are essentially lipid-free
(Levin et al., 1974), this activity is probably due to the presence
of detergent, rather than to the presence of lipid in either
preparation. Preliminary experiments in our laboratory (A. Y. H.
Lu, unpublished observations) indicate that under certain well-de-
fined incubation conditions, several detergents including Emulgen
911 can partially replace the lipid fraction in the reconstituted
system.

B. N-demethylation of Benzphetamine. The N-demethylation of
benzphetamine by the reconstituted system using varying concentrations of cytochrome P-450 purified through Step III or purified
through Step IV (B fraction) is shown in Figure 8. The metabolism
of benzphetamine was measured in the presence of a fixed amount of
lipid and NADPH-cytochrome c reductase while varying the concentration of cytochrome P-450. At low concentrations of hemeprotein,
the rate of the reaction was proportional to the hemeprotein concentration, reaching a plateau at approximately 0.3-0.4 nmole of
cytochrome P-450 when the reductase became limiting in the reaction.
It can be seen that the Step IV B fraction, which contains small
amounts of Emulgen 911 (0.05 mg per nmole cytochrome P-450) is at
least as active as the Step III preparation or the Step III preparation plus Emulgen (0.05 mg per nmole cytochrome P-450). Thus,
the amount of the nonionic detergent present in the more purified

TABLE III

Requirement for NADPH- cytochrome c reductase and lipid in the metabolism of benzphetamine by cytochrome P-450

	Benzphetamine N-demethylation	
Fractions	Step III	Step IV B
	(nmoles NADPH oxidized/min)	
P-450 (0.2 nmole)	0.0	0.0
P-450 + lipid	0.0	0.0
P-450 + reductase	0.32	0.08
P-450 + reductase + lipid	4.44	5.16
P-450 (1.0 nmole)	0.32	0.0
P-450 + lipid	1.69	0.0
P-450 + reductase	1.21	0.48
P-450 + reductase + lipid	7.90	8.39

The reaction mixture, in a final volume of 1.0 ml contained 0.2 or 1.0 nmole Step III or Step IV B cytochrome P-450, 0.1 mg lipid, 195 units of NADPH-cytochrome c reductase, and other necessary cofactors as previously described (Lu et al, 1972). The high concentration of cytochrome P-450 was used to demonstrate its lack of contamination with other components. Hydroxylation of benzphetamine was only proportional to concentrations of cytochrome P-450 in the range of 0.2-0.4 nmoles. Thus, the rate of metabolism of benzphetamine using 0.2 nmole was approximately 60% of the rate obtained using 1 nmole of hemeprotein.

cytochrome P-450 fraction has no apparent inhibitory effect on catalytic activity. However, when the concentration of Emulgen per nmole cytochrome P-450 exceeds 0.25 mg/nmole, a significant inhibition of benzphetamine metabolism occurs (Table IV). This may explain the poor catalytic activity of the A fraction cytochrome P-P-450 towards benzphetamine (Table IV) since the Emulgen content per nmole hemeprotein is 0.35 mg/nmole.

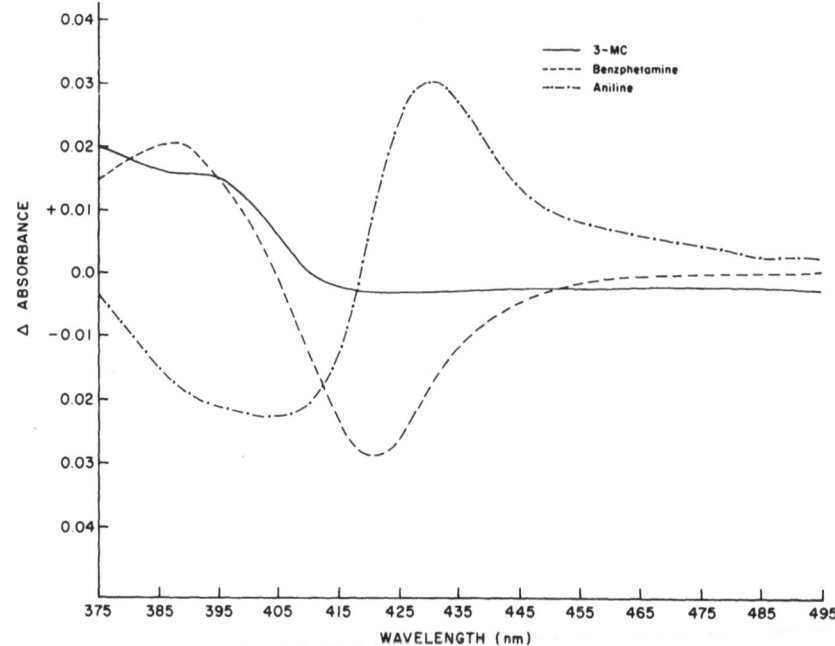

Fig. 7 a and b. Spectral changes (difference spectra) caused
by the addition of substrates to partially purified (Step IV,
fraction B) cytochrome P-450 (a) and P-448 (b). Hemeproteins
were diluted to 0.8 nmoles/ml in 0.1 M potassium phosphate buffer
(pH 7.4). Benzphetamine, aniline, or 3-methylcholanthrene (3-MC)
were added to the sample cuvette at a final concentration of 1 mM,
5mM, and 5 µM, respectively. 3-Methylcholanthrene was dissolved
in acetone, and an equal volume of acetone was added to the reference cuvette.

C. Hydroxylation of Benzo[a]pyrene. Detailed studies on the
catalytic activity of the cytochrome P-448 in fractions A and B
isolated from 3-methylcholanthrene treated rats suggest that the
high Emulgen content in fraction A may not be responsible for the
poor catalytic activity of this cytochrome fraction, at least when
benzo[a]pyrene is used as the substrate (Table V). As can be seen,
the fraction A cytochrome P-448 metabolizes benzo[a]pyrene at 5-10%
of the rate of the fraction B cytochrome P-448. Addition of
fraction A to fraction B at a ratio of 2 to 1 inhibits the catalytic activity of the B fraction by only 13%. Increasing the ratio
of A to B to 10 to 1 results in a 64% inhibition of activity.
While the inhibition observed would be expected by an increased
detergent concentration, this result could also be explained as due
to a competition between the two hemeprotein fractions for NADPH-
cytochrome c reductase. West and Lu (1972) have shown that cytochrome P-450 and P-448 do compete for NADPH-cytochrome c reductase

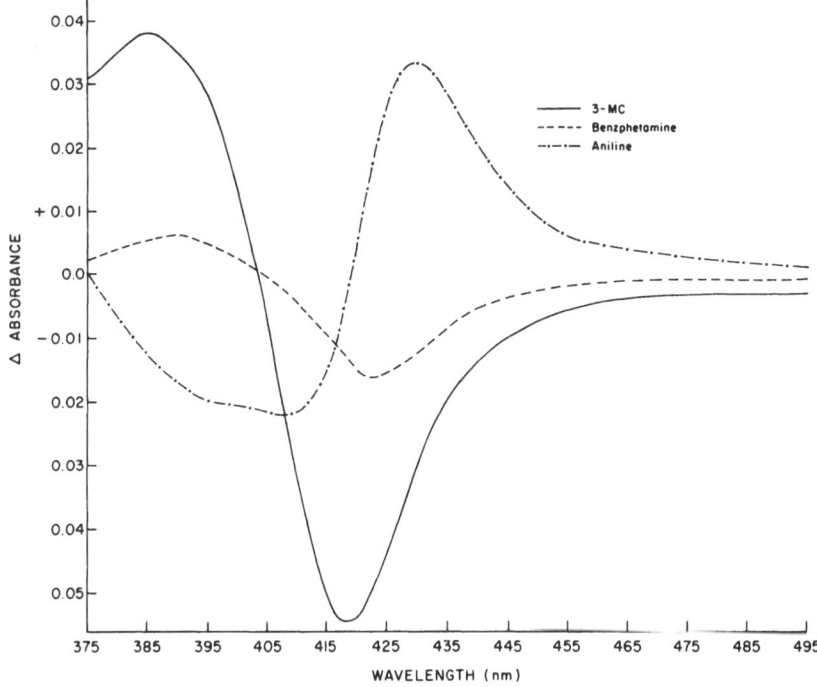

Fig. 7b.

in the metabolism of benzo[a]pyrene when the two hemeproteins are
added simultaneously. If the inhibition of benzo[a]pyrene hydroxylase activity of fraction B cytochrome P-448 by fraction A is indeed due only to detergent, then addition of a boiled fraction A
should still result in inhibition of metabolism. In fact, the
opposite result is obtained; i.e., boiled fraction A stimulates
the benzo [a]pyrene hydroxylase activity of fraction B cytochrome
P-448 by 37-45%. If, indeed, fraction A cytochrome P-448 is a
different hemeprotein as indicated by the spectral studies presented earlier, the effect of Emulgen on the catalytic activity
of this hemeprotein may be quite different from its effect on the
more purified cytochrome P-448 (B fraction). Thus, no definitive
conclusions can be drawn at this time concerning the poor catalytic activity of the cytochrome P-450 and P-448 A fractions
described.

     B. Hydroxylation of Naphthalene. An arene oxide has been
demonstrated to be an obligatory intermediate in the metabolism of
napthalene to 1-naphthol and naphthalene 1,2-dihydrodiol (Jerina et
al., 1970; Jerina and Daly, 1974). This arene oxide is formed by
the monooxygenase system and either undergoes a non-enzymatic rearrangement to form the corresponding phenol (1-naphthol) or is

TABLE IV

Effect of Emulgen 911 on the metabolism of benzphetamine by the
reconstituted system

Cytochrome P-450	Emulgen Concentration (mg/nmole hemeprotein)	Benzphetamine N-demethylation (nmoles NADPH oxidized/min)
Step III (0.2 nmole)	--	5.2
Step IV B (0.2 nmole)	.05	5.2
	.15	5.7
	.25	6.1
	.45	2.1
	.85	0.6
Step IV A (0.15 nmole)	0.35	1.1
(0.50 nmole)	0.35	1.3

Cytochrome P-450 was partially purified from phenobarbital-pretreated rats and
assayed for benzphetamine metabolism. The Step III preparation contains no
Emulgen. The Step IV B preparation of the hemeprotein contains 0.05 mg Emulgen/
nmole cytochrome P-450 after removal of the excess detergent as described in
Methods. Additional Emulgen was added to the hemeprotein before incubating with
reductase, lipid, and cofactors. The Step IV A preparation contains 0.35 mg/Emulgen/
nmole cytochrome P-450 which cannot be removed from the preparation.

hydrated enzymatically by epoxide hydrase to naphthalene-1,2-
dihydrodiol. Table VI shows the effect of epoxide hydrase on the
metabolism of naphthalene by the Step IV B cytochrome P-450 system
which is essentially free of epoxide hydrase. In the absence of
epoxide hydrase, the predominant product of metabolism was the
phenol (1-naphthol). This is in contrast to results obtained with
liver microsomes where more than 50% of the product was the dihydrodiol of naphthalene. Addition of small amounts of the partially
purified epoxide hydrase from the Step IV A fraction did not influence the rate of total naphthalene metabolism but did result in
a significant change in the ratio of dihydrodiol to phenol formed.
Although the epoxide hydrase was contaminated with cytochrome P-450

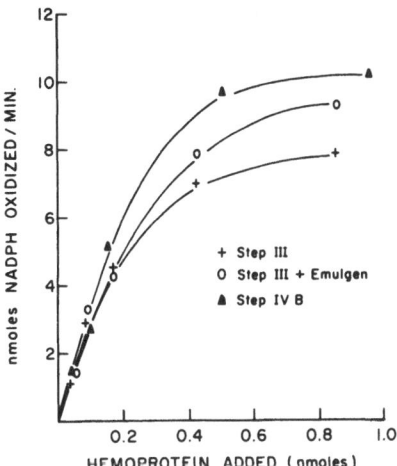

Fig. 8. N-demethylation of benzphetamine as a function of hemeprotein concentration in the presence of fixed amount of NADPH-cytochrome c reductase and lipid. The incubation contained 100 μmoles potassium phosphate buffer (pH 7.4), 1 μmole benzphetamine, 0.1 μmole NADPH, 0.1 mg lipid, 200 units of reductase, and Step III cytochrome P-450, Step III cytochrome P-450 + Emulgen (0.05 mg/nmole P-450), or Step IV B cytochrome P-450 (which contains 0.05 mg Emulgen/nmole P-450).

(fraction A), the purification of hydrase per nmole of hemeprotein was great enough that addition of epoxide hydrase to the reaction mixture results in the addition of only 20% more hemeprotein. Moreover, this cytochrome fraction has poor catalytic activity for the metabolism of naphthalene, and the rate of total naphthalene metabolism did not change upon addition of epoxide hydrase, which supports the suggestion that neither the detergent present in fraction A nor the cytochrome P-450 in the A fraction affect the metabolism of naphthalene by the more purified and enzymatically active cytochrome P-450. Further studies are in progress to completely separate the epoxide hydrase and cytochrome P-450 in the A fraction. The complete separation and recombination of monooxygenase and epoxide hydrase activities may be a valuable tool in studies on the activation of polycyclic hydrocarbons to reactive ultimate carcinogens.

SUMMARY

1. Partial purification of liver microsomal cytochrome P-450 results in the separation of two forms of cytochrome P-450 from phenobarbital-treated rats and two forms of cytochrome P-448 from 3-methylcholanthrene-treated rats.

TABLE V

Metabolism of benzo[a]pyrene by the partially purified cytochrome
P-448 containing reconstituted system

Cytochrome P-448	Benzo[a]pyrene metabolism (nmole equivalents 3-OH benzo[a]pyrene formed per 5 min)
Step IV A (.1 nmole)	.07
(.2 nmole)	.09
(.5 nmole)	.09
(1 nmole)	.09
Step IV B (.1 nmole)	1.22
(.2 nmole)	1.78
(.5 nmole)	2.28
Step IV B (.1 nmole) + Step IV A (.2 nmole)	1.06
Step IV B (.1 nmole) + Step IV A (1 nmole)	0.44
Step IV B (.1 nmole) + boiled Step IV A (.2 nmole)	1.67
Step IV B (.1 nmole) + boiled Step IV A (1 nmole)	1.77

Cytochrome P-448 was partially purified from 3-methylcholanthrene

pretreated rats and assayed for benzo[a]pyrene hydroxylase activity

in the presence of reductase and lipid as previously described

(Levin et al, 1974).

    2. Each of the four cytochrome fractions has different
spectral properties (absolute spectra, CO difference spectra, and
ethylisocyanide difference spectra).

    3. The hemeprotein fractions which elute from a DEAE-cellulose
column at 100 mM KCl (fraction IV B) are more highly purified
than the hemeproteins (fraction IV A) that elute in the column
volume.

    4. The more highly purified cytochrome fractions (IV B) con-
tain 9-11 nmoles of cytochrome P-450 or P-448 per mg protein (an
approximately 5-7 fold purification over microsomes) and are

TABLE VI

Metabolism of napthalene by liver microsomes and partially purified cytochrome P-450

Preparation	1-Napthol	Dihydrodiol	Total	Ratio of 1-Naphthol to Dihydrodiol
	(nmoles formed/min/nmole hemeprotein)			
Microsomes	2.5	3.5	6.0	0.7
Step IV B	6.3	0.4	6.7	15.8
Step IV B + epoxide hydrase	3.2	3.8	7.0	0.8

Liver microsomes and partially purified cytochrome P-450 were prepared from phenobarbital-pretreated rats. The incubation and analysis of naphthalene metabolism were as previously described (Dansette et al., 1974). The amount of epoxide hydrase (Step IV A) added to the incubation was 0.03 mg protein (7 units of epoxide hydrase).

enzymatically active in the metabolism of a variety of substrates when combined with lipid and NADPH-cytochrome c reductase. These hemeprotein fractions are free of cytochrome b_5 and NADPH-cytochrome c reductase, and the hemeproteins are purified approximately 100-fold with respect to phospholipid. The cytochrome P-450 and P-448 are virtually free of epoxide hydrase.

ACKNOWLEDGEMENTS

We thank Mrs. Cathy Chvasta for her assistance in the preparation of this manuscript.

REFERENCES

Alvares, A. P., Schilling, G., Levin, W. and Kuntzman, R. 1967. Studies on the induction of CO-binding pigments in liver microsomes by phenobarbital and 3-methylcholanthrene. Biochem. Biophys. Res. Commun. 29: 521-526.

Alvares, A. P., Schilling, G., Levin, W. and Kuntzman, R. 1968. Alteration of the microsomal hemoprotein by 3-methylcholanthrene:

effects of ethionine and actinomycin D. J. Pharmacol. Exp. Therap. 163: 417-424.

Bligh, E. G. and Dyer, W. T. 1959. A rapid method of total lipid extraction and purification. Can. J. Biochem. Physiol. 37:911-917.

Chaplin, M. D. and Mannering, G. J. 1970. Role of phospholipids in the hepatic microsomal drug-metabolizing system. Molec. Pharmacol. 6:631-640.

Comai, K. and Gaylor, J. L. 1973. Existence and separation of three forms of cytochrome P-450 from rat liver microsomes. J. Biol. Chem. 248: 4947-4955.

Conney, A. H. 1967. Pharmacological implications of microsomal enzyme induction. Pharmacol. Rev. 19: 317-366.

Dansette, P. M., Yagi, H., Jerina, D. M., Daly, J. W., Levin, W., Lu, A. Y. H., Kuntzman, R. and Conney, A. H. 1974. Assay and partial purification of epoxide hydrase from rat liver microsomes. Arch. Biochem. Biophys., in press.

Eling, T. E. and DiAugustine, R. P. 1971. A role of phospholipids in the binding and metabolism of drugs by hepatic microsomes. Use of the fluorescent hydrophobic probe 1-anilinonaphthalene-8-sulphonate. Biochem. J. 123: 539-549.

Fujita, T. and Mannering, G. J. 1973. Electron transport components of hepatic microsomes. Solubilization, resolution, and recombination to reconstitute aniline hydroxylase activity. J. Biol. Chem. 248: 8150-8156.

Gaylor, J. L. and Delwiche, C. V. 1969. Removal of nonionic detergents from proteins by chromatography on Sephadex LH-20. Anal. Biochem. 28: 361-368.

Imai, Y., and Sato, R. 1966. Evidence for two forms of P-450 hemeprotein in microsomal membranes. Biochem. Biophys. Res. Commun. 23: 5-11.

Jerina, D. M. and Daly, J. W. 1974. Formation, toxicity and carcinogenicity of arene oxides: A new aspect of drug metabolism. Science, in press.

Jerina, D. M., Daly, J. W., Witkop, B., Zaltzman-Nirenberg, P. and Udenfriend, S. 1970. 1,2 Naphthalene oxide as an intermediate in the microsomal hydroxylation of naphthalene. Biochemistry 9: 147-156.

Leibman, K. C. and Estabrook, R. W. 1971. Effects of extraction with isooctane upon the properties of liver microsomes. Molec. Pharmacol. 7: 26-32.

Levin, W., Lu, A. Y. H., Ryan, D., West, S., Kuntzman, R. and Conney, A. H. 1972. Partial purification and properties of cytochromes P-450 and P-448 from rat liver microsomes. Arch. Biochem. Biophys. 153: 543-553.

Levin, W., Ryan, D., West, S. and Lu, A. Y. H. 1974. Preparation of a partially purified, lipid-depleted cytochrome P-450 and reduced nicotinamide adenine dinucleotide phosphate-cytochrome c reductase from rat liver microsomes. J. Biol. Chem. 249: 1747-1754.

Lowry, O. H., Rosebrough, N. J., Farr, A. L. and Randall, R. J. 1951. Protein measurement with the Folin phenol reagent. J. Biol. Chem. 193: 265-275.

Lu, A. Y. H. and Coon, M. J. 1968. Role of hemoprotein P-450 in fatty acid ω-hydroxylation in a soluble enzyme system from liver microsomes. J. Biol. Chem. 243: 1331-1332.

Lu, A. Y. H. and Levin, W. 1972. Partial purification of cytochromes P-450 and P-448 from rat liver microsomes. Biochem. Biophys. Res. Commun. 46: 1334-1339.

Lu, A. Y. H., Kuntzman, R., West, S., Jacobson, M. and Conney, A. H. 1972. Reconstituted liver microsomal enzyme system that hydroxylates drugs, other foreign compounds, and endogeneous substrates. II. Role of the cytochrome P-450 and P-448 fractions in drug and steroid hydroxylation. J. Biol. Chem. 247: 1727-1734.

Lu, A. Y. H., Levin, W., West, S. B., Jacobson, M., Ryan, D., Kuntzman, R. and Conney, A. H. 1973. Reconstituted liver microsomal enzyme system that hydroxylates drugs, other foreign compounds, and endogeneous substrates. VI. Differential substrate specificities of the cytochrome P-450 fractions from control and phenobarbital-treated rats. J. Biol. Chem. 248: 456-460.

Oesch, F., Jerina, D. M. and Daly, J. W. 1971. Substrate specificity of hepatic epoxide hydrase in microsomes, and in a purified preparation: Evidence for homologous enzymes. Arch. Biochem. Biophys. 144: 253-261.

Omura, T., and Sato, R. 1964a. The carbon-monoxide-binding pigment of liver microsomes. II. Solubilization, purification and properties. J. Biol. Chem. 239: 2379-2385.

Omura, T. and Sato, R. 1964b. The carbon-monoxide-binding pigment of liver microsomes. I. Evidence for its hemoprotein nature. J. Biol. Chem. 239: 2370-2378.

Phillips, A. H., and Langdon, R. G. 1962. Hepatic triphosphopyridine nucleotide-cytochrome c reductase: isolation, characterization and kinetic studies. J. Biol. Chem. 237: 2652-2660.

Sladek, N. E. and Mannering, G. J. 1966. Evidence for a new P-450 hemoprotein in hepatic microsomes from methylcholanthrene-treated rats. Biochem. Biophys. Res. Commun. 24: 668-674.

Stern, J. O., Levin, W. and Lu, A. Y. H. 1974. Studies on the spin state of 3-methylcholanthrene induced microsomal cytochrome P-450. Fed. Proc. 33: 1387.

Strobel, H. W., Lu, A. Y. H., Heidema, J. and Coon, M. J. 1970. Phosphatidylcholine requirement in the enzymatic reduction of hemoprotein P-450 in fatty acid, hydrocarbon, and drug hydroxylation. J. Biol. Chem. 245: 4851-4854.

van der Hoeven, T. A. and Coon, M. J. 1974. Preparation and properties of partially purified cytochrome P-450 and NADPH-cytochrome P-450 reductase from rabbit liver microsomes.

J. Biol. Chem., in press

Vore, M., Hamilton, J. G. and Lu, A. Y. H. 1974a. Organic solvent extraction of liver microsomal lipid. I. The requirement of lipid for 3,4-benzpyrene hydroxylase. Biochem. Biophys. Res. Commun. 56: 1038-1044.

Vore, M., Lu, A. Y. H., Kuntzman, R. and Conney, A. H. 1974b. Organic solvent extraction of liver microsomal lipid. II. Effect on the metabolism of substrates and binding spectra of cytochrome P-450. Mol. Pharmacol., in press.

West, S. B. and Lu, A. Y. H. 1972. Reconstituted liver microsomal enzyme system that hydroxylates drugs, other foreign compounds and endogenous substrates. Archiv. Biochem. Biophys. 153: 298-303.

BIOCHEMICAL CHARACTERIZATION OF HIGHLY PURIFIED CYTOCHROME P-450

AND OTHER COMPONENTS OF THE MIXED FUNCTION OXIDASE SYSTEM OF LIVER

MICROSOMAL MEMBRANES*

Minor J. Coon, Theodore A. van der Hoeven, David A.
Haugen,‡ F. Peter Guengerich, Janice L. Vermilion,+ and
David P. Ballou, Department of Biological Chemistry,
Medical School, The University of Michigan
Ann Arbor, Michigan 48104

I.   RECONSTITUTION OF THE MIXED FUNCTION OXIDASE SYSTEM OF
     LIVER MICROSOMES: REQUIRED COMPONENTS

The cytochrome P-450-containing enzyme system of liver microsomes is of particular interest to biochemists, pharmacologists,
and toxicologists because of its ability to hydroxylate or otherwise modify not only steroids and fatty acids, but also drugs,
insecticides, carcinogens, and a variety of other foreign compounds. Our laboratory accomplished the solubilization, resolution, and reconstitution of this enzyme some years ago (Lu and
Coon, 1968; Coon and Lu, 1969) and has subsequently been concerned with the purification and biochemical characterization of
the individual components.

When the liver microsomal enzyme system was solubilized with
deoxycholate in the presence of glycerol and other protective
agents and then resolved by column chromatography on DEAE-cellulose,
two fractions, designated A and B, were found to be required for
hydroxylation activity (Lu and Coon, 1968). The first fraction
contained both cytochrome P-450 (A_1) and NADPH-cytochrome P-450
reductase (A_2), which could be partially separated because the
latter was eluted from the column at higher ionic strength, and
the second fraction contained a heat-stable, chloroform-soluble
component which was called Factor B (Lu et al., 169a). B was
subsequently shown to contain phospholipids, and phosphatidylcholine was identified as the active component and shown to be
required along with A_1 and A_2, as well as NADPH and molecular
oxygen, for the hydroxylation of hydrocarbons, drugs, and fatty
acids (Strobel et al., 1970; Lu et al., 1970). More recently,

25

evidence has been obtained that cytochrome P-450 purified to a
state where it is free of significant amounts of other known
electron acceptors such as cytochrome b_5, flavins, nonheme iron,
or other metals accepts two electron per molecule of heme from
dithionite (Ballou et al., 1974; van der Hoeven and Coon, 1975).
This finding was unexpected since other hemoproteins such as
cytochrome c take up only one electron. The unidentified acceptor,
which is referred to as Factor C, appears to have an oxidation-
reduction potential that is close to that of cytochrome P-450, but
how it is associated with the hemoprotein and whether it plays an
essential role in the hydroxylation reactions catalyzed by this
enzyme system remain to be determined. This report will summarize
our recent studies on the purification and properties of these
microsomal components and the manner in which they function to
accomplish substrate hydroxylation.

II.   LIVER MICROSOMAL CYTOCHROME P-450 (P-450$_{LM}$)

A. Solubilization and Purification

Although P-450$_{LM}$ has proved to be an unusually difficult
protein to purify, the use of ionic and nonionic detergents along
with suitable protective agents has finally resulted in significant
progress. Several procedures have recently been reported by our
own and other laboratories for the partial purification of
P-450$_{LM}$ (Autor et al., 1973a; Coon et al., 1973a; Levin et al.,
1972, 1974; Fujita et al., 1973; Sato et al., 1973), and this
protein has recently been obtained in highly purified form from
phenobarbital-induced rabbit liver microsomes by van der Hoeven
and Coon (1974).

Microsomes prepared from male rabbits induced with phenobarbital were prepared and submitted to pyrophosphate extraction
with the results shown in Table I. The resulting preparation was
almost entirely free of hemoglobin and contained P-450$_{LM}$ and
NADPH-cytochrome c reductase as well as cytochrome b_5 at a somewhat elevated level. The extracted microsomal membranes were then
solubilized with cholate and fractionated by precipitation with
polyethylene glycol (PEG) 6000, DEAE-cellulose column chromatography in the presence of Renex 690 (a nonionic detergent), and
hydroxylapatite column chromatography in the presence of Renex,
with the results shown in Table II. In the particular preparation
shown P-450$_{LM}$ was purified to a concentration of 15.2 nmoles per mg
of protein. The values in parentheses indicate the concentrations
obtained at each step when the procedure was carried out on several
batches of microsomes. The best P-450 preparation obtained was at
a concentration of 17.5 nmoles per mg of protein.

TABLE I

Effect of Pyrophosphate Extraction on Microsomal Components[a]

Preparation	Protein	Cyt. P-450	Hemoglobin	Cyt. b_5	NADPH-cyt. c reductase
	mg	nmoles per mg protein	nmoles per mg protein	nmoles per mg protein	Specific activity (nmoles per min per mg protein)
Liver microsomes	1,610	2.7	0.75	1.2	70
Extracted, sonicated microsomes	1,130	3.4	0.01	1.4	130

[a] Protein was estimated by the method of Lowry et al. (1951), cytochrome P-450 as described by Omura and Sato (1964), hemoglobin as the CO complex using an extinction coefficient of 154 mM^{-1} cm^{-1} at 418 nm, cytochrome b_5 by enzymatic reduction with NADH (Omura and Sato, 1964) in the presence of Triton X-100-solubilized NADH-cytochrome b_5 reductase (Spatz and Strittmatter, 1971) and deoxycholate (100 µg per ml, final concentration). NADPH-cytochrome P-450 reductase was assayed by its activity toward cytochrome c in 0.3 M potassium phosphate buffer, pH 7.7, at 30° as described elsewhere (van der Hoeven and Coon, 1974).

TABLE II

Purification of Liver Microsomal Cytochrome P-450

from Phenobarbital-induced Rabbits[a]

Preparation	Protein	Cytochrome P-450	Yield
	mg	nmoles per mg protein	%
Extracted, sonicated microsomes	3,020	2.7 (2.6-3.6)	100
Polyethylene glycol precipitate (8-13%)	1,360	6.4 (5.4-7.0)	100
DEAE-cellulose column eluate (0.5% Renex)	304	8.9 (8.9-12.0)	33
Hydroxylapatite column eluate (0.1% Renex)	63	15.2 (13.0-17.5)	11

[a]The purification procedure is that of van der Hoeven and Coon (1974)

with slight modifications and with the addition of a hydroxylapatite

column chromatography step.

B.   Properties of Highly Purified P-450$_{LM}$

As shown in Figure 1, SDS-polyacrylamide gel electrophoresis
was used to determine the polypeptides present in microsomes and in
the various cytochrome P-450 preparations. A comparison of normal
microsomes (Gel F) with microsomes from phenobarbital-induced
animals (Gels A and G) shows that both contain a group of major
polypeptides with mobilities of 0.5 to 0.7, along with a number
of other less prominent bands. In the induced microsomes one of
the bands, as shown by the arrow, is present in significantly
greater amount. Of special interest, the same polypeptide is the
one obtained upon purification with polyethyleneglycol (Gel B),
DEAE-cellulose (Gel C), and hydroxylapatite (Gels D, E, and H).
It is apparent that the purified P-450$_{LM}$ contains only a single
major polypeptide, and densitometer tracings indicate that it may
account for as much as 90% of the protein in such preparations.

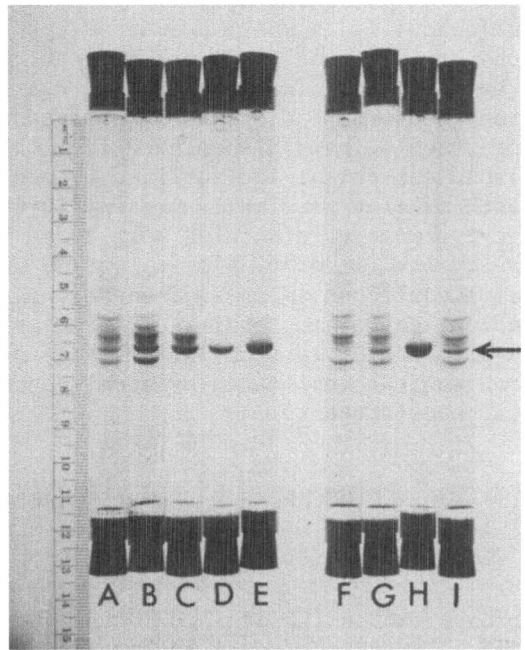

Fig. 1. SDS-polyacrylamide gel electrophoresis of cytochrome P-450 and other microsomal proteins. A discontinuous buffer
system was used as described by Laemmli (1970) under the conditions reported previously (van der Hoeven and Coon, 1974).
Migration was from top to bottom as shown in the figure; the line
at the bottom, marked by a stainless steel wire, indicates the
position of the tracking dye band. The following samples were
used at the protein level indicated: A, pyrophosphate-extracted
microsomes from phenobarbital-induced animals, 10 µg; B, PEG
precipitate, 15 µg; C, DEAE-cellulose eluate, 10 µg; D and E,
hydroxylapatite eluate, 3 and 10 µg; F, pyrophosphate-extracted
microsomes from normal animals, 10 µg; G, pyrophosphate-extracted
microsomes from phenobarbital-induced animals, 10 µg; H,
hydroxylapatite eluate, 10 µg; and I, pyrophosphate-extracted
microsomes from normal animals (10 µg) mixed with hydroxylapatite
eluate (1.3 µg).

When the purified P-450 was combined with normal microsomes the
same band was intensified (Gel I), as would be expected. In
other experiments with calibrated gels the major polypeptide in
the purified P-450$_{LM}$ preparation was found to have a molecular
weight of about 49,000 daltons.

Analytical data on a typical purified preparation of P-450$_{LM}$ are presented in Table III. A small amount of cytochrome P-420 resulting from the breakdown of P-450 was present, but no cytochrome b_5 could be detected. The heme content was slightly higher than would be predicted from the cytochromes present, perhaps because of the presence of heme dissociated from these proteins, and the phospholipid content of the purified cytochrome P-450 was less than one-tenth of that originally present in the microsomes. NADPH- and NADH-cytochrome c reductases were found to be present at only very low levels. In other similar enzyme preparations no labile sulfide or significant amounts of copper, manganese, molybdenum, scandium, selenium, cobalt, or chromium were found. Furthermore, the presence of an iron-sulfur protein can be ruled out since the iron content, estimated by atomic absorption, is close to the total heme content.

III.   NADPH-CYTOCHROME P-450 REDUCTASE

A.   Solubilization and Purification

Our earlier experiments (Lu et al., 1969a) showed that the lipase-solubilized NADPH-cytochrome c reductase described by Williams and Kamin (1962), and kindly furnished by Dr. Kamin, was incapable of replacing our detergent-solubilized reductase as a component of the reconstituted hydroxylation system. Obviously, the ability of a reductase preparation to transfer electrons to an artificial electron acceptor such as cytochrome c, which is not a normal microsomal component, provides no information as to its ability to function as an electron carrier to cytochrome P-450 for the purpose of substrate hydroxylation.

We have pursued the purification of the detergent-solubilized NADPH-cytochrome P-450 reductase from liver microsomes of phenobarbital-treated rats with the results shown in Table IV. The microsomes were extracted with deoxycholate in the presence of glycerol, dithiothreitol, KCl, EDTA, and citrate buffer as described earlier (Lu and Coon, 1968), and the extract was then submitted to column chromatography on ion exchange columns under various conditions. Finally, column chromatography on hydroxylapatite in the presence of Renex gave an enzyme preparation purified over 100-fold from the starting material. Of particular interest, the ratio of the activities toward cytochrome c and toward cytochrome P-450 (measured by benzphetamine hydroxylation in the reconstituted enzyme system with the reductase as the rate-limiting component) remained constant throughout purification.

TABLE III

Properties of Purified Cytochrome P-450$_{LM}$

Component[a]	Concentration or activity (per mg protein)	
Cytochrome P-450	13.0	nmoles
Cytochrome P-420	0.3	nmole
Cytochrome b_5	0	
Heme	14.4	nmoles
Phospholipid	50	nmoles
NADPH-cytochrome $\underline{c}$ reductase	0.25	nmole per min
NADH-cytochrome $\underline{c}$ reductase	0.43	nmole per min
Renex	0.43	mg

[a]Cytochrome P-420 and total heme were determined by published procedures (Omura and Sato, 1964). Phospholipids were extracted by the method of Bligh and Dyer (1959) and estimated by determination of the phosphorus content (King, 1932). Renex was extracted into ether, and after evaporation of the solvent the residue was dissolved in water and the absorbance at 278 nm was determined.

TABLE IV

Purification of Rat Liver Microsomal

NADPH-Cytochrome P-450 Reductase

Preparation	Specific activity (nmoles min^{-1} mg^{-1})		Ratio
	Cyt. c reduction	Hydroxylation[a]	
Microsomes	305		
DEAE-cellulose column eluate (0.05% deoxycholate)	1,880	112	16.8
DEAE-cellulose column eluate (0.4% Renex)	15,100	890	16.9
DEAE-Sephadex A-50 column eluate (0.1% Renex)	21,500	1,220	17.6
Hydroxylapatite column eluate (0.1% Renex)	33,200	2,060	16.1

[a]The ability of the reductase to reduce cytochrome P-450 was estimated from the rate of NADPH disappearance, followed at 340 nm, in a reaction mixture containing benzphetamine, purified cytochrome P-450, phosphatidyl-choline, deoxycholate, the reductase as the rate-limiting component, and NADPH as the final addition.

B. Properties of Highly Purified Reductase

SDS-polyacrylamide gel electrophoresis of the purified reductase showed a single major band accounting for about 70% of the protein, along with some minor bands. Electrophoresis on calibrated gels indicated that the major polypeptide has a molecular weight of about 80,000 daltons. A typical purified NADPH-cytochrome P-450 reductase preparation had the properties shown in Table V. Neither cytochrome P-450 nor cytochrome b_5 could be detected, but FMN and FAD were present in the amounts of 0.65 and 0.74 nmole, respectively, per 80,000 ng of Since the reductase preparation is about 70% pure, it appears likely that the enzyme contains one molecule of each of these flavins per polypeptide chain. Iyanagi and Mason (1973) have previously reported the presence of both flavin nucleotides in detergent-solubilized and partially purified reductase preparations from rabbit and pig liver microsomes. Their maximal activity toward cytochrome c was reported to be 12,400 nmoles min^{-1} mg^{-1}, but no evidence was presented for cytochrome P-450 reduction.

It should be noted that our detergent -solubilized NADPH-cytochrome P-450 reductase from rat liver microsomes is active not only toward cytochrome P-450 from the same source (Lu and Coon, 1968), but also that from rabbit (Lu et al., 1969b), mouse (Nebert et al., 1973), and human liver (Kaschnitz and Coon, 1972, 1974) and from a yeast (Lebeault et al., 1971; Duppel et al., 1973) as well as the carcinogen-inducible cytochrome P-448 from mouse (Nebert et al., 1973) and rat liver (Lu et al., 1972).

IV. EVIDENCE FOR AN UNIDENTIFIED ELECTRON ACCEPTOR

Anaerobic titrations of partially purified cytochrome $P-450_{LM}$ revealed the uptake of two electrons per molecule during the linear phase, whereas the titration of $P-450_{cam}$ (kindly donated by Dr. I. G. Gunsalus) and cytochrome c showed the expected uptake of only one electron per molecule (Table VI). The titration curves from a typical experiment with a more purified preparation of $P-450_{LM}$ are presented in Fig. 2. Upon successive additions of reductant the Soret band at 416 nm disappeared with the appearance of the expected peak at 450 nm, along with several isosbestic points. The plot of these results in Fig. 3 shows that 1.05 nmoles of dithionite were consumed per molecule of cytochrome P-450 converted to the reduced CO complex. Since dithionite is a two-electron donor, it is evident that 2.1 electrons were taken up for each molecule of $P-450_{LM}$ reduced.

The experiments summarized in Table VII provide additional

TABLE V

Properties of Purified NADPH-cytochrome P-450 Reductase

Component	Concentration or activity
NADPH-cyt. c reductase	30,400 nmoles per min per mg protein
FMN	0.65 nmole per 80,000 ng protein
FAD	0.74 nmole per 80,000 ng protein
Cytochrome b_5	0
Cytochrome P-450	0

evidence for the presence of this electron acceptor, which we are calling Factor C. In Expt. 1, $P\text{-}450_{LM}$ was reduced by dithionite, oxidized by dichlorophenolindophenol (DCPIP), and again reduced by dithionite. In Expt. 2, the reduction was carried out photochemically, and the reduced P-450 was then oxidized by DCPIP, and in each instance a 2-electron transfer was observed. Two electrons were also consumed when NADPH served as the electron donor in the presence of catalytic amounts of the reductase, as shown in Expt. 3.

In another study the cytochrome P-450 was converted to P-420 under various conditions and then titrated. As shown in Table VIII, approximately two electrons were taken up from dithionite whether P-420 was made from P-450 in the presence of a mercurial, or urea, or a combination of alkylating agents and urea. Finally, P-420 prepared by the treatment of $P\text{-}450_{LM}$ with urea was reduced photochemically, exposed to iodoacetamide, and then oxidized in air. After dialysis the resulting preparation still consumed about two electrons per molecule of P-420. This experiment rules out the involvement of a redox-active disulfide group as the unknown electron acceptor, since the sulfhydral groups formed upon reduction would presumably have been alkylated by iodoacetamide and would therefore no longer be capable of participating in electron transfer.

The ability of $P\text{-}450_{LM}$ to catalyze substrate hydroxylation when reduced either by dithionite or photochemically and then exposed to air is shown in Table IX. Since autoxidation of the P-450 competes with the hydroxylation reaction, the theoretical

TABLE VI

Electron Uptake by Various Hemoproteins Upon Titration

with Dithionite Under Anaerobic Conditions[a]

Hemoprotein titrated	Electrons consumed per molecule	
	Lag phase	Linear phase
Cytochrome c	0	1.0
P-450$_{cam}$	$\leq$ 0.1	1.1
P-450$_{cam}$ (phosphatidylcholine and glycerol present)	$\leq$ 0.1	1.1
P-450$_{LM}$ (glycerol present)	0.5	$\approx$2.0
P-450$_{LM}$ (phosphatidylcholine and glycerol present)	0.5	2.0$\pm$0.2

[a]The titrations were carried out in an anaerobic apparatus (Burleigh et al., 1969) with carbon monoxide containing less than 0.5 ppm of oxygen as the gas phase.

yield of product was not expected. The relative activities of the substrates correspond roughly to their activities in the presence of the usual complete enzyme system containing NADPH and the reductase. Such experiments show that artificially reduced cytochrome P-450 is capable of catalyzing substrate hydroxylation or, in other words, that the reductase apparently does not play a facilitating role other than that of electron transfer. On the other hand, the results do not indicate whether the second electron acceptor (Factor C) functions directly in the hydroxylation reaction.

V. PROPERTIES OF THE RECONSTITUTED ENZYME SYSTEM

A. Substrate Specificity

The activity toward various substrates by microsomes, by the reconstituted enzyme system before purification, and by the reconstituted system containing highly purified P-450$_{LM}$ and

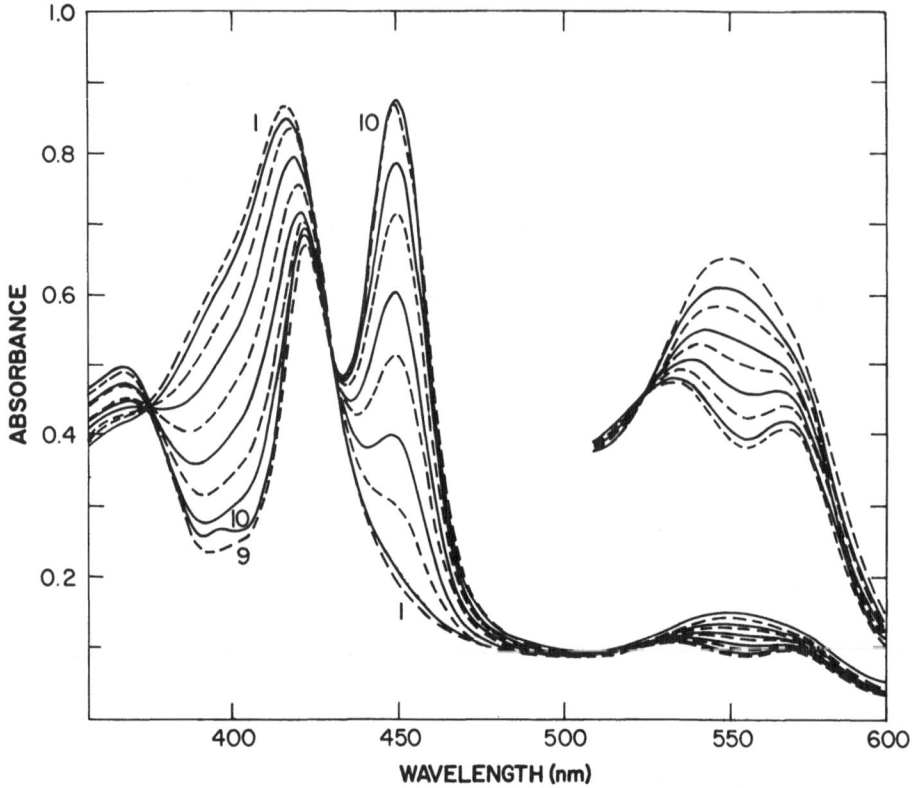

Fig. 2. Spectra observed during stepwise titration by
dithionite of cytochrome P-450 purified to a concentration of
13.3 nmoles per mg of protein. A catalytic amount of methyl
viologen was present.

reductase is shown in Table X. Since the results are expressed as
turnover numbers, it should be noted that such values are highly
meaningful only in the reconstituted system, in which cytochrome
P-450 is readily made the rate-limiting component. In contract,
cytochrome P-450 may not necessarily be the rate-limiting component
in microsomal suspensions. The results show that both the solubilized but unpurified P-450 and the highly purified P-450 retain
activity toward a variety of substrates. The highest turnover
numbers toward benzphetamine and cyclohexane have been obtained
with the most purified preparations. Despite the broad substrate
specificity observed for the highly purified cytochrome P-450
from microsomes of phenobarbital-induced rabbits, it would be

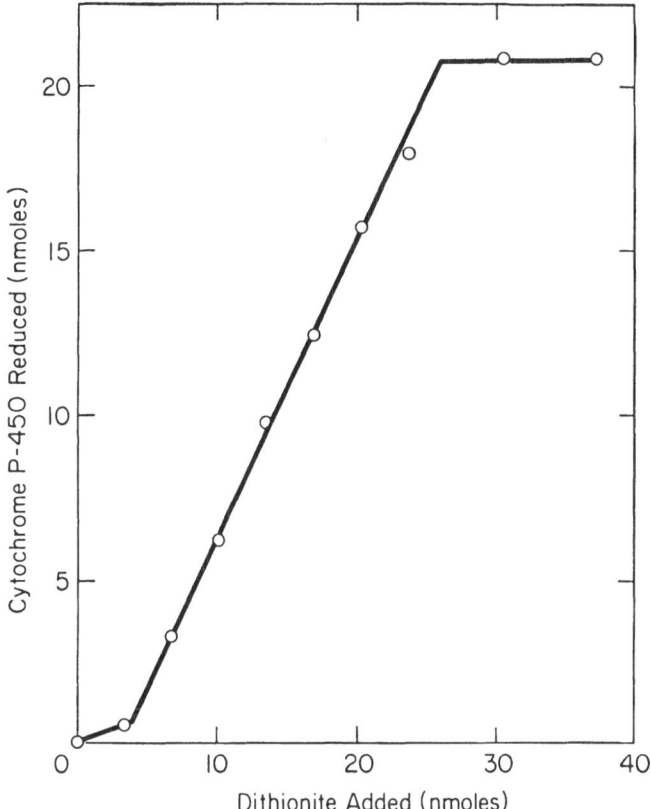

Fig. 3. Cytochrome P-450 reduction as a function of dithionite added.

surprising if a single species of P-450 were to account for all of the activities associated with microsomes and attributed to this pigment. Our more recent studies indicate that liver microsomal P-450 exists in multiple forms differing in molecular weight and substrate specificity (van der Hoeven et al., 1974).

Mention should also be made of the physical properties of the reconstituted enzyme system. It appears likely that a dissociable complex containing cytochrome P-450, NADPH-cytochrome P-450 reductase, and phosphatidylcholine functions in catalysis, but no evidence could be obtained for the formation of aggregates or membrane-like structures upon the addition of phospholipid to the other components of the enzyme system (Autor et al., 1973a, 1973b). The techniques employed were sedimentation, gel exclusion chromatography, and electron microscopy. It remains to be

TABLE VII

Electron Transfer in Oxidation and Reduction

of Cytochrome P-450

Experimental Conditions	Electrons accepted per molecule P-420
P-420 (5 x 10^{-3} $\underline{M}$ $\underline{p}$-mercuribenzoate)	1.8
P-420 (8 $\underline{M}$ urea)	1.9
P-420 (10^{-2}$\underline{M}$ iodoacetate and 8 $\underline{M}$ urea)	2.3
P-420 (10^{-2}$\underline{M}$ iodoacetamide and 8 $\underline{M}$ urea)	2.2
P-420 (8 $\underline{M}$ urea) reduced photo-chemically, exposed to 10^{-2} $\underline{M}$ iodoacetamide, oxidized and dialyzed	2.4

established what effect phospholipid would have on these hydrophobic proteins in the absence of detergents. However, contrary to a report from another laboratory (Shoeman et al., 1970), hydroxylation activity definitely does not require that P-450$_{LM}$ occur in a membranous structure. Much remains to be learned about the physical properties of these proteins from liver microsomal membranes, and their present availability in highly purified form should facilitate such studies.

B.   Proposed Scheme for Mechanism of Cytochrome P-450-catalyzed Hydroxylation Reactions

The reaction sequence presented in Fig. 4 indicates the manner in which electrons are believed to flow from NADPH to molecular oxygen in this enzyme system during substrate hydroxylation. The reductase, now known to contain both FMN and FAD, is first reduced

TABLE VIII

Electron Uptake by Cytochrome P-420 from Dithionite

Experiment	Conditions	Electrons accepted or donated per molecule P-450
1	Reduction by dithionite	+2.0
	Oxidation by DCPIP	-2.0
	Reduction by dithionite	+2.1
2	Photochemical reduction, then oxidation by DCPIP	-2.1
3	Reduction by NADPH in presence of reductase	+1.9

and then transfers electrons to cytochrome P-450. The scheme shows the involvement of Factor C as well as the heme iron atom, although, as already stated, the function of this factor in hydroxylation needs to be clarified. The presence of phosphatidylcholine is necessary for the rapid flow of electrons to cytochrome P-450 (Strobel et al., 1970), but its involvement in other steps, such as oxygen insertion, is not ruled out. The formation of a complex containing P-450, substrate, and superoxide anion is shown as a possible step in the formation of the hydroxylated substrate and water. Although evidence was obtained earlier from an effect of superoxide-generating and -dismuting agents on benzphetamine hydroxylation in the unpurified, reconstituted enzyme system containing cytochrome P-450 solubilized with deoxycholate and fractionated by DEAE-cellulose column chromatography (Strobel and Coon, 1971), such effects were not seen with cytochrome P-450 solubilized with cholate and fractionated with ammonium sulfate (Coon et al., 1973b). Accordingly, further investigation will be necessary to clarify whether superoxide serves as an activated form of oxygen in this enzyme system.

TABLE IX

Substrate Hydroxylation by Reduced P-450

Method of reduction	Substrate	Yield of product (% of theoretical yield based on amount of reduced P-450$_{LM}$ present)
Dithionite	Laurate	4
Photochemical	Laurate	6
Dithionite	Aminopyrine	12
Photochemical	Benzphetamine	50

Cytochrome P-450 was reduced chemically or photochemically, indicated, under anaerobic conditions in the presence of substrate. Air was then admitted to the system, and after 3 min product formation was determined. The hydroxylation of radioactive laurate was determined according to Kusunose et al. (1964), and ^{14}C-labeled formaldehyde liberated by the hydroxylation of radioactive aminopyrine or benzphetamine was estimated according to Poland and Nebert (1973). The radioactive aminopyrine was generously provided by Dr. D. W. Nebert.

TABLE X

Substrate Hydroxylation in Rabbit Liver Enzyme System

Substrate	Turnover number (nmoles per nmole P-450 per min)		
	Microsomes	Crude reconstituted system	Purified reconstituted system
Benzphetamine	9.4	22.0	41.8
Cyclohexane		23.2	40.3
Hexane		30.8	35.7
Octane			14.5
Hexobarbital			7.8
Ethylmorphine	1.7		3.1
Laurate	1.6	1.8	0.9
Aniline	0.3		0.5

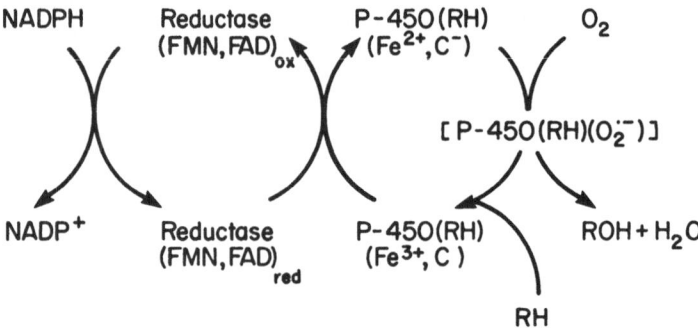

Fig. 4.   Properties for an unidentified electron acceptor:
         Factor C

VI.   SUMMARY

Cytochrome P-450 has been purified from liver microsomes of
phenobarbital-induced rabbits in the presence of ionic and non-
ionic detergents to concentrations over 17 nmoles per mg of protein.   The purified cytochrome P-450$_{LM}$ gives a single major band
on SDS-polyacrylamide gel electrophoresis representing about 90%
of the total protein.   The polypeptide chain has a molecular
weight of about 49,000 daltons.

NADPH-cycochrome P-450 reductase has been purified from
liver microsomes of phenobarbital-induced rats in the presence of
ionic and nonionic detergents to a stage where it catalyzes the
reduction of 33,000 nmoles of cytochrome c per min per mg of protein.   The ratio of activities toward cytochrome P-450 and cytochrome c is constant throughout purification.   The purified reductase contains equimolar amounts of FMN and FAD and gives a
single major band on SDS-polyacrylamide gel electrophoresis
accounting for about 70% of the total protein; the molecular weight
is about 80,000 daltons.

The purified cytochrome P-450 is free of cytochrome b$_5$ but
contains another electron acceptor, provisionally called Factor C,
which is equivalent in amount to the heme present.   Two electrons
are taken up per molecule of cytochrome P-450 from dithionite or
from NADPH in the presence of catalytic amounts of the reductase,
and both electrons are readily transferred from the reduced cytochrome P-450 to molecular oxygen or artificial electron acceptors.

The reconstituted enzyme system containing purified cytochrome
P-450, purified NADPH-cytochrome P-450 reductase, and phosphatidylcholine retains the ability to catalyze the hydroxylation of

drugs, fatty acids, hydrocarbons, and aniline in the presence of NADPH and molecular oxygen.

FOOTNOTES

* This research was supported by Grant GB-30491X from the National Science Foundation and Grant AM-10339 from the United States Public Health Services.
+ Postdoctoral Fellow, United States Public Health Service.
‡ Predoctoral Trainee, United States Public Health Service, Grant GM-00187.

REFERENCES

Autor, A. P., Kaschnitz, R. M., Heidema, J. K., van der Hoeven, T. A., Duppel, W., and Coon, M. J. 1973a. Role of phospholipid in the reconstituted liver microsomal mixed function oxidase system containing cytochrome P-450 and NADPH-cytochrome P-450 reductase. Drug Metab. Disp. 1:156-161.

Autor, A. P., Kaschnitz, R. M., Heidema, J. K., and Coon, M. J. 1973b. Sedimentation and other properties of the reconstituted liver microsomal enzyme system containing cytochrome P-450, reduced triphosphopyridine nucleotide-cytochrome P-450 reductase, and phosphatidylcholine. Mol. Pharmacol. 9:93-104.

Ballou, D. P., Veeger,. C., van der Hoeven, T. A., and Coon, M. J. 1974. Properties of partially purified liver microsomal cytochrome P-450: acceptance of two electrons during anaerobic titration. FEBS Letters 38:337-340.

Bligh, E. G., and Dyer, W. J. 1959. A rapid method of total lipid extraction and purification. Can. J. Biochem. Physiol. 37:911-917.

Burleigh, B. D., Jr., Foust, G. P., and Williams, C. H., Jr. 1969. A method for titrating oxygen-sensitive organic redox systems with reducing agents in solution. Anal. Biochem. 27:536-544.

Coon, M. J., and Lu, A. Y. H. 1969. Fatty acid ω-oxidation in a soluble microsomal enzyme system containing P-450, pp. 151-166. In J. R. Gillette et al. (ed.). Microsomes and Drug Oxidations. Academic Press, New York.

Coon, M. J., van der Hoeven, T. A., Kaschnitz, R. M., and Strobel, H. W. 1973a. Biochemical studies on cytochrome P-450 solubilized from liver microsomes: partial purification and mechanism of catalysis. Ann. N. Y. Acad. Sci. 212:449-457.

Coon, M. J., Strobel, H. W., and Boyer, R. F. 1973b. On the mechanism of hydroxylation reactions catalyzed by cytochrome P-450. Drug Metab. Disp. 1:92-97.

Duppel, W., Lebeault, J. M., and Coon, M. J. 1973. Properties of
    a yeast cytochrome P-450-containing enzyme system which
    catalyzes the hydroxylation of fatty acids, alkanes and drugs.
    Eur. J. Biochem. 36:583-592.
Fujita, T., Shoeman, D. W., and Mannering, G. J. 1973. Differ-
    ences in P-450 cytochromes from livers of rats treated with
    phenobarbital and with 3-methylcholanthrene. J. Biol. Chem.
    248:2192-2201.
Iyanagi, T., and Mason, H. S. 1973. Some properties of hepatic
    reduced nicotinamide adenine dinucleotide phosphate-cyto-
    chrome c reductase. Biochemistry 12:2297-2308.
Kaschnitz, R. M., and Coon, M. J. 1972. Solubilized human liver
    cytochrome P-450: phospholipid requirement in hydroxylation
    reactions. Abstracts, Fifth International Congress on
    Pharmacology, San Francisco, p. 120.
Kaschnitz, R. M., and Coon, M. J. 1974. Drug, fatty acid, and
    hydrocarbon hydroxylation by solubilized human liver cytochrome
    P-450: phospholipid requirement. Biochem. Pharmacol. in
    press.
King, E. J., 1932. The colorimetric determination of phosphorus.
    Biochem. J. 26:292-297.
Kusunose, M., Kusunose, E., and Coon, M. J. 1964. Enzymatic ω-
    oxidation of fatty acids. I. Products of octanoate, decanoate,
    and laurate oxidation. J. Biol. Chem. 239:1374-1380.
Laemmli, U. K. 1970. Cleavage of structural proteins during the
    assembly of the head of bacteriophage T4. Nature 227:680-685.
Lebeault, J. M., Lode, E. T., and Coon, M. J. 1971. Fatty acid and
    hydrocarbon hydroxylation in yeast: role of cytochrome P-450
    in Candida tropicalis. Biochem. Biophys. Res. Commun. 42:
    413-419.
Levin, W., Lu, A. Y. H., Ryan, D., West, S., Kuntzman, R. and
    Conney, A. H. 1972. Partial purification and properties of
    cytochromes P-450 and P-448 from rat liver microsomes. Arch.
    Biochem. Biophys. 153:543-553.
Levin, W., Ryan, D., West, S., and Lu, A. Y. H. 1974. Preparation
    of partially purified, lipid depleted cytochrome P-450 and
    reduced nicotinamide adenine dinucleotide phosphate-cytochrome
    c reductase from rat liver microsomes. J. Biol. Chem. 249:
    1747-1754.
Lowry, O., Rosebrough, N. J., Farr, A. L., and Randall, R. J. 1951.
    Protein measurement with the Folin phenol reagent. J. Biol.
    Chem. 193:265-275.
Lu, A. Y. H., and Coon, M. J. 1968. Role of hemoprotein P-450
    in fatty acid ω-hydroxylation in a soluble enzyme system from
    liver microsomes. J. Biol. Chem. 243:1331-1332.
Lu, A. Y. H., Junk, K. W., and Coon, M. J. 1969a. Resolution of
    the cytochrome P-450-containing ω-hydroxylation system of
    liver microsomes into three components. J. Biol. Chem.
    244:3714-3721.

Lu, A. Y. H., Strobel, H. W., and Coon, M. J. 1969b. Hydroxylation of benzphetamine and other drugs by a solubilized form of cytochrome P-450 from liver microsomes: lipid requirement for drug demethylation. Biochem. Biophys. Res. Commun. 36:545-551.

Lu, A. Y. H., Strobel, H. W., and Coon, M. J. 1970. Properties of a solubilized form of the cytochrome P-450-containing mixed function oxidase of liver microsomes. Mol. Pharmacol. 6:213-220.

Lu, A. Y. H., Kuntzman, R., West, S., Jacobson, M., and Conney, A. H. 1972. Reconstituted liver microsomal enzyme system that hydroxylates drugs, other foreign compounds, and endogenous substrates. II. Role of cytochrome P-450 and P-448 fractions in drug and steroid hydroxylations. J. Biol. Chem. 247:1727-1734.

Nebert, D. W., Heidema, J. K., Strobel, H. W. and Coon, M. J. 1973. Genetic expression of aryl hydrocarbon hydroxylase induction. Genetic specificity resides in the fraction containing cytochrome P_{448} and P_{450}. J. Biol. Chem. 248:7631-7636.

Omura, T., and Sato, R. 1964. The carbon monoxide-binding pigment of liver microsomes. I. Evidence for its hemeprotein nature. J. Biol. Chem. 239:2370-2378.

Poland, A. P., and Nebert, D. W. 1973. A sensitive radiometric assay of aminopyrine N-demethylation. J. Pharmacol. Exp. Ther. 184:269-277.

Sato, R., Satake, H., and Imai, Y. 1973. Partial purification and some spectral properties of hepatic microsomal cytochrome P-450. Drug Metab. Disp. 1:6-13.

Shoeman, D. W., White, J. G., and Mannering, G. J. 1970. Role of lipid induced aggregation in the function of "soluble" cytochrome P-450. Pharmacologist 12:260.

Spatz, L., and Strittmatter, P. 1971. A form of cytochrome b_5 that contains an additional hydrophobic sequence of 40 amino acid residues. Proc. Nat. Acad. Sci. U. S. 68:1042-1046.

Strobel, H. W., Lu, A. Y. H., Heidema, J., and Coon, M. J. 1970. Phosphatidylcholine requirement in the enzymatic reduction of hemoprotein P-450 and in fatty acid, hydrocarbon, and drug hydroxylation. J. Biol. Chem. 245:4851-4854.

Strobel, H. W., and Coon, M. J. 1971. Effect of superoxide generation and dismutation on hydroxylation reactions catalyzed by liver microsomal cytochrome P-450. J. Biol. Chem. 246:7826-7829.

van der Hoeven, T. A., and Coon, M. J. 1974. Preparation and properties of partially purified cytochrome P-450 and NADPH-cytochrome P-450 reductase from rabbit liver microsomes. J. Biol. Chem. 249: in press.

van der Hoeven, T. A., Haugen, D. A., and Coon, M. J. 1974. Separation and characterization of multiple forms of liver

microsomal cytochrome P-450. Pharmacologist 16: in press.
Williams, C. H., Jr., and Kamin, H. 1962. Microsomal triphospho-
   pyridine nucleotide-cytochrome c reductase of liver. J.
   Biol. Chem. 237:587-595.

IMMUNOCHEMICAL AND COMPOSITIONAL COMPARISON OF CYTOCHROME P-450$_{cam}$

OF PSEUDOMONAS PUTIDA AND P-450$_{LM}$ OF PHENOBARBITAL-INDUCED RABBIT

LIVER MICROSOMES*

Karl Dus, William J. Litchfield, and Anne G. Miguel
Biochemistry Department
University of Illinois, Urbana, Illinois

and

Theodore A. van der Hoeven, David A. Haugen, William
L. Dean, and Minor J. Coon
Department of Biological Chemistry
The University of Michigan, Ann Arbor, Michigan

During the last two years my laboratory has been concerned with the determination of the structure and sequence of cytochrome P-450$_{cam}$ of Pseudomonas putida. One of our more recent lines of investigations is to compare this hemeprotein to P-450 cytochrome of higher organisms. The main objective of this investigation is to identify characteristic structural features common to P-450 type hemeproteins, and to demonstrate the value of the readily soluble and accessible, pure cytochrome P-450$_{cam}$ as a structural model.

As an initial approach, the microsome fraction of phenobarbital-treated rats was isolated and tested for immunological cross reactivity with anti-cytochrome P-450$_{cam}$ antisera. These experiments as well as all the subsequent immunochemical cross reactions were carried out by radioimmunoassay using a competitive binding assay with ^{125}I-labeled cytochrome P-450$_{cam}$ which had about 50% of its tyrosines monoiodinated. The response of this microsome preparation in our assay was marginally positive and seemed to level off at about 15% (see Figure 1A). This result was promising because it implied that not only was microsomal P-450 able to cross react with anti-P-450$_{cam}$ antibodies but also that a measurable amount of this P-450 was actually located at or very close to the surface of the microsomes. In order to pursue this problem further we

selectively cleaved cytochrome P-450$_{LM}$ with BrCN under mild conditions. This technique had previously yielded a small heme-containing peptide of P-450$_{cam}$ which was so tightly structured that
it retained the heme in stoichiometric amounts throughout purification. Furthermore, a significant portion of the antigenic
determinants of the native protein seemed to be still associated
with this peptide (Dus et al., 1973). Initially we proceeded with
BrCN digests of whole liver microsomes rather than via isolation of
highly purified P-450$_{LM}$ and its subsequent cleavage with BrCN.
From the resulting peptide mixture we recovered after many purification steps a heme-containing peptide fraction, Fraction 3 in
Figure 1A, which was still quite complex but showed a Soret maximum between 390 and 400 nm. It also exhibited twice as much cross
reactivity with anti-P-450$_{cam}$ antibodies as did the undegraded
microsomes. But most importantly, we demonstrated with this
experiment that microsomal P-450 was indeed related to P-450$_{cam}$
and that some of the P-450$_{LM}$ molecules were probably somewhat
exposed on the surface of the microsomes.

After these exploratory experiments, it was most important to
pursue the comparison at the level of the pure, functional heme
proteins and the pure hemepeptides derived from them. Therefore
highly purified preparations of P-450$_{LM}$ of phenobarbital-induced
rabbits (van der Hoeven and Coon, 1974) were tested for their
ability to cross react with purified anti-cytochrome P-450$_{cam}$
antibodies. A representative competitive binding assay using a
preparation of a specific P-450$_{LM}$ content of 13.3 nmoles/mg protein
is shown in Figure 1B while Figure 2 demonstrates that anti-P-450$_{cam}$
antibodies are also specific and potent inhibitors of microsomal
hydroxylation reactions using benzphetamine as substrate. In each
case inhibitions of 60-70% were obtained. These findings clearly
point to a substantial similarity of antigenic site structures of
these two hemeproteins.

In Table I we show a comparison of the amino acid compositions
of cytochromes P-450$_{cam}$ of Pseudomonas putida and P-450$_{LM}$ of
phenobarbital-induced rabbits. These data were obtained by a series
of analyses using both acid and enzymic hydrolysis; they are calculated in terms of residues assuming a molecular weight of approximately 45,000 daltons for both cytochromes. It should be noted
that, while this analysis refers to a preparation of a specific
P-450$_{LM}$ content of 13.3 nmoles/mg protein which corresponds to less
than complete purity, data from SDS acrylamide gel electrophoresis,
shown to you by Dr. Coon just a few minutes ago, clearly demonstrate
that only one major protein band can be seen in this preparation
together with traces of contaminants. Thus, we conclude that the
major contaminant present in this preparation is the apoprotein of
P-450$_{LM}$, and that it is therefore quite justified to use this
preparation for comparative amino acid analyses. This comparison

TABLE I

Amino Acid Compositions

Amino Acids	P-450$_{cam}$		P-450$_{LM}$	
	Protein	Hemepeptide	Protein	Hemepeptide
CySO$_3$H	6	1	6	1
Asp	27		21	
Asn	9	> 3	14	> 4
Thr	19	2	23	2
Ser	21	3	26	2
Glu	42		24	
Gln	13	> 3	19	> 4
Pro	27	3	24	2
Gly	26	8	30	3
Ala	34	6	23	2
Val	24	4	27	2
Met	9	-	8	-
Ile	24	2	19	2
Leu	40	3	46	4
Tyr	9	1	11	1
Phe	17	2	28	2
His	12	1	11	1
Lys	13	2	19	2
Trp	1	-	1	-
Arg	24	2	29	2
HSer	--	1	--	1
Total	397	47	409	37
Heme	1	1	1	1

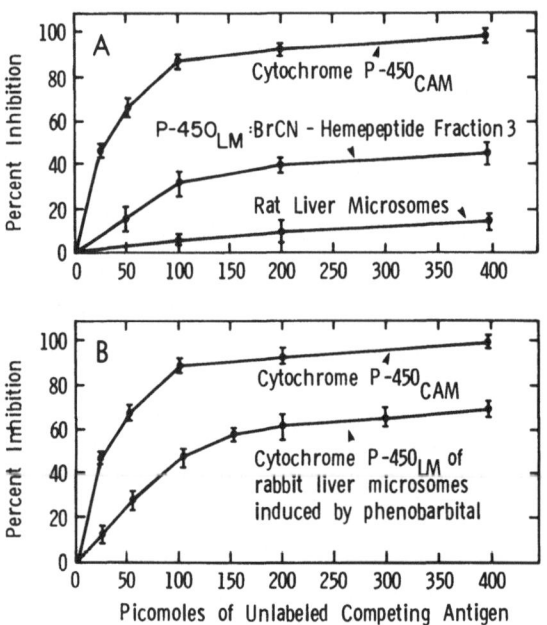

Fig. 1. Competitive Binding Assay Using ^{125}I-labeled
Cytochrome P-450$_{cam}$ and Anti-P-450$_{cam}$ Antibodies. Aliquots of
unlabeled antigen were mixed with 0.15 M NaCl-50 mM sodium
phosphate buffer (pH 7.5) and a constant volume of antibody
solution to give a total volume of 300 μl. The mixture was
incubated at room temperature for 2 hr before an aliquot of
labeled antigen, sufficient to saturate the antibody at equivalence,
was added. After incubation for 48 hr at 10°, the precipitates
were spun down at 3,000 x g, washed once with cold buffer, dissolved in 200 μl of 0.2 N NaOH, and counted with the ^{14}C window
of a Beckman Model LS-30 liquid scintillation counter. The per
cent inhibition by purified P-450$_{LM}$ is based on maximal binding
between P-450$_{cam}$ and its antibodies normalized to 100%.

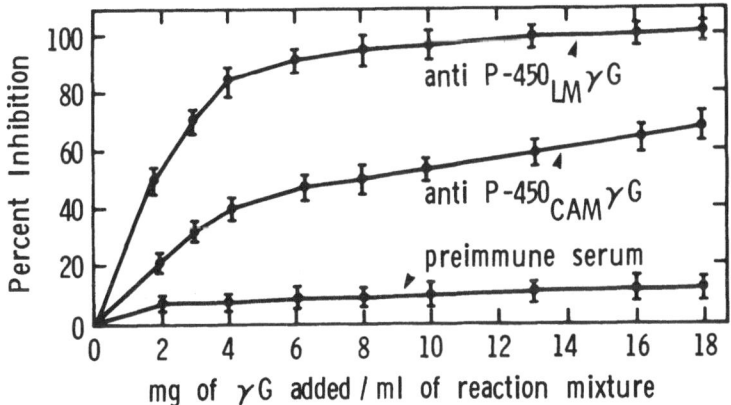

Fig. 2.   Inhibition of Benzphetamine Hydroxylation.   The
effect of antibodies upon substrate-dependent NADPH oxidation by
a reconstituted P-450$_{LM}$ system was monitored with a Gilford 2000
spectrophotometer at 340 nm.

tells us that we are dealing with closely related proteins of predominantly hydrophobic character which, among other features, contain one heme group, six half-cystines, and a carbohydrate unit
(Dus et al., 1970). Although the precise molecular weight of
phenobarbital-induced P-450$_{LM}$ is not yet known our comparison
suggests that this hemeprotein may be slightly larger than
P-450$_{cam}$ because of additional amino acid residues and a larger
carbohydrate unit. But while P-450$_{cam}$ has a pI of 4.5, which is
in agreement with the excess of acidic over basic residues (Dus
et al., 1970), P-450$_{LM}$ has apparently fewer acidic than basic
residues and thus should have a basic pI. Nevertheless,
preliminary electrofocusing experiments indicated a pI for
P-450$_{LM}$ that is even more acidic than that of P-450$_{cam}$. This
could be due to the presence of acidic carbohydrate units.

Finally, we selectively degraded purified cytochrome P-450$_{LM}$
with BrCN under mild conditions and isolated a small, pure hemepeptide. The amino acid composition of this peptide is compared in
Table I to that of the respective small hemepeptide obtained from
P-450$_{cam}$. In addition to a general similarity in composition, it
is worth noting that both hemepeptides contain one residue each of
cysteine and histidine, which probably represent the groups
chelating to the heme iron. We are now ready to proceed with the
determination of the amino acid sequences of these hemepeptides
which should yield information more readily interpretable in terms
of homology of structure and function.

Thus, in closing I would like to point out that the experimental approach we have chosen makes it possible to compare corresponding portions of both cytochromes. These portions also happen
to be the centers of the catalytic action of these proteins, which
is of interest to biochemists and pharmacologists. This work,
which should give detailed information on how the active site of
each of these proteins is constructed, can be done without determining the complete amino acid sequence of such large polypeptides,
and even without having access to completely homogeneous preparations of membrane-bound P-450 cytochromes.

SUMMARY

Although highly purified cytochrome P-450 of Pseudomonas putida (P-450$_{cam}$), and that from phenobarbital-induced rabbit liver microsomes (P-450$_{LM}$), differ markedly in their catalytic and physical properties, they show immunological cross reaction by competitive binding and inhibition of catalytic activity, and are of similar amino acid composition. Upon treatment with cyanogen bromide they yield small heme-containing peptides of highly similar amino acid composition.

Footnotes

*   Supported in part by Grants GM-18902 to K. D. and AM-10339 to M. J. C. from the National Institutes of Health, and GB-30419X to M. J. C. from the National Science Foundation.

REFERENCES

Dus, K., Katagiri, M., Yu, C. A., Erbes, D. L., and Gunsalus, I. C. 1970. Chemical Characterization of Cytochrome P-450$_{cam}$. Biochem. Biophys. Res. Commun. 40: 1423-1430.

Dus, K., Miguel, A. G., Smith, P. C. Litchfield, W. J., and Harrison, J. E. 1973. Hemepeptides Derived from Cytochrome P-450$_{cam}$. 9th International Congress of Biochemistry, Stockholm, Abstracts, p. 341.

van der Hoeven, T. A., and Coon, M. J. 1974. Preparation and Properties of Partially Purified Cytochrome P-450 and NADPH-cytochrome P-450 Reductase from Rabbit Liver Microsomes. J. Biol. Chem. 249: in press.

IMMUNOCHEMICAL AND FUNCTIONAL SIMILARITIES AND DIFFERENCES AMONG IRON-SULFUR PROTEINS INVOLVED IN MAMMALIAN STEROIDOGENESIS[1]

Jeffrey Baron
Department of Pharmacology
The Toxicology Center
The University of Iowa
Iowa City, Iowa 52242

In the mitochondrial fraction of steroidogenic tissues, cytochrome P-450 mediates the initial step in the biosynthesis of the steroid hormones, the oxidative cleavage of the cholesterol side-chain to form pregnenolone. In addition, cytochrome P-450 also mediates the 11β- and 18- hydroxylations of steroids which occur in adrenocortical mitochondria. While cytochrome P-450 plays a central role in these mixed-function oxidations, the ability of this hemoprotein to function as the terminal oxidase depends on the sequential transfer of two reducing equivalents from NADPH to it by means of specific electron transport sequences. The identification of the components of these electron transport sequences was greatly advanced by the studies of Omura and his colleagues in 1966. These investigators resolved the 11β-hydroxylase system of bovine adrenocortical mitochondria into three specific protein fractions: cytochrome P-450; an NADPH-dependent flavoprotein dehydrogenase; and an iron-sulfur protein, adrenodoxin. Based on the results of reconstitution experiments, it was suggested that the iron-sulfur protein serves as an electron transfer intermediate between the flavoprotein dehydrogenase and both cytochrome P-450 and cytochrome c. Subsequently, the cholesterol side-chain cleavage systems of adrenocortical mitochondria (Simpson and Boyd, 1967), placental mitochondria (Mason and Boyd, 1971), and ovarian mitochondria (Sulimovici and Boyd, 1968), were also resolved into three similar protein fractions. Although the electron transport sequence associated with the cholesterol side-chain cleavage reaction occurring in testicular mitochondria has resisted resolution, these mitochondria do contain cytochrome P-450 and an iron-sulfur protein which is capable of replacing adrenodoxin in the reconstituted 11β-hydroxylase system of bovine adrenocortical mitochondria (Kimura and Ohno, 1968).

These observations have led to the assumption that the electron
transport sequences involved in cytochrome P-450-mediated steroid
hydroxylations in the mitochondria of all steroidogenic tissues are
similar.

In an attempt to elucidate the possible similarities or differences which may exist among these electron transport sequences,
we have employed immunochemical techniques (Masters et al., 1971);
Baron et al., 1972; Baron, 1973). For these studies, rabbits or a
goat were immunized with adrenodoxin, isolated and purified to homogeneity from bovine adrenocortical mitochondria. Highly specific
inhibitory antibodies against the iron-sulfur protein were obtained. The present report describes the results of recent studies
on the effects of the goat antibody to adrenodoxin on steroid hydroxylation and electron transport activities catalyzed by adrenal
mitochondria from a variety of species and by ovarian, testicular
and placental mitochondria.

STUDIES WITH MAMMALIAN ADRENAL MITOCHONDRIA

Since a variety of species has been employed for studies on
the mechanism and control of adrenal steroidogenesis, the immunochemical and functional similarities among adrenodoxin(s) present
in adrenal mitochondria prepared from a number of these species were
investigated. Figure 1 demonstrates the interaction on Ouchterlony
double immunodiffusion plates between the goat anti-adrenodoxin
immunoglobulin (Ig) and the adrenal iron-sulfur proteins. It can
readily be seen that the antibody to bovine adrenodoxin interacts
with the iron-sulfur proteins present in bovine and sheep adrenocortical mitochondria and in rat, mouse, cat, dog, guinea pig,
rabbit and human adrenal mitochondria. Although there is considerable cross-reactivity, the presence of spurs where the precipitin
lines cross indicates that, while these adrenal iron-sulfur proteins appear to be similar, they are not immunochemically identical.

Having demonstrated that the goat antibody to adrenodoxin could
interact with the adrenal iron-sulfur proteins of these species, the
effects of this antibody on activities catalyzed by these adrenal
mitochondria were examined. Initial experiments were performed
with bovine adrenocortical mitochondria, the sources of the
adrenodoxin employed for immunizations. As seen in Figure 2, the
addition of goat anti-adrenodoxin Ig to sonicated bovine adrenocortical mitochondria resulted in the inhibition of the NADPH-
cytochrome c reductase activity catalyzed by these mitochondria.
This titration curve is similar to those reported previously (Baron
et al., 1972; Baron, 1973) and again demonstrates the requirement
for the iron-sulfur protein in the NADPH-dependent reduction of
cytochrome c catalyzed by bovine adrenocortical mitochondria. It

TABLE I

Effect of goat anti-adrenodoxin Ig on NADPH-cytochrome c reductase
activities catalyzed by sonicated mammalian adrenal mitochondria.[a]

Adrenal mitochondria	mg Ig protein/mg mitochondrial protein	nmoles cytochrome c reduced/min/mg protein	Percent of control activity
Sheep	0	70.7	100
	2 (immune)	24.2	34
	2 (pre-immune)	69.8	99
Rat	0	55.6	100
	2 (immune)	7.3	13
	2 (pre-immune)	56.7	102
Mouse	0	43.5	100
	2 (immune)	7.4	17
	2 (pre-immune)	43.4	100
Cat	0	24.1	100
	2 (immune)	6.5	27
	2 (pre-immune)	25.3	105
Dog	0	38.3	100
	2 (immune)	9.6	25
	2 (pre-immune)	40.6	106
Guinea pig	0	27.8	100
	2 (immune)	7.2	26
	2 (pre-immune)	30.0	108
Rabbit	0	13.3	100
	2 (immune)	3.9	29
	2 (pre-immune)	14.6	110

[a]The conditions were the same as those described for Figure 2.

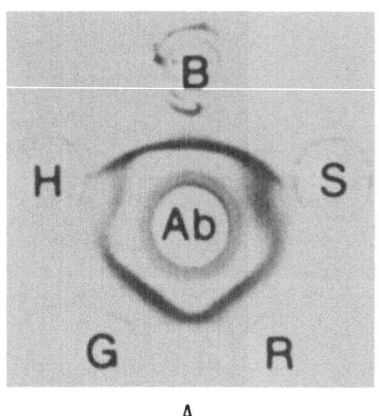
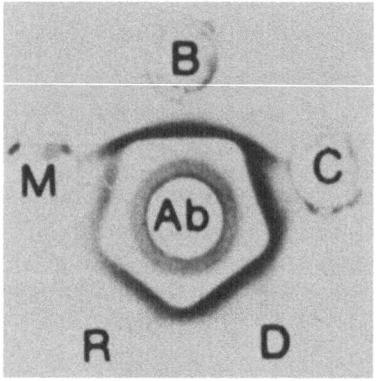

                    A   B

Fig. 1. Ouchterlony double immunodiffusion of goat anti-
adrenodoxin Ig against mammalian adrenal mitochondria. The center
wells of both plates contained goat anti-adrenodoxin Ig (Ab). In
plate "a," the outer wells contained sonicated mitochondria prepared from bovine (B) and sheep (S) adrenal cortex and from rabbit
(R), guinea pig (G) and human (H) adrenals. In plate "b," the
outer wells contained sonicated mitochondria prepared from cat
(C), dog (D), rat (R) and mouse (M) adrenals and from bovine
(B) adrenal cortex.

should be noted that no inhibition of this enzymic activity was
observed in the presence of the pre-immune Ig. When comparable
experiments were performed with adrenal mitochondria from other
species, similar results were obtained. The data presented in
Table I show that the goat antibody to bovine adrenodoxin was also
capable of inhibiting the NADPH-dependent reduction of cytochrome
c catalyzed by sonicated sheep adrenocortical mitochondria and
sonicated rat, mouse, cat, dog, guinea pig and rabbit adrenal
mitochondria. Furthermore, the complete titration curves of
NADPH-cytochrome c reductase activities obtained with these mitochondrial preparations were quite similar to that seen when

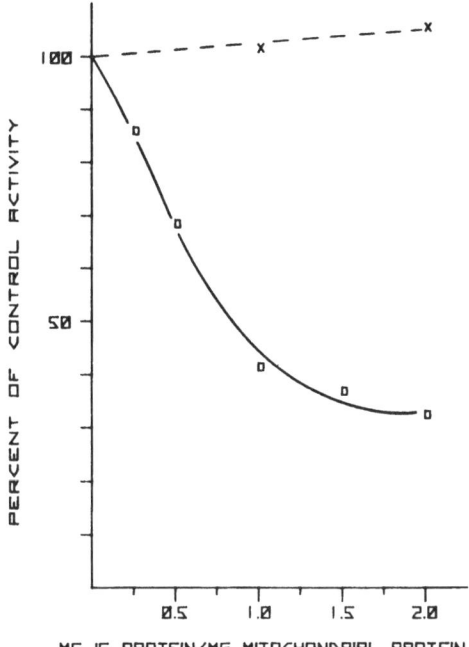

Fig. 2.   Titration of NADPH-cytochrome c reductase activity of bovine adrenocortical mitochondria by goat anti-adrenodoxin Ig. Bovine adrenocortical mitochondria were sonicated for 4 minutes. NADPH-cytochrome c reductase activity was determined as previously described (Baron et al., 1972). Anti-adrenodoxin Ig and pre-immune Ig were substituted for buffer in the indicated concentrations. The solid line represents activities in the presence of anti-adreno-doxin Ig, and the dashed line represents activities in the presence of the pre-immune Ig. In the absence of Ig, NADPH-cytochrome c reductase activity was 57.1 nmoles cytochrome c reduced min^{-1} (mg protein)$^{-1}$.

sonicated bovine adrenocortical mitochondria were studied (Figure 2). Of greatest importance, however, the results shown in Figure 3 demonstrate that the anti-adrenodoxin Ig also inhibited the NADPH-dependent reduction of cytochrome c catalyzed by soni-cated human adrenal mitochondria. Thus, the iron-sulfur proteins involved in NADPH-dependent electron transport activities of adrenal mitochondria from a number of species, including man, appear to be both immunochemically and functionally similar.

Although the goat antibody to adrenodoxin exerted profound in-hibitory action against the NADPH-cytochrome c reductase activities

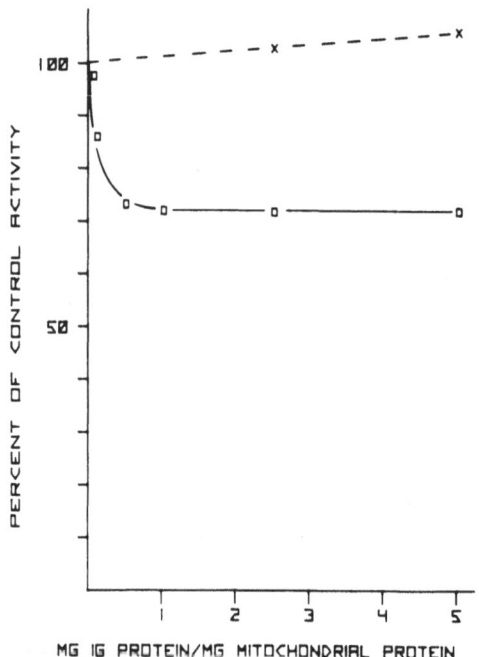

Fig. 3.   Titration of NADPH-cytochrome c reductase activity
of human adrenal mitochondria by goat anti-adrenodoxin Ig.   The
conditions were the same as those described for Figure 2.   In the
absence of Ig, NADPH-cytochrome c reductase activity was 20.8
nmoles cytochrome c reduced min^{-1} (mg protein)$^{-1}$.

catalyzed by these sonicated adrenal mitochondria, the antibody
was without effect on the NADH-dependent reduction of cytochrome
c catalyzed by these preparations (Table II). Since mitochondrial
NADH-cytochrome c reductase activity does not involve adrenodoxin
(Omura et al., 1966), these results demonstrate that the antibody
is specific for adrenodoxin and are in agreement with previous observations (Masters et al.,1971; Baron et al.,1972; Baron, 1973).

We have previously shown (Baron et al., 1972) that rabbit
antibody to bovine adrenodoxin inhibited the 11β-hydroxylation of
deoxycorticosterone catalyzed by frozen and thawed bovine adrenocortical mitochondria.   Consistent with this, the data presented
in Figure 4 show that goat anti-adrenodoxin Ig inhibited the
NADPH-dependent reduction of cytochrome c and the 11β-hydroxy-
lation of deoxycorticosterone catalyzed by a solubilized preparation derived from sonicated bovine adrenocortical mitochondria.
Furthermore, the conversion of cholesterol to pregnenolone

TABLE II

Lack of effect of goat anti-adrenodoxin Ig on NADH-cytochrome c
reductase activities catalyzed by sonicated mammalian mitochondria[a].

Adrenal mitochondria	mg Ig protein/mg mitochondrial protein	nmoles cytochrome c reduced/min/mg protein	Percent of control activity
Bovine	0	378.5	100
	2 (immune)	401.2	106
	2 (pre-immune)	382.3	101
Sheep	0	168.7	100
	2 (immune)	168.0	100
	2 (pre-immune)	169.7	100
Rat	0	59.0	100
	2 (immune)	59.8	101
	2 (pre-immune)	56.6	96
Mouse	0	250.3	100
	2 (immune)	247.8	99
	2 (pre-immune)	237.8	95
Cat	0	64.7	100
	2 (immune)	66.6	103
	2 (pre-immune)	62.1	96
Dog	0	44.6	100
	2 (immune)	42.4	95
	2 (pre-immune)	38.4	86
Guinea pig	0	127.0	100
	2 (immune)	123.2	97
	2 (pre-immune)	119.4	94
Rabbit	0	35.8	100
	2 (immune)	33.7	94
	2 (pre-immune)	32.9	92
Human	0	85.7	100
	2 (immune)	79.7	93
	2 (pre-immune)	80.8	94

[a]Conditions were the same as those described for Figure 2. NADH-cytochrome c
reductase activities were determined as previously described (Baron et al., 1972).

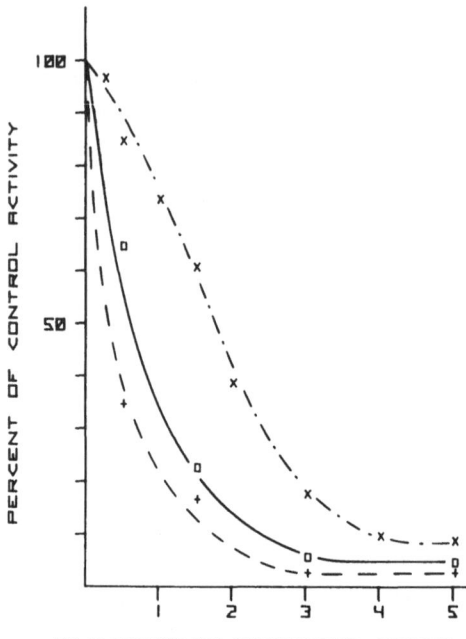

Fig. 4. Titration of NADPH-cytochrome c reductase, 11β-
hydroxylase and cholesterol side-chain cleavage activities by
goat anti-adrenodoxin Ig. Supernatant fraction S_1 was prepared
from sonicated bovine adrenocortical mitochondria according to
the method of Omura et al., (1966). The 11β-hydroxylation of
deoxycorticosterone (+) and NADPH-cytochrome c reductase activity
(x) were determined as previously described (Baron et al., 1972).
The cholesterol side-chain cleavage activity (□) was determined
by a modification of the method of Simpson and Boyd (1967).
Control activities were NADPH-cytochrome c reductase, 249.3
nmoles cytochrome c reduced min^{-1} (mg protein)$^{-1}$; 11β-hydroxylase,
1.25 nmoles corticosterone formed min^{-1} (mg protein)$^{-1}$; and
cholesterol side-chain cleavage, 9% conversion to pregnenolone
in 60 min.

catalyzed by this preparation was also inhibited in the presence
of the antibody. The antibody was also found to inhibit the
cholesterol side-chain cleavage activities of frozen and thawed
mitochondria derived from the adrenals of all of the species
discussed above.[2] These results thus demonstrate the requirement
for adrenodoxin in steroid hydroxylation catalyzed by cytochrome
P-450 in the mitochondria of mammalian adrenals. The differences
seen in Figure 4 between the titration curve for NADPH-cytochrome

c reductase activity and those for the two steroid hydroxylations
may be related to the fact that adrenodoxin is required for the
donation of the two electrons to cytochrome P-450 needed for
overall hydroxylation (Baron et al., 1972; Cooper et al., 1973),
while only one electron is needed for the reduction of cytochrome
c (Kamin et al., 1965). Thus, the steroid hydroxylations should
be more sensitive to the inhibitory action of the antibody. In
addition, the differences between the titration curves observed
for the two steroid hydroxylation reactions may reflect the differential solubilization of the two forms of cytochrome P-450
present in bovine adrenocortical mitochondria, each of which is
specific for one of the hydroxylations (Jefcoate et al.,1972).

COMPARISON OF ADRENAL, OVARIAN, TESTICULAR AND PLACENTAL MITOCHONDRIA

Since the goat anti-adrenodoxin Ig was capable of inhibiting
cholesterol side-chain cleavage and other NADPH-dependent activities involving the iron-sulfur protein in adrenal mitochondria,
the effects of this antibody on comparable enzymic activities
occurring in the mitochondria of other mammalian steroidogenic
tissues were examined. As can be seen in Figure 5, the NADPH-
dependent reduction of cytochrome c catalyzed by sonicated rat
adrenal and rat ovarian mitochondria were both inhibited in the
presence of the goat anti-adrenodoxin Ig. Indeed, the two titration curves are remarkably similar. No inhibition of this enzymic activity was observed in the presence of the goat pre-immune
Ig. To investigate the role of the mitochondrial iron-sulfur
proteins in the cholesterol side-chain cleavage activities of
these and other steroidogenic tissues, mitochondria were subjected to repeated freezing and thawing. This procedure for
altering the mitochondrial membrane permeability to the antibody
and reduced pyridine nucleotides does not significantly decrease
the conversion of cholesterol to pregnenolone and progesterone
while sonication does. Decreases in cholesterol side-chain
cleavage activity seen after sonication are presumably related
to the liberation of unlabeled cholesterol from the mitochondrial
membranes which would be capable of diluting the labeled cholesterol substrate, thereby causing an apparently lowered conversion
of labeled cholesterol to pregnenolone and progesterone. The
data in Table III show that the addition of the goat anti-
adrenodoxin Ig to frozen and thawed rat adrenal and ovarian mitochondria resulted in the inhibition of both NADPH-cytochrome c
reductase activity and the conversion of cholesterol to pregnenolone and progesterone. Thus, the iron-sulfur proteins of
adrenal and ovarian mitochondria appear to be immunochemically
and functionally similar. The stimulation of cholesterol side-
chain cleavage activity in the presence of the goat pre-immune

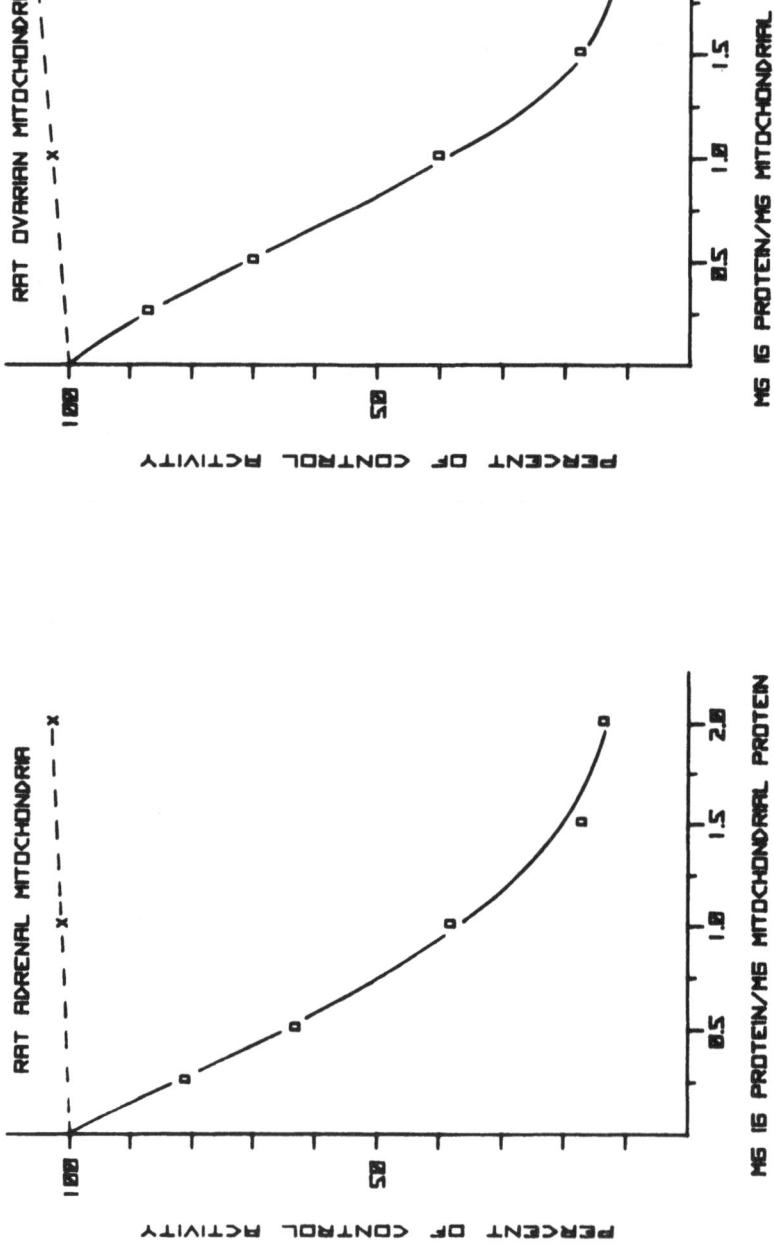

Fig. 5. Titration of NADPH-cytochrome c reductase activities of rat adrenal and ovarian mitochondria by goat anti-adrenodoxin Ig. The conditions were the same as those described for Figure 2. In the absence of Ig, NADPH-cytochrome c reductase activities were: adrenal, 55.6 nmoles cytochrome c reduced min^{-1} (mg protein)$^{-1}$; ovary, 64.1 nmoles cytochrome c reduced min^{-1} (mg protein)$^{-1}$.

TABLE III

Effects of goat anti-adrenodoxin Ig on activities of rat adrenal, ovarian and testicular mitochondria.

Mitochondria[a]	mg Ig protein/mg mitochondrial protein	NADH-cyt. c reductase[b]		NADPH-cyt. c reductase[b]		cholesterol side-chain cleavage[c]	
		nmoles/min/mg	% control	nmoles/min/mg	% control	% conversion	% control
Adrenal	0	59.0	100	113.4	100	65.4	100
	5 (immune)	57.2	97	23.8	21	3.3	5
	5 (pre-immune)	53.7	91	123.6	109	63.4	97
Ovary	0	63.7	100	67.0	100	22.0	100
	5 (immune)	64.3	101	10.7	16	1.0	5
	5 (pre-immune)	65.5	103	75.7	113	27.1	141
Testis[d]	0	86.0	100	7.8	100	4.8	100
	5 (immune)	80.8	94	6.0	77	0.7	15
	5 (pre-immune)	84.6	98	8.5	109	7.4	154

[a] Mitochondria were subjected to repeated freezing and thawing (5X).

[b] The reduction of cytochrome c was determined as described previously (Baron et al., 1972).

[c] The % conversion of cholesterol to pregnenolone in 60 minutes was determined by a modification of the method of Simpson and Boyd (1967).

[d] Mature male rats were treated with human chorionic gonadotropin (500 IU/rat, subcutaneously) at 0, 48 and 96 hours and were sacrificed at 120 hours.

Ig is probably due to the presence of albumin in the Ig fraction
which could bind and solubilize the cholesterol. The inability
of the goat antibody to adrenodoxin to inhibit the NADH-dependent
reduction of cytochrome c catalyzed by ovarian mitochondria indicates that this ovarian activity, similar to the comparable activity of adrenal mitochondria, does not involve the iron-sulfur
protein.

When similar studies were performed with testicular mitochondria prepared from mature rats, the NADPH-dependent enzymic
activities were extremely low and were not consistently inhibited
by the antibody. However, when testicular mitochondria prepared
from rats which had been treated with human chorionic gonadotropin
(HCG) were employed, enzymic activities were elevated and the
goat anti-adrenodoxin Ig did inhibit both NADPH-cytochrome c reductase activity (Figure 6) and the conversion of cholesterol to
pregnenolone and progesterone (Table III). These results are consistent with the observation of Mason et al., (1973) that, although testicular mitochondria of untreated rats contain little or
no adrenodoxin-like iron-sulfur protein, the treatment of rats
with HCG results in the increased synthesis of such a protein.
It is readily apparent that the degree of inhibition of the NADPH-
cytochrome c reductase activity of testicular mitochondria produced
by the antibody is considerably less than that seen with adrenal
and ovarian mitochondria (Figure 5). However, as seen in Table
III, although the NADPH-dependent reduction of cytochrome c catalyzed by HCG-treated rat testicular mitochondria is inhibited by
only 20 to 30%, at the same time the cholesterol side-chain cleavage activity is almost completely inhibited. The residual NADPH-
cytochrome c reductase activity may thus be due, in part, to either
microsomal contamination or to the presence of another pathway
which does not require the iron-sulfur protein. Thus, the iron-
sulfur protein of testicular mitochondria involved in the oxidative
cleavage of the cholesterol side-chain appears to be immunochemically and functionally similar to the adrenal and ovarian iron-
sulfur proteins. It should be noted that, analogous to adrenal
and ovarian mitochondria, the NADH-cytochrome c reductase activity
of testicular mitochondria was not inhibited by the antibody to
adrenodoxin.

Unlike its inhibitory effects on the NADPH-dependent activities of adrenal, ovarian and testicular mitochondria, the goat
antibody to adrenodoxin, even at relatively high concentrations,
did not inhibit the NADPH-dependent reduction of cytochrome c
catalyzed by sonicated mitochondria prepared from the sonicated
human term placental mitochondria (Table IV). Moreover, no inhibition of the cholesterol side-chain cleavage activities of
these preparations was observed in the presence of the antibody.[2]
Although an iron-sulfur protein has been isolated from human term

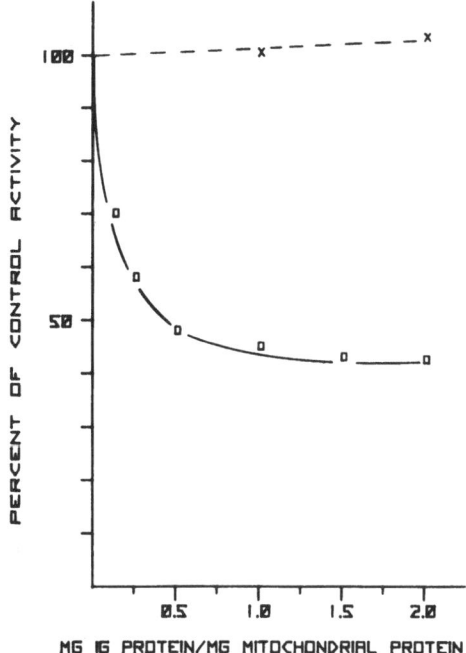

Fig. 6. Titration of NADPH-cytochrome c reductase activity of HCG-treated rat testicular mitochondria by goat anti-adrenodoxin I_g. The conditions were the same as those described for Figure 2. Rats were treated with HCG (500 IU/rat, subcutaneously) at 0, 48 and 96 hours and were sacrificed at 120 hours. In the absence of Ig, NADPH-cytochrome c reductase activity was 9.6 nmoles cytochrome c reduced min^{-1} (mg protein)$^{-1}$.

placental mitochondria, Mason and Boyd (1971) observed that the ratio of iron to sulfur in this protein was 2 to 1, rather than 1 to 1 as found in adrenodoxin, and Billiar and Little (1969) could detect no acid-labile sulfur in this protein. These reports suggest that the placental iron-sulfur protein differs from adrenodoxin. Consistent with these findings, no interaction was observed between the goat anti-adrenodoxin Ig and rat term placental mitochondria on Ouchterlony double immunodiffusion (Figure 7), while considerable interaction was observed between the antibody and rat adrenal, ovarian and testicular mitochondria. Thus, the iron-sulfur protein present in term placental mitochondria appears to differ both chemically and immunochemically from the adrenal, ovarian and testicular iron-sulfur proteins involved in mitochondrial steroid hydroxylations mediated by cytochrome P-450.

TABLE IV

Lack of effect of goat anti-adrenodoxin Ig on NADPH-cytochrome c reductase activities of rat and human term placenta.

Preparation	mg Ig protein/ mg protein	NADPH-cytochrome c reductase[a]	
		nmoles cytochrome c reduced/min/mg	Percent of control activity
Rat placental mitochondria[b]	0	5.3	100
	10 (immune)	7.0	132
	20 (immune)	7.0	132
	10 (pre-immune)	7.2	136
	20 (pre-immune)	8.2	155
Human placental mitochondria[b]	0	14.0	100
	10 (immune)	15.4	110
	20 (immune)	15.7	112
	10 (pre-immune)	16.9	121
	20 (pre-immune)	17.1	122
Human placental mitochondrial - S_1[c]	0	24.7	100
	10 (immune)	25.4	103
	20 (immune)	23.7	96
	10 (pre-immune)	28.9	117
	20 (pre-immune)	28.4	115

[a]Determined as previously described (Baron et al., 1972).

[b]Mitochondria were sonicated for 4 minutes.

[c]Supernatant fraction prepared from sonicated mitochondria according to the method of Omura et al. (1966).

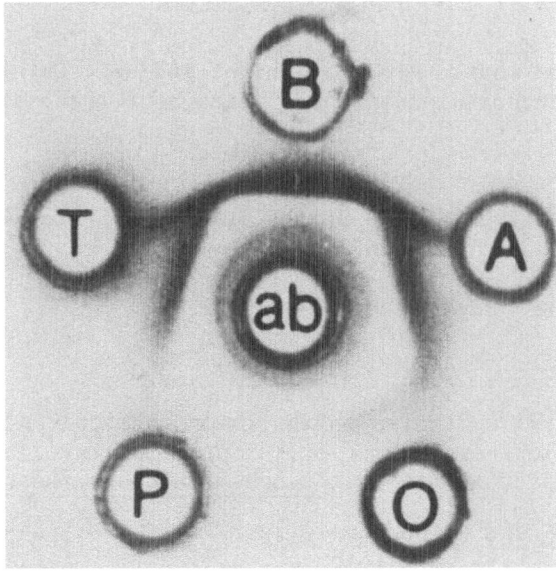

Fig. 7. Ouchterlony double immunodiffusion of goat anti-
adrenodoxin Ig against mitochondria from rat steroidogenic
tissues. The center well contained goat anti-adrenodoxin
(Ab), and the outer wells contained sonicated mitochondria prepared from bovine adrenal cortex (B), rat adrenal(A), rat ovary
(O), HCG-treated rat testis (T) and rat term placenta (P).

SUMMARY

An antibody prepared against bovine adrenodoxin has been
employed to study possible similarities or differences among the
iron-sulfur protein components of electron transport sequences
associated with cytochrome P-450 function in the mitochondria of
mammalian steroidogenic tissues. Although they are not identical,
the "adrenodoxins" in mitochondria from the adrenals of a number
of species, including man, are both immunochemically and functionally similar. The ability of the antibody to inhibit adrenal
11β-hydroxylase and cholesterol side-chain cleavage activities
demonstrates the requirement for adrenodoxin in these mixed-
function oxidations. The results presented also demonstrate
the immunochemical and functional similarities among the iron-
sulfur proteins which are involved in the oxidative cleavage of
the cholesterol side-chain occurring in adrenal, ovarian and
testicular mitochondria. The iron-sulfur protein involved in
cholesterol side-chain cleavage activity in term placental mitochondria, however, appears to be immunochemically different from
these other proteins.

ACKNOWLEDGEMENTS

The author thanks Mrs. Joy Shimek and Mrs. Cathy Hayford for their expert technical assistance throughout these studies.

FOOTNOTES

1   This research was supported by United States Public Health Service Grant GM 12675.

2   J. Baron, unpublished observations.

REFERENCES

Baron, J., 1973. Studies on the Immunochemical and functional similarities among iron-sulfur proteins involved in mammalian steroid hydroxylations. Biochem. Biophys. Res. Commun. 54: 764-769.
Baron, J., Taylor, W. E. and Masters, B. S. S. 1972. Immunochemical studies on electron transport chains involving cytochrome P-450. The role of the iron-sulfur protein, adrenodoxin, in mixed-function oxidation reactions. Arch. Biochem. Biophys. 150:105-115.
Billiar, R. B. and Little, B. 1969. Isolation of placental and liver "mitochondrial" fractions which support steroid 11β-hydroxylation. Biochim. Biophys. Acta 187:243-249.
Cooper, D. Y., Schleyer, H., Levin, S. S. and Rosenthal, O. 1973. Studies on the partially purified heme protein P-450 from the adrenal cortex. Ann. N. Y. Acad. Sci. 212:227-240.
Jefcoate, C. R., Hume, R., and Boyd, G. S. 1970. Separation of two forms of cytochrome P-450 from adrenal cortex mitochondria. FEBS Letters 9:41-44.
Kamin, H., Masters, B. S. S., Gibson, Q. H. and Williams, C. H. 1965. Microsomal TPNH-cytochrome c reductase. Fed. Proc. 24: 1164-1171.
Kimura, T. and Ohno, H. 1968. Preparation of testis non-heme iron protein and substitution for adrenodoxin by various non-heme iron proteins in steroid 11β-hydroxylation. J. Biochem. 63: 716-724.
Mason, J. I. and Boyd, G. S. 1971. The cholesterol side-chain cleavage enzyme system in mitochondria of human term placenta. Eur. J. Biochem. 21: 308-321.
Mason, J. I., Estabrook, R. W. and Purvis, J. L. 1973. Testicular cytochrome P-450 and iron-sulfur protein as related to steroid metabolism. Ann. N. Y. Acad. Sci. 212:406-419.
Masters, B. S. S., Baron, J., Taylor, W. E., Isaacson, E. L. and LoSpalluto, J. 1971. Immunochemical studies on electron transport chains involving cytochrome P-450. I. Effects of antibodies to pig liver microsomal reduced triphosphopyridine

nucleotide-cytochrome c reductase and the non-heme iron protein from bovine adrenocortical mitochondria. J. Biol. Chem. 246:4143-4150.

Omura, T., Sanders, E., Estabrook, R. W., Cooper, D. Y. and Rosenthal, O. 1966. Isolation from adrenal cortex of a non-heme iron protein and a flavoprotein functional as a reduced triphosphopyridine nucleotide-cytochrome P-450 reductase. Arch. Biochem. Biophys.117:660-673.

Simpson, E. R., and Boyd, G. S. 1967. Partial resolution of the mixed-function oxidase involved in the cholesterol side-chain cleavage reaction in bovine adrenal mitochondria. Biochem. Biophys. Res. Commun. 28:945-950.

Sulimovici, S. and Boyd, G. S. 1968. The effect of ascorbic acid in vitro on the rat ovarian cholesterol side-chain cleavage enzyme system. Steroids 12: 127-149.

GENERAL DISCUSSION

Session I - Part 1

PEISACH: While sitting in my seat listening to Dr. Coon, I did some "fast numerology" with the mechanism proposed by Dr. Coon on a two-electron reduction of cytochrome P-450, and the numbers just didn't seem to add up. Before performing this numerology, I shall have to define terms. Let us assume that ferric heme is at oxidation state III, ferrous heme is at oxidation state II, molecular oxygen at oxidation state IV (because four electrons are required to reduce oxygen to water), superoxide anion at oxidation state III and peroxide at oxidation state II. The standards for this system, atomic iron and water, are each defined as oxidation state zero. To begin with, if we are to believe Dr. Coon's mechanism, it means that cytochrome P-450 in the resting state starts in the reaction cycle at oxidation state III. We know that is true since the protein has the EPR of an S = 1/2 low spin ferric heme. If two reducing equivalents are added (as Dr. Coon proposes) this means that the protein is now at the formal oxidation state of one. If molecular oxygen having an oxidation state of IV is now added (we know that oxygen is required for this reaction), the formal oxidation state of the hypothetical protein-oxygen complex is V. What heme protein compounds that we know of can be at oxidation state V? There are two examples that I can bring to mind. One is compound I of horseradish peroxidase. This complex is green, has a much depressed Soret and an unusual spectrum in the visible. Nothing having these properties appears during the turnover of cytochrome P-450. The second protein which can be at state V is cytochrome c peroxidase (of yeast). With hydroperoxides it forms a red compound which contains ferryl heme (at oxidation state IV) which is EPR silent and a free radical (at oxidation state I) which does exhibit an EPR.

Since nobody has been able to demonstrate the formation of free radicals during the turnover of cytochrome P-450, this type of heme protein intermediate is not likely. Although Dr. Coon's mechanism does explain the stoichiometry of reaction, it doesn't really explain the mechanism involving intermediates. I think that the numbers just don't seem to add up.

COON: I don't follow your numerology at all. The overall reaction is very simple. Two electrons are taken up, one by the ferric heme and presumably one by the unknown factor which we are calling C. These two electrons must eventually reduce one oxygen atom to water. You have overlooked that the other oxygen atom is not reduced to H_2O but is inserted into the resulting product. Everything adds up. We have looked with EPR at the purified P-450 during various stages of reduction. We see no free radicals either forming or disappearing. One possibility is that spin coupling prevents our seeing the free radicals.

PEISACH: One of the rules of mathematics is that when an odd number is added to an even number, the result is an odd number. Ferric heme has an odd number of electrons, one in this case $(S = 1/2)$. Adding two electrons, an even number, must lead to an odd spin situation. If there is an odd number of spins, then, in principle at least, an EPR spectrum must be observed. Something doesn't add up still.

JERINA: Dr. Peisach, I am puzzled by your bringing oxygen into the argument whatsoever, since this is an anaerobic reduction of Dr. Coon's preparation.

PEISACH: The reason I brought oxygen into the numerology is because Dr. Coon shows in his cycle the formation of an oxygenated cytochrome P-450 complex in which he says oxygen is bound as superoxide anion and the iron in an unspecified oxidation state. If two electrons are added to the cytochrome, then the protein is formally at oxidation state I. Adding four electrons from oxygen, the number of spins still remains odd.

JERINA: That's assuming that the iron is taking up both electrons, isn't it?

COON: Yes.

PEISACH: The result is similar if the iron atom in cytochrome P-450 only takes up one electron. The second electron must be taken up by some other hypothetical species in the protein that we can call "X." Adding two electrons to the protein, one goes to ferric heme and leads to ferrous heme while the second electron will end up in X which now becomes $X^{\overline{}}$. Here $X^{\overline{}}$ is paramagnetic and an EPR spectrum should be observed, but isn't.

SCHENKMAN: My question concerns the possibility of detergent action, that is, the requirement for a lipid factor. Since detergents will displace each other from proteins, there's a possibility that the lipid requirement is merely there to displace detergent.

LEVIN: I think that possibility is very unlikely. Dr. Mary Vore (Biochem. Biophys. Res. Commun. 56: 1038, 1974) has used organic solvent-extracted microsomes to demonstrate the requirement of lipid for microsomal hydroxylation. These microsomes, in which 80-90% of the lipid has been removed, have a decreased activity towards the hydroxylation of benzo[α]pyrene which can only be restored upon the addition of crude lipid or synthetic phosphatidylcholine. Since these preparations had not been in contact with any detergents, the requirement for phospholipid cannot be attributed to removal of detergent from the system. In addition, Dr. Lu has results with the reconstituted system showing that under certain conditions, several detergents can almost completely replace phosphatidylcholine for the metabolism of benzphetamine (Biochem. Biophys. Res. Commun., in press). I think these results clearly demonstrate that lipid is a required component of the system.

COON: To answer Dr. Schenkman, it's highly unlikely, from several lines of evidence, that the role of the phospholipid is to overcome detergent inhibition.

PETERSON: I would like to ask both Dr. Coon and Dr. Levin a question. I was fascinated, Dr. Coon, by your titration spectra, titration with dithionite of your purified P-450 which has a specific concentration of 15 nmoles P-450/mg protein. You still have a very large Soret band left when you have all the P-450 reduced. I noticed in Dr. Levin's slide that he did not have such a Soret band and his specific concentration was only 11 nmoles P-450/mg protein. What do you attribute this remaining Soret band to when all of your P-450 and all of your heme is supposed to be either P-450 or P-420?

COON: We're not absolutely certain. This may in part be an effect of the particular detergent used on the heme. I think that in this enzyme preparation some P-420 was present; we are puzzled ourselves by the observations in the Soret region.

BALLOU: I believe that Dr. Peisach's comments concerning the "numerology" of a two electron uptake by the heme are pointing at the question of what happens to the EPR spectrum during the course of this reduction. Since the resting state of P-450$_{LM}$ is certainly the ferric state and the Fe^{+1} state is virtually unknown, how do we envision putting in two electrons per heme? To put it in another light, if one begins the experiment with an EPR signal (e. g., Fe^{+3}) and adds one electron, it is very possible to wipe out the signal. However, the addition of two electrons to an already existing EPR signal must preserve at least potentially some form of signal.

To explain both the fact that two electrons are taken up per heme and the fact that we start with a signal, Fe^{+3} (low spin), and after titration end up with no observable signal, we (in collaboration with Dr. Richard Sands) propose the following three possible models (Figure 1) which include an additional electron acceptor, "C."

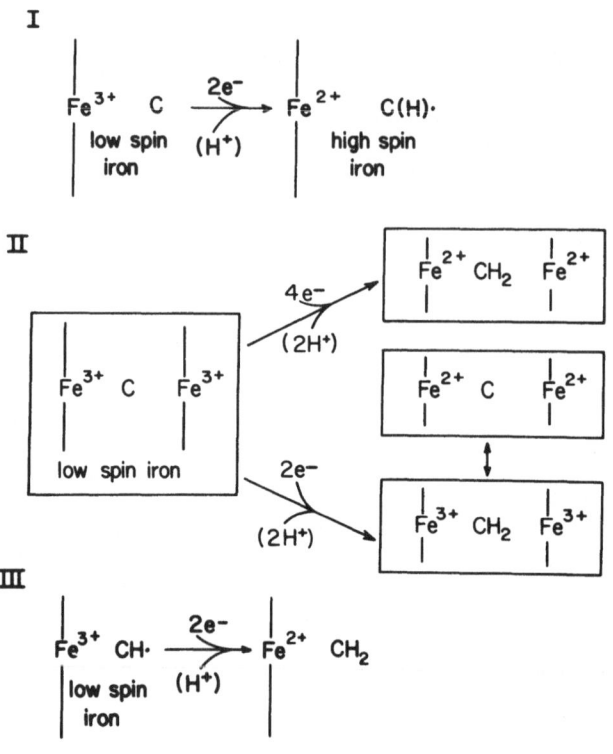

Figure 1.

The first model and the one we consider most likely, proposes that we begin the experiment with ferric heme primarily in low spin form which is tightly coupled electrochemically with the group C. Electrons are taken up two at a time to produce a high spin ferrous heme still coupled tightly to the C(H)·radical. Although this would be expected to exhibit the EPR signal to a free radical, we believe that the high spin ferrous (S=2) can have strong spin orbit coupling to the radical, thus greatly enhancing the radical relaxation rate and broadening the signal to the point where it is not observable, even at 7°K. That the P-450$_{cam}$ exists in a high spin ferrous state (1) makes this possibility quite likely. The principal difficulty arises in the explanation of the chemical nature of the one electron-acceptor C. A combination of Mössbauer spectroscopy (to see if it is high spin) and magnetic suscepti-bility (to see if the free radical exists) could provide a reason-able test for this model.

The second model suggests that two hemes act in concert with

one two-electron acceptor, C. In this model, no free radicals
would be observed. One-electron reduced species would interact
intermolecularly to produce two-electron reduced species (via
dismutation). This might be treated by studying the kinetics of
the intermolecular rearrangement.

The third possibility is that the low spin ferric is present
along with a free radical. Addition of 2 electrons would make
both signals diamagnetic. If this were the case the only signal
one might see is that from species which have lost C(H)· during
purification. The obvious test of this is to integrate the EPR
spectra to see if there is 1.0 electron per heme or, perhaps,
only 0.1 electron per heme.

KAMIN: I've gotten a lot of very sophisticated explanations
here but I'd like to first ask for some simple ones from Dr. Coon.
Your titration data certainly seem to be of very high quality and
are very interesting, but I think that before we get into electron
counting, I want to ask you whether you've gotten certain types of
data: 1) the proposed picture says that the first electron goes
in and doesn't need oxygen, the second electron goes into an oxygenated compound. The first question I have to ask you is, how
sure are you that you do not have oxygen? Oxygen is a dangerous
contaminant, even though I recognize that anaerobiosis at the
University of Michigan is better than in almost any other institution.

COON: We could not be titrating oxygen in the anaerobic experiments since it is present in' less than 0.5 parts per million.
I should add, incidentally, that we have not seen the proposed
oxygenated complex described by Estabrook et al. in 1971. (Bioch.
Biophys. Res. Com. 42, 132-139.) We have looked very carefully
for it under a variety of conditions.

KAMIN: That's one factor taken care of. The other is: how
good a search have you been able to make for another reducible
group in the protein? Rational chemistry suggests that there is
another group, possibly aromatic, possibly a tryptophane, which
could be a second electron acceptor. Have you analyzed for total
iron in your system as against heme iron? Are there other metals?
Things are complicated enough with two electrons: in the present
picture perhaps you could get three, and then even worse numerology
could set in.

COON: We've shown with neutron activation analysis and
emission spectroscopy that there's no metal present with redox
capability other than iron. In our better preparations the total
iron content agrees with the heme content. Therefore, there is no
significant amount of nonheme iron in the preparations.

There appear to be no disulfides. There is nothing in the visible
or UV spectrum which shows cofactors other than heme.

PEISACH: Dr. Ballou wants to have his cake and eat it, too.
On the one hand, he enjoys the possibility of an unpaired spin
species in his cytochrome P-450 preparation and on the other hand,
wants to explain away the lack of EPR signal after a two electron
addition. He therefore presents a series of models none of which
is correct from both theoretical and practical reasons, in order
to reconcile the experimental findings.

To begin with, the first model that he proposes, I assume, is
the one of cross relaxation of the radical resonance with that of
the iron. This mechanism is part of the folklore of solid state
physics often used to explain the unusual broadening sometimes
observed of a resonance line. Let us assume that in order to
"obliterate" a resonance signal of a radical species, that the
amount of broadening is roughly 1000 gauss. This would mean that
there would be a few thousand megacycles of interaction between
the iron and the radical. This would require that the relaxation
time would be correlated with the broadening. This phenomenon is
theoretically possible, but since $1/T_1$ is a temperature-dependent
phenomenon, then at least over the range of temperatures used to
study the EPR, a radical signal would be observed at some point.
As was reported today, this is not the case.

The second hypothesis suggests a dimerization property of
cytochrome P-450 during the course of reaction. Again, although
possible, there is no evidence for this phenomenon.

The third possibility is ruled out since the position in
magnetic field where one would observe a free radical is so different from that for cytochrome P-450. Furthermore, the EPR line
shape for a radical is so obvious to any investigator that it
surely would be seen in a spectrum. On the other hand, if there
is weak dipolar coupling between the iron and a radical, then
a half-field resonance would be observed in the fully oxidized
protein. For a radical coupled to $S = 1/2$ iron, this spectrum
would be so obviously different in the region to slightly lower
field than $g = 2$, that it would be clearly recognizable. Finally,
if the coupling were large, then the EPR of $S = 1/2$ iron would
not be observed at all in the fully oxidized protein.

The numbers still don't add up.

PETERSON: Dr. Baron, I was kind of surprised when you
mentioned your cross-reactivity of your iron sulfur proteins that
you didn't talk about any work we did several years ago, particularly with respect to the work that Karl Dus presented this morning,

in that there is no cross-reactivity between the putidaredoxin isolated in the bacterial system and the adrenoxodin antibody.

BARON: That is correct. When we began our studies with the antibody to adrenodoxin, we examined its effects on the EPR spectrum of reduced adrenodoxin. The antibody was found to prevent the reduction of adrenodoxin, as evidenced by the lack of appearance of the g 1.94 signal. Moreover, when the antibody was added to samples containing reduced adrenodoxin, the g 1.94 signal was greatly decreased. These effects were observed at ratios of 5 to 10 mg lgG protein per mg of enzyme protein, and were interpreted as indicating that the antibody affected, either directly or indirectly, the active site of adrenodoxin. The next logical step at that point was to examine the effect of the antibody to adrenodoxin on putidaredoxin. We found that this antibody had absolutely no effect on either the EPR spectrum of reduced putidaredoxin or the reduction of cytochrome c catalyzed by the Pseudomonas system, even at ratios up to 100 mg lgG protein per mg enzyme protein.

SCHENKMAN: Dr. Coon, concerning this stimulatory factor that you saw in the oxidation of ethylmorphine or benzphetamine, or Dr. Mannering's stimulatory factor, Dr. Cinti and I have been looking at it a bit. It seems to be just a protection of the enzyme activity, for some reason. The reaction seems to be linear in the presence of this protein, but in the absence of this protein it curves off very fast. And so it may not be a stimulatory factor, but protection of the enzyme.

PEISACH: Dr. Levin, when you state that there's a difference in the CO difference spectrum, what do you take the difference spectrum against?

LEVIN: This was measured as a CO-difference spectrum in the usual manner (CO + dithionite versus dithionite). We have also observed the peak shift in the absolute CO spectrum.

PEISACH: The reason I brought up this question is that unless you take an absolute spectrum, you can't tell whether the difference observed in the CO spectrum is due to an inherent difference in the CO spectrum itself or a difference in the spectrum of the reduced protein which is used as a reference. This raises another point. Many workers insist on using the technique of difference spectroscopy on purified preparations where that is really not necessary. Here, you may lose data by resorting solely to this technique of spectroscopy to study materials optically.

LEVIN: We agree with you to a certain extent and that is why we always check for spectral differences in both the absolute and difference spectra. However, since some of these fractions that I described are far less pure than our other hemeprotein fractions,

we do not like to use absolute spectra exclusively since other unknown contaminants in the preparations may have absorbance in the
visible range which could alter the absolute spectra.

MASTERS: I have a comment to make in regard to the FMN and FAD
content of the reductase which we have also corroborated. We have
repeated earlier experiments using a new fluorometric method of
Faeder and Segal to determine the FMN and FAD content. I've talked
with Dr. Sato and he sees the same thing. So I think the answer
is that NADPH-cytochrome c reductase does contain both FMN and
FAD in equimolar amounts. Nobody knows why the earlier data did
not show that there was FMN there, especially in view of the fact
that FMN has 10 times or more the fluorescence of FAD. So it's
a very peculiar phenomenon, but it's there nonetheless.

The second comment I want to make is with regard to the cytochrome c reductase activity of your preparation. I'm glad to see
that cytochrome c reduction has been vindicated - that we're getting
very close to the published value that I put in the literature, i.
e., 38,100 nmoles/(min x mg protein), a specific activity which
has also been reported by Omura and his co-workers. So we're
getting there - you're almost at 30,000 nmoles/(min x mg protein).

COON: Our specific activity is about 33,000, but I see no
reason our protein and yours should have the same specific activity
towards cytochrome c. Ours is a different protein which has not
been subjected to proteolytic cleavage and retains activity toward
cytochrome P-450 as well as cytochrome c.

MASTERS: I agree, and your reductase preparation supports
hydroxylation.

COON: Our P-450 reductase has a polypeptide chain of higher
molecular weight since it has not been subjected to proteolysis,
as is the case with your cytochrome c reductase.

MASTERS: That's possible, but my point was made with regard
to the nature of cytochrome c reduction by the NADPH-cytochrome c
reductase. I think earlier it might have been said that cytochrome
c reduction was an artifactual activity of the enzyme and I really
think that this catalytic property has been vindicated in that
sense. It is artifactual in the sense that there is no cytochrome
c in liver microsomal membranes.

COON: I agree that cytochrome c reduction is artifactual in
that sense.

REFERENCE (see p. 74)

1.   Sharrock, M., Münck, E., Debrunner, P., Lipscomb, J. D.,Mar-
shall, V. and Gunsalus, I.C., Proc. Nat. Acad. Sci.U.S.A.70 (1972).

COMPARISON OF THE INDUCTION COURSE, BIOPHYSICAL CHEMICAL INTERACTIONS

AND PHOTOCHEMICAL ACTION SPECTRA OF PHENOBARBITAL AND 3-METHYLCHOLAN-

THRENE INDUCED HEPATIC MICROSOMAL P-450

D. Y. Cooper*, H. Schleyer, J. H. Thomas, H. M. Vars and
O. Rosenthal
Harrison Department of Surgical Research and Johnson
Research Foundation, School of Medicine, University of
Pennsylvania
Philadelphia, Pennsylvania 19174

INTRODUCTION

That the P-450-dependent microsomal mixed function oxidase systems induced in rat livers by polycyclic hydrocarbons (3-methylcholanthrene (3-MC)) are different from those induced by phenobarbital (PB) was first reported by Sladek and Mannering (1966). More recently Fouts (1973) has observed that 3-MC is much less effective than phenobarbital in inducing the microsomal electron transport system necessary for reduction of P-450. It is likely, therefore, that during the course of induction of enzyme activity with polycyclic hydrocarbons and the decline of this activity to preinduction levels there would be not only differences in the type of P-450 from that in PB-induced animals, but also changes in the reductive capacity of the microsomal electron transport system.

In order to assess the best conditions for a systematic study of the light reversal of CO-inhibition of various arylhydroxylase catalyzed reactions, it appeared necessary to determine the physical and chemical properties of the P-450 and the capacity of the electron transport system induced by polycyclic hydrocarbons throughout the time course of induction and decay of enzyme activity. Comparative studies of these properties in 3-MC- and PB-treated rats are the subject of this paper.

MATERIAL AND METHODS

A. <u>Animals</u>. Male Sprague Dawley rats of three age groups
(starting weight ≈ 100 g, 150 g and 200 g) were used in these
studies. The rats were housed in individual cages, fed a Nutritional Biochemical Normal Protein Test Diet and weighed daily.
Food intake of all animals was equalized to approximately that of
the animals having the lowest food consumption.

After a 2 - 3 day equilibrium period the animals were injected intraperitoneally with saline (0.5 ml), corn oil (0.5 ml),
phenobarbital (80 mg/kg) or 3-methylcholanthrene (20 mg/kg) each
day for four days. The animals were killed by decapitation after
12 hours of starvation at appropriate intervals during and after
the 4 day injection period as indicated in the text. Hemoglobin
was removed from the liver by perfusing the organ <u>in situ</u> after
canulating the superior vena cava through the right atrium, tying
the inferior vena cava just below the liver and cutting the portal
vein. Approximately 100 ml of 0.15 M cold KCl and 100 ml of cold
0.25 M sucrose were used in each liver perfusion. The microsomes
were prepared by the method of Schenkman et al. (1967).

B. <u>Chemicals</u>. The chemicals used in this study were obtained from the following sources: napthalene, anthracene, pyrene,
phenanthrene, benzanthracene, 9,10-dimethylbenzanthracene, benzo(a)-
pyrene, 3-methylcholanthrene, from Eastman Kodak; benzene, sodium
dithionite, aniline, from Fisher Scientific Co.; phenobarbital, from
Merck and Co.; hexobarbital, from Winthrop Stearns; codeine, from
Mallinckrodt Chemical Works; DPN, TPN, TPNH, DPNH and cytochrome
c, from Sigma Chemical Co. SKF-525A was a gift of the Smith Kline
and French Laboratories. Metopirone was kindly supplied by Dr. J.
J. Chart of the Ciba Pharmaceutical Company, Summit, New Jersey.

C. <u>Enzyme Assays</u>. Microsomal P-450 was estimated by difference spectroscopy with a Yang-Chance scanning spectrophotometer.
The differential millimolar absorption coefficient $\Delta\epsilon = 91$ mM^{-1}cm^{-1}
(450 - 490 nm, $Fe^{2+} \cdot CO - Fe^{2+}$) reported by Omura and Sato (1964b)
was used. Cytochrome b$_5$ was estimated as described by Omura and
Sato (1964a). Utilization of TPNH and DPNH was measured fluoro-
metrically by the method of Estabrook (1962). TPNH-cytochrome c
and TPNH-P-450 reductases were measured as described by Omura et
al., (1966). Aniline hydroxylation and codeine demethylation
activities were determined by the method described by Schenkman
et al., (1967).

RESULTS

A. <u>Time Course of Induction of Hepatic Microsomal P-450
in 200 g Rats Treated with PB or 3-MC</u>. The time course of induction

of microsomal P-450 by PB or 3-MC is given in Figure 1. The specific concentration of P-450 is recorded as nmole P-450(Fe^{2+})·CO per mg protein. As reported by others, both PB and 3-MC increased microsomal P-450 content in the 4 day treatment period. During this period, dithionite ($Na_2S_2O_4$) and TPNH were equally effective in reducing the P-450 of both types of microsomes as indicated by the approximately equal amounts of P-450 estimated by either of the two methods of reduction. The P-450 content of the microsomes from PB-induced animals started to decrease shortly after cessation of treatment and had fallen to the control values within 6 days. The type of reductant used was of no consequence in quantitating P-450.

In the 3-MC microsomes, when P-450 was quantitated by using TPNH as reductant, it was also found that the specific P-450 concentration had fallen to control values by the 6th day after cessation of the injections. On the other hand, when $Na_2S_2O_4$ was used to reduce the heme P-450 a strikingly different time course was observed, with the microsomal P-450 concentration still rising up to 11 days after the injections had been discontinued.

The estimates of the specific concentration of P-450 on the day of maximal induction by PB (day 3, after 3 daily injections) and by 3-MC (day 17, the 13th day after 4 daily injections) are compared in Table I. Several points are brought out in this table. On the 17th day of the experiment, control animals treated with corn oil had approximately doubled their microsomal P-450 content. By the 17th day, the microsomal P-450 of the PB-treated animals had fallen below the control value. With 3-MC-induced microsomes, TPNH was slightly more effective than dithionite in reducing P-450 from animals treated 3 days, but on the 17th day, TPNH could reduce only 25% of the P-450 that was reducible with $Na_2S_2O_4$.

B. Time Course of Induction of Hepatic Microsomal P-450 in 100 g Rats Treated with PB and 3-MC. The situation is more complex in rats with starting weight of 100 g. A difference in the ability of TPNH and dithionite to reduce P-450 for formation of the CO compound was already observed in the corn oil treated control animals (Figure 2). Dithionite reduction yields estimates of the specific concentration of microsomal P-450 of 3 to 6 times the estimates obtained with TPNH.

The time course of induction of microsomal P-450 in 100 g rats treated with PB (Figure 3) is essentially the same as in 200 g rats, but TPNH could only reduce ≳ 30% of the P-450 present in these microsomes. However, the relative increases in P-450(Fe^{2+})·CO with reference to the corresponding control values were nearly identical with both types of reductant.

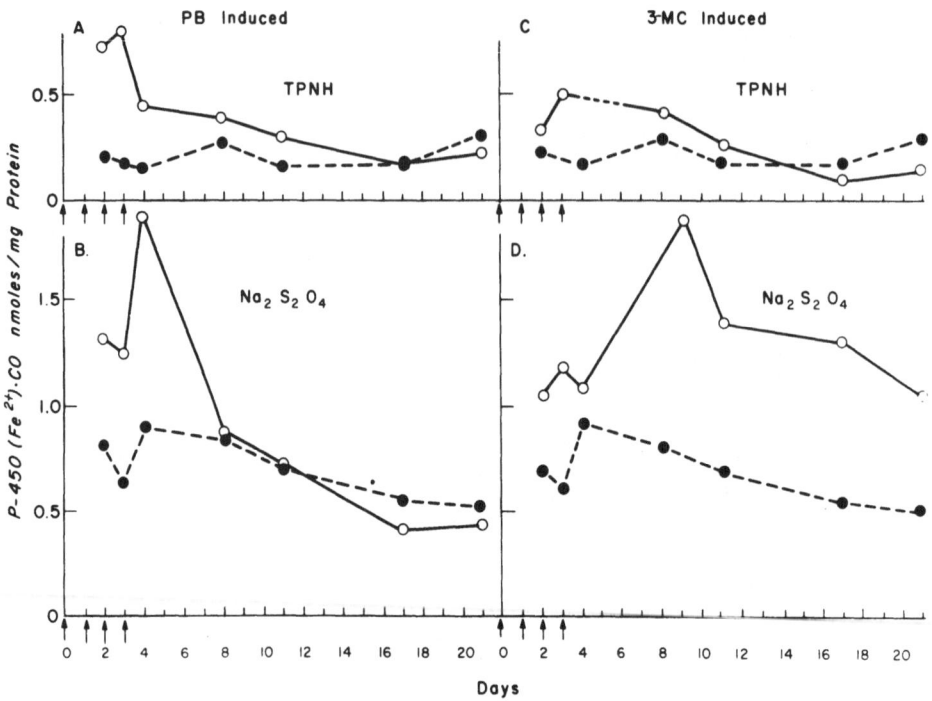

Figure 1.   Time course of induction of hepatic microsomal
P-450 by PB- or 3-MC-treatment of rats weighing 200 - 300 g.   For
each experimental point microsomes prepared from the pooled livers
of 2 rats were used.   0.5 ml of a microsome suspension (protein
range 6-13 mg) was diluted with 50 mM potassium phosphate buffer,
pH 7.4, to 6 ml, divided equally between 2 optical cells of 1 cm
light path, and a base line of equal light absorption established.

In the dithionite experiments, $Na_2S_2O_4$ was added to the experimental cell and the difference spectrum reduced-oxidized (total
reducible pigments) recorded. $Na_2S_2O_4$ was then added to the reference cell, the experimental cell was bubbled with CO for 30 sec.
and the difference spectrum of P-450$(Fe^{2+})\cdot$CO was recorded.

In the TPNH reduction experiments, hexobarbital (140 μM) was
first added to the experimental cell to record the substrate-pro-
duced difference spectrum which was thereafter blanked out by
adding hexobarbital to the reference cell.   This was followed by
the addition of DPNH (43 μM) to the experimental cell to obtain

(Legend cont'd next page)

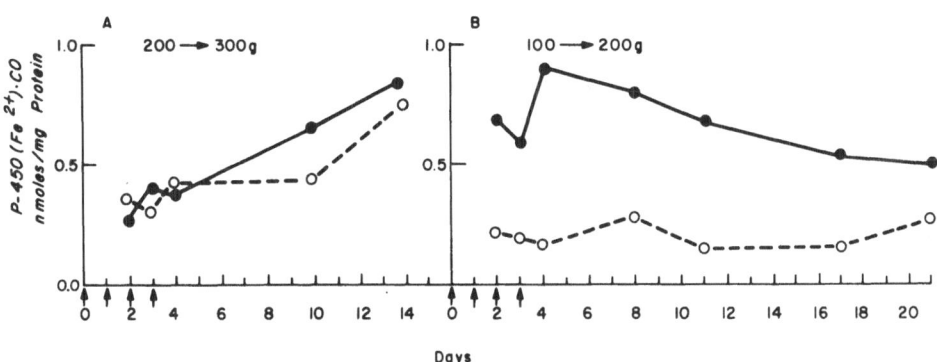

Figure 2. P-450 levels in microsomes prepared from control rats (corn oil treated). The animal weights ranged either from 200 to 300 g (A) or from 100 to 200 g (B). Experimental conditions were as described in legend to Figure 1. o— — —o P-450 reduced with TPNH; ●——● $Na_2S_2O_4$. ↑ indicates the days of injection with corn oil (Mazola).

the difference spectrum of cytochrome b_5. After compensation of this spectrum by adding DPNH to the reference cell, TPNH was added to both cells and the difference spectrum of P-450(Fe^{2+})·CO obtained as after $Na_2S_2O_4$ reduction. o——o = rats injected with inducer in corn oil. ●— — —● = simultaneous controls injected with corn oil only. Days of injection are indicated by arrows. Graphs A + B and C + D represent respectively PB- and 3-MC-treated groups. Reductant is TPNH in A and C and $Na_2S_2O_4$ in B and D.

TABLE I

Comparison of P-450 concentration estimated as $P\text{-}450(Fe^{2+}) \cdot CO$ by TPNH and $Na_2S_2O_4$ reduction on days 3 and 17. Animal weights 200 – 300 g. Experimental conditions as described in legend of Figure 1.

	$P\text{-}450(Fe^{2+}) \cdot CO$ (nmoles/mg protein)					
	Control		PB		3-MC	
Day	TPNH	$Na_2S_2O_4$	TPNH	$Na_2S_2O_4$	TPNH	$Na_2S_2O_4$
3*	0.302	0.401	1.632	1.44	1.501	1.11
17 +	0.721	0.811	0.366	0.673	0.549	2.30

* Day 3:   first day after 3 daily injections
+ Day 17:   13th day after 4 daily injections.

     The time course of P-450 induction by 3-MC was essentially the same as in the 200 g animals except that the maximum was reached earlier (on the 8th day) and the subsequent decrease was slower.

     C. Properties of the Difference Spectrum of $P\text{-}450(Fe^{2+}) \cdot CO$ of Microsomes of 3-MC Treated Animals (8 days after 4 injections of 3-MC). When microsomes from control animals and PB-treated animals of the 200 to 300 g weight range had been reduced with either TPNH or $Na_2S_2O_4$ the usual difference spectrum of $P\text{-}450(Fe^{2+}) \cdot CO$ appeared with a maximum at 451 nm, a small shoulder around 420 nm and a deep trough at 409 nm. A different picture was seen, however, when the microsomes prepared from animals 8 days after 4 days of injection with 3-MC were reduced with TPNH (Figure 4, curve A). Only a small fraction (about 25%) of the $P\text{-}450(Fe^{2+}) \cdot CO$ was detected that was formed when $Na_2S_2O_4$ was used as reductant, but a large absorption band in the 420 nm region

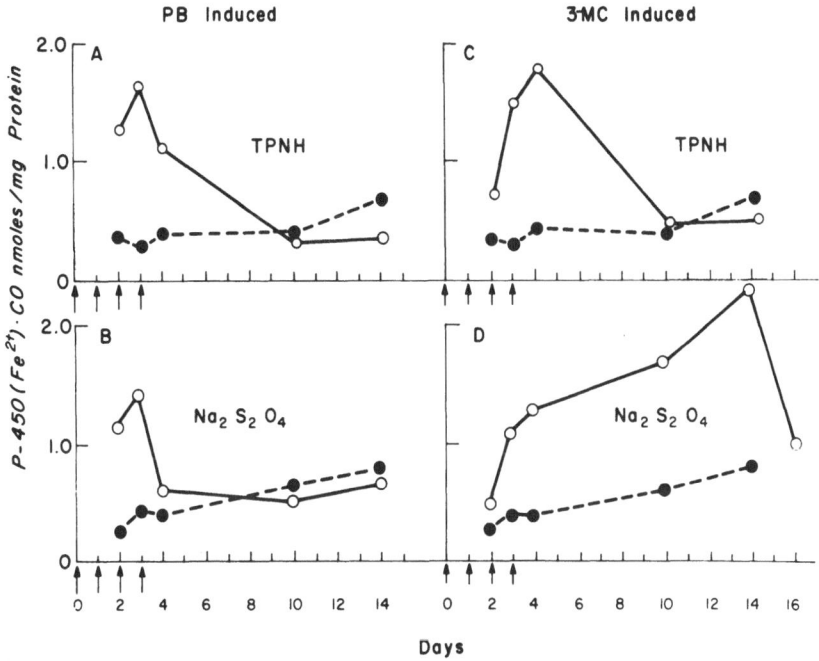

Figure 3. Time course of induction of hepatic microsomal
P-450 by PB and 3-MC in rats weighing 100 to 200 g. Experimental
conditions as described in legend to Figure 1. A. PB-induced,
reduced with TPNH; B. PB-induced, reduced with $Na_2S_2O_4$; C. 3-MC-
induced, reduced with TPNH; D. 3-MC-induced, reduced with $Na_2S_2O_4$.
o——o treated animals; ● - - ● control animals.

indicative of P-420$(Fe^{2+})\cdot$CO was present. This band seemed to
disappear after addition of $Na_2S_2O_4$ to both cuvettes (Figure 4,
curve B). If, however, the positive absorptions in the 420 nm
region are corrected for the negative absorptivity of P-450·CO
at 420 nm $[\Delta\varepsilon(420 - 490\ nm) = -41\ cm^{-1}\ mM^{-1}]$ according to the
procedure of Omura and Sato (1967), one arrives at nearly identical
values for the P-420 concentration with both types of reductant.
In other words, P-420 can be virtually fully reduced with TPNH as
contrasted to the 25% reduction of P-450, a result in accord with
the higher redox potential of the former ($E_o^{'} = -20$ mV vs. ≈ -400 mV).

These experiments were carried out with microsome suspensions
in air-saturated buffer solution in open cuvettes and the commercial carbon monoxide was known to contain at least 3% oxygen. When
these experiments were repeated with scrubbed gases in special
cuvettes, our system designed for titrations under strictly O_2-
free conditions, nearly stoichiometric quantities of TPNH sufficed

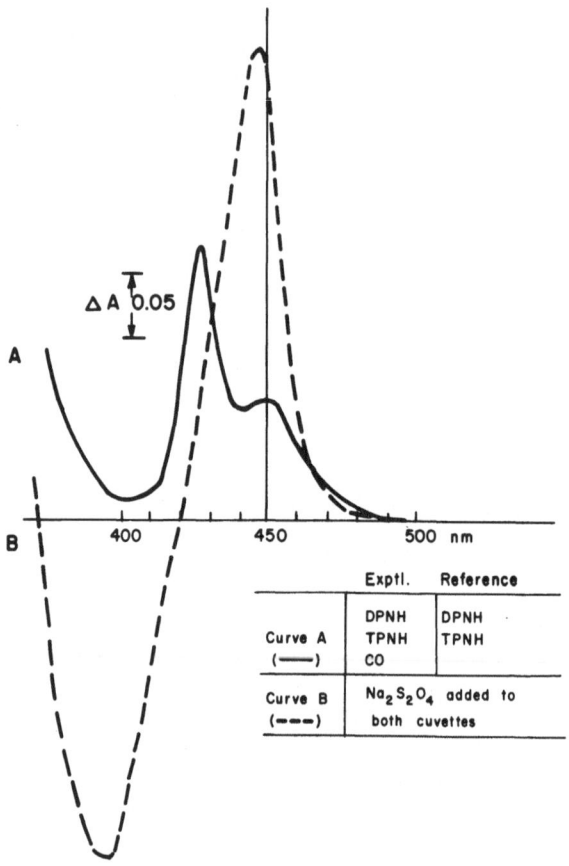

Figure 4.   Spectra of P-450(Fe^{2+})·CO of 3-MC microsomes (8 days after 4 injections) reduced with TPNH and $Na_2S_2O_4$. Experimental conditions as described in Figure 1. Animal weights 200 to 300 g. ———— TPNH reduction; ---- $Na_2S_2O_4$ added after TPNH.

to develop the full P-450(Fe^{2+})·CO spectrum that was detected with dithionite. These experiments and additional ones at intermediate O_2 concentrations indicate clearly that our problems in reducing microsomes with TPNH were caused by the presence of varying levels of O_2 and for unknown reasons the microsomes of 3-MC-induced animals and of younger animals in all groups appear more sensitive to these conditions.

D. <u>Spectral Response Observed after Addition of Hexobarbital.</u> Addition of hexobarbital to microsomes from control and PB treated animals produced the characteristic Type I spectral change as previously described whereas a reverse Type I (modified Type II) spectral change was found on addition to 3-MC induced microsomes (Figure 5). Aniline produced a Type II spectral change with both PB and 3-MC induced microsomes (Figure 5). These spectral changes are essentially the same as those reported from other laboratories (cf. Orrenius and Ernster 1974).

The magnitude of the spectral changes (ΔA/mg protein) elicited by hexobarbital during and after injection of the two inducers is recorded in Figure 6. It is evident that the magnitude of these changes followed the time course of P-450 induction as shown in Figure 1. The same relationship held true for the spectral response to aniline (not shown).

E. Drug Metabolism during the Time Course of Enzyme Induction by PB and 3-MC. The rate of aniline hydroxylation (Figure 7A) was increased during the injection period and fell slowly to control values after the injections were stopped. The time course of induction of aniline hydroxylation was the same for PB- and 3-MC-induced microsomes. It is interesting to note that this is the same time course as that for induction of the P-450 that can be reduced by TPNH (see Figure 1).

The rate of codeine demethylation was increased only by PB induction (Figure 7B). This increase was most marked during the injection period, but the rate remained elevated above the controls through the 10th day. 3-MC induction resulted in a decrease of the rate of codeine demethylation which began after the third dose, and the rate remained depressed throughout the course of the experiment.

F. Photochemical Action Spectra in PB- and 3-MC-Induced Microsomes. The photochemical action spectrum was measured as previously described by Rosenthal and Cooper (1967). Action spectra for codeine demethylation by PB-induced microsomes and for aniline hydroxylation by 3-MC-induced microsomes are depicted in Figure 8. The maxima of light reversal of the CO-inhibition for both substrates and both types of inducers are listed in Table II.

The action spectrum for codeine demethylation by microsomes from PB-induced animals gave a single light reversal peak whose maximum was at 450 nm. The action spectrum for aniline hydroxylation by PB-induced microsomes showed some light reversal of the CO inhibition in the 400 to 450 nm spectral range with a shoulder at 450 nm and maxima around 430 and 415 nm.

The 3-MC-induced microsomes yielded anomalous action spectra with both substrates. In these microsomes a main light reversal

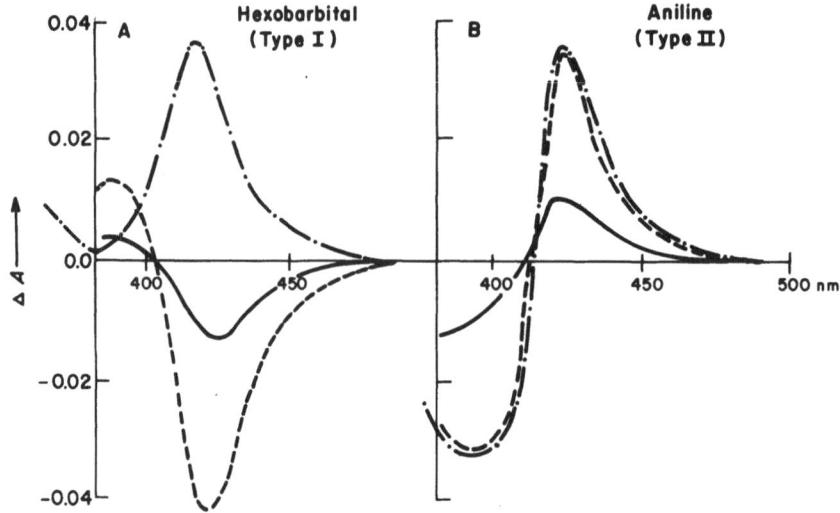

Figure 5. Spectral changes produced by addition of hexobarbital (Type I and Reverse Type I or Modified Type II) and aniline (Type II) to microsomes prepared from 200 - 300 g animals
treated with corn oil, PB and 3-MC. A. Hexobarbital addition
$(1.4 \times 10^{-4}M)$; B. Aniline addition $(5 \times 10^{-4}M)$. ——— corn oil
control, [P-450] = 0.38 nmoles/mg protein; --- PB, [P-450] =
1.6 nmoles/mg protein; ——·——·—— 3-MC, [P-450] = 2.3 nmoles/
mg protein.

band at 440 to 450 nm and a lower band at 420 to 430 nm were found
for codeine demethylation. With aniline hydroxylation there was
little light reversal with wavelengths of light around 450 nm, but
there was a broad band in the 440 to 410 nm region. Note that in
the action spectra for aniline 430 nm was used as reference instead
of 450 nm.

These experiments aimed at comparing a substrate that is demethylated with one that is hydroxylated by microsomes induced by
PB or 3-MC. Codeine is satisfactory for studying demethylation by
PB-induced microsomes, but is unsuitable for studies of 3-MC-in-
duced microsomes since 3-MC induction depresses codeine demethylase
activity.

Aniline is not a good substrate for these studies because it
not only is hydroxylated at a very slow rate (0.5 to 1.0 nmole
min^{-1}(mg protein)$^{-1}$) but it also appears to depress the rate of
electron transfer to P-450 or to interfere with the combination of
P-450(Fe^{2+}) with CO.

TABLE II

Action Spectra Maxima

[nm]

Substrate	Inducer	
	PB	3-MC
Codeine	450	440-450; 420-430
Aniline	450(sh+); 430; 415	420-440

+sh = shoulder.

G. Induction of P-450 by Other Polycyclic Hydrocarbons. The series of polycyclic hydrocarbons given in Figures 9A and 9B were tested for their ability to induce TPNH- and $Na_2S_2O_4$-reducible P-450. No significant induction of P-450 measured by either reductant was found with benzene, napthalene, anthracene, phenanthrene or benzanthracene. The most effective inducer comparable to phenobarbital was 3-methylcholanthrene. During the four day induction period 3-MC and PB induced P-450 systems that could be reduced with TPNH as well as with $Na_2S_2O_4$. It is interesting that pyrene, benzo(a)pyrene and 9,10-dimethylbenzanthracene also induced P-450 in the four day test period, but the induced pigments were only reducible by $Na_2S_2O_4$. This inability of TPNH to reduce P-450 induced by certain polycyclic hydrocarbons raises the question whether there is a qualitative and/or quantitative difference in the metabolism of certain polycyclic hydrocarbons.

As evident from Table III the ability of a compound to induce P-450 was correlated with molecular size and order of carcinogenicity.

H. Reaction of Metopirone and SKF 525A with P-450(Fe^{3+}) and P-450(Fe^{2+}). It has been reported from several laboratories that SKF 525A gives a Type I spectral change while metopirone produces a Type II spectral change when added to microsomes of PB-induced

TABLE III

Order of carcinogenicity, tryptophan resonance overlap integral, incumbrance area (Å^2), $\Delta\varepsilon_{420-490}$ and % increase in P-450 concentration with arylhydrocarbons of increasing size.

Arylhydrocarbon	Order of* Carcinogenicity	Overlap* Integral	Incumbrance** Area (Å^2)	$\Delta\varepsilon_{420-490}$ nm	$\dfrac{\Delta p\text{ind}+450}{p\text{con}450}$ %
Benzene	0	0	32.0	-0.15	- 45
Napthalene	0	0	48.0	-0.28	- 16
Anthracene	0	1,550	61.5	-0.01	+ 7
Phenanthrene	0	0	64.0	-0.02	0
Pyrene	?	-	79.4	-0.004	+174
1,2-Benzanthracene	+	1,350	99.0	-0.004	+ 26
9,10-Dimethyl-1,2-Benzanthracene	++++	3,400	105.0	0.000	+ 74
3,4-Benzpyrene	++++	6,000	94.0	+0.003++	+158
3-Methylcholanthrene	++++	2,200	99.0	+0.015++	+340

* Birks, 1961.

** Arcos and Arcos, 1955; Arcos et al., 1961.

\+ $\Delta p\text{ind}_{450}$ = specific concentration (nmoles/mg protein) of P-450 in induced microsomes minus its specific concentration (PC_{450}) in microsomes of corn oil controls.

\+\+ Inverse Type I spectral change.

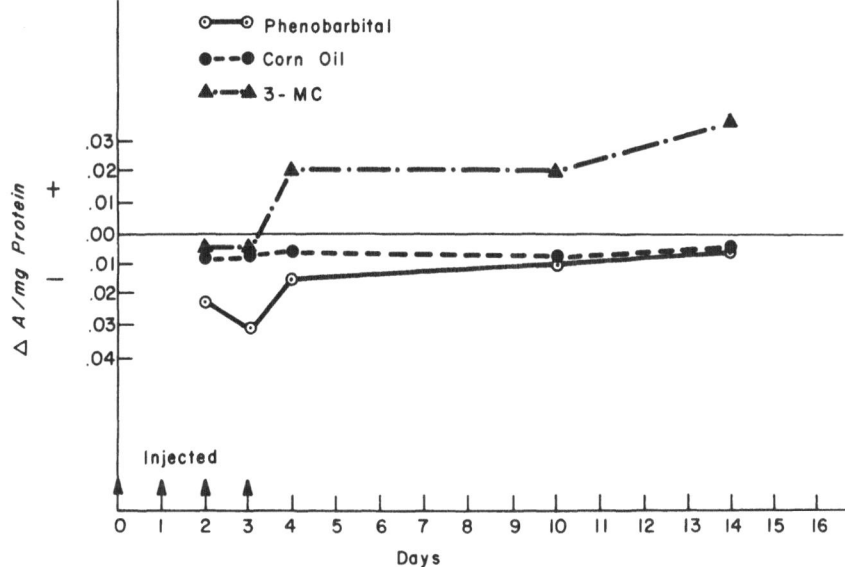

Figure 6. Time course of spectral response observed on addition of hexobarbital to microsomes from animals treated with corn oil, PB or 3-MC. Experimental conditions as described in the legend to Figure 1.

animals. Hildebrandt (1972) has described the spectral change he observed when metopirone is added to reduced PB microsomes. He found an absorption maximum at 444 nm. The maxima of the spectral changes observed by us when metopirone and SKF 525A were reacted with microsomes of PB and 3-MC induced animals at times of maximal inductions are summarized in Table IV. The corresponding P-450(Fe^{2+})·CO absorption maxima are also given for comparison.

SKF 525A did not give a definite spectral change when added to PB-induced microsomes reduced by $Na_2S_2O_4$, but it did give a rather large, broad spectral change with a peak at 434 nm and shoulders at 421 and 451 nm when added to dithionite-reduced 3-MC-induced microsomes. With metopirone when added to reduced 3-MC microsomes a maximum at 443 nm and a shoulder at 425 nm were found, while in PB-induced microsomes the spectrum had the appearance of a P-450(Fe^{2+})·CO difference spectrum whose Soret band had been shifted to 444 nm.

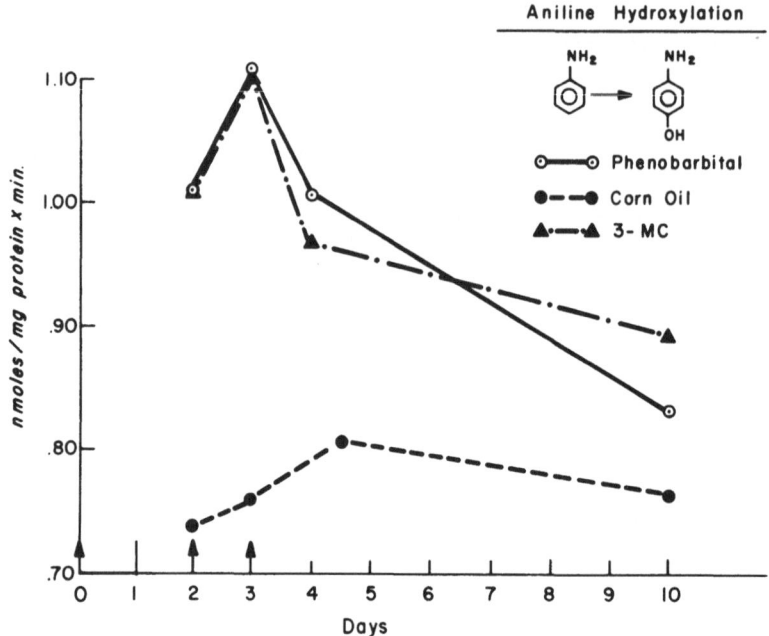

Figure 7A-B. Aniline hydroxylation (A) and codeine demethylation (B) by microsomes from livers of 200 gram rats. The complete
incubation systems, in a total volume of 3.0 ml, contained: 2.4
ml Tris-HCl buffer, 50 mM, pH 7.4; 2 - 5 mg of microsomal protein;
and final concentrations of 5 mM aniline or codeine, 37 x 10^{-3}M
MgCl$_2$; 1.1 x 10^{-3}M TPN; and 4.4 x 10^{-3}M glucose-6-phosphate; with
3 E.U. of glucose phosphate dehydrogenase. Incubations were
carried out for 15 min at 37°C, the reaction being initiated by
addition of the TPNH generating system. The reaction was stopped
with appropriate volumes of trichloroacetic acid.

DISCUSSION

The results reported here confirm those of other laboratories
in showing that the microsomal P-450 system induced by 3-MC is
different from those induced by PB or of non-induced controls, the
latter two being qualitatively identical. The two types of induced
systems differ in their reactions with hexobarbital, metopirone,
SKF 525A and ethyl isocyanide. The time course of induction and
the reducibility of the oxygenase systems induced by the two agents
are entirely different, as well as the CO combining properties
and their light reversibility. It is interesting that the least
distinguishing feature of these two systems is the small 3 nm difference in the absorption maxima of their CO difference spectra.

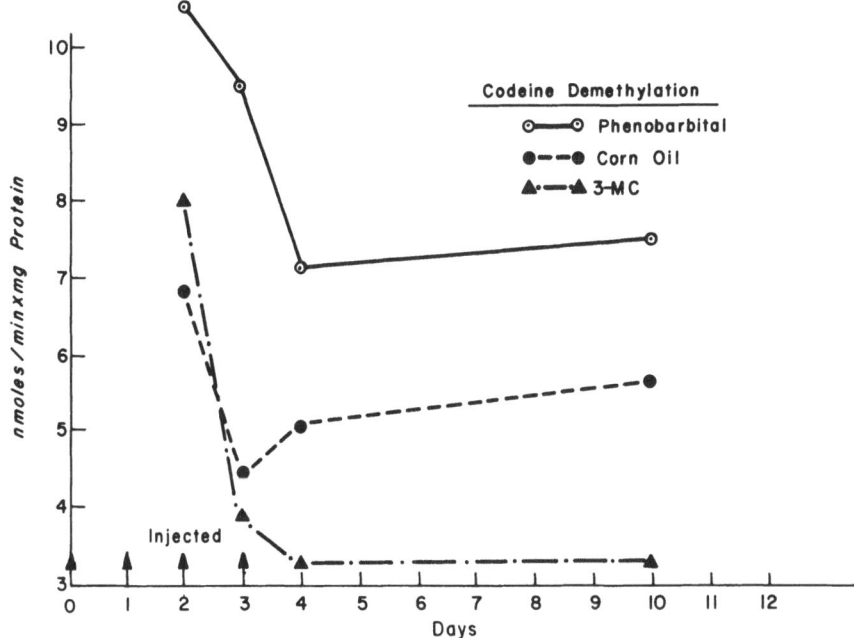

Figure 7B.

There is no evidence that the PB type system can be converted
to the 3-MC system or vice versa by the addition of Type I or
Type II substrates. The absorption coefficients of the
P-450(Fe^{2+})·CO of these two types of P-450 are not very different,
which is in agreement with the values of Fujita et al.,(1973), who
reported millimolar absorption coefficients of 58 and 78 for the PB
and 3-MC-induced microsomes, respectively.

The observation that the P-450 of microsomes from all younger
animals as well as from older animals 8-10 days after 4 treatments
with 3-MC does not react readily with carbon monoxide when TPNH
is used as reductant is of significance. Not only is this finding
an interesting property of microsomal P-450 but it also raises
important questions regarding the design of further experiments to
define the mechanism by which P-450 activates oxygen in hydroxylations catalyzed by mixed function oxidases.

Work of our laboratory (Cooper et al., 1973) and of Waterman
and Mason (1970) has demonstrated that P-450 has a very low oxidation reduction potential, in the range of -380 mV to -400 mV.
These studies have also shown that under strictly O_2-free conditions an excess of TPNH does not convert all of the
P-450(Fe^{3+}) to P-450(Fe^{2+}). This is reflected by the failure of

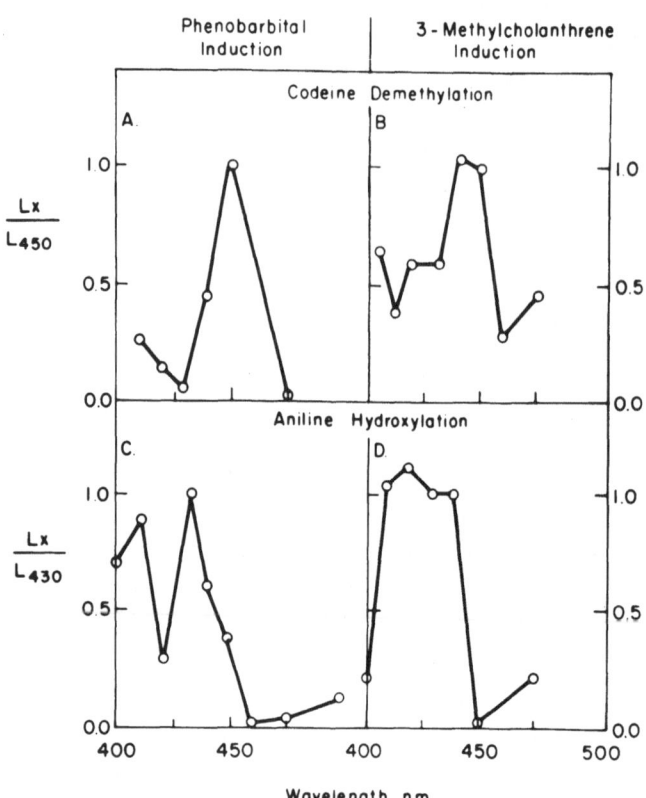

Figure 8. Relative photochemical action spectra for codeine demethylation and aniline hydroxylation. The hepatic microsomes were prepared from rats weighing approximately 150 g after 4 single daily injections of the inducer. The incubation system has been described in the legend to Figure 7. The reaction vessels were equilibrated with gas mixtures containing 4% O_2 or 8% CO + 4% O_2, the balance being made up in both cases with N_2. Warburg's partition constants, K, for the distribution of P-450 between CO and O_2 in these reactions, were as follows: PB microsomes - codeine 0.79, aniline 1.66; 3-MC microsomes - codeine 2.0, aniline 4.38. The relative photochemical action of the monochromatic light bands in reversing the CO inhibition is expressed as the ratio L_x/L_{ref}, where $L = 1/i \cdot (\Delta K/K_d)$, i = (mole quanta) $\cdot$ cm^{-2} $\cdot$ min^{-1} and $\Delta K = K_1 - K_d$, the irradiation-produced increase in K. Reference wavelength (L_{ref}) is 450 nm for codeine and 430 nm for aniline. Regarding details of the method and derivation of the constants, cf. Rosenthal and Cooper, 1967.

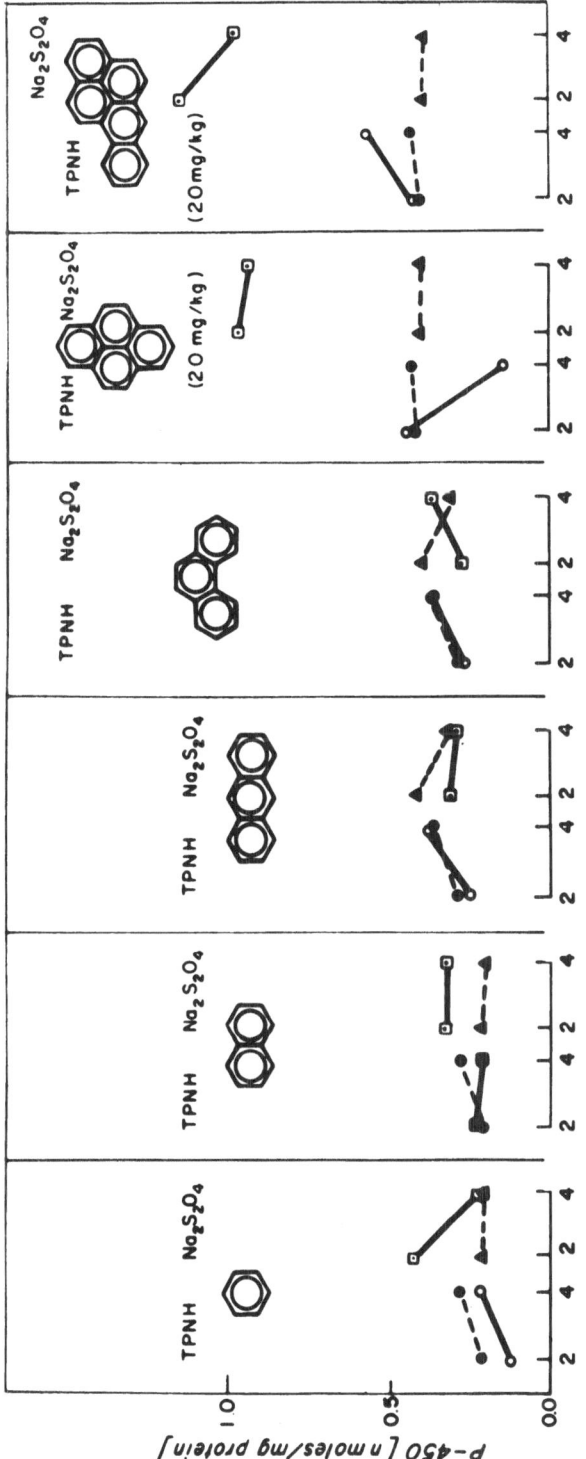

Figure 9A-B. Induction of P-450 by a series of aryl hydrocarbons of increasing molecular size, in 200-300 g rats. P-450 measured by formation of P-450(Fe2+)·CO using TPNH and Na₂S₂O₄ as reductants. Experiments carried out as described in the legend to Figure 1. The dosages of the various hydrocarbons tested were adjusted to be approximately equimolar to the 20 mg/kg/day dosage of 3-MC. Injections were given for 2 or 4 consecutive days and the animals were sacrificed 24 hrs. after the last injection.

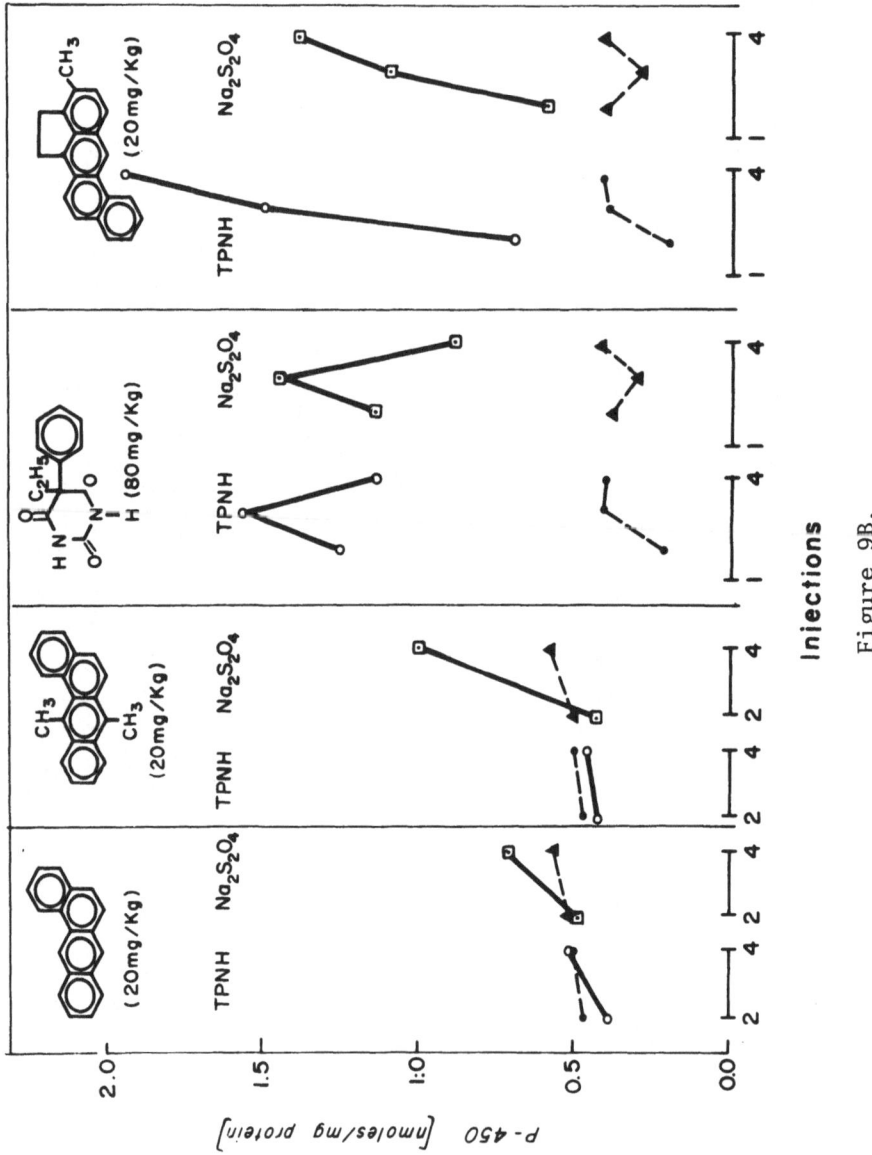

Figure 9B.

TABLE IV

Optical Difference Spectra Maxima ΔA (+ reagent - reagent)

$$\left[nm\right]$$

Inducer →	PB		3-MC	
Reagent	Oxid.	Red.	Oxid.	Red.
SKF-525 A	420*	-	423	434 (421 sh.) (451 sh.)
Metopirone	425	444	425	443 (425 sh.)
CO	-	450	-	448

* negative ΔA (Type I spectrum)

TPNH to remove completely the low spin ferric hemoprotein signal (2.42, 2.24, 1.91 g values) of the EPR spectrum unless excess CO is present, It follows, therefore, that under any condition that would allow hydroxylation to proceed <u>in vitro</u>, i.e., atmospheric O_2 concentration of at least 4%, P-450 would be only detectable in the Fe^{3+} form. This is what one would expect from the oxidation reduction potentials of the known components of the microsomal electron transport system:

1) $TPN^+ + e^- \rightleftharpoons TPNH + H^+$ (-320 mV)

2) $Fp + e^- \rightleftharpoons FpH$ (-250 mV)

3) $FpH + e^- \rightleftharpoons FpH_2$ (-320 mV)

4) $P\text{-}450(Fe^{3+}) + e^- \rightleftharpoons P\text{-}450(Fe^{2+})$ (-400 mV)

There is a 100 mV gap between P-450 and its electron source.

In order for hydroxylation to occur there should be a

mechanism by which electrons are "pulled" toward P-450. This could be accomplished by the rapid reaction of P-450(Fe^{2+}) with a ligand to which it has a high affinity, such as O_2. We may imagine the following reaction steps for this temporary "fixing" of electrons: 1) Formation of the species P-450(Fe^{3+})·O_2^{1-} by an internal electronic rearrangement of P-450(Fe^{2+})·O_2; 2) Introduction of a second electron in the region of O_2 by an as yet unknown microsomal enzyme yielding the species P-450(Fe^{3+})·O_2^{2-}; 3) In the presence of a suitable substrate, the oxygen molecule is cleaved in a concerted reaction into an oxygen atom which is inserted into the hydroxylation substrate, and O^{2-} which attracts 2 protons to form H_2O.

In the absence of an hydroxylation substrate P-450(Fe^{3+})·O_2^{2-} decomposes to P-450(Fe^{3+}) and peroxide anion which reacts with 2 protons to form H_2O_2.

CO is even more effective in fixing the electrons in the P-450(Fe^{2+})·CO complex because the rate of dissociation of CO from P-450 is slow.

SUMMARY

A comparison has been made of the physical and chemical properties of hepatic microsomal P-450 and associated enzyme systems from rats treated with phenobarbital or with 3-methylcholanthrene and other polycyclic aryl hydrocarbons. The results of these studies, though preliminary in nature, indicate clearly that the aryl-induced mixed-function oxidase systems differ significantly from the PB-induced ones in time course of induction, spectral properties, hydroxylase and demethylase activities, CO-inhibition of these reactions and light-reversal of the inhibition. The results support and extend the findings of other investigators regarding the differential biophysical and biochemical properties of aryl-induced systems and provide an experimental design for studying these properties in greater depth at the maximum of aryl induction.

FOOTNOTES

* Dr. Cooper is the recipient of Research Development Award 5 KO3 HL 25132

REFERENCES

Arcos, J. C. and Arcos, M., 1955. Size and Carcinogenic Activity of the Polycyclic Hydrocarbons. Naturwiss 43: 608.
Arcos, J. C., Conney, A. H. and Buu-Hoi, Ng. Pk., 1961. Induction of Microsomal Enzyme Synthesis by Polycyclic Hydrocarbons of

Different Molecular Sizes. J. Biol. Chem. 236: 1291-1299.
Birks, J. B., 1961. A Physical Theory of Carcinogenesis by Aro-
    matic Hydrocarbons. Nature 190: 232-235.
Cooper, D. Y., Schleyer, H., Levin, S. S. and Rosenthal, O.,
    1973. Studies on the Partially Purified Heme Protein P-450
    from the Adrenal Cortex. Ann. N. Y. Acad. Sci. 212: 227-240.
Estabrook, R. W., 1962. Fluorometric Measurement of Reduced Pyri-
    dine Nucleotide in Cellular and Subcellular Particles. Anal.
    Biochem. 4: 231-245.
Fouts, J. R., 1973. Some Selected Studies on Hepatic Microsomal
    Drug-Metabolizing Enzymes - Environment Interactions, p. 380-
    385, in Microsomes and Drug Oxidations, R. W. Estabrook,
    J. R. Gillette, K. C. Leibman, eds., Williams and Wilkens,
    Baltimore, M. D.
Fujita, T., Shoeman, D. W. and Mannering, G. J. 1973. Differences
    in P-450 Cytochromes from Livers of Rats Treated with Pheno-
    barbital and with 3-Methylcholanthrene. J. Biol. Chem. 248:
    2192-2201.
Hildebrandt, A., 1972. The Binding of Metyrapone to Cytochrome
    P-450 and its Inhibitory Action on Microsomal Hepatic Mixed
    Function Oxidation Reactions, pp. 79-102, in Biological
    Hydroxylation Mechanisms, Biochemical Soc. Symposia No. 34,
    G. Boyd and R. M. S. Smellie, eds., Academic Press, London
    and New York.
Omura, T., Sanders, E., Estabrook, R. W., Cooper, D. Y. and
    Rosenthal, O., 1966. Isolation from Adrenal Cortex of a
    Nonheme Iron Protein and a Flavoprotein Functional as a
    Reduced Triphosphopyridine Nucleotide-Cytochrome P-450
    Reductase. Arch. Biochem. Biophys. 117: 660-673.
Omura, T. and Sato, R., 1964a. The Carbon Monoxide Combining Pig-
    ment of Liver Microsomes. I. Evidence for its Heme Protein
    Nature. J. Biol. Chem. 239: 2370-2376.
Omura, T. and Sato, R., 1964b. The Carbon Monoxide Combining Pig-
    ment of Liver Microsomes. II. Solubilization, Purification
    and Properties. J. Biol. Chem. 239: 2379-2385.
Omura, T. and Sato, R., 1967. Isolation of Cytochromes P-450 and
    P-420, pp. 556-561, in Methods in Enzymology, Vol. X, R. W.
    Estabrook and M. Pullman, eds., Academic Press, New York
    and London.
Orrenius, S. and Ernster, L., 1974. Microsomal Cytochrome P-450-
    Linked Monoxygenase Systems in Mammalian Tissues, pp. 215-
    244, in Molecular Mechanisms of Oxygen Activation, O.
    Hayaishi, ed., Academic Press, New York and London.
Rosenthal, O. and Cooper, D. Y., 1967. Methods of Determining the
    Photochemical Action Spectrum, pp. 616-629, in Methods in
    Enzymology, Vol. X, R. W. Estabrook and M. Pullman, eds.,
    Academic Press, New York and London.
Schenkman, J., Remmer, H. and Estabrook, R. W., 1967. Spectral
    Studies of Drug Interaction with Hepatic Microsomal Cytochrome

Mol. Pharmacol. 3:   113-123.

Sladek, N. E. and Mannering, G. J., 1966.   Evidence for a New
    P-450 Hemoprotein in Hepatic Microsomes from Methylcholanthrene-
    Treated Rats. Biochem. Biophys. Res. Commun. 24: 668-674.

Waterman, M. R. and Mason, H. S., 1970. The Redox Potential of
    Liver Cytochrome P-450. Biochem. Biophys. Res. Commun. 39:
    450-454.

CYTOCHROME P-450 IN THE ACTIVATION AND INACTIVATION OF CARCINOGENS

Helmut Greim+, Peter Czygan++, Anthony J. Garro,
Ferenc Hutterer, Fenton Schaffner, Hans Popper
Stratton Laboratory for the Study of Liver Disease,
Department of Pathology, and Department of Microbiology,
Mount Sinai School of Medicine of The City University
of New York, New York, New York 10029
Otto Rosenthal, David Y. Cooper
Harrison Department of Surgical Research, University of
Pennsylvania, Philadelphia, Pennsylvania 19104

SUMMARY

The capacity of isolated mouse liver microsomes to alter the
mutagenicity for bacteria of the primary carcinogen N-methyl-N'-
nitro-N-nitrosoguanidine (MNNG) and the secondary one dimethylnitrosamine (DMN) was studied. Microsomal activation of DMN and inactivation of MNNG were decreased by protein- and protein-choline-
deficient diets and were increased by pretreatment with microsomal
enzyme inducers. The decrease and increase paralleled the content
of cytochrome P-450 present in the different microsomal preparations.
With human liver microsomes of differing cytochrome P-450 contents
similar correlation was obtained, whereas normal rat liver microsomes did not activate or inactivate DMN or MNNG. Oxidative demethylation of DMN by mouse liver microsomes and the activation of
DMN to a mutagen followed similar kinetics. Both reactions were
inhibited by carbon monoxide and the inhibition was maximally
reversed by monochromatic light at 450 nm. These observations indicate that at least some carcinogens are activated or inactivated

+ Presently: Institut für Toxikologie, D-74 Tübingen,
  Wilhelmstrasse 56

++ Presently: Ludolf-Krehl-Klinik, D-68 Heidelberg,
  Bergheimer Strasse 58, Germany

by the unspecific cytochrome P-450 dependent enzyme system, suggesting that the extent of this biotransformation may be one factor influencing human carcinogenesis.

INTRODUCTION

Several studies indicate that the microsomal cytochrome P-450 dependent biotransformation system is involved in chemical carcinogenesis. These include the observations that environmental factors, drugs, and nutrition (McLean and McLean, 1966; Shargel and Mazel, 1968; Marshall and McLean, 1969; Conney et al., 1971; Mannering, 1971; Alvares et al., 1973; Mgbodlie et al., 1973), which influence the activity of this enzyme system, also affect tumor formation in carcinogen-treated animals (Tannenbaum and Silverstone, 1953; Miller et al., 1958; Marugami et al., 1967; Kunz et al., 1969; Falk, 1971; Magour and Nievel, 1971; McLean and Marshall, 1971; Peraino et al., 1971; Rogers and Newberne, 1971). Consequently the status of microsomal enzyme activity should affect carcinogenesis induced by chemicals that are metabolized by these enzymes. However, a precise relationship between microsomal enzyme activity and the biological or mutagenic activity of carcinogens has not been established. Since most carcinogenic compounds are mutagenic to microorganisms (Miller and Miller, 1971; Ames et al., 1973a), the ability of isolated mouse liver microsomes to alter the mutagenic activities of two chemical carcinogens was studied. Furthermore, the involvement of cytochrome P-450 in the activation of the secondary carcinogen was investigated. As example of a primary carcinogen MNNG (N-methyl-N'-nitro-N-nitrosoguanidine) was chosen (Schoental, 1966), that is inactivated by microsomal enzymes (Popper et al., 1973). As a secondary one DMN (dimethylnitrosamine) was used (Magee and Barnes, 1967) that is activated to a mutagen by oxidative demethylation forming monomethylnitrosamine, subsequently degrading to highly reactive carbonium ions (Miller and Miller, 1965; Malling, 1971).

MATERIALS AND METHODS

Male Swiss-Webster mice, Camm Research Institute, Wayne, N.J., weighing 20 g, were randomized in groups of 24 animals and kept under constant environmental conditions at 22°. The animals were fed ad libitum until 12 hr before sacrifice. To induce microsomal enzyme activity one group each received a single intraperitoneal injection of 20mg/kg 3-methylcholanthrene or 500 mg/kg of the polychlorinated biphenyls Arochlor 1254 (PCB) (Monsanto, Chicago, Ill.) or 1% sodium-phenobarbital in the drinking water. The mice were sacrificed 2 days, 4 days or 7 days after beginning of the treatment, respectively. To reduce microsomal enzyme activity, groups of mice were fed semisynthetic diets, containing 30, 10, 10 and choline-deficient, 3 or 0% protein. The 0% protein diet was given for 7 days, all others were given for 28 days. After a 12 hr fast, the animals were killed by decapitation.

Human liver microsomes were isolated from specimen taken by surgical biopsies for histological examination. The methods used for the preparation of hepatic microsomes, the measurement of cytochrome P-450, oxidative demethylation of DMN and aminopyrine have been described previously (Czygan et al., 1973a). Mutagenicity of DMN and MNNG was determined by use of a strain of Salmonella typhimurium TA 1535 (generously provided by Professor B. N. Ames, Dept. of Biochemistry, University of California at Berkeley, Calif.), which carries the his G 46 mutation and requires histidine for growth (Ames et al., 1973b). Bacteria, microsomes, an NADPH generating system and the mutagen were incubated as described previously (Popper et al., 1973). Samples were withdrawn at 0, 15 and 30 minutes and plated for total colony forming units (CFU) and his$^+$ CFU on appropriate selective media. The mutation frequency was expressed as his$^+$ CFU/10^8 CFU The reactions were linear over 30 minutes, with microsomal protein concentrations of 2.5-12.5 mg/assay in 2.7 ml. As controls, the system was incubated without NADPH, without microsomes, or without the mutagen.

RESULTS AND DISCUSSION

In the control system without microsomal enzyme activity, (NADPH omitted), the primary carcinogen MNNG is highly mutagenic and does not require microsomal activation (Table 1, "control"). When MNNG is incubated with liver microsomes from untreated mice, (Table 1, "untreated"), its mutagenic effect is reduced. Incubation with microsomes from mice pretreated with 3-methylcholanthrene, phenobarbital or Arochlor 1254 with elevated microsomal enzyme activities further reduced mutagenicity of MNNG. The extent of this reduction correlates with the amount of cytochrome P-450 and the drug metabolizing enzyme activities of the different microsomal preparations.

By contrast, when DMN is incubated without microsomes, no mutation of the bacterial strain is observed. DMN requires metabolic activation, indicating that it is a secondary carcinogen. Mutagenic activity again is dependent on the activity status of the microsomal preparations as shown by a moderately increased mutagenic activity when exposed to microsomes isolated from mice pretreated with phenobarbital or 3-methylcholanthrene.

TABLE I

Induction of Microsomal Cytochrome P-450 and
Mutagenicity of Primary (MNNG) and Secondary (DMN)
Carcinogen

PRE-TREATMENT	nmoles P-450: mg Protein	HIS$^+$ per 10^8 CFU	
		DMN	MNNG
Control	0.71 : 0.15	10	11500
Untreated	0.71 : 0.15	250	8000
3-MC	1.39 : 0.19	1000	4000
PB	2.22 : 0.27	1100	2400
PCB	2.98 : 0.36	3800	1500

Mutagenicity is highly increased after exposure to microsomes
from mice pretreated with the strong microsomal enzyme inducer
Arochlor 1254.

On the other hand, with microsomes of low cytochrome P-450
content obtained from mice fed different low protein diets the
same correlation between microsomal enzyme activity and mutagenicity was found (Table 2). In the control experiment without
metabolically active microsomes, mutagenicity of MNNG is highest
whereas secondary DMN has no such effect. Microsomes with normal
content in cytochrome P-450 per mg protein inactivate MNNG and
activate DMN. With decreasing microsomal cytochrome P-450 contents activation of DMN and inactivation of MNNG decrease corres-
pondingly (Czygan et al., 1974).

Correlation between microsomal enzyme activity and biological
activity of mutagenic carcinogens can be further demonstrated by
measuring the kinetics of the reactions involved (Fig. 1). Oxidative demethylation of DMN detected by formaldehyde formation
has a K_m of 40 mM, both with normal and induced mouse liver
microsomes. With induced microsomes V_{max} is increased fourfold
identical to a fourfold higher content in cytochrome P-450. When
mutation frequency was measured, K_m is 20 mM, almost similar to
that of the formaldehyde formation from DMN, and V_{max} of the
mutation frequency again is four- to fivefold greater with microsomes of high cytochrome P-450 content (Fig. 2). Both the

Table II

Correlation between Dietary Protein, Microsomal
Cytochrome P-450 and Mutagenicity of a Primary(MNNG)
and a Secondary (DMN) Carcinogen

Dietary Protein	nmoles P-450/ mg Protein	HIS^+ per 10^8 CFU	
		DMN	MNNG
Control	0.68 : 0.13	10	16000
30%	0.68 : 0.13	2800	4500
10%	0.48 : 0.11	1800	6000
3%	0.21 : 0.10	750	10000
0%	0.22 : 0.12	750	14500

oxidative demethylation of DMN and its activation to a mutagen are
progressively inhibited in the presence of increasing $CO:O_2$ ratios
in the gas phase of the incubation flask. 50% inhibition occurs
with 2% CO and 4% oxygen in the nitrogen atmosphere. This inhibitory effect is reversed by irradiating these samples with
monochromatic light, and maximal reversal occurred in both reactions at 450 nm (Fig. 3).

With rat liver microsomes, however, this correlation between
the amount of cytochrome P-450 and the metabolism of the carcinogens
could not be demonstrated. Normal rat liver microsomes obtained
from untreated rats do not activate DMN, whereas activation of DMN
by microsomes from phenobarbital pretreated rats is similar to
that with microsomes from mice with similar cytochrome P-450 content. This inability of normal rat liver microsomes to activate
DMN seems to be due to an inhibitor, since dilution of normal rat
liver microsomes by induced ones reduced inhibition of DMN activation less than the amount of cytochrome P-450 present would
suggest (Fig. 4).

Human liver microsomes, however, with different microsomal
cytochrome P-450 contents metabolize the carcinogens (Czygan et al.,
1973b). Activation of DMN to a mutagen closely depends on the
amount of cytochrome P-450 present (Fig. 5), and inactivation of

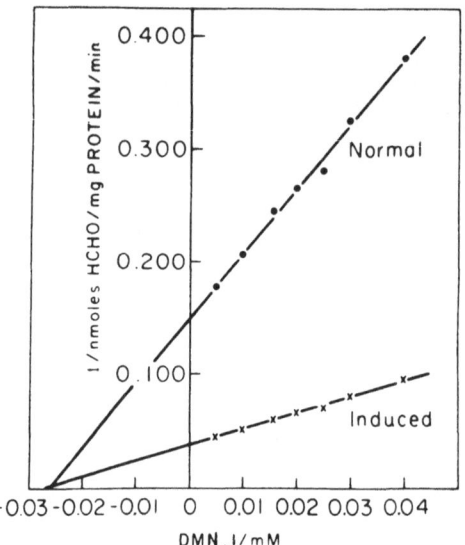

Fig. 1.   Lineweaver-Burk plots of oxidative demethylation of
DMN as measured by formaldehyde formation by normal and induced
mouse liver microsomes.   The incubation system contained 6 mg
microsomal protein and an NADPH generating system in 2.7 ml
0.1 M phosphate buffer, pH 7.4.   The incubation was carried out
for 15 min.   The cytochrome P-450 content of the normal and induced microsomes was 0.7 nmole/mg protein and 2.9 nmoles/mg protein, respectively.

the primary carcinogen MNNG also correlates with the amount of
cytochrome P-450 of the different microsomal preparations
(Fig. 6). Cytochrome P-450 content of surgical human liver microsomes varies widely, ranging in 29 cases from 4 to 50 nmoles/g
tissue or 0.1 to about 0.9 nmoles/mg microsomal protein (Fig. 7).

Considering the involvement of cytochrome P-450 in the metabolism of at least some of the carcinogens, this great variation
possibly may be one of the reasons why susceptibility of man to
chemical carcinogens differs. However, the degree to which these
observations generally apply to human carcinogenesis is a matter of
conjecture, especially considering that carcinogenesis is a complex
process not only involving numerous host factors but also the
nature of the reactants formed that may be different from those
formed by microsomes in vitro. This also applies to the question
whether such metabolizing test systems used in this investigation

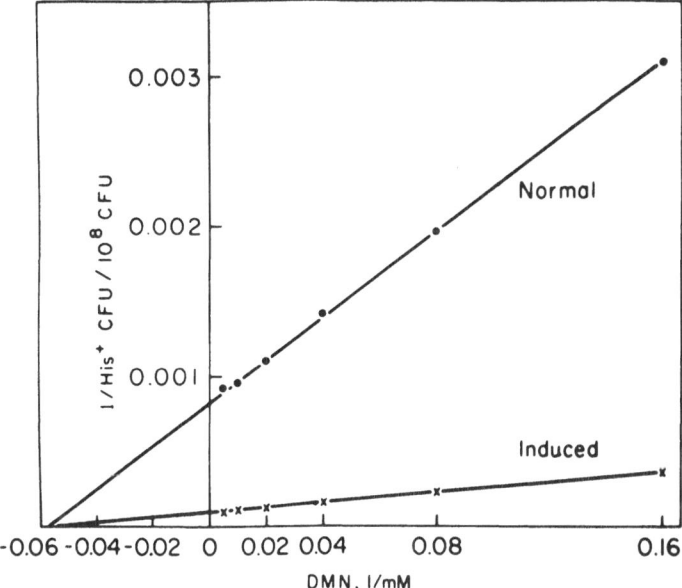

Fig. 2. Activation of DMN to a mutagen. Cells (6 to 9×10^8)
of Salmonella typhimurium TA 1535 were suspended in the incubation system described in Fig. 1. The rate of DMN activation
was monitored by measuring the frequency of <u>his</u>* revertants after
15 min incubation.

can be utilized to screen for carcinogenic compounds (Ames et al.,
1973a and b). The nature of the reactants produced by the microsomes and the nature of bacterial lipopolysaccharide layers of the
cell membrane and bacterial metabolism of the reactants are
factors affecting sensitivity of the test. Additionally, species
differences in microsomal capability to activate or inactivate
carcinogenic compounds influence the outcome of the screening
procedure. Consequently, relevancy of such tests is limited.
They are suitable to investigate particular problems such as those
presented here. But in view of the varied ways in which chemical
carcinogens are metabolized and interact with cell components,
these test systems are not yet appropriate to screen for carcinogens. Moreover, each test system must be adapted to each
problem and to each group of chemicals to be investigated. One
unifying fact, however, is the involvement of cytochrome P-450 in
the metabolism of several of such potential carcinogens.

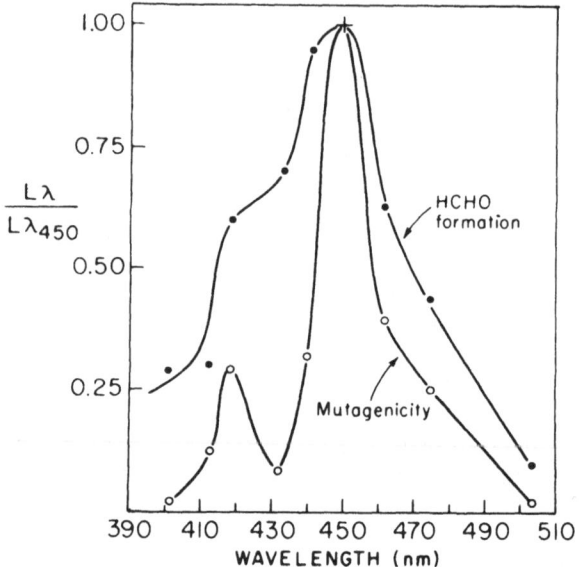

Fig. 3. Photochemical activation of the oxidative demethylation of DMN as measured by formaldehyde formation (●) and the activation of DMN to a mutagen as measured by the mutation frequency (0). The incubation systems are identical to those described in Figure 1 and 2. The incubates were gassed with a $CO:O_2$ ratio of 2:1 and 0.5:1, respectively, for 5 min prior to the addition of DMN. Incubation continued for 15 min during which the incubation flasks were irradiated with monochromatic light of different wavelengths. Light sensitivity is defined by $L = (1/i) \times (\Delta K/K_d)$, where i is light intensity in terms of mole quanta·cm^{-2}·min^{-1}, K_d the distribution constant in darkness, and ΔK the increase in K produced by light of a given wavelength. The partition constant K between oxygen and CO at which 50% inhibition occurred was calculated from Warburg's partition equation (Warburg, 1949).

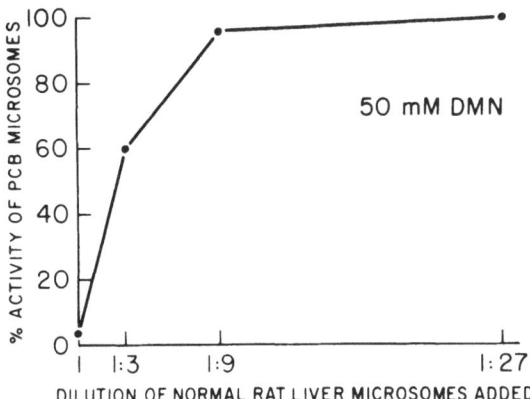

Fig. 4 The effect of diluting normal rat liver microsomes
by microsomes isolated from rats pretreated with 500 mg/kg
Arochlor 1254 on the activation of DMN to a mutagen. Protein
content of the incubate of 2.7 ml was maintained at 6 mg.
Cytochrome P-450 content of normal and induced liver microsomes
was 0.8 and 2.4 nmoles/mg protein, respectively. Incubation system and the procedure to determine mutagenicity are described in
Figures 1 and 2.

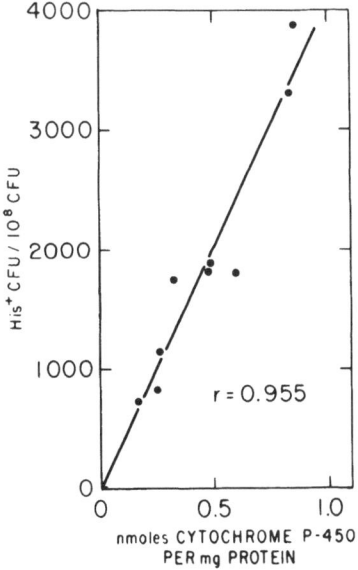

Fig. 5. Relationship between cytochrome P-450 content of
human liver microsomes and ability of these microsomes to activate
mutagenicity of DMN during 30 min incubation.

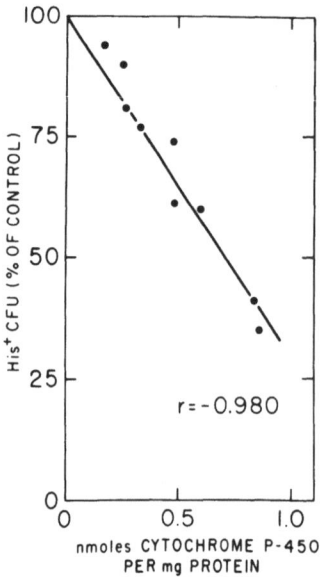

Fig. 6. Relationship between cytochrome P-450 content of human liver microsomes and ability of these microsomes to inactivate MNNG during 30 min incubation. One-hundred percent indicates mutagenicity of MNNG when incubated without metabolically active microsomes.

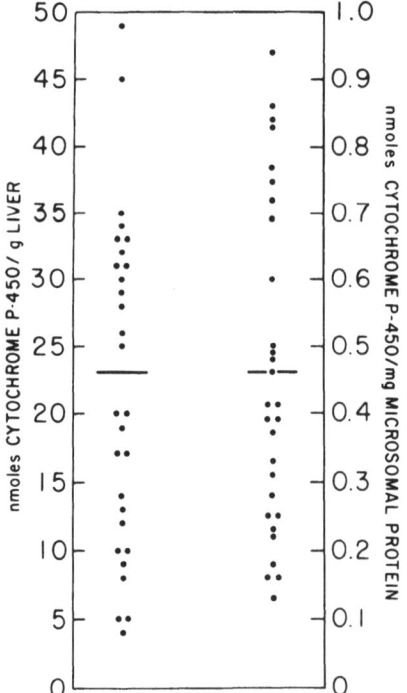

Fig. 7. Hepatic cytochrome P-450 content in 29 patients undergoing abdominal surgery. Horizontal bars represent the means which were 24±14 nmoles cytochrome P-450/g liver and 0.46±0.23 nmoles cytochrome P-450/mg microsomal protein.

References

Alvares, A. P., Bickers, D. R., and Kappas, A. 1973. Polychlor-
    inated biphenyls: A new type of inducer of cytochrome P-448
    in the liver. Proc. Natl. Acad. Sci. USA 70:1321-1325.

Ames, B. N., Durston, W. E., Yamasaki, E., and Lee, F. D. 1973a.
    Carcinogens are mutagens: A simple test system combining
    liver homogenates for activation and bacteria for detection.
    Proc. Natl. Acad. Sci. USA 70:2281-2285.

Ames, B. N., Lee, F. D., and Durston, W. E. 1973b. An improved
    bacterial test system for the detection and classification of
    mutagens and carcinogens. Proc. Natl. Acad. Sci. USA
    70:782-786.

Conney, A. H., Welch, R., Kuntzman, R., Poland, A., Poppers,
    P. J., Finster, M., Wolff, J. A., Munro-Faure, A. D., Peck,
    A. W., Bye, A., Chang, R., and Jacobson, M. 1971. Effects of
    environmental chemicals on the metabolism of drugs, car-
    cinogens, and normal body constituents in man. Ann. N. Y.
    Acad. Sci. 179:155-172.

Czygan, P., Greim, H., Garro, A. J., Hutterer, F., Schaffner, F.,
    Popper, H., Rosenthal, O., and Cooper, D. Y. 1973a. Micro-
    somal metabolism of dimethylnitrosamine and the cytochrome
    P-450 dependency of its activation to a mutagen. Cancer
    Res. 33:2983-2986.

Czygan, P., Greim, H., Garro, A. J., Hutterer, F., Rudick, J.,
    Schaffner, F., and Popper, H. 1973b. Cytochrome P-450 content
    and the ability of liver microsomes from patients undergoing
    abdominal surgery to alter the mutagenicity of a primary and
    a secondary carcinogen. J. Natl. Cancer Inst. 51:1761-1764.

Czygan, P., Greim, H., Garro, A., Schaffner, F., and Popper, H.
    1974. The effect of dietary protein deficiency on the ability
    of isolated hepatic microsomes to alter the mutagenicity of a
    primary and a secondary carcinogen. Cancer Res. 34:119-123.

Falk, H. L. 1971. Anticarcinogenesis-An alternative. Progr.
    Exptl. Tumor Res. 14:105-137.

Kunz, W., Schaude, G., and Thomas, C. 1969. Die Beeinflussung
    der Nitrosamincarcinogenese durch Phenobarbital und
    Halogenkohlenwasserstoffe. Z. Krebsforsch. 72:291-304.

Magee, P. N., and Barnes, J. M. 1967. Carcinogenic nitroso com-
    pounds. Advan. Cancer Res. 10:163-256.

Magour, S., and Nievel, J. G. 1971. Effect of inducers of drug
    metabolizing enzymes on diethylnitrosamine metabolism and
    toxicity. Biochem. J. 123:8-9p.

Malling, H. V. 1971. Dimethylnitrosamine. Formation of mutagenic
    compounds by interaction with mouse liver microsomes.
    Mutation Res. 13:425-429.

Mannering, G. J. 1971. Properties of cytochrome P-450 as affected
    by environmental factors: Qualitative changes due to adminis-
    tration of polycyclic hydrocarbons. Metabolism 20:228-245.

Marshall, W. J., and McLean, A.E.M. 1969. The effect of oral phenobarbitone on hepatic microsomal cytochrome P-450 and demethylation activity in rats fed normal and low protein diets. Biochem. Pharmacol. 18:158-167.

Marugami, M., Ito, N., Konishi, Y., Hsia, Y., and Farber, E. 1967. Influence of 3-methylcholanthrene on liver carcinogenesis in rats ingesting DL-ethionine, 3-methyl-4-dimethyl-aminoazobenzene, and 2-fluorenylacetamide. Cancer Res. 27: 2011-2019.

McLean, A. E. M., and Marshall, A. 1971. Reduced carcinogenic effects of aflatoxin in rats given phenobarbitone. Brit. J. Exptl. Pathol. 52:322-329.

McLean, A. E. M., and McLean, E. K. 1966. The effect of diet and 1, 1, 1-trichloro-2, 2-bis(p-chlorophenyl)ethane (DDT) on microsomal hydroxylating enzymes and on sensitivity of rats to carbon tetrachloride poisoning. Biochem. J. 100:564-571.

Mgbodlie, M. U. K., Hayes, J. R., and Campbell, T. C. 1973. Effect of protein deficiency on the inducibility of the hepatic microsomal drug metabolizing enzyme system. II. Effect of enzyme kinetics and electron transport system. Biochem. Pharmacol. 22:1125-1132.

Miller, J. A., and Miller, E. C. 1965. Metabolism of drugs in relation to carcinogenicity. Ann. N. Y. Acad. Sci. 123:125-140.

Miller, E. C., and Miller, J. A. 1971. The mutagenicity of chemical carcinogens, pp. 83-119. In A. Hollaender (ed). Chemical Mutagens. Principles and Methods for Their Detection. Plenum Press, New York-London.

Miller, E. C., Miller, J. A., Brown, R. R., and McDonald, J. C. 1958. On the protective action of certain polycyclic aromatic hydrocarbons against carcinogenesis by aminoazo dyes and 2-acetylaminofluorene. Cancer Res. 18:469-477.

Peraino, C., Frey, R. J. M., and Staffelt, E. 1971. Reduction and enhancement by phenobarbital of hepatocarcinogenesis induced in the rat by 2-acetylaminofluorene. Cancer Res. 31:1506-1512.

Popper, H., Czygan, P., Greim, H., Schaffner, F., and Garro, A. J. 1973. Mutagenicity of primary and secondary carcinogens altered by normal and induced hepatic microsomes. Proc. Soc. Exp. Biol. and Med. 142:727-729.

Rogers, A. E., and Newberne, P. M. 1971. Nutrition and aflatoxin carcinogenesis. Nature 229:62-63.

Shargel, L., and Mazel, P. 1968. Phenobarbital and 3-methyl-cholanthrene induction of microsomal azoreductase in riboflavin deficient rats. Fed. Proc. 27:302.

Schoenthal, R. 1966. Carcinogenic activity of N-methyl-N'-nitro-N-nitrosoguanidine. Nature 209:726-727.

Tannenbaum, A., and Silverstone, H. 1953. Nutrition in relation
    to cancer. Advan. Cancer Res. 1:451-501.
Warburg, O. 1949. Heavy prosthetic groups and enzyme action.
    Oxford University Press, London.

EFFECT OF CYCLIC AMP ON THE PHENOBARBITAL INDUCED INCREASE IN

CYTOCHROME P-450 AND HYPERTROPHY OF THE ENDOPLASMIC RETICULUM OF

THE RAT LIVER

Ferenc Hutterer, Kenneth Dressler, Helmut Greim, Peter
Czygan, Fenton Schaffner and Hans Popper
The Stratton Laboratory for the Study of Liver Disease
Mount Sinai School of Medicine of the City University
of New York
New York, New York 10029

It has been known for almost a decade that catecholamines depress hepatic drug metabolism (Kato and Gillette, 1965). The action
of these hormones is mediated by cyclic adenosine -3', 5'- monophosphate (cAMP) and the cAMP level in the liver becomes elevated
after hormone administration (Exton et al., 1971). However, there
is relatively little information available concerning the direct
involvement of cAMP in hepatic drug metabolism. Weiner et al.,
(1972a) have recently shown that cAMP, or rather its dibutyryl derivative, $N^6,0^{2'}$-dibutyryl cAMP (DBcAMP), increases sleeping time
after hexobarbital administration. Subsequently it was demonstrated
that DBcAMP treatment decreases the activities of hepatic aniline
hydroxylase and aminopyrine demethylase and the concentration of
cytochrome P-450 (Ross et al., 1973) and partially inhibits the
induction of cytochrome P-450 stimulated by phenobarbital (Dressler
et al., 1973).

We will present here the results of further studies on the
effects of DBcAMP on phenobarbital induction, more specifically
its effect on cytochrome P-450 and b_5, NADPH-cytochrome c reductase, NADPH-cytochrome P-450 reductase, and on the hypertrophy of
the smooth endoplasmic reticulum of the liver.

In our experiments we used Sprague-Dawley male rats with
body weights ranging from 120 to 140 g. The animals were divided
into four groups: one group received saline injections and served
as controls; the second group was injected i.p. with phenobarbital;

the third group received phenobarbital together with DBcAMP; and
the fourth group received DBcAMP alone. The daily dosage of phenobarbital was 100 mg per kg body weight and that of the DBcAMP was
0.2 mmoles per kg body weight. The animals were injected on five
consecutive mornings so that the phenobarbital treatment would not
interfere with their eating habits. Their food intake was adjusted,
each animal consuming 14-15 g per day. Twenty-four hours after
the last injections the animals were killed by decapitation, their
livers homogenized, and the microsomes isolated by ultracentrifugation. The concentrations of cytochromes P-450 and b_5 were determined on the Aminco-Chance spectrophotometer and the activities of
NADPH-cytochrome P-450 reductase and NADPH-cytochrome c reductase
were measured on the same spectrophotometer equipped with an
Aminco-Morrow rapid mixing device (Hutterer et al., 1970). The
microsomal protein concentration was measured by the method of
Lowry et al., (1951). The phospholipid was extracted from the
isolated microsomes as described by Folch et al., (1956). The
extraction was made complete by ultrasonication of the microsomes
in chloroform:methanol. Phosphorus was assayed as described by
Bartlett (1958).

Phenobarbital treatment resulted in enlargement of the liver
and increased microsomal protein content. DBcAMP given simultaneously with phenobarbital further enhanced the liver weight and
microsomal protein content. DBcAMP treatment alone did not significantly alter either of these parameters (Table I).

Phenobarbital increased hepatic cytochrome P-450 content by
nearly 500 nmoles. The cytochrome P-450 content was also higher
in the group treated with phenobarbital and DBcAMP, but the increment was less than 200 nmoles. DBcAMP treatment alone decreased
the cytochrome P-450 content of the liver to 60% of control values
(Table II and Figure 1). The change in cytochrome b_5 content was
in each group similar in direction but lesser in degree than the
change in P-450 (Figure 2).

The activites of NADPH-cytochrome P-450 and NADPH-cytochrome
c reductase increased after phenobarbital treatment to a degree
comparable to the increase in P-450 content. DBcAMP decreased
the activities of the reductases slightly more than it decreased
the P-450 content.

The phospholipid content of the total liver increased in the
phenobarbital treated rats to 294 μmoles from the control value
of 223 μmoles. Simultaneous DBcAMP treatment further enhanced the
phospholipid content to 318 μmoles. DBcAMP treatment alone did
not significantly alter the phospholipid content of the microsomes;
the phospholipid content in this group was 252 μmoles (Figure 3).

TABLE I

Effect of DBcAMP on the body weight, liver weight, and microsomal
protein content of control and phenobarbital (PB) treated rats.

	Body weight	Liver weight	Microsomal protein
	g	g/100 g b.w.	mg/liver weight
Control	146	4.4 (0.6)[a]	294 (18)
PB	155	5.9 (0.5)	335 (13)
PB + DBcAMP	154	7.0 (0.4)	380 (56)
DBcAMP	145	4.9 (0.6)	305 (15)

[a]Standard deviation in parentheses, N = 8. These differences are
not statistically significant: in liver weight between the control group and the DBcAMP group; in microsomal protein content
between the control group and the DBcAMP group and between the
PB group and the PB + DBcAMP group. All other differences are
statistically significant.

    Therefore, the number of nanomoles of cytochrome P-450
associated with each 1000 nmoles of phospholipid was 1.14 in the
control group, 2.43 in the phenobarbital treated group, 1.43 in
the phenobarbital plus DBcAMP treated group, and 0.61 in the group
treated with DBcAMP alone.

    Electron microscopic examination showed an increase in the
smooth endoplasmic reticulum in both phenobarbital and phenobarbital plus DBcAMP treated groups, though the endoplasmic reticulum
was more vesicular and less tubular in form in the DBcAMP treated
animals.

    The data presented here indicate that DBcAMP slightly enhances
the hypertrophy produced in the liver by phenobarbital. This is
reflected by the increase in liver weight, microsomal protein and
phospholipid content, and is further confirmed by electron micro-
scopic examination. In contrast, however, DBcAMP decreases the
concentration of cytochrome P-450 and other enzymes in the biotransformation system in the hypertrophied membrane: assuming
an even distribution, the number of nmoles of P-450 inserted into

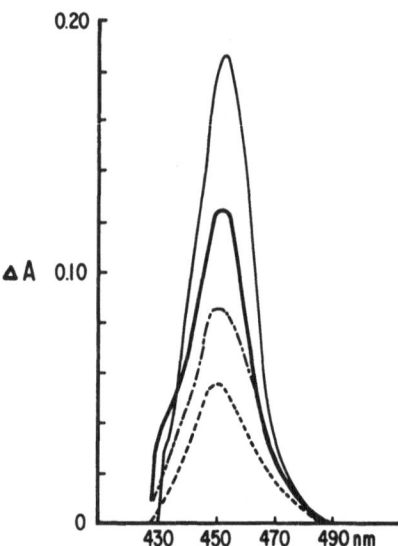

Fig. 1. Difference spectra of dithionite reduced and CO saturated microsomes versus dithionite reduced microsomes. From top
to bottom the peaks represent the phenobarbital treated, phenobarbital plus DBcAMP treated, control and DBcAMP treated groups,
respectively.

the membrane per 1000 nmoles of phospholipid is 2.4 in the phenobarbital treated rats, but only 1.4 after phenobarbital plus
DBcAMP; assuming an uneven distribution, namely that the increment
in cytochrome P-450 is only associated with the newly formed phospholipid, the number of nmoles of P-450 per 1000 nmoles of phospholipid becomes 6.4 in the phenobarbital treated group and 2.0 in
the phenobarbital plus DBcAMP group. Thus the administration of
DBcAMP to phenobarbital treated rats produces a hypertrophic but
relatively "empty" endoplasmic reticulum. The nature of the increased microsomal protein is not known. DBcAMP, while decreasing
the cytochrome P-450 content, does stimulate the synthesis of many
other enzymes, among them the microsomal glucose-6-phosphatase
(Kacew and Singhal, 1974; Schwartz et al., 1974).

The mechanism by which cAMP affects the biotransformation
system is not known. Weiner et al., (1972b) suggested that cAMP
stimulates the production or release of an inhibitor of drug
handling enzyme activities. Such a mechanism could be operational

TABLE II

Effect of DBcAMP on the hepatic cytochrome P-450 content of control and phenobarbital (PB) treated rats.

	Cytochrome P-450	
	nmoles/liver weight	Δ nmoles
Control	257 (31)[a]	
PB	714 (90)	+457
PB + DBcAMP	456 (24)	+199
DBcAMP	153 (26)	-104

[a]Standard deviation in parentheses, N = 8. Comparison of the cytochrome P-450 content of any one group with any other reveals a statistically significant difference.

very shortly after cAMP administration, but it doesn't explain the decrease in concentration of the enzymes of the biotransformation system. In fact, the decrease in enzyme concentration itself is sufficient to explain the hypoactivity of the biotransformation system after cAMP treatment without the assumption of an inhibitor.

In the following we shall try to correlate our findings with some of the known reactions of cAMP (Figure 4).

cAMP may interfere with heme synthesis by decreasing the activity of δ-ALA syntehtase (Amruthavalli and Ramasarma, 1973). Interference with heme synthesis can indeed lead to the formation of hypertrophic smooth endoplasmic reticulum with low cytochrome P-450 content, as is the case when aminotriazole and phenobarbital are administered together (Raisfeld et al., 1970). However, amino-triazole does not depress the NADPH-cytochrome c reductase activity (Baron and Tephly, 1969) while DBcAMP does. We have no information as to whether cAMP interferes with flavokinase, an enzyme necessary in the synthesis of FAD. We have no evidence that the synthesis of heme and FAD are controlled by the same mechanism and therefore it is unlikely that cAMP could depress both syntheses simultaneously and to the same degree.

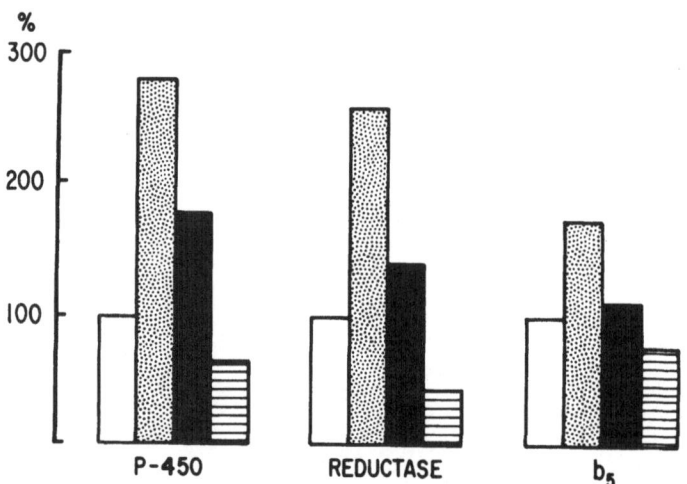

Fig. 2. Relative changes in cytochrome P-450 and cytochrome
b_5 contents and in the activity of NADPH-cytochrome c reductase.
Empty bar represents the control group; dotted bar, the phenobarbital
treated group; black bar, the phenobarbital plus DBcAMP treated
group; bar with horizontal lines, the DBcAMP treated group.

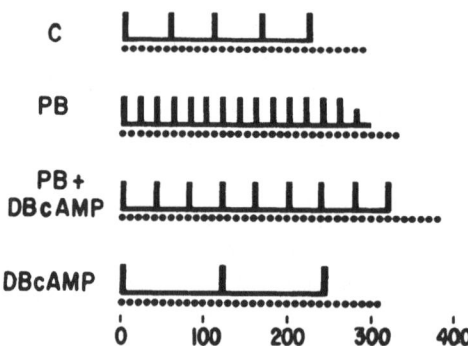

Fig. 3. Correlation of changes in microsomal protein, phospholipid, and cytochrome P-450 contents. Dotted line represents
protein, mg/total liver/100 g body weight; solid line (horizontal)
represents phospholipid, μmoles/total liver/100 g body weight;
vertical lines each represent 50 nmoles of cytochrome P-450. C =
control group; PB = phenobarbital treatment.

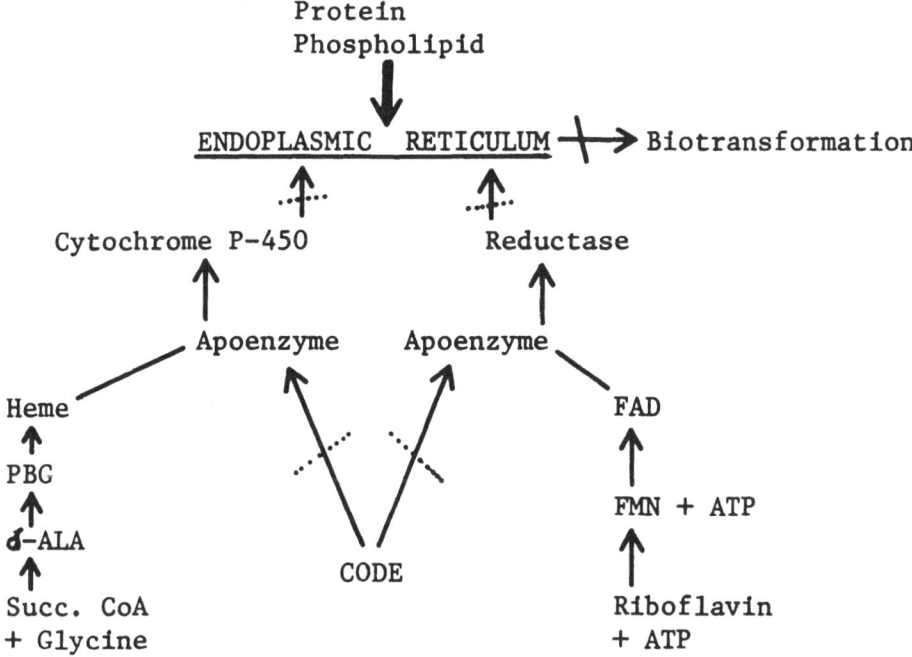

Figure 4.   Possible sites of action of cAMP on the microsomal
biotransformation system.   Heavy arrow indicates increased microsomal protein and phospholipid contents; arrow crossed by solid
line indicates proven decrease in biotransformation activities; arrows crossed by dotted lines indicate possible sites of action of
cAMP, namely decreased apoenzyme synthesis or decreased insertion
of the enzymes into the endoplasmic reticulum.

cAMP stimules the formation and/or release of proteolytic
enzymes from lysosomes (Deter and de Duve, 1967).   This may lead
to increased degradation of the enzymes of the biotransformation
system.   In view of the increased microsomal protein content of
the liver, this is unlikely unless we assume a greatly increased
susceptibility of the enzymes of the biotransformation system to
the action of the lysosomal proteases, an assumption which has no
experimental support.   Increased lipolysis, stimulated by cAMP
(Rizack, 1964), might also account for the loss of enzymes from
the microsomal membranes.   However, the observed increase in microsomal phospholipid makes this an improbable explanation.

cAMP may decrease the rate of synthesis of the apoenzymes of
both the cytochromes and the flavoproteins of the biotransformation

system. This might explain the parallel decrease in these enzymes,
provided that their synthesis is under the same genetic control.
In spite of the fact that cAMP stimulates the synthesis of a great
number of hepatic enzymes (Jost et al., 1969; Wicks et al., 1974)
rather than depressing it, this mechanism is worth studying.

Activation of kinases, the so-called "master reaction" of
cAMP (Jost and Rickenberg, 1971), may provide a suitable explanation
and working hypothesis for the action of cAMP on the biotransformation system. Recent observations indicate that the enzymes of the
biotransformation system are inserted into the preformed endoplasmic
reticulum membrane by hydrophobic interaction (Enomoto and Sato,
1973; Rogers and Strittmatter, 1974). It was also shown in the
case of cytochrome b_5 that the enzyme molecule has a hydrophobic
tail which can be removed by trypsin with the loss of the ability
of the enzyme to bind to the membrane. The hydrophobic segment
comprises 25% of the molecule and contains a number of serine
residues. We speculate that cAMP, which stimulates endogenous phosphorylation of the endoplasmic reticulum membrane of the rat liver
(Jergil and Ohlsson, 1974), may also stimulate phosphorylation of
serine residues on the hydrophobic segment of the cytochrome.
This would cause an increase in the number of charged groups on the
surface of the membrane and/or enzyme, an increase in membrane
potential (Friedmann and Dambach, 1973), and a possible weakening
of hydrophobic interaction. Thus cAMP may exert its influence on
the biotransofrmation system by inhibiting the insertion of the
enzymes of the biotransformation system into the endoplasmic reticulum membrane. This hypothesis is eminently susceptible to testing
since Enomoto and Sato (1973) have demonstrated that solubilized
and purified cytochrome b_5 can be reinserted into the microsomal
membranes in vitro and this reaction can be followed quantitatively
(Cinti, 1975).

Whatever is the mechanism, it is clear that cAMP decreases
the content of cytochrome P-450 and other enzymes of the biotransformation system of the endoplasmic reticulum, both in normal liver
and during induction by phenobarbital. Whether the converse is
true, namely that induction of cytochrome P-450 requires decreased
hepatic cAMP concentrations has not yet been documented directly,
but it has been shown that the cAMP concentration in the liver decreases significantly after barbiturate administration (Kimura et
al., 1974).

Whether the effect of cAMP on the biotransofrmation system is
a pharmacological action or part of an endogenous regulating
mechanism also remains to be established.

REFERENCES

Amruthavalli, E. and Ramasarma, T. 1973. Induction of δ-Amino-
    laevulinate synthetase under environmental-stress conditions.
    Biochem. J. 136: 1091-1096.
Baron, J. and Tephly, T. R. 1969. Effect of 3-amino-1,2,4-
    triazole on the stimulation of hepatic microsomal heme syn-
    thesis and induction of hepatic microsomal oxidases produced
    by phenobarbital. Molec. Pharmacol. 5: 10-20.
Bartlett, G. R. 1958. Phosphorus assay in column chromatography.
    J. Biol. Chem. 234: 466-468.
Cinti, D. L. 1975. The role of cytochrome b_5 in mixed function
    oxidations: Effect of microsomal binding of the hemoprotein
    on hepatic N-demethylations, in this publication.
Deter, R. L. and de Duve, C. 1967. Influence of glucagon, an
    inducer of cellular autophagy, on some physical properties
    of rat liver lysosomes. J. Cell. Biol. 33: 437-449.
Dressler, K., Czygan, P., Skews, T., Greim, H. and Hutterer, F.
    1973. Effect of dibutyryl cyclic AMP on the phenobarbital
    induced increase of cytochrome P-450 in the liver. Fed.
    Proc. 32: 865.
Enomoto, K. and Sato, R. 1973. Incorporation in vitro of purified
    cytochrome b_5 into liver microsomal membranes. Biochem. and
    Biophys. Res. Comm. 51(1)1-7.
Exton, J. H., Robison, G. A., Sutherland, E. W. and Park, C. R.
    1971. Studies on the role of adenosine 3',5'-monophosphate
    in the hepatic actions af glucagon and catecholamines. J.
    Biol. Chem. 246: 6166-6177.
Folch, J., Lees, M. and Stanley, G. H. S. 1956. A simple method
    for the isolation and purification of total lipids from ani-
    mal tissues. J. Biol. Chem. 226: 497-509.
Friedmann, N. and Dambach, G. 1973. Effects of glucagon, 3',5'-
    AMP, and 3',5'-GMP on ion fluxes and transmembrane potential
    in perfused livers of normal and adrenalectomized rats.
    Biochim. Biophys. Acta. 307: 399-403.
Hutterer, F., Bacchin, P. G., Denk, H., Schenkman, J. B., Schaffner,
    F. and Popper, H. 1970. Mechanism of cholestasis. 2.
    Effect of bile acids on the microsomal electron transfer sys-
    tem in vitro. Life. Sci. 9: 1159-1166.
Jergill, B. and Ohlsson, R. 1974. Phosphorylation of proteins in
    rat liver. Endogenous phosphorylation and dephosphorylation
    of proteins from smooth and rough endoplasmic reticulum and
    free ribosomes. Eur. J. Biochem. 46: 13-25.
Jost, J. P. and Rickenberg, H. V. 1971. Cyclic AMP. Ann. Rev.
    Biochem. 40: 741-774.
Jost, J. O., Hsie, A. W. and Rickenberg, H. V. 1969. Regulation
    of the synthesis of rat liver serine dehydratase by adenosine

3',5'-cyclic monophosphate. Biochem. Biophys. Res. Comm. 34: 748-754.

Kacew, S. and Singhal, R. L. 1974. Role of cyclic adenosine 3',5'-monophosphate in the action of 1,1,1-trichloro-2,2-bis-(p-chlorophenyl)-ethane (DDT) on hepatic and renal metabolism. Biochem. J. 142: 145-152.

Kato, R. and Gillette, J. R. 1965. Sex differences in the effects of abnormal physiological states on the metabolism of drugs by rat liver microsomes. J. Pharmacol. Exp. Ther. 150: 285-291.

Kimura, H., Thomas, E. and Murad, F. 1974. Effects of decapitation, ether and pentobarbital on guanosine 3',5'-phosphate and adenosine 3',5'-phosphate levels in rat tissues. Biochim. Biophys. Acta. 343: 519-528.

Lowry, O. H., Rosebrough, N. J., Farr, A. L. and Randall, R. J. 1951. Protein measurement with the Folin phenol reagent. J. Biol. Chem. 193: 265-275.

Raisfeld, I. H., Bacchin, P., Hutterer, F. and Schaffner, F. 1970. The effect of 3-amino-1,2,4-triazole on the phenobarbital-induced formation of hepatic microsomal membranes. Molec. Pharmacol. 6: 231-239.

Rizack, M. A. 1964. Activation of an epinephrine-sensitive lipolytic activity from adipose tissue by adenosine 3',5'-phosphate. J. Biol. Chem. 239: 392-395.

Rogers, J. J. and Strittmatter, P. 1974. The binding of NADH cytochrome b_5 reductase to liver microsomes. Fed. Proc. 33: 1254.

Ross, W. E., Simrell, C. and Oppelt, W. W. 1973. Sex-dependent effects of cyclic AMP on the hepatic mixed function oxidase system. Res. Comm. Chem. Path. Pharmacol. 5: 319-332.

Schwartz, A. L., Raiha, N. C. R. and Rall, T. W. 1974. Effect of dibutyryl cyclic AMP on glucose-6-phosphatase activity in human fetal liver explants. Biochim. Biophys. Acta. 343: 500-509.

Weiner, M., Buterbaugh, G. G. and Blake, D. A. 1972a. Inhibition of hepatic drug metabolism by cyclic 3',5'-adenosine monophosphate. Res. Comm. Chem. Path. Pharmacol. 3: 249-263.

Weiner, M., Buterbaugh, G. G. and Blake, D. A. 1972b. Studies on the mechanism of inhibition of drug biotransformation by cyclic adenosine nucleotides. Res. Comm. Chem. Path. Pharmacol. 4: 37-50.

Wicks, W. D., Barnett, C. A. and McKibbin, J. B. 1974. Interaction between hormones and cyclic AMP in regulating specific hepatic enzyme synthesis. Fed. Proc. 33: 1105-1111.

EVIDENCE FOR THE ACTIVATION OF 3-METHYLCHOLANTHRENE AS A

CARCINOGEN IN VIVO AND AS A MUTAGEN IN VITRO BY P_1-450 FROM

INBRED STRAINS OF MICE

Daniel W. Nebert and James S. Felton
Section on Developmental Pharmacology, Laboratory of
Biomedical Sciences
National Institute of Child Health and Human Development
National Institutes of Health
Bethesda, Maryland 20014

SUMMARY

Genetic differences in aromatic hydrocarbon "responsiveness"
exist among various mouse strains. New formation of cytochrome
P_1-450 and the induction of aryl hydrocarbon (benzo[a]pyrene)
hydroxylase (as well as numerous other monooxygenase activities)
appear to be associated ultimately with genes that cosegregate at
a small number of genetic loci. By comparing "responsive" and
"nonresponsive" siblings, we can evaluate the susceptibility of
each individual to various mutagenic chemicals in vitro or carcinogenic agents in vivo.

3-Methylcholanthrene, benzo[a]pyrene, 7,12-dimethylbenz[a]-
anthracene, benz[a]anthracene, and dimethylnitrosamine were tested
for their metabolic activation to frameshift mutagens in vitro
by liver fractions from various inbred strains of mice treated
previously with 3-methylcholanthrene, β-naphthoflavone, or phenobarbital. The mutagenicity of 3-methylcholanthrene, but not of the
other four compounds, in Salmonella histidine mutant tester
strains TA1537 or TA1538 is expressed as an additive trait and is
closely associated with the genetically mediated induction of aryl
hydrocarbon hydroxylase activity and new cytochrome P_1-450
formation, both of which appear to be expressed as an autosomal
dominant trait. 3-Methylcholanthrene in vitro in the presence of
a liver homogenate fraction from phenobarbital-treated mice is
less mutagenic--per molecule of CO-binding hemoprotein--than that
from 3-methylcholanthrene-treated "responsive" mice. Tumorigenesis
caused by the subcutaneous administration of 3-methylcholanthrene--

but not by similar treatment with benzo[a]pyrene or 7,12-dimethyl-
benz[a]anthracene--is highly correlated with aromatic hydrocarbon
"responsiveness."

We suggest these results represent differences between cytochrome P_1-450 and other cytochromes P-450 in their respective
production of certain reactive metabolic intermediates with MC
as the substrate. The other four compounds in this study (i)
might not require metabolism for mutagenesis and carcinogenesis,
(ii) might be associated more closely with metabolic activation
by some cytochrome P-450 species other than P_1-450, (iii) might
be metabolically activated to the same extent by either cytochrome
P_1-450 or other cytochromes P-450, or (iv) might be associated
with a metabolic pathway other than the monooxygenase system.

A simple, inexpensive, and very sensitive test for the
detection of chemical compounds as mutagens has been recently developed (1, 2). The detection of mutations involves a simple
back-mutation test: the reversion from histidine requirement in
Salmonella typhimurium auxotrophs to growth on minimal medium.
Additional mutations (1-3) include a defective excision repair
system for DNA and a low lipopolysaccharide content in the
bacterial cell wall. Consequently, extremely hydrophobic molecules
may enter these bacteria rather easily, and the bacteria will be
far more sensitive to any mutagen, because intercalation and/or
covalent binding with DNA will not be readily repaired. Thus,
DNA repair as a factor can be separated from the mechanism of DNA
damage. With this sytem, 18 carcinogens--including MC,[1] BP,
DMBA, aflatoxin B_1, 2-acetylaminofluorene, benzidine, and
dimethylamino-trans-stilbene--all have been shown (3) to be metabolically activated by rat liver homogenates to form potent frameshift mutagens. The structural features which these compounds
have in common are (i) a ring system sufficiently planar for a
stacking interaction with DNA base pairs, and (ii) a portion of
the molecule capable of being metabolized to a reactive group, so
that covalent bonding to DNA is likely to occur.

Definition of frameshift mutation. The scientific principle
involved in frameshift mutation is, we feel, of such potential
fundamental importance as to justify a few introductory remarks
and simple illustrations. Figure 1 shows the six possible DNA
base-pair combinations. It is theorized that a molecule sufficiently planar for intercalation with DNA may interact specifically, for steric reasons, with only one of these six possible
base-pair combinations. A less specific interaction might also be
possible with several, or all, of these six combinations. One
interaction which has been demonstrated (4) by 3-dimensional computer analysis is the specific intercalation of the actinomycin
D molecule with the -G-C- base-pairing. The end result is a
                          -C-G-

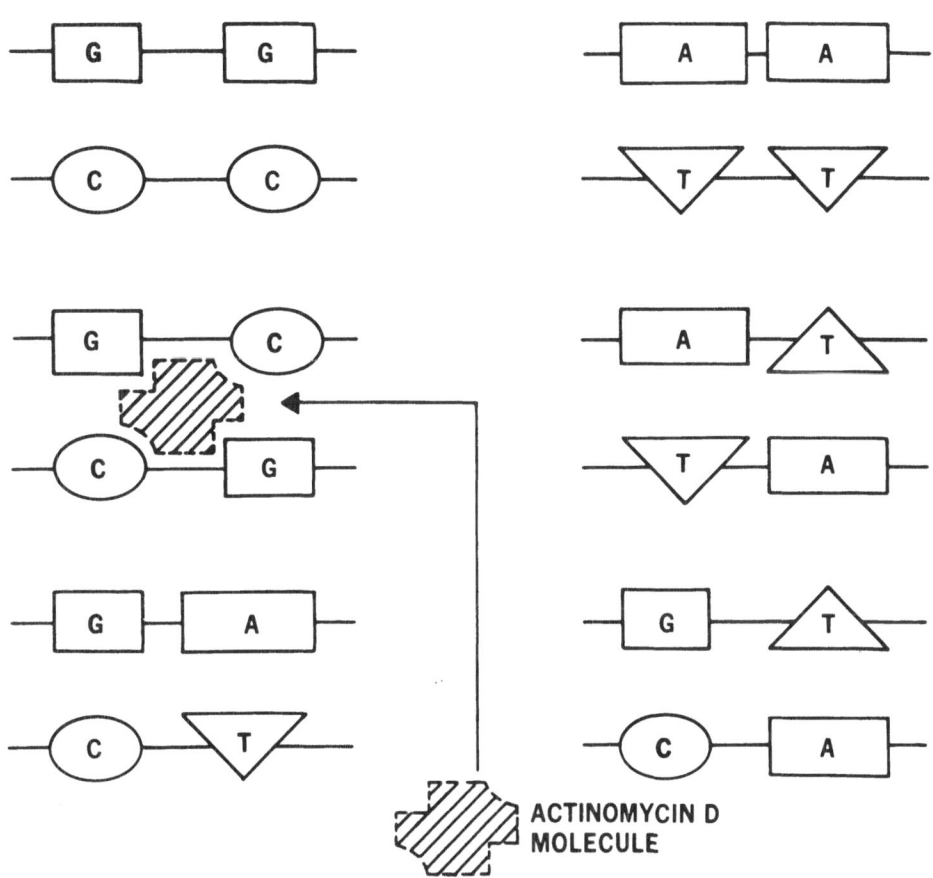

Fig. 1. The six possible DNA base-pair arrangements. The
precise steric "fit" for the molecule of actinomycin D into the
-G-C base pairing is illustrated. Genes containing larger numbers
-C-G-
of this -G-C sequence will presumably be more susceptible to the
       -C-G-
intercalation of actinomycin D and its subsequent effects.Other
planar molecules may likewise interact sterically with one or more
of the base-pair combinations in a similar specific manner. G =
guanine; C = cytosine; A = adenine; T = thymine.

covalent binding of actinomycin D to the nitrogen-7 of guanine
(4, 5), thereby resulting in an irreversible block in transcription
of DNA into RNA. It should be emphasized that the set of tester
strains (1-3) for detecting frameshift mutagens is not yet complete. For example, strain TA1538 contains the CGCGCGCG sequence;
strains TA1535, TA1536, and TA1537 have other "hot spot" sequences
in one of the histidine genomes (1-3). Not until all six possible
DNA base-pair combinations are sufficiently developed will this
frameshift mutagen "tester kit" (1-3) be complete. Therefore, a
chemical could be shown to be a mutagen only in the presence of
one tester strain, or the compound could be mutagenic with as many
as all six tester strains. The lack of mutagenicity for a particular chemical with the current set of only 4 of the 6 possible
bacterial tester strains therefore does not necessarily lead to
the conclusion that the compound is not mutagenic for at least two
reasons: (i) metabolic or transport conditions of the compound
in vivo are not duplicated in the in vitro system; (ii) the mutagenic agent reacts only with one of the DNA base-pair combinations
not yet developed. All of the negative results with compounds used in
this preliminary report will certainly require further investigation
as the complete bacterial "tester kit" becomes more fully developed.

    The mechanism of mutation by compounds which intercalate in
the DNA base-pair stack is thought to be that the intercalation
distorts the DNA backbone so that a mispairing during DNA repli-
cation, repair, or recombination causes the addition or deletion
of a base (6). These chemicals are called frameshift mutagens
because the reading frame of the messenger RNA is shifted (Fig. 2),
resulting in synthesis of a "nonsense" peptide. In the case of
each Salmonella tester strain (1-3), a frameshift mutation has
already been introduced into one of the genes of the histidine
operon; it is the intercalation of the frameshift mutagens that
corrects the reading frame so that a functional protein in the
histidine biosynthetic pathway is now synthesized.

    Metabolic activation of mutagens by monooxygenases. The
possibilities that certain compounds may require metabolic activation to the proximal or ultimate carcinogen have been recently
reviewed (7-9). It is now commonly accepted (Fig. 3) that the
oxidative metabolism of numerous hydrophobic compounds --
especially polycyclic aromatic hydrocarbons--proceeds via reactive arene oxide (epoxide) intermediates, which can isomerize to
phenols, be converted enzymatically to trans-dihydrodiols or
gluthathione conjugates, or become covalently bound to cellular
nucleic acids and proteins. A prominent candidate for such important arene oxide formation is the microsomal monooxygenase activity,
aryl hydrocarbon (benzo[a]pyrene) hydroxylase, a NADPH-dependent
cytochrome P-450-mediated enzyme system (cf. ref. 9-11).

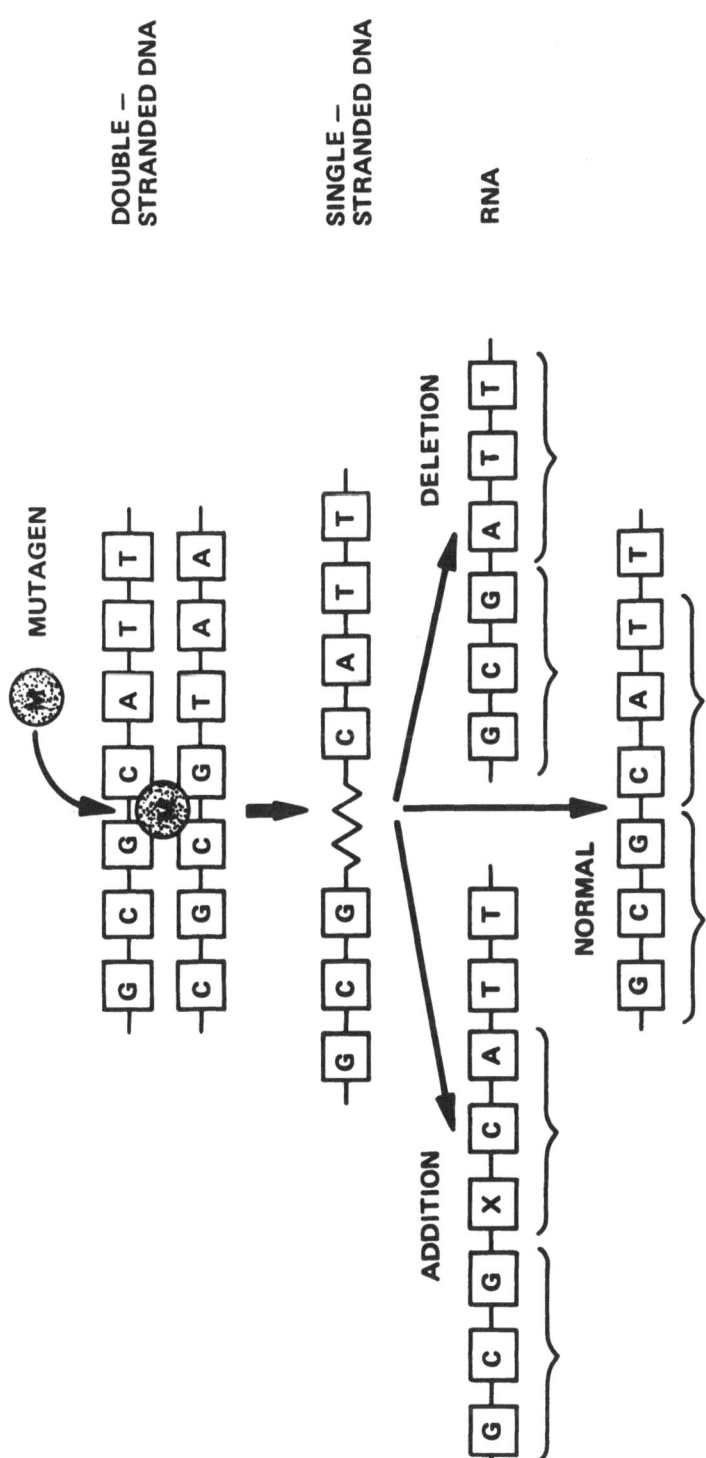

Fig. 2. Illustrated mechanism of frameshift mutation, whereby the intercalation of a mutagen (M) into a certain DNA base-pair causes in some way either addition (X = purine or pyrimidine) or deletion of a base in the transcribed RNA. Abbreviations of the bases are the same as in Figure 1, plus U = uracil. This change in base sequence affects the "readout" of triplets when the RNA is translated into a peptide, thereby resulting in a "nonsense protein," which can be detected as a mutation.

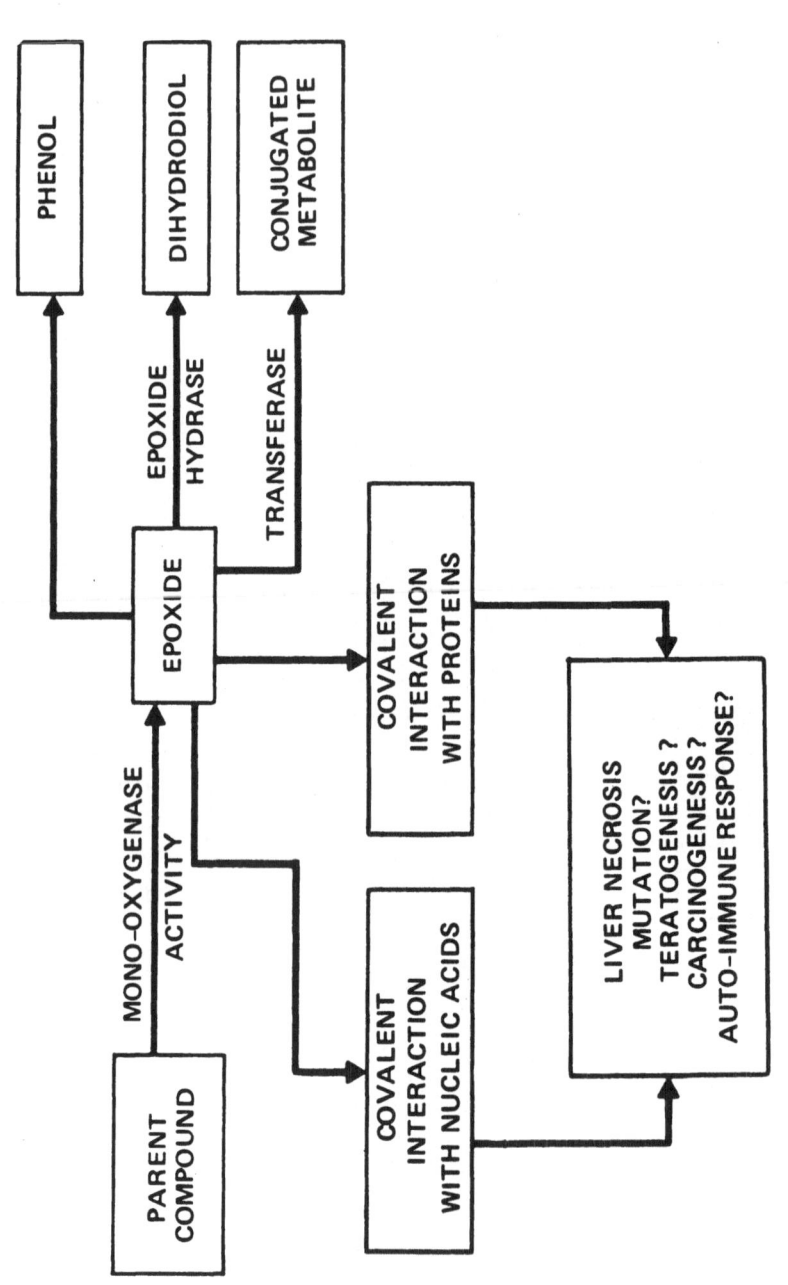

Fig. 3. Possible metabolic pathways for reactive epoxide (arene oxide) intermediates, subsequent to oxygenation of the hydrophobic aromatic parent substrates by the cytochrome P-450-mediated monooxygenase system(s).

Genetic differences in the monooxygenase system. In spite of
the extreme complexity of the membrane-bound multicomponent monooxygenase system (9-11), the response of mice to aromatic hydrocarbon inducers is controlled by a surprisingly small number of
genetic loci. A single gene difference was initially thought to
regulate aryl hydrocarbon hydroxylase induction by aromatic hydrocarbons (12-15), and the Ah locus for aromatic hydrocarbon "re-
sonsiveness" was suggested (16). However, recent studies indicate that at least two, and probably more, nonlinked loci may be
involved (11, 17). The stimulation of at least 11 enzyme activities plus new formation [2] of cytochrome P_1-450 is a constant feature
among mice from numerous genetic backgrounds, including recombinant
inbred sublines.[3] Although the (B6D2)F_1 heterozygote can in fact
(11) be distinguished from the inbred dominant B6 parent, we feel
it is still valid for the sake of this report to present the genetic scheme shown in Figure 4. The allele carried by the "responsive" B6 mouse is Ah^b, and the allele carried by the "nonresponsive" D2 mouse is Ah^d. From the two lower boxes, it can be
seen that the F_2 generation and the offspring from the (B6D2)F_1 X
D2 backcross result in 75% and 50%, respectively, of the mice
being "responsive."

A "responsive" animal has higher inducible hydroxylase activities than a "nonresponsive" animal not only in liver but also in
numerous nonhepatic tissues as well (11, 12, 16, 17). Genetic
differences in the CO-binding hemoprotein associated with increased
aromatic hydrocarbon hydroxylation quite likely will be associated
with increased amounts of the reactive arene oxide intermediates
(Figure 3). This experimental model system thus allows one to
study--among siblings--the susceptibility of each individual to
carcinogenic or toxic environmental chemicals. These genetic
differences among siblings offer an especially powerful probe in
the research areas of toxicology, chemical carcinogenesis, and
pharmacology, because many test substances produce nonspecific
toxicity, malnutrition, or other disruptions of normal physiology.

F_1 Ah^b/Ah^b X Ah^d/Ah^d ↓ Ah^b/Ah^d	Ah^b/Ah^d X Ah^b/Ah^b ↓ Ah^b/Ah^b:Ah^b/Ah^d
F_2 Ah^b/Ah^d X Ah^b/Ah^d ↓ Ah^b/Ah^b:Ah^b/Ah^d:Ah^b/Ah^d:Ah^d/Ah^d	Ah^b/Ah^d X Ah^d/Ah^d ↓ Ah^b/Ah^d:Ah^d/Ah^d

Fig. 4. Simplified genetic scheme for aromatic hydrocarbon
"responsiveness" in the mouse.

Understanding the mechanisms whereby polycyclic hydrocarbons initiate cancer is, to say the least, extremely complex. By simplifying some of the experimental variables, however, we feel it is possible to gain useful information. In this report we combine the bacterial mutagenesis assay in vitro and the genetic differences in the mouse hepatic cytochrome P-450-mediated monooxygenase system. With the mutagenesis assay, the variable of DNA repair is eliminated. With the genetic differences among inbred strains of mice, the metabolism of a chemical to its mutagenically or carcinogenically active form may be different between cytochrome P-450 and P_1-450. We show here that the activation of MC to a mutagen in vitro and to a carcinogen in vivo is highly correlated with the genetically mediated aromatic hydrocarbon "responsiveness" and its associated increase in cytochrome P_1-450 concentration, whereas the activation of other carcinogens such as BP, DMBA, benz[a]anthracene, and dimethylnitrosamine in vitro does not appear to be closely associated with the induced monooxygenase activities and cytochrome P_1-450 content.

Comparison of different bacterial tester strains. Following treatment with aromatic hydrocarbons such as MC or BNF, the inducible hepatic hydroxylase activity in the "responsive" inbred B6 mouse rises 3- to 8-fold, whereas the enzyme activity in the "nonresponsive" D2 inbred strain does not increase (11-17). In the presence of either tester strain TA1537 or TA1538 (Tables 1 and 2), a striking genetic difference between liver S-9 from MC- or BNF-treated B6 and D2 mice was found in the activation of MC in vitro: 92 versus 22 colonies with bacterial tester strain TA1537 and 162 versus 15 (MC) and 176 versus 28 (BNF) with tester strain TA1538. The colony counts shown are the average of duplicate determinations, which always varied by less than 10%; experiments performed on different weeks gave the same relative differences, although the absolute colony counts sometimes varied by as much as 50%. Small differences between B6 and D2 mice were observed with BP or DMBA as the activated mutagen in vitro, but these increases in histidine reversion with B6 were always less than twice that found with D2.

The histidine revertant rate with S-9 plus the solvent dimethylsulfoxide without any compound added in vitro was higher than the revertant rate with a compound added in vitro but without the S-9 fraction; this effect was slight with bacterial tester strain TA1537, but marked with tester strain TA1538. We do not understand the mechanism for this observation, which also had been seen by Ames and coworkers but not commented upon (3); however, we believe that the difference, for example, between 162 and 15 colonies for B6 and D2, respectively, is significantly a more than 10-fold difference and not a 2- to 3-fold difference--although 50 to 70 colonies are observed in the presence of S-9 but without

TABLE 1

Comparison of three different bacterial tester strains in determining activation of MC to a mutagen in vitro by liver S-9 fractions from MC-treated C57BL/6N or DBA/2N inbred mice.

μg of MC added	Histidine revertants per plate								
	Strain TA1535			Strain TA1537			Strain TA1538		
	B6	D2	No S-9	B6	D2	No S-9	B6	D2	No S-9
500	19	11		112	20				
200	20	28		94	13				
100			9	104	15	12	190	33	18
25				24	9				
10				19					
0	4	6		9	7		44	64	8

The environment of the animal room, feeding of the mice, and prior treatment with MC (single intraperitoneal dose of 80 mg kg^{-1} in corn oil 40 hours before sacrifice) have been previously described (12, 14). The mouse liver homogenate was centrifuged for 10 min at 9,000 x g, and the supernatant fraction was decanted and saved; this is the "S-9 fraction" (2, 3). The mutagenesis test with the S-9 fraction was carried out exactly as described (3). To 2 ml of molten top agar at 45° were added 0.1 ml of the bacterial culture (2-3 x 10^9/ml), 100 μl of dimethylsulfoxide or water containing the compound to be tested, and 0.5 ml of the S-9 "Mix" (which contains 0.3 ml of S-9 fraction, 8 μmoles MgCl$_2$, 33 μmoles KCl, 5 μmoles glucose-6-phosphate, 4 μmoles NADP, and 100 μmoles sodium phosphate buffer, pH 7.4, per ml). The colonies on each plate (histidine revertants) were counted after a 2-day incubation at 37° (18).

TABLE 2

Activation of various polycyclic hydrocarbons to mutagens in vitro by liver S-9 fractions from C57BL/6N or DBA/2N inbred mice treated with MC, β-naphthoflavone, or phenobarbital.

In vivo pretreatment of mice[a]	Compound[b] added in vitro		Histidine revertants per plate					
			Strain TA1537			Strain TA1538		
	Name	Amount (µg)	B6	D2	No S-9	B6	D2	No S-9
MC	MC	100	92	22	9	162	15	10
	BP	50	21	16	10			
		10				182	156	28
		5	46	14				
	DMBA	100	52	26	9	60	44	30
		10	17	10				
	BA	100	21	15	8	74	40	18
	DMN	100				95	101	15
	NONE	–	14	28		50	68	
BNF	MC	100				176	28	18
	NONE	–				31	58	
Pheno-barbital	MC	100				85	92	17
	BP	10				182	185	10
	DMBA	100				65	62	15
	BA	100				125	112	
	DMN	100				130	129	
	NONE	–				57	65	
None	MC	100				28	25	15
	BP	10				169	164	
	DMBA	100				43	32	
	BA	100				40	55	
	DMN	100				93	73	
	NONE	–				45	69	

a   Previous treatment with MC or BNF consisted of a single intra-peritoneal dose (80 mg kg^{-1}) of either compound in corn oil 40 hr before sacrifice (18). Phenobarbital was administered in 0.85% NaCl as an intraperitoneal dose of 60 mg kg^{-1} for 4 consecutive days, and the livers were removed 24 hr after the last dose. Controls received either corn oil or 0.85% NaCl alone.

b   Abbreviations in this Table include BA, benz[a]anthracene; and DMN, dimethylnitrosamine.

any compound added in vitro (vide infra in discussion of
Figure 6B).

Comparison of various inducers of monooxygenase activities
in vivo and various mutagens in vitro. This striking genetic difference with MC between the B6 and D2 inbred strains was not found
in mice treated previously with phenobarbital or in control mice
(Table 2), nor was a large difference found when either BP, DMBA,
benz[a]anthracene, or dimethylnitrosamine was used as the mutagen
in vitro . The induction of hepatic hydroxylase activity by
phenobarbital is similar in B6 and D2 mice (12, 14). In the
activities of MC in vitro, the facts--(i) that a difference exists
between B6 and D2 mice treated previously with either MC or BNF
and (ii) that no difference is seen between these mouse strains
following either no treatment or phenobarbital treatment--suggest
that MC activation might be related to aromatic hydrocarbon "responsiveness."

Comparison of other inbred strains and F_1 hybrids. Table 3
shows a highly significant correlation between the mutagenicity
of MC in vitro and the hepatic hydroxylase activity in several
"responsive" and "nonresponsive" MC-treated inbred strains and in
MC-treated F_1 hybrids having high, intermediate, or low levels of
the hydroxylase activity. We found no evidence for a diffusible
cellular product, since the mixing of S-9 fractions from MC-treated
"responsive" and "nonresponsive" mice was purely additive (data
not shown). Further, the MC-treated (C3H/HeN) (DBA/2N)F_1 hybrid
displayed intermediate values for both the revertant rate in vitro
and the hydroxylase activity (Table 3). With respect to MC activation in vitro among MC-treated (B6D2)F_1 hybrids and offspring
from the B6D2 x D2 backcross (Figure 5) and F_2 progeny (data not
illustrated), an interesting genetic relationship emerged. Whereas the inducible hydroxylase activity segregated as an autosomal
dominant trait, the histidine reversion rate--although associated
exclusively with the inducible enzyme activity--segregated co-
dominantly, or in an additive fashion.

Relationship between cytochrome P_1-450 and mutagenicity by
MC. It is likely (11, 19-22) that different forms of monooxygenase
activities and "cytochromes P-450" exist. Dissimilarities in the
hydroxylase activity and CO-binding cytochrome from (i) control
mice or MC-treated "nonresponsive" mice, (ii) phenobarbital-
treated mice, and (iii) MC-treated "responsive" mice have been
demonstrated (11) by various techniques. If one plots the specific hydroxylase activity as a function of CO-binding hemoprotein
(Figure 6A), these three groups (triangles, squares, and circles,
respectively) are clearly seen. Figure 6B shows that MC in vitro--
in the presence of the S-9 fraction from phenobarbital-treated
B6 or D2 mice--is less mutagenic, per molecule of CO-binding
cytochrome, than that from MC-treated B6 mice.

TABLE 3

Relationship between the hepatic hydroxylase activity and the
activation of MC to a mutagen in vitro by liver S-9 fractions
from various MC-treated inbred or hybrid mice.

"Responsiveness" to aromatic hydrocarbons[a]	Source of S-9 fraction	Histidine revertants per plate[b]	Hydroxylase specific activity[c]
"Responsive"	C57BL/6N	114	2080
	C57BL/6J	91	1660
	C3H/HeN	74	1630
	(C57BL/6N)(DBA/2N)F$_1$	93	1950
"Intermediate"	(C3H/HeN)(DBA/2N)F$_1$	53	1010
"Nonresponsive"	DBA/2N	31	380
	AKR/N	43	600
	(C57BL/6N)(AKR/N)F$_1$	36	340

All mice received a single dose of MC 40 hr prior to sacrifice (18).
The livers from 3 mice were combined for each inbred or hybrid
sample. One hundred µg was the dosage of MC as the mutagen in vitro
and the bacterial tester strain was TA1538. The correlation coefficient r for these data is 0.91 (P<0.001).

[a]For detailed discussions involving classification of these inbred
strains and these genetic crosses, see refs. 11, 12, 14 and 17.

[b]In this table and in the subsequent figures, a background revertant rate of about 18--when no S-9 was added to the reaction mix-
ture--has already been subtracted.

[c]The determinations of hydroxylase activity and protein content
were carried out exactly as previously described (12, 14). In
this table and subsequently in this report, one unit of aryl
hydrocarbon hydroxylase activity is defined (12, 14) as that amount
of enzyme activity catalyzing per min at 37⁰ the formation of
hydroxylated product equivalent to that of 1 pmole of the recry-
stallized 3-hydroxybenzo[a]pyrene standard. Specific activity
denotes units per mg of liver microsomal protein.

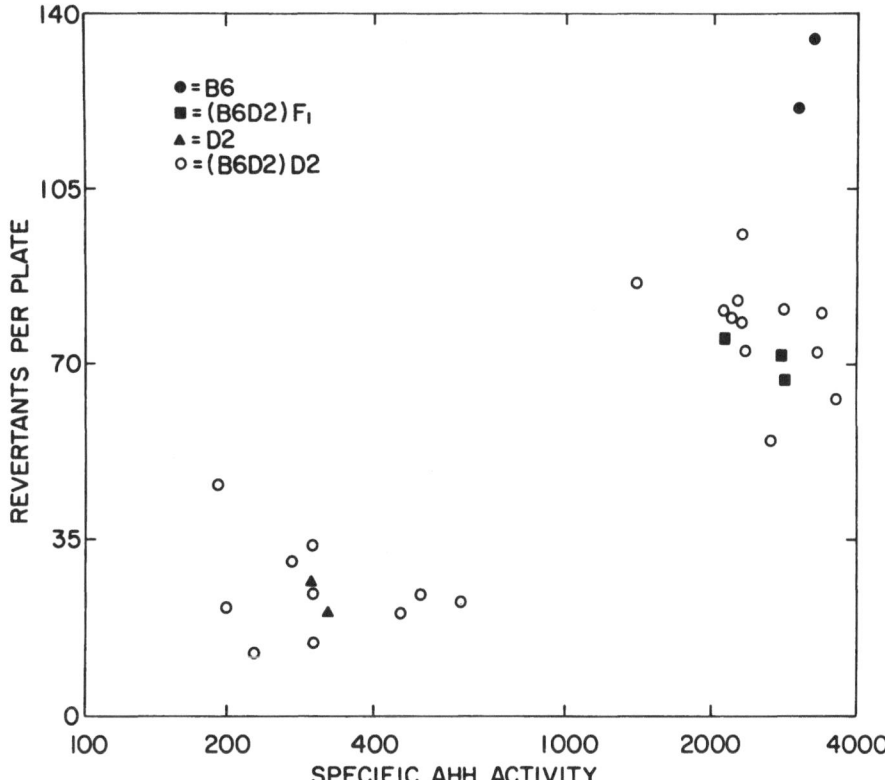

Fig. 5. Relationship between MC as a mutagen <u>in vitro</u> and aryl hydrocarbon hydroxylase (AHH) activity in liver S-9 fractions from MC-treated C57BL/6N (●), DBA/2N (▲), (C57BL/6N) (DBA/2N)F₁ (■), and offspring from the (C57BL/6N)-(DBA/2N) x DBA/2N backcross (O). In this figure and the next, 100 μg of MC as the mutagen in dimethylsulfoxide was added per plate, and the bacterial tester strain TA1538 was used; the background revertant rate--ranging in each experiment between 7 and 30 when no S-9 was added to the reaction mixture--is always subtracted (18).

A linear relationship exists between the values from MC-treated D2 mice and values from B6 mice treated with MC <u>in vivo</u> for 6 to 48 hours. We feel that this finding supports our suggestion above that about 20 colonies is a real number for the S-9 from MC-treated D2 mice in the presence of MC <u>in vitro</u> and the 50 to 70 colonies--when S-9 is present without any compound

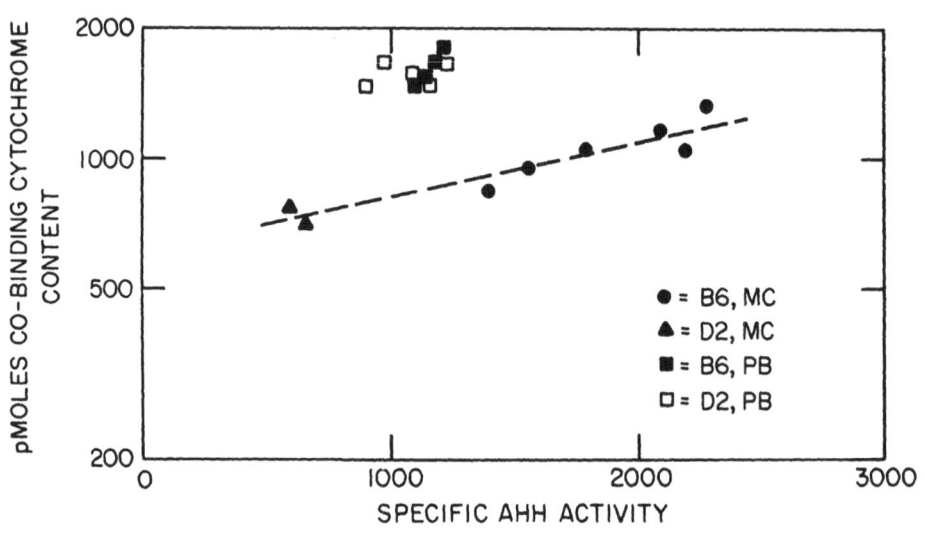

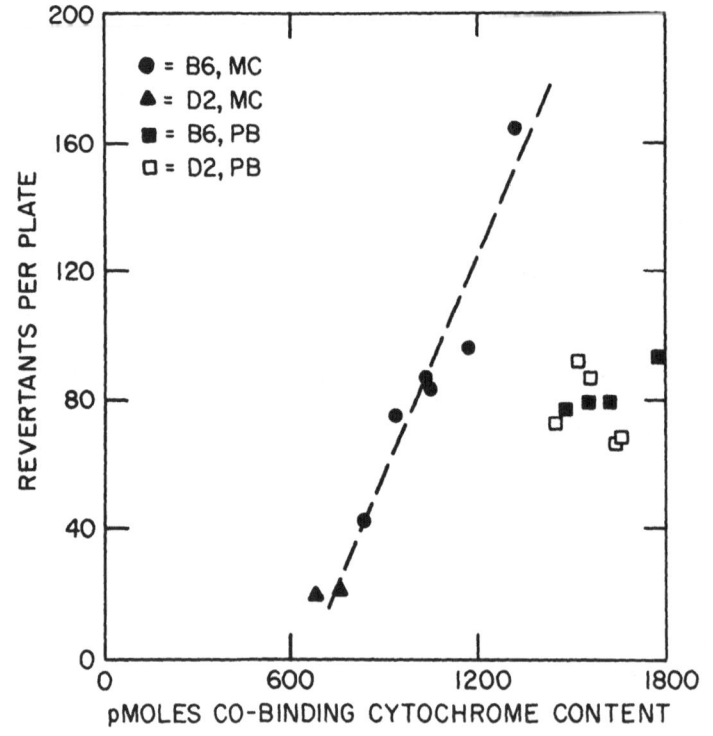

Figure 6a and 6b. Relationships between aryl hydrocarbon hydroxylase (AHH) activity, microsomal CO-binding cytochrome content, and MC as a mutagen in vitro, in liver from MC-treated (●) or phenobarbital-treated (■) C57BL/6N and MC-treated (Δ) or phenobarbital-treated (□) DBA/2N inbred mice (18). B6 mice were sacrificed 6, 9, 12, 18, 20, and 48 hours following MC administration As expected, the revertant rate and specific hydroxylase activity both increased in proportion to the length of MC treatment in the B6 mice. Each closed circle or triangle represents the liver combined from two MC-treated mice. Each closed or open square denotes an individual phenobarbital-treated mouse. The dashed lines drawn between the closed triangles and circles were calculated with the Monro-matic computer program for least-squares analysis. The CO-binding cytochrome content is expressed in pmoles per mg of microsomal protein. The CO difference spectral method (23) was used for determining the concentrations of cytochromes P-450 or P_1-450. The absorption coefficient of 91 $mM^{-1}cm^{-1}$ was used for the difference in absorbance between the Soret maximum and the 490 nm baseline for the hemoprotein-CO complex reduced with dithionite (23). There is (16) no marked difference between the absorption coefficient of hepatic cytochrome P-450 in aromatic hydrocarbon-treated "nonresponsive" mice and that of cytochrome P_1-450 in aromatic hydrocarbon-treated "responsive" mice.

added in vitro represents a falsely elevated background rate which is an artifact that we are as yet unable to explain. In further support of this statement, different concentrations of MC were used with S-9 present, and a straight line decreasing toward the origin was seen (18) down to a concentration of 0.1 μg of MC per plate. At this MC concentration, the number of colonies was the same for S-9 from control mice or from MC-treated D2 mice or for plates in which the S-9 had been omitted. Only with the addition of less than 0.1 μg of MC per plate does the mutation rate rise (18). Also, the falsely elevated background reversion rate is not caused by toxicity of the polycyclic hydrocarbons to the bacteria (18).

Differences in "metabolite profile" between cytochromes P_1-450 and P-450. Comparing MC versus phenobarbital as an inducer (Figure 7),various laboratories have demonstrated that hydroxylations may occur in different chemical positions on the molecule for such substrates as biphenyl (24), testosterone (25), bromobenzene (26), and n-hexane (27). Such differences in the metabolite profile of a polycyclic hydrocarbon might result in marked differences in the reactivity of intermediates and therefore might result in marked dissimilarities in the carcinogenicity of a given compound.

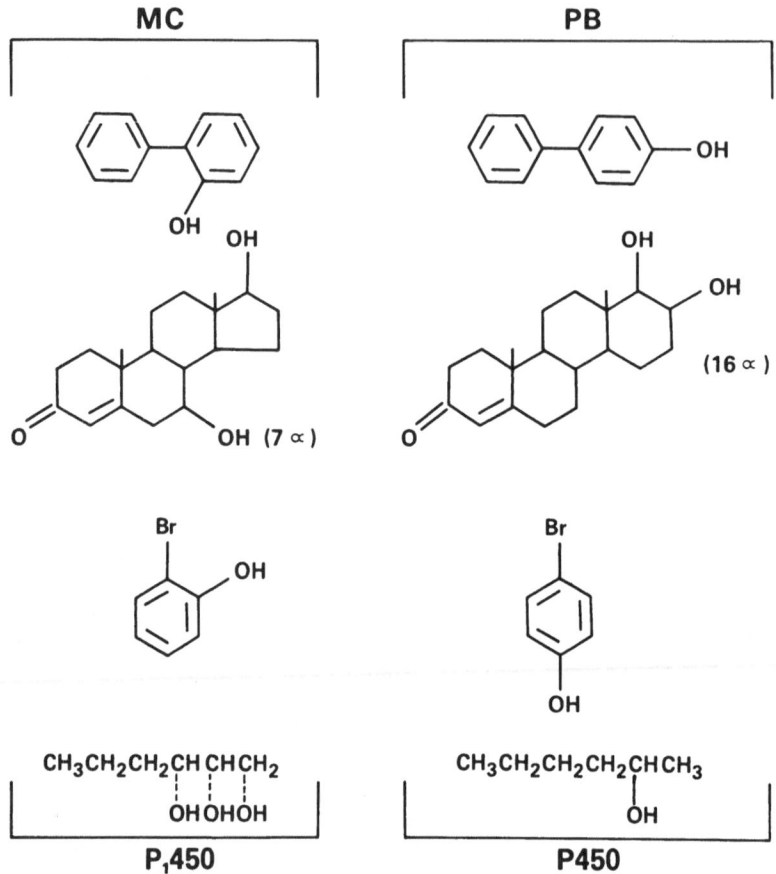

Fig. 7.   Chemical structures of different metabolites which
we suggest are formed by cytochrome P$_1$-450 in the liver of rats
treated previously with MC and by another form of P-450 in the
liver of rats treated previously with phenobarbital.

Quite likely different reactive intermediates may intercalate
with different DNA base-pairs, thereby damaging or activating
different genomes.

Genetic differences in MC as a carcinogen. Figure 8 shows
the carcinogenic index[4] as a function of hepatic aryl hydrocarbon
hydroxylase induction for 14 inbred strains of mice treated with
MC. Whereas the correlation coefficient r was 0.90 (P<0.001) for
tumors produced by a 150- μg subcutaneous dose of MC, r was 0.28
and 0.22 for tumors produced by the same subcutaneous dose of DMBA
and BP, respectively (P>0.20 for both). This correlation between

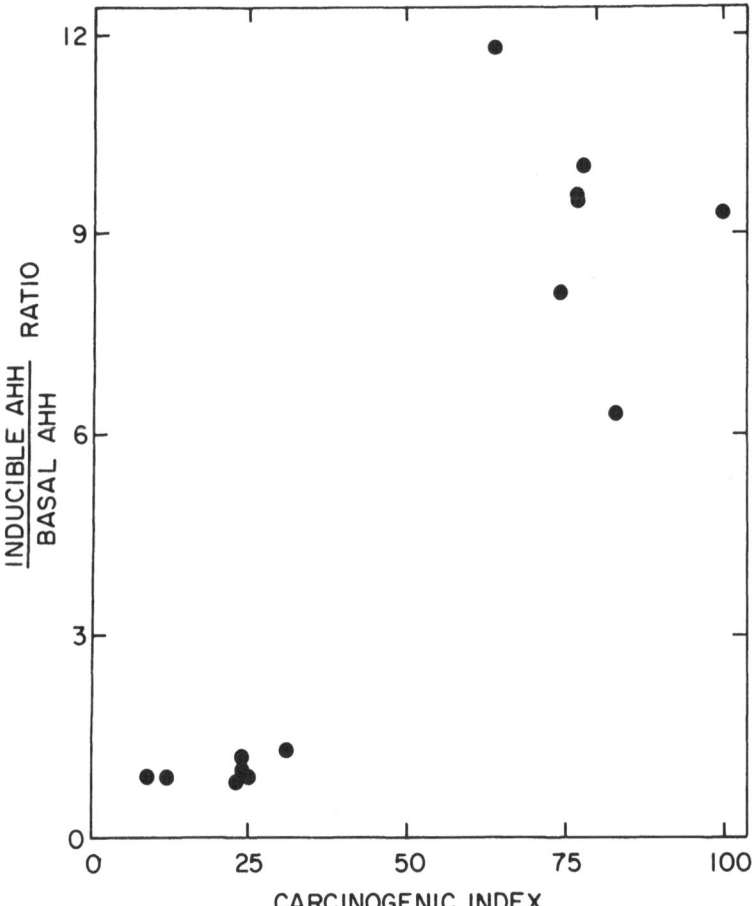

Fig. 8. Relationship between the carcinogenic index and the genetically mediated induction of aryl hydrocarbon hydroxylase activity by MC for each of 14 inbred mouse strains (29); the correlation coefficient r is 0.90 (P<0.001). Each closed circle represents the average result from a group of 30 inbred mice of a certain strain. The carcinogenic index was evaluated after a subcutaneous dose of 150 µg of MC had been given to a minimum of 30 weanling mice of each strain (32). The "inducible AHH/basal AHH ratio" reflects the mean of hepatic hydroxylase activity in MC-treated mice divided by the mean hepatic enzyme activity in control mice (N ≥5 for each of the two groups). Whether the MC-inducible hydroxylase activity in the nonhepatic tissues appears to segregate as a single gene with the inducible hepatic enzyme activity has not been examined for many of these strains (11,17).

MC-produced tumorigenesis and inducible hydroxylase activity has been shown (28) to remain true in all possible backcrosses and intercrosses between B6 and D2 mice. These data for DMBA and BP (29-34), therefore, indicate that important differences exist in the metabolic activation of the three carcinogens, DMBA, BP, and MC. These results might represent different metabolite "profiles," as the result of at least 2 different monooxygenase active-sites (11). A metabolite of MC generated by cytochrome P_1-450 might therefore be more carcinogenic than a metabolite of MC formed by cytochrome P-450. Such a hypothesis fits our data in Figure 8; the results from the laboratories of Kouri and Nebert (31-33) suggest that the same is not true for DMBA or BP.

CONCLUSIONS

We realize that in this study we have not examined the possibly important contributions of epoxide hydrases (9, 35), glutathione concentrations (36), and the epoxide-glutathione transferases (37)--all present in the S-9 fraction--to the mechanism of MC mutagenesis in vitro. Yet, our results with strain TA1538 in vitro are in support of the earlier in vivo studies (29-34), implicating an association between genetically mediated hydroxylase induction and MC-initiated tumorigenesis in these mice.

Our present study does not show conclusively that the mutagenicity demonstrated by these other carcinogens in vitro is mediated by some form of CO-binding hemoprotein other than cytochrome P_1-450 nor have we ruled out the possibility that metabolism of these other compounds to the ultimate carcinogen is carried out at least in part by cytochrome P_1-450. However, the mutagenicity of BP, DMBA, benz[a]anthracene, and dimethylnitrosamine does not appear to be as closely related as MC to the genetically mediated hydroxylase induction and concomitant new cytochrome P_1-450 formation. The mechanism by which a compound becomes an ultimate carcinogen may be highly specific yet quite different for each compound: (i) metabolism might not be required for carcinogenesis (38); (ii) metabolic activation other than arene oxide formation might be important (39-45); (iii) formation of the arene oxide might be necessary (46-48); or (iv) metabolic activation might not be mediated by cytochrome P_1-450 but by some other CO-binding hemoprotein species (49,50).

ACKNOWLEDGEMENT

We appreciate very much the help and encouragement of Dr. Bruce N. Ames and Edith Yamasaki in setting up the bacterial system, and the collaboration with Dr. Richard E. Kouri (29) in certain tumorigenesis studies with mice.

FOOTNOTES

1   Abbreviations: MC, 3-methylcholanthrene; BNF, β-naphtho-flavone; BP, benzo[a]pyrene; DMBA, 7,12-dimethylbenz[a]anthracene; the hydroxylase, aryl hydrocarbon (benzo[a]pyrene) hydroxylase; B6, the inbred C57BL/6N mouse strain; D2, the inbred DBA/2N strain.

2   We refer to "cytochrome P_1-450" as that species of CO-binding hemoprotein which increases in concentration in response to aromatic hydrocarbon treatment.

3   Steven A. Atlas, Daniel W. Nebert, and Benjamin A. Taylor, manuscript in preparation.

4   The carcinogenic index (C.I.) provides a means for relating the latency period to tumor incidence (28) by the equation C.I. = $\frac{P}{T}$ x 100 where P = per cent of mice developing tumors within 8 months and T = average latency period (in days) for a tumor of 1.5- to 2.0-cm dimensions to develop.

REFERENCES

1   Ames, B. N. (1971) Chemical mutagens: principles and methods for their detection, Vol. 1, pp. 267-282. In A. Hollaender (ed.). Plenum Press, New York.

2   Ames, B. N., Lee, F. D. and Durston, W. E. (1973) An improved bacterial test for the detection and classification of mutagens and carcinogens. Proc. Nat. Acad. Sci. USA 70, 782-786.

3   Ames, B. N., Durston, W. E., Yamasaki, E. and Lee, F. D. (1973) Carcinogens are mutagens: a simple test system combining liver homogenates for activation and bacteria for detection. Proc. Nat. Acad. Sci. USA 70, 2281-2285.

4   Sobell, H. B. and Jain, S. C. (1972) Stereochemistry of actinomycin binding to DNA. II. Detailed molecular model of actinomycin-DNA complex and its implications. J. Molec. Biol.68, 21-34.

5   Reich, E., Franklin, R. M., Shatkin, A. J. and Tatum, E. L. (1962) Action of actinomycin D on animal cells and viruses. Proc. Nat. Acad. Sci. USA 48, 1238-1245.

6   Drake, J. W. (1970) The molecular basis of mutation, 273 pages. Holden-Day: San Francisco.

7   Miller, E. C. and Miller, J. A. (1971) Chemical mutagens: principles and methods for their detection, Vol. 1, pp. 83-119. In A. Hollaender (ed.). Plenum Press, New York.

8 Heidelberger, C. (1973) Advances in cancer research, Vol. 18,
      pp. 317-366. In G. Klein and S. Weinhouse (eds.).
      Academic Press, New York.
9 Jerina, D. M., and Daly, J.W. (1974) Arene oxides: A new
      aspect of drug metabolism. Science 185, 573-582.
10 Gillette, J. R., Davis, D. C. and Sasame, H. A. (1972) Cyto-
      chrome P-450 and its role in drug metabolism. Annu.Rev.
      Pharmacol. 12, 57-84.
11 Nebert, D. W., Robinson, J. R., Niwa, A., Kumaki, K. and
      Poland, A. P. (1975) Genetic expression of aryl hydro-
      carbon hydroxylase activity in the mouse. J. Cell.
      Physiol., in press.
12 Gielen, J. E., Goujon, F. M. and Nebert, D. W. (1972) Genetic
      regulation of aryl hydrocarbon hydroxylase induction.
      II. Simple Mendelian expression in mouse tissue in vivo.
      J. Biol. Chem. 247, 1125-1137.
13 Nebert, D. W., Goujon, F. M. and Gielen, J. E. (1972) Aryl
      hydrocarbon hydroxylase induction by polycyclie hydro-
      carbons; Simple autosomal dominant trait in the mouse.
      Nature New Biol. 236, 107-110.
14 Nebert, D. W. and Gielen, J. E. (1972) Genetic regulation of
      aryl hydrocarbon hydroxylase induction in the mouse.
      Fed. Proc. 31, 1315-1325.
15 Thomas, P. E., Kouri, R. E. and Hutton, J. J. (1972) The
      genetics of aryl hydrocarbon hydroxylase induction in
      mice: a single gene difference between C57BL/6J and
      DBA/2J. Biochem. Genet. 6, 157-168.
16 Nebert, D. W., Gielen, J. E. and Goujon, F. M. (1972) Genetic
      expression of aryl hydrocarbon hydroxylase induction.
      III. Changes in binding of n-octylamine to cytochrome
      P-450. Mol. Pharmacol. 8, 651-666.
17 Robinson, J. R., Considine, N. and Nebert, D. W. (1974)
      Genetic expression of aryl hydrocarbon hydroxylase in-
      duction. Evidence for the involvement of other genetic
      loci. J. Biol. Chem., in press.
18 Felton, J. S. and Nebert, D. W. (1975) Association of geneti-
      cally mediated increases in aryl hydrocarbon hydroxylase
      and cytochrome P_1-450 with activation of certain car-
      cinogens to mutagens in vitro. Manuscript submitted
      for publication.
19 Alvares, A. and Siekevitz, P. (1973) Gel electrophoresis of
      partially purified cytochromes P_{450} from liver micro-
      somes of variously-treated rats. Biochem. Biophys. Res.
      Commun. 54, 923-929.
20 Welton, A. R. and Aust, S. D. (1974) Multiplicity of cytochrome
      P_{450} hemoproteins in rat liver microsomes. Biochem.
      Biophys. Res. Commun. 56, 898-906.
21 Levin, W., Lu, A. Y. H., Ryan, D., West, S., Kuntzman, R. and
      Conney, A. H. (1975) Partial purification and separation

of multiple forms of cytochrome P-450 and cytochrome
P-448 from rat liver microsomes. Adv. Exp. Med. Biol.,
in press.

22 Coon, M. J., Haugen, D. A. and van der Hoeven, T. A. (1975)
Properties of purified cytochrome P-450 and NADPH-cyto-
chrome P-450 reductase from rabbit liver microsomes.
Adv. Exp. Med. Biol., in press.

23 Omura, T. and Sato, R. (1964) The carbon monoxide-binding pigment of liver microsomes. I. Evidence for its hemoprotein nature. J. Biol. Chem. 239, 2370-2378.

24 Creaven, P. J. and Parke, D. V. (1966) The stimulation of
hydroxylation by carcinogenic and non-carcinogenic compounds. Biochem. Pharmacol. 15, 7-16.

25 Kuntzman, R., Levin, W., Jacobson, M. and Conney, A. H.
(1968) Studies on microsomal hydroxylation and the demonstration of a new carbon monoxide binding pigment in
liver microsomes. Life Sci. 7, 215-224.

26 Zampaglione, N., Jollow, D. J., Mitchell, J. R. Stripp, B.,
Hamrick, M. and Gillette, J. R. (1973) Role of detoxi-
fying enzymes in bromobenzene-induced liver necrosis.
J. Pharmacol. Exper. Ther. 187, 218-227.

27 Frommer, U., Ullrich, V. and Orrenius, S. (1974) Influence of
inducers and inhibitors on the hydroxylation pattern of
n-hexane in rat liver microsomes. FEBS. Lett., in press.

28 Iball, J. (1939) The relative potency of carcinogenic compounds. Amer. J. Cancer 35, 188.

29 Nebert, D. W., Benedict, W. F. and Kouri, R. E. (1974) Aromatic hydrocarbon-produced tumorigenesis and the genetic
differences in aryl hydrocarbon hydroxylase induction.
In Chemical Carcinogenesis, (P. O. P. Ts'o and J. A.
Dipaolo, eds.), Marcel-Dekker Inc.: New York, pp. 271-288.

30 Kouri, R. E., Ratrie, III, H. and Whitmire, C. E. (1974)
Genetic control of susceptibility to 3-methylcholan-
threne-induced subcutaneous sarcomas. Int. J. Cancer 13,
714-720.

31 Nebert, D. W., Benedict, W. F., Gielen, J. E., Oesch, F. and
Daly, J. W. (1972) Aryl hydrocarbon hydroxylase, epoxide
hydrase, and 7,12-dimethylbenz[a]anthracene-produced
skin tumorigenesis in the mouse. Mol. Pharmacol. 8,
374-379.

32 Kouri, R. E., Salerno, R. A. and Whitmire, C. E. (1973) Relationships between aryl hydrocarbon hydroxylase inducibility and sensitivity to chemically induced subcutaneous
sarcomas in various strains of mice. J. Nat. Cancer
Inst. 50, 363-368.

33 Benedict, W. F., Considine, N. and Nebert, D. W. (1973)
Genetic differences in aryl hydrocarbon hydroxylase induction and benzo[a]pyrene-produced tumorigenesis in

the mouse. Mol. Pharmacol. 9, 266-277.

34   Kouri, R. E., Ratrie, H. and Whitmire, C. E. (1973) Evidence
        of a genetic relationship between susceptibility to 3-
        methylcholanthrene-induced subcutaneous tumors and in-
        ducibility of aryl hydrocarbon hydroxylase. J. Nat.
        Cancer Inst. 51, 197-200.

35   Oesch, F. (1972) Mammalian epoxide hydrases: Inducible en-
        zymes catalyzing the inactivation of carcinogenic and
        cytotoxic metabolites derived from aromatic olefinic
        compounds. Xenobiotica 3, 305-340.

36   Mitchell, J. R., Jollow, D. J., Potter, W. Z., Gillette,
        J. R. and Brodie, B. B. (1973) Acetaminophen-induced
        hepatic necrosis. IV. Protective role of glutathione.
        J. Pharmacol. Exp. Ther. 187, 211-217.

37   Sims, P. (1973) Epoxy derivatives of aromatic polycyclic
        hydrocarbons. The preparation and metabolism of epoxides
        related to 7,12-dimethylbenz[a]anthracene. Biochem. J.
        131, 405-413.

38   Wattenberg, L. W. (1972) Dietary modification of intestinal
        and pulmonary aryl hydrocarbon hydroxylase activity.
        Toxicol. Appl. Pharmacol. 23, 741-748.

39   Fried, J. and Schumm, D. E. (1967) One electron transfer
        oxidation of 7,12-dimethylbenz[a]anthracene, a model
        for the metabolic activation of carcinogenic hydrocarbons.
        J. Amer. Chem. Soc. 89, 5508-5509.

40   Dipple, A., Lawley, P. D. and Brookes, P. (1968) Theory of
        tumor initiation by chemical carcinogens: dependence
        of activity on structure of ultimate carcinogen. Eur.
        J. Cancer 4, 493-506.

41   Wilk, M. and Girke, W. (1969) Jerusalem symposia on quantum
        chemistry and biochemistry. Vol. 1, pp. 91-105. In
        E. D. Bergmann, and B. Pullman (eds.). Israel Academy
        of Science and Humanities, Jerusalem.

42   Flesher, J. W., and Sydnor, K. L. (1971) Carcinogenicity of
        derivatives of 7,12-dimethylbenz[a]anthracene. Cancer
        Res. 31, 1951-1954.

43   Thorgeirsson, S. S., Jollow, D. J., Sasame, H. A. Green, I.
        and Mitchell, J. R. (1973) The role of cytochrome P-450
        in N-hydroxylation of 2-acetylaminofluorene. Mol.
        Pharmacol. 9, 398-404.

44   Baird, W. M., Dipple, A., Grover, P. L., Sims, P. and Brookes,
        P. (1973) Studies on the formation of hydrocarbon-
        deoxyribonucleoside products by the binding of deriva-
        tives of 7-methylbenz[a]anthracene to DNA in aqueous
        solution and in mouse embryo cells in culture. Cancer
        Res. 33, 2386-2392.

45   Fahmy, O. G. and Fahmy, M. J. (1973) Oxidative activation of
        benz[a]anthracene and methylated derivatives in muta-
        genesis and carcinogenesis. Cancer Res. 33, 2354-2361.

46   Grover, P. L., Sims, P., Huberman, E., Marquardt, H. Kuroki,
      T. and Heidelberger, C. (1971) In vitro transformation
      of rodent cells by K-region derivatives of polycyclic
      hydrocarbons. Proc. Nat. Acad. Sci. USA 68, 1098-1101.

47   Huberman, E., Aspiras, L., Heidelberger, C., Grover, P. L.
      and Sims, P. (1971) Mutagenicity to mammalian cells of
      epoxides and other derivatives of polycyclic hydrocarbons.
      Proc. Nat. Acad. Sci. USA 68, 3195-3199.

48   Huberman, E., Kuroki, T., Marquardt, H., Selkirk, J. K.,
      Heidelberger, C., Grover, P. L. and Sims, P. (1972)
      Transformation of hamster embryo cells by epoxides and
      other derivatives of polycyclic hydrocarbons. Cancer
      Res. 32, 1391-1396.

49   Czygan, P., Greim, H., Garro, A. J., Hutterer, F., Schaffner,
      F., Popper, H., Rosenthal, O. and Cooper, D. Y. (1973)
      Microsomal metabolism of dimethylnitrosamine and the
      cytochrome P-450 dependency of its activation to a
      mutagen. Cancer Res. 33, 2983-2986.

50   Czygan, P., Greim, H., Garro, A., Schaffner, F. and Popper, H.
      (1974) The effect of dietary protein deficiency on the
      ability of isolated hepatic microsomes to alter the muta-
      genicity of a primary and a secondary carcinogen.
      Cancer Res. 34, 119-123.

INCREASED TRANSLATION AS A RESULT OF ELEVATED INITIATION FACTOR

ACTIVITY AFTER ADMINISTRATION OF 3-METHYLCHOLANTHRENE

Edward Bresnick, John Hopkinson and P. M. Prichard
Department of Cell and Molecular Biology
Medical College of Georgia
Augusta, Georgia 30902

SUMMARY

Protein synthesis initiation factors present in a crude 0.5 M KCl microsomal wash fraction were isolated from the livers of immature rats that had been injected either 2 or 17 hours earlier with the polycyclic hydrocarbon, 3-methylcholanthrene (3MC), and then were tested for their ability to stimulate natural mRNA-directed protein synthesis in vitro. After purification of the initiation factors by means of ammonium sulfate fractionation, and DEAE-cellulose and Sephadex G-200 chromatography, M_2A and M_2B, but not M_3 or M_1, from the livers of 3MC-pretreated rats were more active than were similarly-prepared control factors in their ability to initiate the synthesis of rabbit globin polypeptides in a highly fractionated cell-free protein synthesizing system derived from rabbit reticulocytes. The greater activity of the M_2A and M_2B preparations from drug-treated rat liver did not appear to be due to differences in the composition of initiation factor protein extracted from the livers of control or experimental rats and occurred very early after administration of 3MC, i.e., 2 hr. The role of the protein synthesis initiation factors in the altered rates of protein synthesis which accompany cytodifferentiation and growth is discussed.

The administration of polycyclic hydrocarbons, e.g., 3-methylcholanthrene (3MC), to rodents leads to a substantial increase in the activity of a number of the mixed function oxidases particularly in liver (reviewed in 1,2). That this elevation in activity indeed represents new enzyme synthesis may be inferred from studies

in which inhibitors of both RNA and protein synthesis were employed (3-5). Furthermore, during the induction process, gene
activation occurs in liver since nuclear RNA polymerase and chroma-
tin template activities are elevated (6-10).

Not only is the action of 3MC apparent in nucleic acid synthesis and function, but protein synthesis is also affected. In
this regard, Gelboin and Blackburn (4) have reported an increased
incorporation of amino acid precursors into protein after administration of the polycyclic hydrocarbon.

More recently (11-13), administration of 3MC has been demonstrated to affect liver initiation factor activity which appears
at least in part to regulate protein synthesis in mammalian
systems (14-17).

Our previous studies suffered from the criticism that the
products of the in vitro all liver protein synthesizing system had
not been characterized. Consequently, we decided to employ the
very-well characterized reticulocyte protein-synthesizing system
where the bulk of the protein elaborated in vitro is indeed hemoglobin. With this system, we will show that the liver initiation
factors, M_2A and M_2B, exhibit greater activity when isolated from
3MC-treated than from corn-oil treated rats in the in vitro synthesis of globin. It is possible, therefore, that pharmacologic
agents, e.g., 3MC, by altering initiation factor activity, may
modulate intracellular rates of protein synthesis.

MATERIALS AND METHODS

Animals: Male Charles River rats, 50-60 g in weight, were
injected, i.p., with either 3MC (20 mg/kg body weight) or corn oil
and were sacrificed at 2 or 17 hr later by decapitation. The
livers were removed, washed in cold 0.9% saline, homogenized to
50% in cold 50 mM Tris pH 7.5 - 250 mM sucrose - 35 mM KCl - 1 mM
dithiothreitol - 0.5 mM EDTA, and both microsomes and a high speed
supernatant fraction were prepared.

Preparation of Rabbit Factors for Protein Synthesis: Rabbit
reticulocyte initiation factors, high speed supernatant fraction,
high-salt washed ribosomes, and transfer RNA were prepared as
previously reported (18-21).

Preparation of Rat Liver Initiation Factors, M_2A, M_2B and M_3:
Rat liver M_2A, M_2B, and M_3 were isolated from the 0.5 M KCl microsomal wash by means of $(NH_4)_2SO_4$ fractionation, DEAE cellulose and
Sephadex G-200 chromatography (22). The high supernatant fraction,
i.e., 124,000 x g for 2 hr, was utilized in the preparation of
elongation factors (EF). A flow diagram of the preparation of the

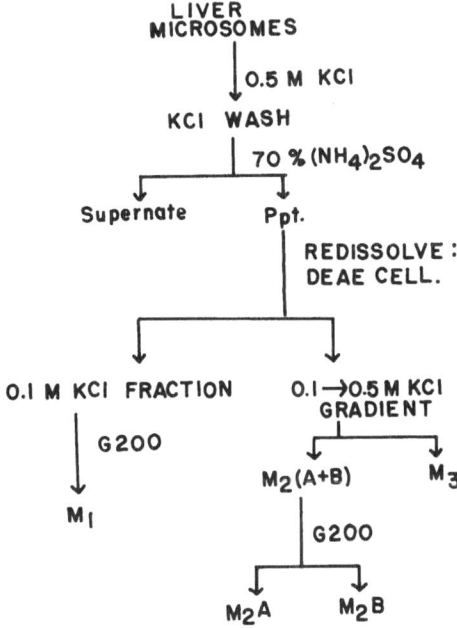

Figure 1.   Flow diagram of purification of initiation factors
from microsomes.

initiation factors is presented in Figure 1. The final purifica-
tions of M_1, M_2A, M_2B and M_3 from rat liver on Sephadex G-200 are
shown in Figures 2-4. From approximately 80 g liver (45 rats), 15
mg M_3 and 3 mg each M_2A and M_2B were obtained. These amounts were
not altered by treatment of the rats with 3MC.

Hemoglobin Synthesis Assay: The assay for rabbit reticulocyte
endogenous mRNA directed hemoglobin synthesis was as described
previously (22). Each 100 µl reaction mixture contained: 20 mM
Tris-HCl, pH 7.5, 80 mM KCl, 4 mM $MgCl_2$, 1 mM ATP, 0.2 mM GTP,
1 mM dithiothreitol, 3 mM PEP,[a] 0.4 I.U. pyruvate kinase, 0.04 mM
19 [^{12}C] -amino acids minus leucine, 0.02 mM ^{14}C leucine (specific
activity, 78 mC/mmole), 0.2 A_{260} units of rabbit reticulocyte salt-
washed ribosomes, 0.12 A_{260} units of rabbit reticulocyte tRNA, 1.4
mg of rabbit reticulocyte supernatant protein [supplemented in some
instances with additional rabbit reticulocyte aminoacyl-tRNA synthetase (27 µg) and EF-1 (44 µg)] and saturating amounts of the
appropriate rabbit reticulocyte initiation factors (M_1,68 µg;
M_2A, 8 µg; M_2B, 5 µg; M_3, 4 µg). Protein synthesis was measured
by the incorporation of ^{14}C leucine into hot TCA - precipitable

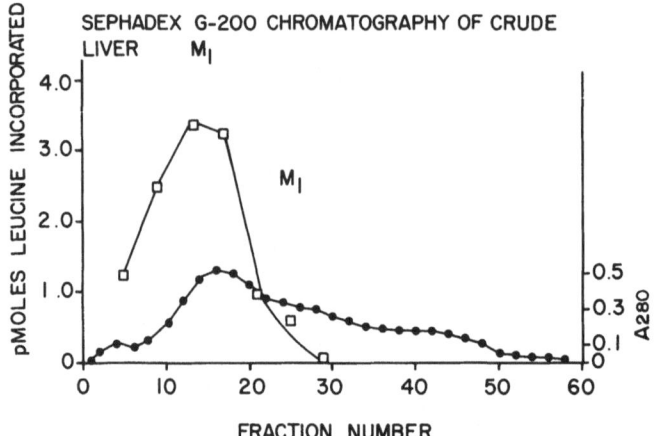

Figure 2. Purification of M_1 on Sephadex G-200. Enzymatic activity (□——□) was determined in the rabbit globin synthesis assay in the presence of saturating amounts of the other factors. ●——●, A_{280}.

material. Reaction mixtures were filtered through Millipore nitro-cellulose filters (type HA, 0.45 µ pore size, 25 mm diameter), washed with cold 5% trichloroacetic acid, dried, and counted in a toluene scintillation fluor which contained 0.5% 2,5-diphenyloxa-zole and 0.03% 1,4-bis-2(4-methyl-5-phenyloxazolyl)-benzene in a Beckman Model LS-230 or LS-150 liquid scintillation spectrometer with an efficiency of 67%.

With initiation factor preparations from both corn oil and 3-methylcholanthrene-pretreated rat liver, incorporation was linear for at least 30 min and was optimal at 4 mM $MgCl_2$ (data not shown).

Chemical determinations. Protein was determined by the method of Lowry et al. Iodoacetate was added to preclude interference from dithiothreitol, and bovine serum albumin served as the ref-erence standard.

RESULTS

As the first step toward establishing an effect of 3MC ad-ministration upon initiation factor activity, we had to establish the validity of employing rat liver factors in a rabbit reticulo-cyte system. Titration curves were determined for rat liver M_1, M_2A, M_2B, and M_3 in the rabbit reticulocyte endogenous mRNA-directed protein synthesis assay. Typical saturation kinetics were observed for all the rat liver initiation factors. Table I

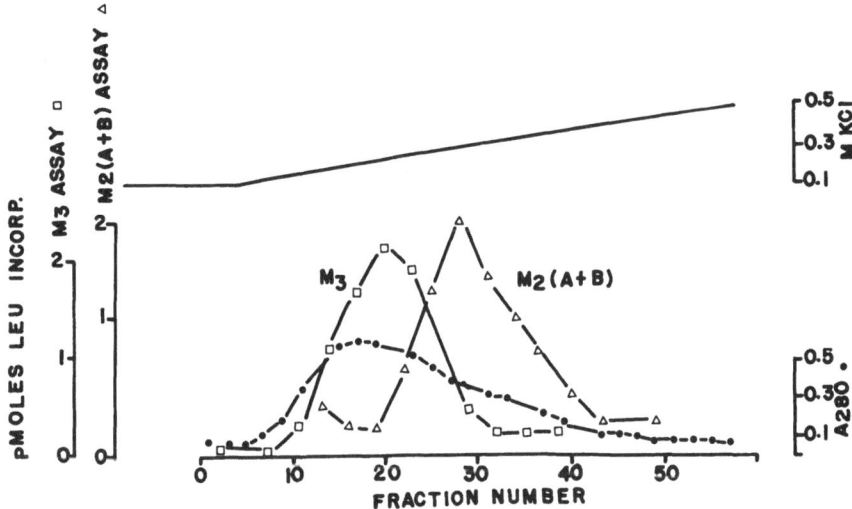

Figure 3. The separation of rat liver initiation factors by DEAE-cellulose column chromatography. Enzymatic activities of rat liver M_3 (□——□) and M_2 (A + B) (△——△) were determined in the rabbit globin protein synthesis assay in the presence of saturating levels of all the other rabbit reticulocyte factors. Blank values of 1.72 and 4.62 pmoles, representing the activity obtained in the absence of M_3 and M_2 (A + B), respectively, have been subtracted from the data. The A_{280} reading of the fractions is indicated as (0——0). The KCl gradient is depicted at the top of the figure as a solid line.

shows that optimal amounts of all these rat liver initiation factors were able to stimulate protein synthesis in a fractionated rabbit reticulocyte system, although not to the same degree as the corresponding rabbit reticulocyte factor.

Having established the utility of this system, we then proceeded to analyze the effects of 3MC administration upon liver factor activity. M_2A and M_2B preparations from the livers of rats injected with 3MC either 2 or 17 hr previously were much more active in stimulating rabbit globin synthesis in a highly fractionated rabbit reticulocyte system (Figures 5A and 5B).

With limiting amounts of the initiation factor, M_2A and M_2B from 2-hour 3MC-pretreated rat liver were 75 and 55%, respectively, more active in stimulating rabbit reticulocyte mRNA-directed protein synthesis than were similarly prepared control factors. M_2A and M_2B from 17-hour 3MC-pretreated rats were 75 and 45% more active, respectively, in this regard (Figures 6A and 6B). With saturating amounts of the initiation factor, M_2A and M_2B prepara-

TABLE I

Exchange of Rabbit Reticulocyte and Rat Liver Initiation Factors
in the Rabbit Reticulocyte Endogenous mRNA-Directed Protein
Synthesis Assay[a]

Assay for:	pmoles (^{14}C) leucine incorporated into protein	fold stimulation
M_1		
control	5.09	
+ reticulocyte M_1	16.77	3.3
+ liver M_1	8.90	1.8
M_2A		
control	6.58	
+ reticulocyte M_2A	15.93	2.4
+ liver M_2A	13.60	2.1
M_2B		
control	5.44	
+ reticulocyte M_2B	15.93	2.9
+ liver M_2B	14.28	2.6
M_3		
control	2.05	
+ reticulocyte M_3	18.40	9.0
+ liver M_3	12.23	6.0

[a] Enzymatic activities of the rat liver initiation factors were determined
in the rabbit reticulocyte endogenous mRNA-directed protein synthesis assay
in the presence of saturating levels of all the other rabbit reticulocyte
factors as detailed in "Materials and Methods". Where indicated, rat liver
factors were added in the following amounts: M_1, 27 µg; M_2A, 13 µg; M_2B, 15 µg;
M_3, 32 µg. In each case, the control contained all the factors except the
one being assayed.

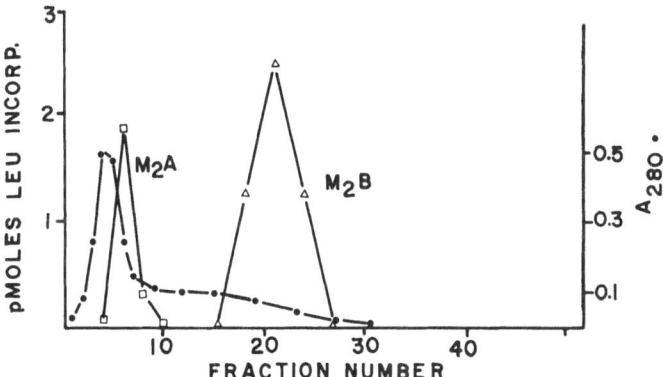

Figure 4. The separation of M$_2$A and M$_2$B by Sephadex G-200 column chromatography. Enzymatic activities of M$_2$A (□——□) and M$_2$B (△——△) were determined as described in Figure 3. Blank values of 3.60 and 4.64 pmoles, representing the activity obtained in the absence of M$_2$A and M$_2$B, respectively, have been subtracted from the data. The A$_{280}$ reading of the fractions is indicated as (●——●).

tions from corn oil-treated rat liver were never able, even at higher protein concentrations, to stimulate the rate of rabbit globin synthesis to a level comparable to that seen when the initiation factor preparations were derived from the livers of 3MC-pretreated rats. M$_3$, on the other hand, did not show any differences between control or 3MC-treated rats (Figure 7). Similarly, no differences in M$_1$ activity were seen between control and experimental rats (data not shown).

Stimulation of rabbit globin synthesis observed when M$_2$A and M$_2$B were isolated from the livers of 3MC-pretreated rats may have been due to a greater ability of these preparations to complete nascent polypeptides rather than to initiate the synthesis of new molecules. To establish which of these mechanisms would prevail, two compounds which specifically inhibit the initiation of cell-free protein synthesis, aurintricarboxylic acid (ATA) (24) and sodium fluoride (25), were tested for their ability to eliminate the greater stimulation seen when the initiation factor preparations were derived from 3MC-pretreated rat liver (Figures 8 and 9). In the absence of either inhibitor, limiting amounts of M$_2$A and M$_2$B from the livers of rats injected with 3MC 2 hr prior to sacrifice were about 75 and 67% more active, respectively, than were similarly prepared control factors. M$_2$A and M$_2$B from 17-hr 3MC-pretreated rats were 97 and 26% more active, respectively, in this regard. Aurintricarboxylic acid (100 µM) and sodium

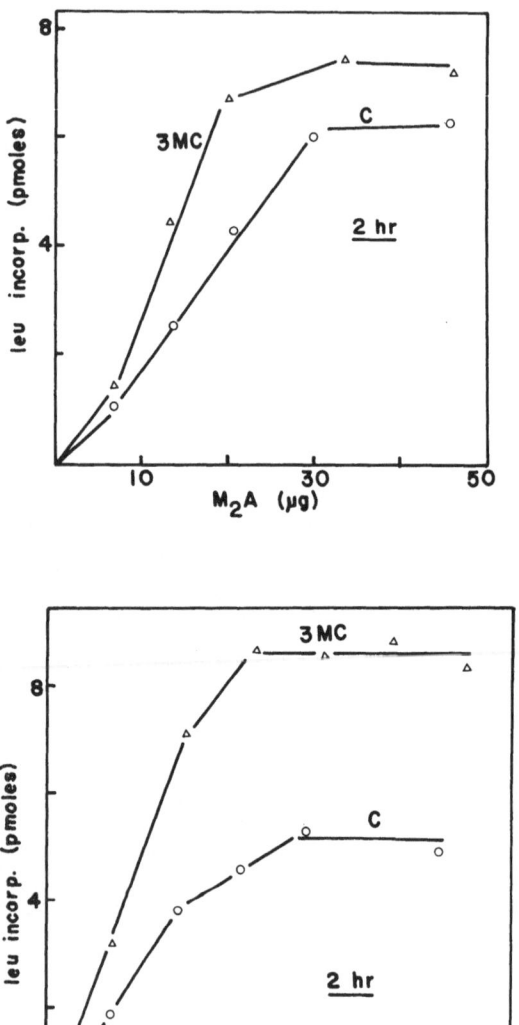

Figure 5A-B. The effect of 2-hr M A$_2$ and M$_2$B on globin mRNA-directed protein synthesis. Increasing amounts of Sephadex G-200 M$_2$A (A) or M$_2$B (B) protein from the livers of rats injected 2 hr earlier with corn oil (o——o) or 3MC (Δ——Δ) were tested for their ability to stimulate rabbit globin synthesis. Blank values of 8.24 or 5.44 pmoles, representing the activity obtained in the absence of M$_2$A or M$_2$B, respectively, have been subtracted from the data.

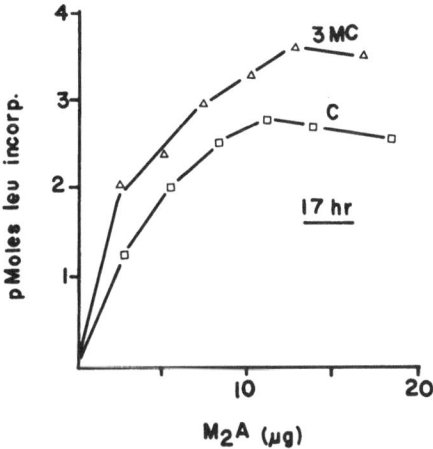

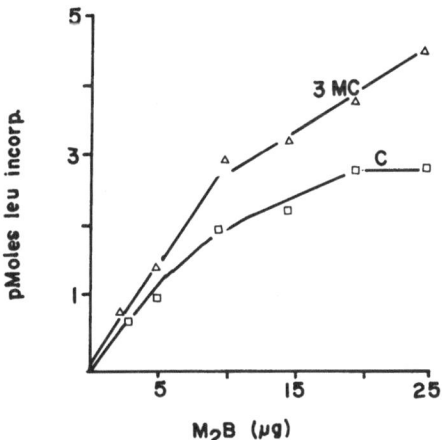

Figures 6A-B. The effect of 17 hour M_2A and M_2B on globin mRNA-directed protein synthesis. Increasing amounts of Sephadex G-200 M_2A (A) or M_2B (B) protein from the livers of rats injected 17 hr earlier with corn oil (□——□) or 3MC (△——△)were tested for their ability to stimulate rabbit globin synthesis as described in "Materials and Methods." Blank values of 5.04 or 7.11 pmoles, representing the activity obtained in the absence of M_2A or M_2B, respectively, have been subtracted from the data.

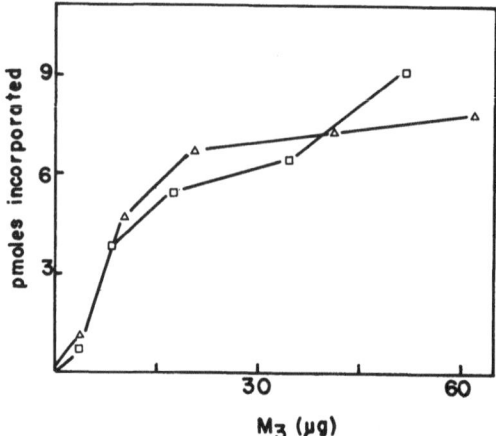

Figure 7.   The effect of M_3 on globin mRNA-directed protein synthesis.   Increasing amounts of DEAE-cellulose M_3 protein from the livers of corn oil (□——□) or 3MC (△——△) pretreated rats were tested for their ability to stimulate rabbit globin synthesis as described in "Materials and Methods."   A blank value of 2.02 pmoles, representing the activity obtained in the absence of M_3, has been subtracted from the data.

fluoride (15 mM) inhibited by 80-90% the initiation and synthesis of the new, complete globin chains which occurred when all of the initiation factors were derived from rabbit reticulocytes.   At these inhibitor concentrations, the greater stimulatory effect of the 3MC-pretreated rat liver M_2A and M_2B was completely abolished.

DISCUSSION

The present study was undertaken to determine how pharmacological agents can modulate intracellular rates of protein synthesis at the translational level.   Lanclos and Bresnick (11,12) had previously shown that the 0.5 M KCl ribosomal wash fraction isolated from the livers of rats injected 15 hr prior to sacrifice with 3MC was more efficacious than the similarly prepared wash fraction from corn oil-pretreated controls in poly(U)-directed protein synthesis and in the binding of methionyl-tRNA$_F$ to the initiator codon, AUG, and to recitulocyte polysomal RNA.   This stimulation appeared to be localized in the combined ($M_3 + M_2$)

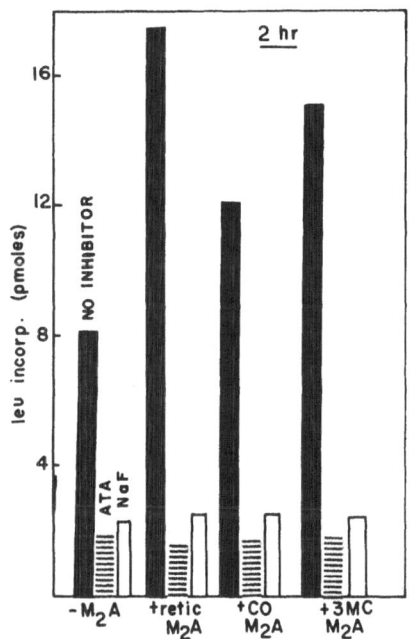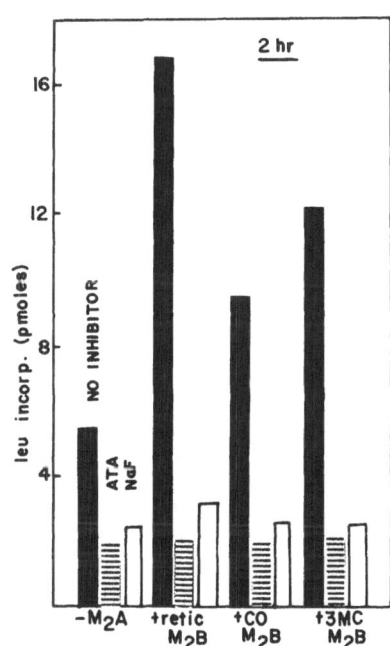

Figures 8A-B. The effect of inhibitors of initiation on the ability of 2-hour M_2A and M_2B to stimulate globin mRNA-directed protein synthesis. Sephadex G-200 M_2A (A) M_2B (B) from rabbit reticulocytes or from the livers of rats injected 2 hours earlier with corn oil (CO) or 3MC were tested for their ability to stimulate rabbit globin synthesis, as described in "Materials and Methods," in the absence of inhibitor ■ or in the presence of 100 μM ATA ▤ or 15 mM NaF □ . The minus M_2A and minus M_2B blanks contain all the initiation factors except the one being tested. Where indicated, 20 μg of rat liver M_2A or 30 μg of rat liver M_2B protein was present.

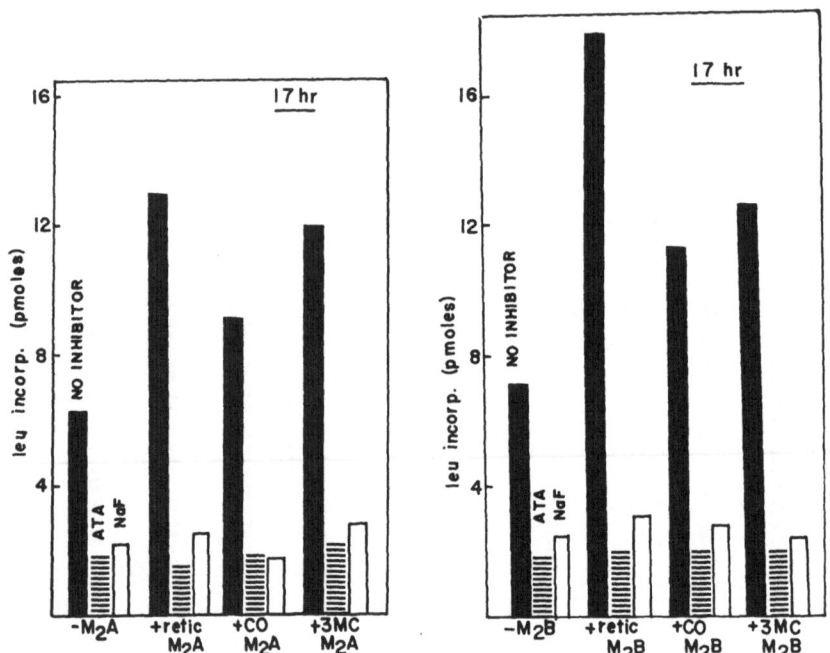

Figures 9A-B. The effect of inhibitors of initiation on the ability of 17-hour M_2A and M_2B to stimulate globin mRNA-directed protein synthesis. Sephadex G-200 M_2A (A) or M_2B (B) from rabbit reticulocytes or from the livers of rats injected 17 hours earlier with corn oil (CO) or 3MC were tested for their ability to stimulate rabbit globin synthesis, as described in "Materials and Methods," in the absence of inhibitor ■ or in the presence of 100 µM ATA ⊟ or 15 mM NaF ☐ . The minus M_2A and minus M_2B blanks contain all the initiation factors except the one being tested. Where indicated, 13 µg of rat liver M_2A or 19 µg of rat liver M_2B protein was present.

components of the ribosomal wash fraction. These effects were analyzed in greater detail in the present study.

This report indicated that, with both limiting and saturating amounts of the initiation factors, M_2A and M_2B, preparations derived from the livers of rats injected either 2 or 17 hr earlier with 3MC were more active than similar factor preparations from control rats in initiating the synthesis of rabbit globin in vitro in a fractionated rabbit reticulocyte, protein synthesizing system. Furthermore, the effect of 3MC administration on the activity of rat liver initiation factors appeared to be selective. Only M_2A and M_2B activities were affected. The increased activity seen with M_2A and M_2B preparations from drug-treated rat liver was not due to differences in the amount or the composition of initiating factor protein extracted from the livers of control and experimental animals. An alternative explanation is that 3MC (or a metabolite) acted in some way as an allosteric effector to increase the activity of M_2A and M_2B. While the direct addition of 3MC to the in vitro system had no effect on M_2A or M_2B activity (unpublished data), this still does not eliminate the activation or metabolism of the parent compound to a more active substance prior to eliciting this response on the initiation factors. Experiments along these lines are underway at this time.

Extrapolations from the results reported in this study rest upon the assumption that changes in the rate of initiation and in initiation factor activity observed in vitro reflect changes in the rate of protein synthesis in the intact cell. Since chain initiation appears to be the rate limiting step in protein biosynthesis in most eukaryotic cells (26), this assumption is probably valid. It is not unreasonable to expect that a multiplicity of steps of peptide chain initiation would be affected as the liver gears up its protein synthesizing machinery in preparation for growth.

The increases in rat liver initiation factor activity at 2 hr after 3MC injection are most interesting because they precede not only all of the reported effects of drug administration on protein synthesis, but also virtually all of the effects of 3MC on transcriptional processes as well. It may be possible that the enhanced initiation factor activity at this time contributes, at least in part, to the very early increases in the synthesis of the hepatic drug-metabolizing enzymes. These observations tend to indicate that pharmacologic agents may exert their effects upon translational events which operate at the level of initiation of protein synthesis.

ACKNOWLEDGEMENTS

We wish to acknowledge the following grants in support of this research: NSF GB-25581 and GB 40633 and N.I.H. GN-18623.

REFERENCES

1   A. H. Conney and J. J. Burns. Adv. in Pharmacol 1: 31 (1962).
2   A. H. Conney. Pharmacol. Rev. 19: 317 (1967).
3. A. von der Decken and T. Hultin. Arch. Biochem. Biophys. 90:
       201 (1967).
4   H. V. Gelboin and N. R. Blackburn. Biochim. Biophys. Acta 72:
       657 (1963).
5   H. V. Gelboin and L. Sokoloff. Science 134: 611 (1961).
6   E. Bresnick. Mol. Pharmacol. 2: 406 (1966).
7   H. V. Gelboin, J. S. Wortham and R. G. Wilson. Nature 214:
       281 (1967).
8   E. Bresnick. Biochim. Biophys. Acta. 217: 204 (1970).
9   J. C. Madix and E. Bresnick. Biochem. Biophys. Res. Commun.
       28: 445 (1967).
10 E. Bresnick and H. Mossé. Mol. Pharmacol. 5: 219 (1969).
11 K. D. Lanclos and E. Bresnick. Res. Commun. Chem Path. and
       Pharmacol. 4: 421 (1972).
12 K. D. Lanclos and E. Bresnick, Drug Met. and Dispos. 1: 239
       (1973).
13 J. Hopkinson, P. M. Prichard and E. Bresnick. Biochim. Bio-
       Phys. Acta, in press.
14 R. L. Miller and R. Schweet. Arch. Biochem. Biophys. 125:
       632 (1968).
15 G. M. Brawerman, M. Ravel, W. Salser and F. Gros. Nature 223:
       957 (1969).
16 S. M. Heywood. Nature 225: 696 (1970).
17 P. M. Prichard, J. M. Gilbert, D. A. Shafritz and W. F.
       Anderson. Nature 226: 511 (1970).
18 J. M. Gilbert and W. F. Anderson. J. Biol. Chem. 245: 2342
       (1970).
19 P. M. Prichard, D. J. Picciano, D. G. Laycock, and W. F.
       Anderson. Proc. Natl. Acad. Sci. 68: 2752 (1971).
20 D. A. Shafritz, P. M. Prichard, J. M. Gilbert, W. C. Merrick
       and W. F. Anderson. Proc. Natl. Acad. Sci. 69: 983
       (1972).
21 D. A. Shafritz and W. F. Anderson. J. Biol. Chem. 245:5553(1970).
22 D. J. Picciano, P. M. Prichard, W. C. Merrick, D. A. Shafritz,
       H. Graf, R. G. Crystal and W. R. Anderson. J. Biol.
       Chem. 248: 204 (1973).
23 O. H. Lowry, N. J. Rosebrough, A. L. Farr and R. J. Randall.
       J. Biol. Chem. 193: 265 (1951).
24 M. L. Stewart, A. P. Grollman and M. T. Hwang. Proc. Natl.
       Acad, Sci. 68: 97 (1971).
25 S. Y. Lin, R. D. Mostellen and B. Hardesty. J. Mol. Biol.
       21: 51 (1966).
26 V. M. Pain and M. J. Clemens. FEBS Letters 32: 205 (1973).

FOOTNOTE: [a)]PEP = phosphoenolpyruvate.

DISCUSSION

Session I - Part 2

HOLTZMAN: I would like to address a question to Dr. Hutterer concerning the effect of cyclic AMP. What has interested me about this is that the original observations on the effect of this agent suggested that it only acts on intact cells. I was wondering whether this could be comparable to the suppression of the synthesis of cytochrome P-450 produced by high carbohydrate diets. The added cyclic AMP produces a high concentration of glucose in the hepatocyte by the well-known mechanisms of the late Dr. Sutherland and may well explain the results you are seeing here.

HUTTERER: Our experiments do not exclude the possibility of the "glucose effect" (the repression of the induction of certain enzymes by glucose). Induction of maltodextrin phosphorylase in E. Coli (Chao and Weatherbee. 1974. J. Bacteriol. 117: 181-188), δ-ALA synthetase in liver cell suspension (Edwards and Elliott. 1974.   J. Biol. Chem. 249: 851-855), and serine dehydratase in rat liver (Jost et al., 1969. Biochim. Biophys. Res. Comm. 34: 748-754),are repressed by glucose but in each case the repression is reversed by cAMP. There are several theories as to how the repression of enzyme induction by glucose occurs. The most plausible one which is consistent with the above examples is that glucose mobilizes insulin which in turn decreases the formation of cAMP. This pathway is circumvented when cAMP is injected in substantial quantities. Therefore it is unlikely that in the case of the cAMP treated animals the glucose effect is responsible for the partial inhibition of the phenobarbital induced increase in cytochrome P-450.

COON: A question for Dr. Cooper. I am intrigued by your action spectra data and I am wondering what causes the results seen at lower wavelength.

COOPER: This is the problem: One of the reasons we undertook these experiments in addition to studying polycyclic hydrocarbon induction was to try to get an idea of what the 420 nm bands are that have been found in some action spectra. We still don't know.

SCHENKMAN: In the action spectra you showed they differed somewhat from your earlier action spectra in that one of the first slides you put up did not have a large peak at 420 nm. This was with the codeine.

COOPER: Codeine was the one substrate that behaves normally. It did in the original work, and it still does now. You don't

seem to get the 420 nm band. But, if you look at the spectra
that Dr. Greim presented for the mutagenicity, if you look at
Dr. Orrenius' spectra, almost all of them, aside from the adrenal
systems, seem to have some effect around   420 nm. The meaning
of this band remains unexplained.

SCHENKMAN: I meant that this action spectra you showed
didn't seem to show the same thing.

COOPER. This one didn't. In the early work we always had
trouble in that region.

SCHENKMAN: Is this a function of your new mirror set up?

COOPER: No. Codeine is a better substrate for studying
photochemical action spectra.

JERINA: In regard to the unusual action spectrum, both Dr.
Peisach and myself were wondering if it's not possible that the
actual chromophore of the heme is really absorbing at the wavelength where you're seeing the action spectrum, since that's
what your data say. Now in principle, if I remember correctly
what you examined, these were all ligandable substrates, substrates that could bind at the heme because of an available lone
pair of electrons. Why not have CO on one side and substrate on
the other side of the heme residue itself allowing this to change
the actual position of the λ max for the tertiary complex of CO,
heme protein and substrate?

COOPER: This possibility has been discussed.

PEISACH: The far-fetched scheme that I suggested to Dr.
Jerina as we were sitting near each other is that there is an
emission from the aniline and a reabsorption of energy. No place
along the way did I suggest that the aniline was binding to the
iron. What I suggested was that there's aniline binding to the
heme by some sort of π overlap mechanism, but I really don't feel
that aniline is actually binding at the iron - I was at least half
misquoted.

REMMER: Concerning this problem. I have only one question.
If, in these experiments, one adds only aniline, and not CO, what
comes out?

SCHLEYER: We have done some experiments along these lines,
but the results are not very clear.

SCHENKMAN: Is there a change in the Type II spectrum?

REMMER: Is there any inhibition of the reaction?

SCHLEYER: To answer Dr. Remmer's question, certain rates are affected by the presence of aniline. As a general rule, we find a slowing down of rates. But these aniline concentrations are very high! So are the reported K_m- and K_s-values; Dr. Schenkman may remember them better than I do.

Disagreeing with Dr. Peisach, I think this could simply mean that we do have the experimental conditions under which aniline just might bind to the ferrous ion because of its extremely high concentration. I don't think I can add any further information at this time, except to state that this is a dangerous concentration range for ligand chemistry.

SCHENKMAN: Dr. Schleyer, under those conditions, wouldn't you expect a 430 nm reversal spectrum?

SCHLEYER: Yes, but I don't think the spectral resolution is that good. The spectrum in the aniline case is actually shifted somewhat towards the suggested 430 nm. But I do not think that this observation tells us anything about the real problem which Dr. Cooper referred to: What happens in the 420 nm region? What gives rise to the features in the 420 nm region of photochemical action spectra of certain microsomal P-450 systems?

CINTI: With reference to this point, has the presence of hemoglobin been considered, since Dr. Gillette reported a number of years ago that aniline oxidation can be catalyzed by hemoglobin. Can the 420 nm peak of the photochemical action spectrum be attributed to hemoglobin or one of its products?

COOPER: Well, we hope hemoglobin has been removed. These animals are carefully perfused and although we can't get all of the hemoglobin out by perfusion, there isn't any evidence for it in the difference spectra. The hemoglobin band would be very sharp at 420 nm and should be easily seen.

JERINA: My comments are directed towards carcinogenesis. What I would like to say is in the way of a word of caution directed to those interested in correlating P-450 and carcinogenic activity in intact animals. I'd like to point out several things which are often ignored, basically because of a desire to implicate P-450. The points that I would like to make are the following: Dr. Wattenberg repeatedly pointed out that induction of microsomal enzymes, more often than not, leads to substantial protection from the action of carcinogens. That point has to be taken into context with the following thoughts. It is not known, nor is it clear to me, what enzymes, in addition to P-450, are being induced by these agents. There are a multitude of potential pathways in a living cell for deactivation of a biochemically induced active

agent capable of binding to some tissue constituent. Thus, it is
quite inadequate to look simply at P-450 and say, "Aha, we're
generating more active intermediate." That's why some people or
some strains of animals are more susceptible to particular types
of carcinogenesis. If, for example, arene oxides are involved,
the following factors are known and necessarily have to be considered: The level of epoxidehydrases, the level of glutathione
transferases, the absolute level of glutathione in the tissues.
What is being dumped by these particular components? This is a
point which must be addressed.

SCHENKMAN: I would like to direct a point to Dr. Nebert concerning a similar thought as Dr. Jerina's. What do you compare
when you measure the carcinogenicity of compounds on subcutaneous
injection of methylcholanthrene and the aryl-hydrocarbon hydroxylase activity? It seems to me you are measuring the latter
in the liver and the former in the topical region. Could you
make some comment on this?

NEBERT: In answer to Dr. Jerina, the data of Dr. Wattenberg
(Toxicol. Appl. Pharmacol.[1972] 23, 741-748) can be interpreted
in any of several ways. The aryl hydrocarbon hydroxylase system
is substrate-inducible. This presumably means that noncarcinogenic
inducers, such as benz[a]anthracene or β-naphthoflavone are in
turn continuously metabolized. A carcinogen such as MC or DMBA
added to such an "induced system" may appear less carcinogenic because the noncarcinogenic inducer competes for "carcinogenesis-
initiating sites" in the cell at any of several levels: (i) the
nonmetabolized parent hydrophobic compound (as Dr. Wattenberg suggests); (ii) the P-450-mediated monooxygenase system from which
the reactive arene oxide intermediate is formed; or (iii) some
step beyond formation of the epoxide (cf. Benedict et al., Int.
J. Cancer [1972] 9, 435-451 for further discussion).

An older study from our laboratory (Nebert et al., Int. J.
Cancer [1970] 6, 470-480) is more in favor of hypothesis (ii) or
(iii). The susceptibility of mice to skin tumorigenesis by DMBA
is diminished during estrus and may be inhibited by estrogens.
We demonstrated that 17-β-estradiol does not effectively block the
hydroxylase induction in fetal cell cultures but that 17-β-estradiol
is an effective competitive inhibitor of the hydroxylase activity
in a cell-free in vitro system. A recent study (Booth et al.,
[1974] Biochem. Pharmacol. 23, 735-744) also corroborates our data.

I agree with Dr. Jerina that numerous potentially detrimental
metabolic pathways other than those involving P-450 might be
important for carcinogenesis. The enzymes that you mention--which
detoxify arene oxides--are obvious extensions of the work I have
just presented and are currently under investigation with this

bacterial mutagenesis assay system (J. S. Felton and D. W. Nebert, manuscript submitted for publication). The whole aspect of DNA repair is also extremely important; I suspect as long as mutations can be excised and the DNA repaired, the organism can remain out of trouble with respect to chemical carcinogenesis. Therefore, in view of the complexities of carcinogenesis, the fact that we have any similarity at all between the bacterial mutagenesis assay in vitro and what we had previously found with tumorigenesis in vivo is surprising to me.

In answer to Dr. Schenkman, at any dose of inducer, the hydroxylase activity is higher in liver, lung, bowel, kidney, and skin from B6 mice than in the corresponding tissue from D2 mice (Gielen et al., [1972] J. Biol. Chem. 247, 1125-1137; D. W. Nebert et al., J. Cell. Physiol., in press). However, we cannot be certain what the hydroxylase activity is in the precise site at which sarcoma forms. Dr. Kouri (Int. J. Cancer [1974] 13, 714-720) finds no significant induction of hepatic hydroxylase activity by the subcutaneous 150 μg dose of MC. It would be interesting to know how much MC remains (metabolized and nonmetabolized) at the site of injection, how much MC reaches other tissues such as the liver, and the fate of MC metabolites from such tissues as the liver and lung.

BRESNICK: In relation to the last comment, we haven't done anything subcutaneously, but we have painted methylcholanthrene on mice. We have also painted the 3-MC-11,12-oxide, and we do get induction of aryl hydrocarbon hydroxylase in lung after a single painting. This was observed not only with 3-MC, but, what is of more interest to us, also with the K-region epoxide of 3-MC, the 11,12-oxide. So it gets there.

REMMER: My comment is of a similar nature; we should be very cautious to interpret these findings. I would like, perhaps, to report about an experiment which nature has actually done for us. What nature has done, our physicians have also done with those patients who received certain compounds throughout their life, for instance phenetidine. Everybody here would agree that phenetidine is metabolized by hydroxylation, and some people have even evidence for an epoxide as intermediate. Now, everybody was thinking that, perhaps, these patients are in real danger from cancer because they got this drug throughout their whole life. Statisticians have made observations on these patients for about 30 years. Even here in Philadelphia we have an institution that has made these observations. And I have heard, just before coming here, from a study in Copenhagen, where scientists have observed these patients for 30 years and have listed any form of cancer which occurred in these patients. When the authors compared these patients with a normal type of population which does not take the drugs, it came out that the incidence of cancer in these patients was even a little lower.

NEBERT: Again, as Dr. Jerina said before and I fully agree,
the levels of enzymes other than cytochromes P-450 are undoubtedly
important. The balance between these various metabolic pathways
and DNA repair must be delicate and no doubt may (i) differ in
each tissue, (ii) be subject to environmental as well as genetic
factors, and (iii) change in response to such factors as hormones,
nutrition, and age. For example, DNA repair is known to become
less efficient with increasing age. I'm sure we all agree that
cancer is complex.

HUTTERER: I wish to comment on Dr. Jerina's remark. We all
realize that multiple factors are involved in the activation of
carcinogens. One of the advantages of the mutagen system is that
mutagenicity is the net result of a series of metabolic steps involved in the activation of carcinogens and thus it reflects the
interplay between cytochrome P-450 and epoxide hydrase or other
enzymes which may participate in the activation. We have some
preliminary indication that the epoxide hydrase inhibitor, when
added to a system containing microsomes, NADPH, a carcinogen, and
bacteria, increased the mutagenicity.

BRESNICK: There's one thing that concerns me about the
mutagenesis studies. It's probably true that all carcinogens are
mutagens, but it sure as heck is not true that all mutagens are
carcinogens; the classic example of that is methyl-methanesulfonate.
This substance is probably as mutagenic as you can get, but as far
as I know, no one has ever gotten a tumor with it. The ultimate
story as to whether a substance is carcinogenic or not, is not
whether it will mutate something but whether you can take that
transformed cell and put it back into an animal and get tumors;
there are a sufficient number of experimental model systems by
which that can be tested.

The second comment: We too have done some experiments with
epoxide hydrase inhibitors, with TCPO (1,1,1-trichloro-propene-
2,3-oxide) which Jerina, Daly and colleagues have reported to be
a potent inhibitor of styrene oxide hydration and which we showed
to be a potent inhibitor of the hydration of the 3-MC-K-region
oxide. And that substance, when painted together with 3-MC
under some stringent conditions which I don't want to go into now,
on susceptible mice, increases the number of tumors, increases
the percentage of tumor-takes and very significantly shortens the
latency period over that seen after painting 3-MC alone.

SCHENKMAN: Dr. Cooper, when you found these differences in
the amount of TPNH-reducible cytochrome P-450, as compared to the
amount of dithionite-reducible cytochrome P-450, I don't know

whether you mentioned it, did you check the reductase activity?
Was there any difference in this?

COOPER: Dr. Schleyer, would you like to answer this? You
have all the data.

SCHLEYER: I think, Dr. Schenkman, I cannot give you a
straight answer to this question, but let me say that there are obvious readily measurable effects on the reductase activities as one
goes through a study of induction and post-induction time course
comparing control animals and induced animals. One finds changes
in the reductase levels when measured in the classical way. But
the data do not fit in a simple scheme which would explain this "reducibility problem." Let me stress this once more: The "reducibility problem" is simply a matter of the oxygen concentration (or
oxygen tension, whatever way you like to look at it) present in the
particular experimental situation.

SCHENKMAN: This is not really an answer! Is there a change in
TPNH-cytochrome c reductase?

SCHLEYER: Let me repeat this and try to make it clear,   If
you measure the rate of cytochrome c reduction under strict exclusion
of oxygen, you obtain a type of "maximal rate" of electron flow to
cytochrome c as acceptor or to any other acceptor used. This is
then the rate you would expect to observe if you do not funnel electrons away to some other member of the electron transport system.
Such a funneling off, however, is precisely what happens in the systems which Dr. Cooper presented. It happens to various degrees
depending on type, dose, and time course of induction, but depends
also - and quite sensitively - on the oxygen tension under the conditions of the experiment. One can convincingly show this by removing
even traces of oxygen from the samples. We do not have all the
answers yet. We do not have an account of all the active
electron transport components in the microsomes, nor do we have the
necessary kinetic information. But even so, it is very clear that
there is usually not a strict one for one relation between the observations and the cytochrome c reductase activity observed with
cytochrome c as artificial acceptor and with all the oxygen
removed from the system. When oxygen is present in the experimental
situation - as it obviously must be in all hydroxylation experiments-
oxygen acts not only as one of the substrates of the hydroxylation reaction, but also as an electron sink for other systems (present to
varying degrees depending on the particular situation).

NARASIMHULU: I would like to ask a question of Dr. Cooper or
Dr. Schleyer. Did you measure the rate of reduction of P-450 in
methylcholanthrene- and phenobarbital-induced rat liver preparations
and, if so, did you measure the effect of substrate on these two
systems?

SCHLEYER: We obtain routinely data on these rates with our microsomal preparations. I don't think, however, that they help us any further at this moment. We could perhaps discuss some of the observations in detail, if you like. But let me just state here for the sake of brevity that there is nothing in these data which, in a simple way, would relate to the rate of P-450 reduction or, more strictly speaking, of formation of P-450(Fe^{2+})·CO. In other words, P-450 reduction can be faster or slower, but this still does not explain the results which Dr. Cooper presented.

NARASIMHULU: Well, what do you find with aniline as a substrate?

SCHLEYER: That has only been studied to a limited extent, with hexobarbital added to the microsomes. There is not a great difference between phenobarbital-induced and methylcholanthrene-induced microsomes on comparable days of induction. With hexobarbital, you do, of course, speed up the TPNH utilization (which we also measure routinely in the fluorometer). But the effects are small and do not explain the findings with the photochemical action spectra. Let me repeat: It is strictly a matter of oxygen concentration in the experiment, what one observes. This is the serious problem one has to worry about when one measures action spectra: One needs the presence of oxygen for the hydroxylation reaction which, let's say, is carried out in an atmosphere containing 4% O_2 and a selected concentration of CO in inert N_2-gas. The role of oxygen as "electron sink" has, therefore, to be appreciated and one has to take this oxygen concentration into account.

ORRENIUS: I would like to ask Dr. Nebert again. Is there any evidence that arene oxides leak out of the tissue where they are formed, and that they may be transported into other tissues?

NEBERT: We have shown (Benedict et al., [1973] Mol.Pharmacol. 9, 266-277) that systemic circulation of some factor--metabolites or proviral or subviral particles--appears to be at least 25 times more effective than topical treatment. When 100 µg of DMBA was administered intraperitoneally, we determined that no more than 4 ng of DMBA or metabolites ever reached the ear; after 4 months of phorbol ester promotion (which prevents DNA repair), tumor incidence was more than 80%. Under the same conditions, 100 ng of DMBA painted directly on the skin of the ear produced no tumors.

BRESNICK: This is a comment to Orrenius' question. I think I alluded to some evidence in answering your question. It is a little indirect, however. If you paint skin with MC-oxide, you do get induction in lung. Now, the 11,12-transdiol of 3-MC - the hydration product of the epoxide-is ineffective as an inducer. The 11-phenol compound of 3-MC is also ineffective as inducer; the epoxide under organ culture conditions has about one-half the efficiency of induction as methylcholanthrene has. So, the only conclusion that can be drawn is that the epoxide is forming something other than the transdiol (the product of the epoxide hydrase reaction) or the phenol, and is reverting back to the hydrocarbon or does get to the lung. We tend to favor the latter possibility.

OPTICAL AND ELECTRON PARAMAGNETIC RESONANCE STUDIES OF PARTIALLY

PURIFIED RABBIT LIVER CYTOCHROME P-450[*]

Charlotte Witmer, Peter Nehls, Peter Krauss, Herbert
Remmer and Robert Snyder, University of Tübingen, West
Germany, and Thomas Jefferson University,
Philadelphia, Pennsylvania, U. S. A.

Hildebrandt, Remmer and Estabrook (1968) studied the "off-
balance" spectrum of oxidized cytochrome P-450 from rabbit liver
microsomes by comparing microsomes from either phenobarbital (PB)
or 3-methylcholanthrene (MC) treated animals in the sample cuvette
with microsomes from control animals in the reference cuvette. By
using the technique of Kinoshita and Horie (1967) in which the
concentrations of cytochrome b_5 are equalized in both cuvettes, they
were able to observe the "absolute spectrum" of the cytochrome P-450
as the difference in absorption between the two cuvettes. When
hexobarbital was added to the sample cuvette containing microsomes
from a polycyclic aromatic hydrocarbon-induced animal, the spectrum
took on the features of hepatic microsomal cytochrome P-450 from
a PB induced animal. By analogy with work on other heme proteins
(George, Bettlestone and Griffith, 1961) it appeared that the
change in spectrum was characteristic of a change in spin state
from the high spin form predominantly observed in the polycyclic
aromatic hydrocarbon-induced cytochrome to the low spin form seen
in the phenobarbital induced type. Thus, it seemed that P-450
could exist in two different spin states which appeared to be
interconvertible on substrate addition and which might represent
the two different forms of a single heme protein. However, this
interpretation did not account for the differences in metabolic
specificity displayed by microsomes from animals induced with
different types of inducing agents, e.g. - polycyclic aromatic
hydrocarbons and barbiturates (Conney, 1967). This discrepancy,
coupled with later studies of Nebert et al., (1973)describing two
genetically different kinds of cytochrome P-450, led us to review,
re-evaluate and expand these studies.

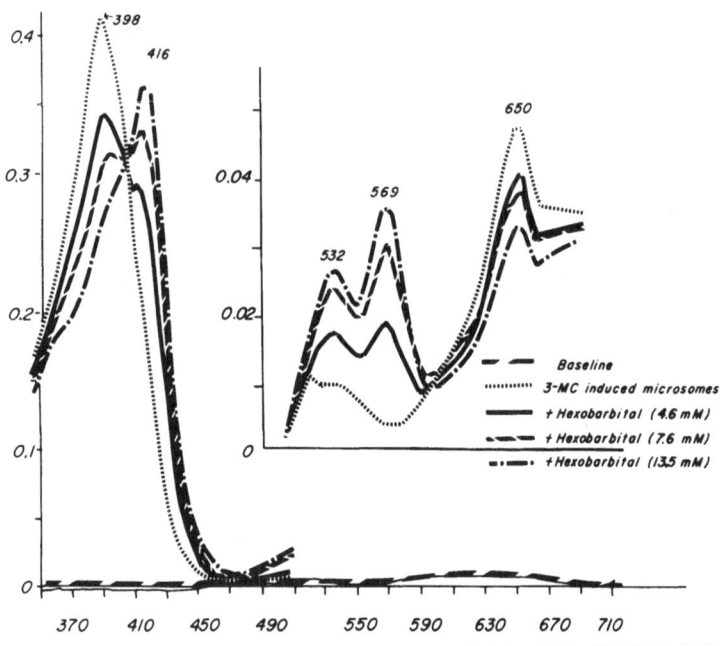

Figure 1. The Effect of Hexobarbital on the Absorption Spectrum of Microsomal P-448. The dotted line indicates the spectrum of microsomal P-448 from rabbit liver, as studied by the "off-balance" technique. Microsomes were isolated after treatment of the rabbit with three daily injections of 3-methylcholanthrene (MC) (25 mg/kg). Food was withheld for 18 hours prior to sacrifice and microsomes were prepared by a method previously described (Snyder et al., 1967). The spectrum of the cytochrome P-448 of microsomes from MC treated animals was determined by recording the spectral difference between sample and reference cuvettes in which the cytochrome b_5 content was balanced. The sample cuvette contained 7.5 nmoles P-450/ml, 3.17 nmoles/ml cytochrome b_5, and 3.20 nmoles/ml of total protein. The reference cuvette contained 3.13 nmoles/ml of cytochrome P-450 from an untreated rabbit, 3.17 nmoles/ml cytochrome b_5 and had a total protein content of 2.40 mg/ml. The two cuvettes were optically balanced at 490 nm prior to the recording of each spectrum. The microsomes were diluted in phosphate buffer 0.1M, pH 7.4 containing 0.15 M KCl. The solid and hatched lines indicate the spectra after the addition of increasing amounts of hexobarbital to the sample cuvette. Equal volumes of buffer were added to the reference cuvette. These spectra were recorded using the Aminco-Chance Split Beam/Dual Wavelength Spectrophotometer in the split beam mode. Note that the absorbance scale is different for the Soret and the higher wavelength regions, and that the baseline has been shifted upward of 500 nm to better display the bands at the higher wavelengths.

Our studies were performed with rabbit liver preparations. Both isolated microsomes and partially purified cytochrome preparations isolated by the method of Lu et al., (1972) were utilized. It should be noted that the isolated preparations were sufficiently devoid of turbidity to allow absolute spectra to be recorded without resorting to the "off-balance" technique.

The spectral characteristics of cytochrome P-450 did not change significantly during purification, indicating that the "off-balance" technique results in a spectrum with features similar to those of the absolute spectrum. Figure 1 shows the "off-balance" spectrum of the heme protein in microsomal suspensions from the livers of MC treated rabbits (dotted line). There is a strong Soret band at 398 nm, a band at 518 nm with a shoulder at 538 nm, and another band in the 640-650 nm region. This spectrum is that of a high spin ferric heme protein (Hill et al., 1970). The solid line shows the spectrum which results upon the addition of hexobarbital or of benzene to the purified preparation. Subsequent additions of hexobarbital increased the 416 nm peak in relation to the 398 nm band, as shown by the hatched lines.

When a sample of the heme protein isolated from the liver of an MC treated rabbit was divided between two cuvettes and a substrate was added to the sample cuvette in small increments, the difference spectra shown in Figure 2 were obtained. The spectra shown are formed on addition of phenobarbital and each is typical of a reverse Type I spectral change (Schenkman, 1972) characterized by an absorption increase which is maximal in the 420-422 nm region, and a decrease appearing as a trough at 385-390 nm. These difference spectra correspond closely to the theoretical spectra calculated by subtracting the absolute spectrum of the MC-induced heme protein from that resulting after the addition of barbiturate.

Cytochrome P-450 isolated from livers of rabbits treated with phenobarbital has a Soret maximum at 417 nm and alpha and beta bands at 568 and 538 nm, respectively. (Figure 3a). Upon the addition of one of the mixed function oxidase substrates, phenobarbital, hexobarbital or benzene, the Soret band shifted toward the blue, with a maximum at 398 nm, the alpha and beta bands took on the appearance of those observed in preparations from 3-methylcholanthrene induced rabbits and a 648 nm band appeared (Figure 3b).

The correlation of optical spectra with the spin state of ferric heme proteins can be ascertained by studying the spin state directly using an electron spin resonance spectrometer.[1]

[1]The terms electron spin resonance (ESR) and electron paramagnetic resonance (EPR) are equivalent terms. EPR is used in this paper except when describing equipment.

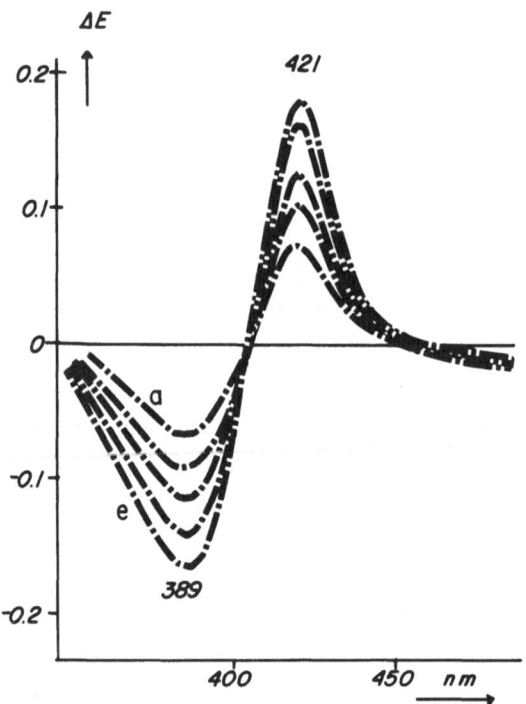

Figure 2. Type R-1 Spectrum with Phenobarbital Addition to
Partially Purified Cytochrome P-448. The cytochrome P-448 was
isolated by the method of Lu et al., (1972); the method of
Schenkman (1970) was used to titrate the cytochrome P-448 with
phenobarbital. Each cuvette contained 8.7 nmoles of cytochrome
P-448. Phenobarbital was added in the following concentrations:
(a) 4.08, (b) 7.87, (c) 9.0, (d) 11.39, (e) 14.66mM. Prior to
the recording of each spectrum, the cuvettes were optically balanced at 490 nm.

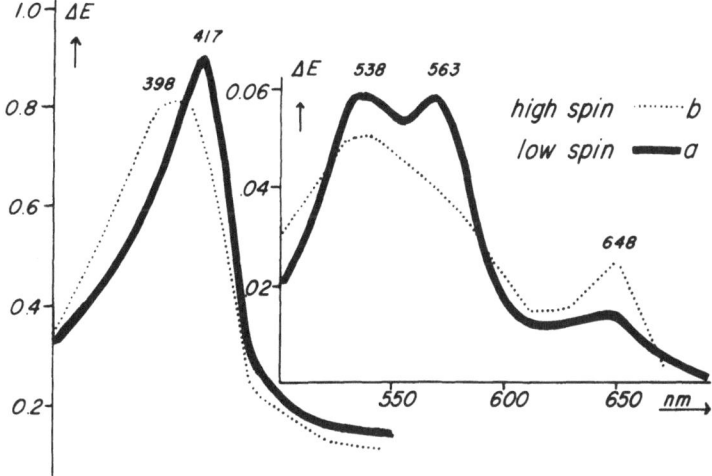

Figure 3. Effect of Barbiturates on the Optical Absorption Spectrum of Isolated Cytochrome P-450. The spectrum shown as (a) is that of cytochrome P-450 isolated from rabbits treated with phenobarbital (PB) (75 mg/kg daily for 3 days). This partially purified hepatic cytochrome is similar to that of microsomal suspensions after PB induction, as studied by the "off-balance" technique (Kinoshita and Horie, 1968). The cytochrome P-450 concentration of this sample was 8.7 nmoles/ml and the diluting buffer was phosphate (100 mM, pH 7.7). The spectrum designated (b) is that recorded after the addition of phenobarbital to a final concentration of 12.2 mM. This is very similar to the spectrum of partially purified cytochrome P-448. Analogous spectral changes also take place if either hexobarbital or benzene is added to the sample cuvette.

Detectability and resolution of the high and low spin signals from Fe^{+3} require that the spectra must be recorded at very low temperatures. Thus these studies were performed using liquid helium as a coolant. Figure 4 shows the first derivative EPR spectrum of a partially purified preparation of heme protein from an MC induced rabbit. The presence of high spin ferric heme protein is indicated by g tensor values at 8.0, 3.7 and 1.73. The signals at 2.42, 2.25 and 1.93 show that the sample also contains some low spin ferric heme P-450. This spectrum was recorded at $4°K$ using a 5000 Gauss sweep; the sample contained 40 mg of total protein

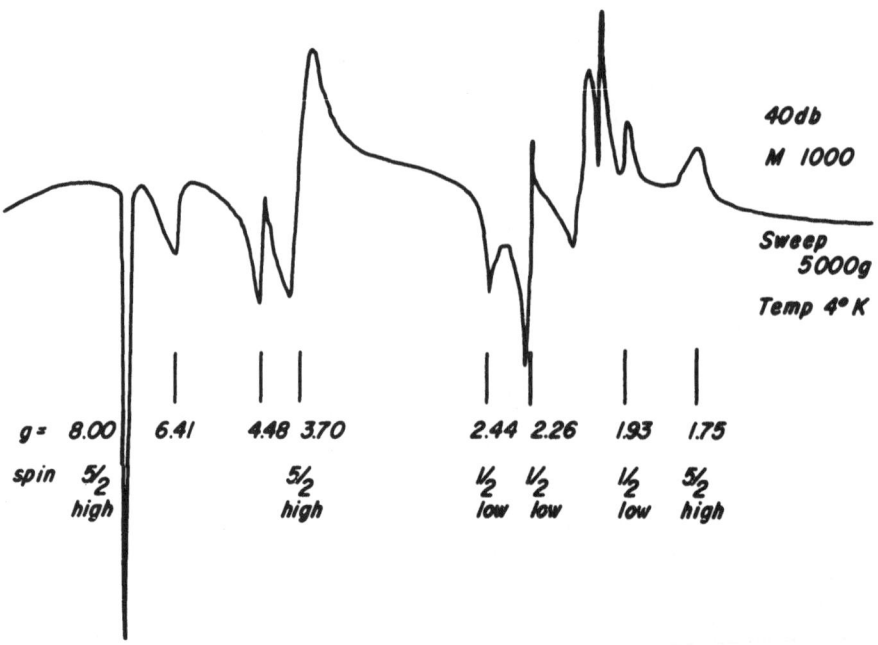

Figure 4. EPR Spectrum of Cytochrome P-448. EPR spectroscopy of a cytochrome P-448 fraction isolated from a 3-methyl-
cholanthrene induced rabbit (3 daily injections, 24 mg/kg) was
carried out using a liquid helium attachment with a Varian E-9
X-band spectrometer. Heme protein concentration was 44 nmoles/ml.
The first derivative spectrum shown here was recorded at 4°K over a
5000 Gauss sweep. The g values indicating either high or low
spin ferric heme iron are identified in the figure. The heme protein was suspended in phosphate buffer (50 mM, pH 7.4) containing
50 mM KCl.

per ml. Double integration of several of these spectra was
carried out by the method of Wyard (1965) to estimate relative
high and low spin concentrations. These calculations indicate
that a high percentage of the heme protein from MC induced rabbits
is in a high spin state.

When EPR spectra were observed of similarly isolated cytochrome preparations from livers of rabbits which had been treated
with phenobarbital, a much lower percentage of the high spin form
of the heme protein was observed (Figure 5). The conditions for
the resonance spectroscopy were the same as those for the MC
induced preparations and the total protein content of the preparations was similar (37 mg/ml in the case of the PB induced heme
protein). Paramagnetic resonance spectra of heme proteins isolated
from untreated rabbits had similar high/low spin ratios to those
of the PB induced cytochrome preparations. Thus it appeared that

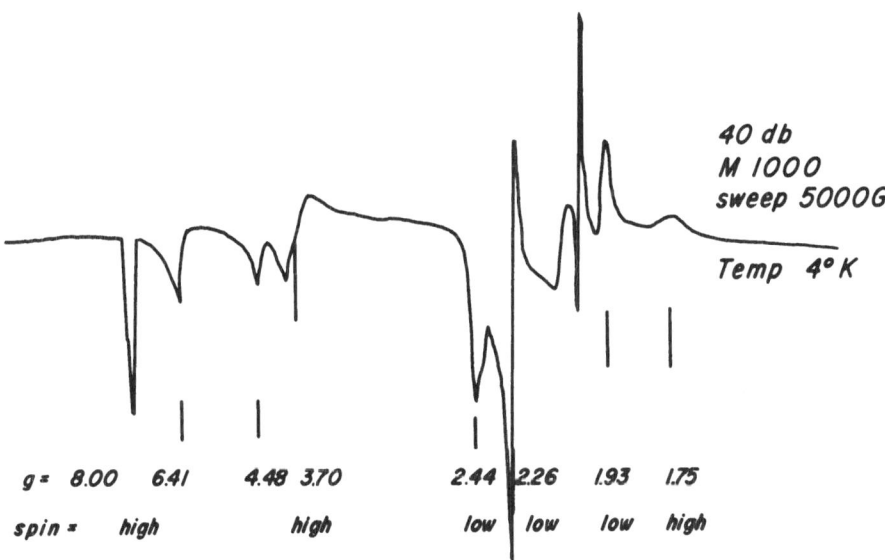

Figure 5. EPR Spectrum of Cytochrome P-450. The conditions for the EPR spectroscopy of the cytochrome P-450 isolated from the liver of a rabbit treated with phenobarbital (3 daily injections 75 mg/kg) were the same as those for the previous figure. The relative amount of high spin heme protein is significantly lower than the relative amount in the sample of cytochrome P-448 (Figure 4). The cytochrome P-450 concentration was 37 nmoles/ml.

3-methylcholanthrene treatment preferentially increased the high spin heme component, and that phenobarbital increased the levels of both the high and low spin components.

Alternately it might be argued that the high spin signals could be the result of covalent binding of 3-methylcholanthrene to the heme protein P-450. Therefore, we treated rabbits with ^{14}C-3-methylcholanthrene and isolated the hepatic microsomal cytochrome. After exhaustively washing the isolated preparations we found that the final ratio of 3-methylcholanthrene to heme protein was about 1:35. Thus 3-methylcholanthrene was not stoichiometrically attached to the cytochrome in a 1:1 ratio either covalently or by any other strong type of bonding. This result indicates that the high spin state is not the result of formation of an enzyme substrate complex but is characteristic of the MC induced cytochrome.

Figure 6 shows the effect of the addition of hexobarbital to high spin heme protein P-450 as measured by EPR spectroscopy. With the addition of hexobarbital there is a decrease in the high

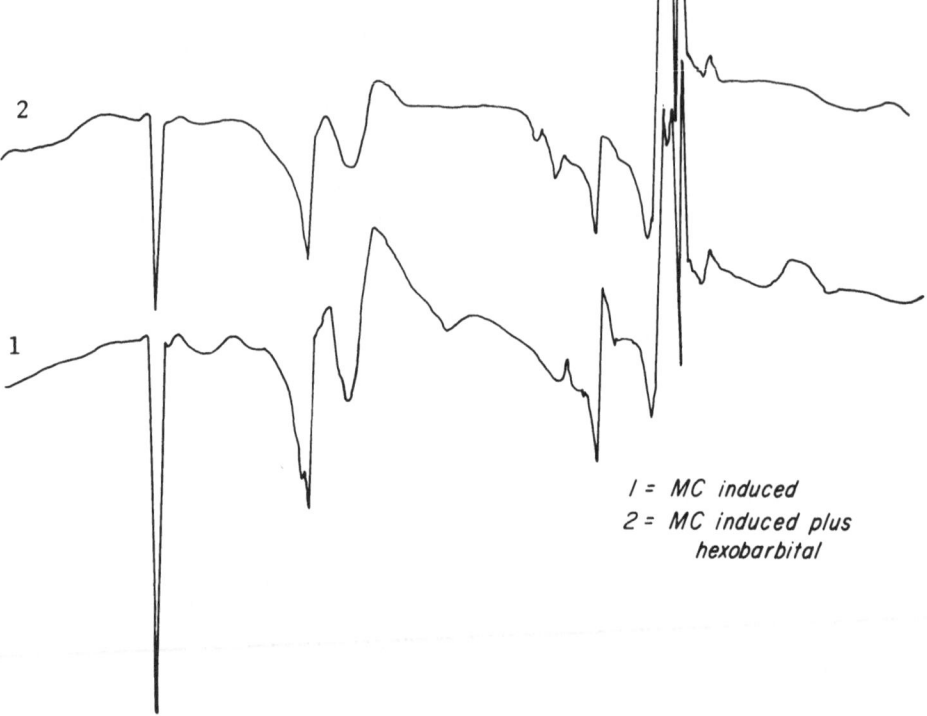

1 = MC induced
2 = MC induced plus
      hexobarbital

Figure 6. The Effect of Hexobarbital on the Spin State of
Cytochrome P-448. Spectrum 1 is the EPR spectrum of the sample
previously shown in Figure 4. To determine the effect of hexobarbital on the spin state of the cytochrome, the sample was thawed
after the spectrum was recorded and hexobarbital was added to a
concentration of 5.15 mM. The sample was rapidly stirred and
immediately refrozen to the temperature of liquid nitrogen. The
temperature was then lowered to that of liquid helium and the spectrum was recorded. The instrument settings and other parameters
are as described in Figure 4. A lower concentration of hexobarbital (1.72 mM) caused a smaller decrease in the high spin
signals (spectrum not shown).

spin signals with a simultaneous increase in the low spin signals,
indicating a spin state conversion. Successive additions of small
amounts of hexobarbital (not shown) indicate that the decrease in
high spin state is dependent on the hexobarbital concentration.
These results are similar to those obtained with optical spectroscopy.

Carbon monoxide binding spectra of preparations from phenobarbital treated rabbits display an absorption maximum at 450 nm
whereas those from 3-methylcholanthrene treated rabbits absorb
maximally at 448 nm. The addition of benzene, hexobarbital or

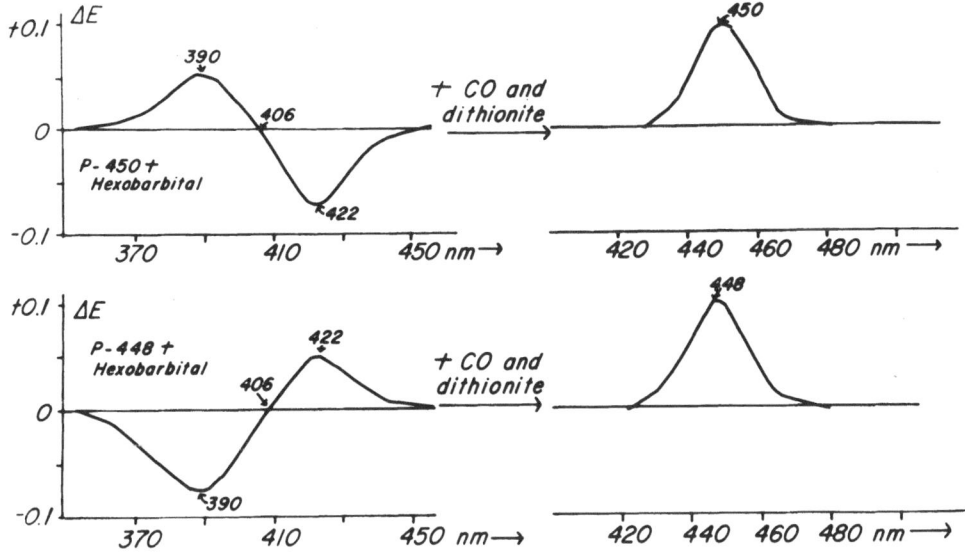

Figure 7.   Reaction of CO with Hexobarbital-Bound Reduced
P-450 and P-448.   This is a schematic diagram of the spectral
changes taking place on addition of hexobarbital (or other substrates) to either cytochrome P-450 (upper diagram) or cytochrome
P-448 (lower diagram), followed by reduction of the heme protein
and saturation with carbon monoxide.]

other substrates to each type of microsomal preparation results
in a characteristic spectral change (Figure 7), i.e., Type I with
the former and Type RI with the latter. When the heme proteins
are then reduced and carbon monoxide is bubbled through the
samples the absorption maxima remain at 450 and 448 nm, respectively. Thus the heme proteins are not interconvertible; the
absorption spectra of the cytochrome-carbon monoxide complexes
are characteristic of the heme proteins and are not a function of
the spin states. It appears that we are justified in terming the
heme protein from phenobarbital treated animals cytochrome P-450
and that from 3-methylcholanthrene, cytochrome P-448. Figure 7
is a schematic summation of these changes.

In order to study preparations which contained high concentrations of both cytochromes P-450 and P-448, we treated rabbits
with both phenobarbital (80 mg/kg for three days) and 3-methyl-
cholanthrene (a single dose of 40 mg/kg on the third day of PB
treatment). The animals were sacrificed 24 hours after the MC

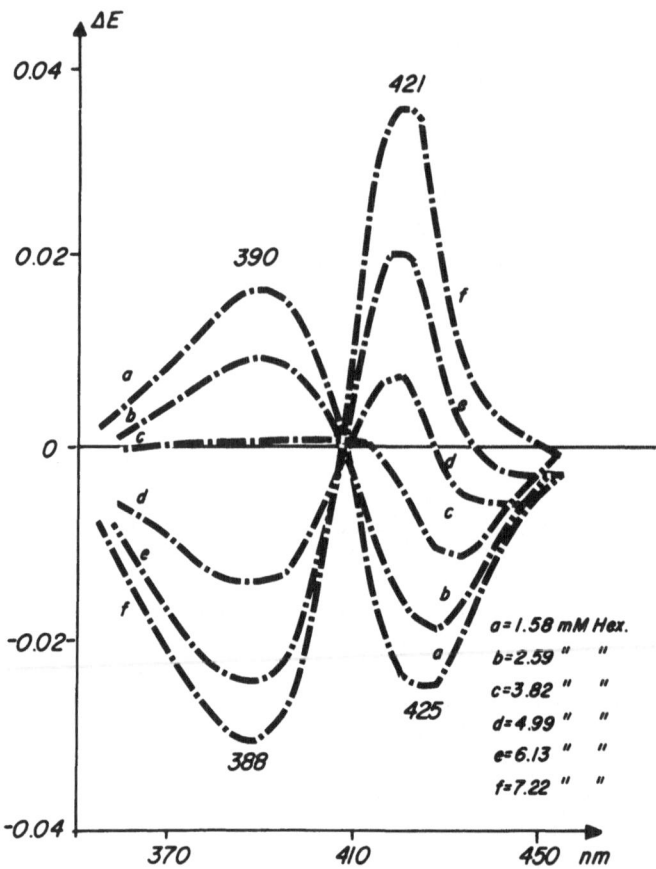

Figure 8. Type I and Type R-I Spectra with PB- and MC-Induced Microsomes. Microsomes were obtained from a rabbit treated with both phenobarbital and 3-methylcholanthrene, as described in the text. Difference spectra were recorded by the method of Schenkman (1974). Each cuvette contained 4.2 nmoles/ml of the mixture of heme-proteins. The final concentrations of hexobarbital are as indicated in the figure. After each addition of hexobarbital, the two cuvettes were optically balanced at 490 nm prior to recording the spectra. Wavelengths are indicated for the curves adjacent to the notations. Microsomes were diluted with phosphate buffer (150 mM, pH 7.5).

treatment and liver microsomes were prepared. Spectral studies with these microsomes showed that the wavelengths of maximal absorption of the carbon monoxide cytochrome complex was indistinct with a broad absorption in the 450 nm region. When the difference spectrum technique was employed to study the interaction of the cytochrome mixture with a substrate, the results shown in Figure 8 were observed. At low concentrations of hexo-

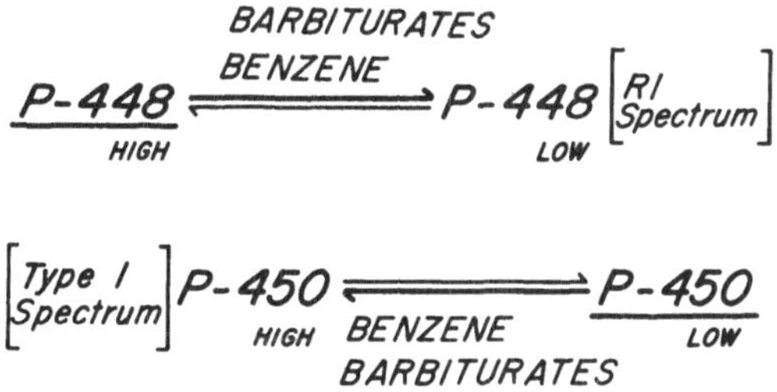

$$P\text{-}448 \underset{HIGH}{\overset{\substack{BARBITURATES \\ BENZENE}}{\rightleftharpoons}} P\text{-}448 \left[\begin{array}{c} RI \\ Spectrum \end{array}\right]_{LOW}$$

$$\left[\begin{array}{c} Type\ I \\ Spectrum \end{array}\right] P\text{-}450 \underset{HIGH}{\overset{}{\rightleftharpoons}} \underset{\substack{BENZENE \\ BARBITURATES}}{} P\text{-}450_{LOW}$$

Figure 9.   Interconversion of Spin States of Cytochromes P-448 and P-450.   Interconversions of spin states of cytochrome P-450 and P-448 are depicted, with the underlined states as the preferred states.   Barbiturates or benzene shift the equilibria as indicated.

barbital (1.5 mM) a Type I spectral change was seen.   At 2.6 mM hexobarbital the Type I spectral change was still observed but with diminished intensity.   When the concentration of hexobarbital was increased to 3.8 mM the Type I spectral change disappeared and the trough of the Type RI spectral change began to appear.   As the concentration of hexobarbital was further increased the Type RI spectral change increased in intensity.

These data suggest that hexobarbital reacts preferentially with cytochrome P-450; thus at lower concentrations of the barbiturate only the Type I spectrum of the P-450 hexobarbital complex is formed.   At higher concentrations the reaction of the hexobarbital with P-448 to form the Type RI spectrum predominates either because the P-450 has been saturated with substrate at lower concentration or the reaction with P-448 masks any further reaction of the P-450 with hexobarbital.   As the Type RI spectrum is indeed the reverse of the Type I, at high concentrations of hexobarbital only the RI spectrum can be observed.

The overlap of the two spectra may explain why some observers have had difficulty in detecting the Type I change after induction with polycyclic hydrocarbons (Schenkman et al., 1967).   Very low concentrations of hexobarbital are required to react exclusively with the small amount of cytochrome P-450 in these microsomes. For the detection of a Type I spectral change, the hexobarbital concentration should be below that at which the transition between the Type I and Type RI spectral change is observed.   This transition point which we have termed S_{crit} appears to be a function of both the hexobarbital and cytochrome concentrations.

The data which we have obtained using both rabbit liver microsomes and partially purified preparations of microsomal heme proteins are consistent with the scheme shown in the last figure (Figure 9). The underlined spin state is the preferred state for each cytochrome. Cytochrome P-450 exists preferentially in the low spin state but upon addition of substrate is partially converted to the high spin state; the change in spin state does not, however, involve conversion to cytochrome P-448. Cytochrome P-448, on the other hand, is partially converted from high to low spin by addition of hexobarbital or other substrates but is not converted to cytochrome P-450. Thus, rabbit liver microsomes appear to contain two separate and distinct heme proteins, cytochrome P-450 and P-448.

ACKNOWLEDGEMENTS

*   This work was supported in part by the Deutsche Forschungsgemeinschaft and by U. S. Public Health Service, NIH Grant ES 00322.

REFERENCES

Conney, A. H. 1967. Pharmacol. Rev. 19, 317-366.

George, P., Beetlestone, J. and Griffith, J. S., in Haematin Enzymes (J. E. Falk, R. Lemberg and R. K. Morton, Eds.), Volume 1, p. 105, Pergamon Press, London, 1961.

Hildebrandt, A., Remmer, H. and Estabrook, R. W. 1968. Cytochrome P-450 of liver microsomes - one pigment or many? Biochem. Biophys. Res. Comm. 30, 607-612.

Hill, H. A. O., Röder, A. and Williams, R. J. P. 1970. The chemical nature and reactivity of cytochrome P-450. Structure and Bonding 8: 123-151.

Kinoshita, T. and Horie, S. 1967. On the absorption of P-450 in rabbit liver microsomes. J. Biochem. (Tokyo) 61: 26-34.

Lu, A. Y. H., Kuntzman, R., West, S., Jacobson, M. and Conney, A. H. 1972. Reconstituted liver microsomal enzyme system that hydroxylates drugs, other foreign compounds, and endogenous substrates. II. Role of the cytochrome P-450 and P-448 fractions in drug and steroid hydroxylations. J. Biol. Chem, 247, 1727-1734.

Nebert, D. W., Heidema, J. K., Strobel, H. W. and Coon, M. J. 1973. Genetic expression of aryl hydrocarbon hydroxylase induction. Genetic specificity resides in the fraction containing cytochromes P-448 and P-450. J. Biol. Chem. 248, 7631-7636.

Schenkman, J. B. 1970. Studies on the nature of the Type I and Type II spectral changes in liver microsomes. Biochemistry 9, 2081-2091.

Schenkman, J. B., Cinti, D. L., Orrenius, S., Moldeus, P. and

Kraschnitz, R. 1972. The nature of the reverse Type I (modified Type II) spectral change in liver microsomes. Biochemistry 11, 4243-4250.

Snyder, R., Uzuki, F., Gonasum, L., Bromfeld, E. and Wells, A. 1967. The metabolism of benzene in vitro. Tox. and Appl. Pharmacol. 11, 346-360.

Wyard, S. J. 1965. Double integration of electron spin resonance spectra. J. Sci. Instr. 42, 769-770.

STUDIES ON THE SPIN STATE OF 3-METHYLCHOLANTHRENE INDUCED

CYTOCHROME P-450 FROM RAT LIVER*

J. O. Stern,** E. Peisach, J. Peisach,*** W. E. Blumberg,
A. Y. II. Lu, S. West, D. Ryan and W. Levin
Departments of Pharmacology and Molecular Biology
Albert Einstein College of Medicine of Yeshiva University
Bronx, New York 10461

Institute for Developmental Studies
New York University, New York, New York 10012

Bell Laboratories, Inc.
Murray Hill, New Jersey 07974

Department of Biochemistry and Drug Metabolism
Hoffman-LaRoche, Inc.
Nutley, New Jersey 07110

SUMMARY

The time course of induction of rat liver microsomal cytochrome P-450 by the polycyclic hydrocarbon 3-methylcholanthrene
was followed by measuring the specific content of cytochrome P-450,
benzpyrene hydroxylase activity, and the percent of cytochrome P-450
existing as the high-spin form (g = 7.9, 3.7 and 1.7) as determined by low temperature EPR spectroscopy. Significant increases
in benzpyrene hydroxylase, cytochrome P-450 and high-spin ferric
hemoprotein are seen twenty-four hours following 3-methylcholanthrene
treatment. Administration of DL-ethionine prior to 3-methylcholan-
threne treatment effectively blocks any increase in benzpyrene
hydroxylase and cytochrome P-450 but not the increase in the levels
of the high-spin species of the hemoprotein normally seen following
3-methylcholanthrene induction. In addition, partially purified
cytochrome P-450 can be isolated from liver microsomes of 3-
methylcholanthrene treated rats as a low-spin ferric hemoprotein
containing essentially no high-spin species (<1%). This partially
purified hemoprotein has the same substrate specificity as the

microsomes from which it was derived. It is therefore concluded
that the appearance of the high-spin form of cytochrome P-450, as
quantitated by EPR, does not correlate with the induction of cytochrome P-450 and benzpyrene hydroxylase activity by 3-methylcholan-
threne.

Liver microsomal cytochromes P-450[1] and the metabolism of
various compounds have been shown to be selectively inducible by
various drugs and polycyclic hydrocarbons (Conney et al., 1959;
Conney, 1967). Microsomes from phenobarbital (PB)-treated rats
have different substrate specificities than microsomes from 3-
methylcholanthrene (3-MC)-treated rats. PB stimulates the hydroxylation of many substrates, while 3-MC only enhances the metabolism
of a limited number of compounds (Alvares et al., 1968a; Rickert
and Fouts, 1970). Reduced liver microsomal cytochrome P-450 from
PB-treated rats exhibits a Soret absorption at 450 nm when ligated to CO while 3-MC treatment causes the formation of a 448 nm
absorbing species (Alvares et al., 1967). Furthermore, ethylisocyanide (EtNC) ligated to reduced cytochrome P-450 leads to the
formation of spectral species showing pH-dependent absorption differences at both 430 and 455 nm (Imai and Sato, 1966; Sladek and
Mannering, 1966) which vary with the drug treatment employed.

From EPR and optical studies of rat liver microsomes, it was
demonstrated that cytochrome P-450 from PB-treated animals is a
low-spin ferric hemoprotein (Miyake et al., 1968). However, under
certain conditions a high-spin form is also observed. Peisach and
Blumberg(1970), employing low-temperature EPR spectroscopy, found
that 3-MC treatment caused an increase in the specific content of
low-spin cytochrome P-450 (g = 2.4, 2.2, and 1.9) in rabbit liver
microsomes compared to control animals and, in addition, discovered
that 3-MC predominantly induces the formation of a newly characterized high-spin form of this cytochrome (g = 7.9, 3.7, and 1.7).
These authors suggested that this high-spin species was a substrate
bound form.

Schenkman et al., (1969) were able to demonstrate that the
apparent absolute spectrum of oxidized, PB-induced microsomal
cytochrome P-450 from rats has a Soret absorption maximum at 420
nm, indicative of a low-spin hemoprotein. However, in the presence
of substrates such as hexobarbital or polycyclic hydrocarbons, the
420 nm Soret absorption decreased and a new absorption at 394 nm
appeared suggesting a conversion to a high-spin form. This 394 nm
Soret peak could also be demonstrated in the apparent absolute
spectrum of oxidized cytochrome P-450 from 3-MC-treated rats in the
absence of any added substrates. As the 394 nm absorbing species
was believed to be a substrate bound form these authors suggested
that the alterations in CO and EtNC difference spectra seen in rats
pretreated with 3-MC were due to the formation of enzyme-substrate

complexes. However, Alvares et al., (1968) demonstrated that ethionine, an inhibitor of protein synthesis, and actinomycin D, an inhibitor of DNA-dependent RNA synthesis, could prevent the changes in the optical spectral properties of the cytochrome P-450 induced by 3-MC suggesting that protein synthesis was required for the changes in the CO and ethyl isocyanide difference spectra of this hemoprotein. They (Alvares et al., 1971) and others (Gnosspelius et al., 1969; Fujita et al., 1973) subsequently demonstrated, under a variety of experimental conditions, that 3-MC and its metabolites were unable to cause the changes in these spectral properties which were ascribed to a newly synthesized hemoprotein.

A frequent misconception is that 3-MC, by some unknown genetic mechanism, directs the synthesis of a new species of cytochrome P-450 which exists, independent of substrate, in a "preferred" high-spin configuration (Jefcoate and Gaylor, 1969; Goujon et al., 1972; Nebert and Kon, 1973; Nebert et al., 1973a). Stern et al., (1973) examining partially purified cytochrome P-450 from 3-MC-induced rats demonstrated that the level of the high-spin species ($g = 7.9$, 3.7, and 1.7), when compared with microsomes from 3-MC-treated rats, was dramatically reduced while the substrate specificity of both remained the same. These authors provided evidence that the $g = 7.9$ high-spin species is, most probably, a substrate bound form and showed that the substrate specificity is not predetermined by the spin-state of the oxidized hemoprotein.

In this communication, we describe those changes in spin-state seen concomitant with induction of cytochrome P-450 following 3-MC treatment and compare these findings for animals where induction has been blocked by ethionine, an inhibitor of protein synthesis.

METHODS

Immature male Long-Evans rats (50-60 g body weight) were divided into six treatment groups, each group consisting of triplicate sets and each set consisting of 3 or 4 animals. The treatment groups were comprised of: (1) Control animals receiving corn oil injected intraperitoneally (i.p.) (2) animals receiving DL-ethionine (500 mg/kg) 60 min and 30 min prior to receiving an injection of corn oil (Alvares et al., 1968b) and sacrificed by decapitation 24 hr later (referred to as ethionine-control), (3) one single injection of 3-MC (25 mg/kg, i.p.) and sacrificed 2 hr later (2 hr, 3-MC), (4) same as group (3) but sacrificed 24 hr after injection of 3-MC (24 hr, 3-MC), (5) animals receiving DL-ethionine 60 min and 30 min prior to receiving an injection of 3-MC (25 mg/kg) and sacrificed 24 hr later

(ethionine-24 hr, 3-MC) and (6) animals receiving daily injections
of 3-MC (25 mg/kg) for 3 days and sacrificed 24 hr after the last
injection (72 hr, 3-MC).

Microsomes used for the EPR studies were prepared by homog-
enizing 3 to 4 livers, obtained from animals within a single set,
in four volumes of 0.05 M Tris buffer (pH = 7.4) containing 1.15%
KCl at 4°C. Homogenates were centrifuged at 10,000 x g for 20
min and the supernatant fractions were then centrifuged at 105,000
x g for 60 min. The microsomal pellets were suspended in 1.15%
KCl and centrifuged again at 105,000 x g for 60 min. Portions of
the microsomal pellets, equivalent to 70-100 mg microsomal protein
(0.5-0.9 g wet weight of microsomes), were transferred into EPR
cavities described previously (Berzofsky et al., 1971), immediately frozen and stored in liquid nitrogen and examined by EPR
within 24 hr. In some cases, microsomes were first suspended in
0.25M sucrose to a protein concentration of 20-30 mg/ml and then
frozen into EPR cavities as a suspension. Partially purified
cytochrome P-450 was always frozen as a solution containing 50-60
nmoles hemoprotein/ml. The percent of high-spin cytochrome P-450
measured either in frozen pellets or frozen suspensions was identical in all cases. The samples, if kept under liquid nitrogen,
were stable for at least 6 months with no losses of either high-
spin or low-spin EPR absorptions. EPR spectra were recorded with
an X-band superheterodyne spectrometer operating near 1.6°K, as
described previously (Feher, 1957). In order to avoid power saturation of EPR absorptions, the incident power on the sample was
70 db below 30 mW when examining high-spin cytochrome P-450, and
80 db below 30 mW when examining the low-spin forms. The relative
levels of high- and low-spin ferric cytochrome P-450 in any preparation were calculated by double integration of the resonance absorption derivative using a computer program written by one of
the authors (W.E.B.) and described previously (Peisach and Blumberg, 1970; Peisach et al., 1973). The error by this method is
±5% for the determination of any value. Furthermore, the sum of
both high- and low-spin cytochrome P-450 is within ±10% of the
total cytochrome P-450 determined optically by the method of Omura
and Sato (1964).

The remaining microsomal pellets, not used for EPR study,
were suspended and diluted in 0.1 M K-PO$_4$ buffer (pH = 7.4),
equivalent to 100 mg wet weight liver/ml and assayed for cytochrome P-450 by difference spectroscopy (Omura and Sato, 1964).
Optical spectra were recorded on an Aminco-Chance Dual-Wavelength
spectrophotometer. 3,4-Benzpyrene hydroxylase activity was
measured by the method of Nebert and Gelboin (1968), with slight
modifications as described by Lu et al., (1972). Protein was determined by the method of Lowry et al., (1951).

Results obtained from the various treatment groups were subjected to statistical analyses (one-way analysis of variance) to
determine if differences between groups were significant. Comparisons between any two groups were done using the method of
Newman-Keuls (Winter, 1962).

The binding of 3-MC to microsomes was studied after injection
of a single dose (25 mg/kg) of [^{3}H]-3-methylcholanthrene (Amer-
sham-Searle Corporation, Arlington Heights, Illinois) having a
specific activity of 14.2 mCi/mmole (12,640 cpm/nmole 3-MC).

Partial purification of cytochrome P-450 from fully-induced,
3-MC-treated rats was achieved using the ionic detergent sodium
cholate with ammonium sulfate fractionation and calcium phosphate
gel adsorption and elution (Step III) by the method of Lu and
Levin (1972). Further purification was achieved with the non-
ionic detergent Emulgen 911 and DEAE-cellulose column chromatography (Step IV) as described by Levin et al., (1974).

RESULTS

EPR spectra of control rat microsomes and fully-induced
(72 hr), 3-MC rat microsomes were identical to those previously
reported (Stern et al., 1973). The most prominent features of
these spectra are absorptions at g = 2.425, 2.254 and 1.915 for
control microsomes, and at g = 2.417, 2.248 and 1.912 for 3-MC
microsomes; these are all ascribed to low-spin (S = 1/2) ferric
cytochrome P-450 (Miyake et al., 1968; Stern et al., 1973). EPR
spectra from ethionine-control microsomes, 2 hr, 3-MC microsomes
and ethionine-24 hr, 3-MC microsomes are qualitatively identical
with control microsomes whereas EPR spectra from the 24 hr, 3-MC
group are qualitatively the same as fully-induced 3-MC microsomes.
Prominent in the spectrum of microsomes from fully-induced 3-MC
rats is the low field absorption feature near g = 8, and other
features to higher field at g = 3.7 and 1.71, indicative of high-
spin (S = 5/2) ferric cytochrome P-450 (Peisach and Blumberg,1970).
In addition, an EPR absorption feature is seen near g = 6 arising
from other high-spin microsomal hemoprotein in a near-axial environment, possibly high-spin cytochrome P-420 (Murakami and Mason,
1967).

Table I summarizes the data from the six experimental groups.
As can be seen following the time course of induction after 3-MC
treatment, an increase in the levels of cytochrome P-450 and a
dramatic increase in the specific activity of benzpyrene hydroxylase were observed 24 hr after 3-MC administration but not at 2 hr
after administration of the hydrocarbon (Table I and Figure 1).

TABLE I

The Effects of In Vivo Administration of Ethionine and 3-Methylchol-
anthrene on the Levels of Hepatic Cytochrome P-450 and Benzpyrene
Hydroxylase

Treatment	Cyt.[a] P-450	Benzpyrene[b] Hydroxylase	Low-spin[c] P-450	High-spin[c] P-450	% High-spin P-450
1. Control	0.88	1.03	0.84	0.04	4.9
	0.86	1.27	0.81	0.05	5.6
	0.97	1.60	0.92	0.05	5.2
2. Ethionine- control	0.63	0.46	0.60	0.03	4.6
	0.56	0.50	0.53	0.03	4.7
	0.54	0.38	0.52	0.03	4.7
3. 2 hr 3-MC	0.93	1.78	0.88	0.05	4.9
	0.90	1.09	0.85	0.05	5.6
	0.90	1.56	0.85	0.05	4.0
4. 24 hr 3-MC	1.27	7.95	1.19	0.08	6.3
	1.40	11.42	1.30	0.10	7.3
	1.37	10.83	1.27	0.10	7.6
5. Ethionine- 24 hr 3-MC	0.66	0.66	0.62	0.04	6.4
	0.70	0.84	0.66	0.04	6.0
	0.66	0.59	0.61	0.05	7.2
6. 72 hr 3-MC	1.48	7.39	1.33	0.15	9.9
	1.56	14.01	1.40	0.16	10.2
	1.93	14.34	1.74	0.19	9.7

[a]Cytochrome P-450 expressed as nmoles/mg protein.

[b]Benzpyrene hydroxylase activity expressed as nmoles hydroxylated/5
min/mg protein.

[c]Low- and high-spin cytochrome P-450 quantitated by double integration
of EPR absorption derivative, nmole/mg protein.

A similar time course of uptake of radioactive 3-MC, as shown in
Figure 1, has been demonstrated by Levine and Singer (1972) using
adult rats of a different strain. Concomitant with these changes,
an increase in the amount (percent of total) of high-spin cytochrome P-450 was observed at 24 and 72 hr after injecting 3-MC but
not at 2 hr after 3-MC administration.

Comparisons of the six treatment groups, with respect to
specific content of cytochrome P-450 (Table II-a) and benzpyrene
hydroxylase activity (Table II-b) show that there are no significant

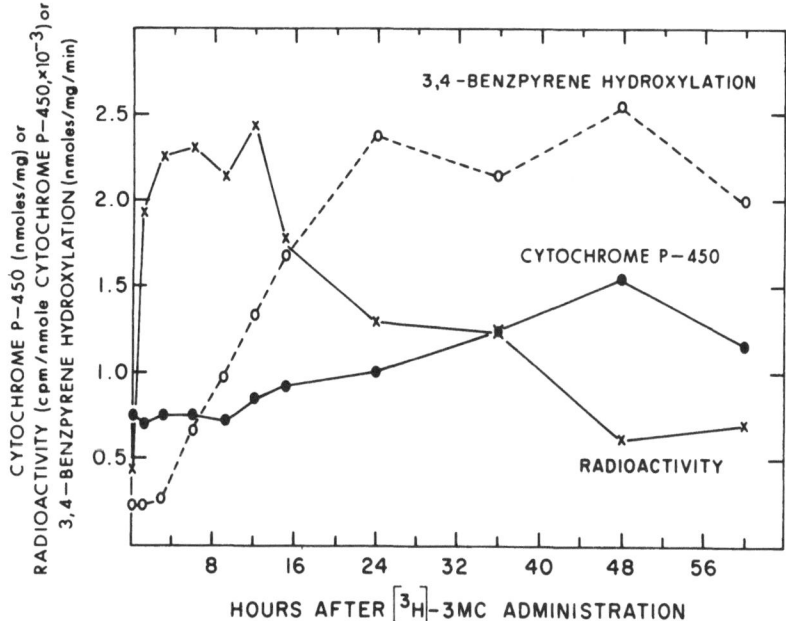

Fig. 1.   Binding of [^{3}H]-3-MC to rat liver microsomes.   Long-Evans male rats were injected intraperitoneally with a single dose (25 mg per kg) of [^{3}H]-3-MC (14.2 mCi/mmole of 3-MC).   Microsomes were prepared from rats sacrificed at the indicated time periods after treatment.   Radioactivity (X), 3,4-benzpyrene hydroxylation (0) and cytochrome P-450 (●) were measured.   Absorbance$_{max}$ = 450 nm, 0-6 hr; 449 nm, 9-15 hr; 448 nm, 24-60 hr.   Three rats were used for each time period.

differences (p < 0.01) among the control, ethionine-control, 2 hr, 3-MC and ethionine-24 hr, 3-MC groups.   Of the six groups only the 24 hr, 3-MC and 72 hr, 3-MC groups are significantly different from the other groups.   Statistical analyses of variance of the six groups, with respect to changes in the levels of high-spin cytochrome P-450 (Table II-c), reveal that groups 1, 2 and 3 (control, ethionine-control, and 2 hr, 3-MC microsomes, respectively) are not significantly different from each other.   These analyses also demonstrate that there is no difference between the 24 hr, 3-MC group and the ethionine-24 hr, 3-MC group in their levels of high-spin cytochrome P-450.   If the same test is applied for the benz-pyrene hydroxylase and cytochrome P-450 data for control, ethionine-control, 2 hr 3-MC and ethionine-24 hr, 3-MC groups, significant differences are observed within this smaller set.   However, both

TABLE II

Newman-Keuls Testing of Significance Between Pairs of Treatment
Groups

Notation: 1 = control group
              2 = ethionine-control
              3 = 2 hr, 3-MC
              4 = 24 hr, 3-MC
              5 = ethionine-24 hr, 3-MC
              6 = 72 hr, 3-MC

(+) denotes significance and (-) lack of significance between a pair.

a) Cytochrome P-450

	1	2	3	4	5	6
1		-	-	+	-	+
2			-	+	-	+
3				+	-	+
4					+	+
5						+

b) Benzpyrene Hydroxylase

	1	2	3	4	5	6
1		-	-	+	-	+
2			-	+	-	+
3				+	-	+
4					+	-
5						+

c) Percent High-spin P-450

	1	2	3	4	5	6
1		-	-	+	+	+
2			-	+	+	+
3				+	+	+
4					-	+
5						+

tests lead to the same conclusion that the percent of cytochrome
P-450 in the high-spin form is neither correlated with the level
of cytochrome P-450 nor with the level of benzpyrene hydroxylase
activity. Statistically valid comparisons can only be made, however, between similarly treated groups, i.e., ethionine-24 hr, 3-MC
versus ethionine-control group and 24 hr, 3-MC group versus control group. These comparisons (using the mean value of each group)
are shown in Figure 2 which clearly demonstrate that the administration of DL-ethionine prior to injecting 3-MC blocks the rise in
levels of cytochrome P-450 and benzpyrene hydroxylase activity
but has no effect in preventing the increase of the high-spin
species of the ferric hemoprotein normally seen with 3-MC induction.

The EPR spectra of the Step III and Step IV partially purified cytochrome P-450 preparations show EPR absorptions ascribable to the low-spin form of the hemoprotein as was seen in microsomes from 3-MC treated rats (Stern et al., 1973). As purification proceeds from microsomes through Step IV the amount of the high-spin species (g = 7.9, 3.7, and 1.7) decreases, as summarized in Table III. In spite of the loss of the high-spin species, the enzymatic activity and the substrate specificity of each of these preparations is the same as of the microsomes from which they were derived (Lu et al., 1973a; Levin et al., 1974). In addition, no g = 6 material is observed in this most purified preparation.

DISCUSSION

The microsomal hemoprotein cytochrome P-450 is known to encompass a heterogenous group of enzymes and the resolution of at least three types of microsomal cytochrome P-450 has been demonstrated by their unique optical, magnetic and enzymic properties (Sladek and Mannering, 1966; Alvares et al., 1967; Hildebrandt et al.,1968; Peisach and Blumberg, 1970; Goujon et al., 1972; Comai and Gaylor, 1973). One of these forms is believed to predominate in control animals, another in PB-treated animals and a third in animals pretreated with polycyclic hydrocarbons (Lu et al., 1973a).

Ferric cytochromes P-450 have also been shown by EPR to exist in two possible spin-states, low-spin and high-spin. Associated with the appearance of the new form of cytochrome P-450 following 3-MC treatment is the increase in hepatic microsomal high-spin cytochrome P-450 (Peisach and Blumberg, 1970; Nebert and Kon, 1973; Stern et al., 1973). The relationship between 3-MC induction and the changes in spin-state has not been resolved but at least five hypotheses have been proposed:

(1) Polycyclic hydrocarbons induce the production of a carrier-type protein which is specific for certain substrates. A carrier protein could bind substrates and then proceed to bind to the cytochrome thereby altering its biochemical and biophysical (e.g. spin-state) properties (Imai and Siekevitz, 1971).

(2) 3-MC treatment induces the formation of some membrane-specific non-heme protein or lipid and this results in an alteration of the conformation and/or configuration of cytochrome P-450 within the membrane matrix (Imai and Siekevitz, 1971; Nebert and Kon, 1973; Nebert et al., 1973a).

(3) The spectral, enzymic and spin-state changes seen following 3-MC treatment are simply caused by direct binding of 3-MC or one of its metabolites to cytochrome P-450 (Hildebrandt et al., 1968; Schenkman et al., 1969).

TABLE III

EPR Analysis of Cytochrome P-450 from Microsomes and Partially
Purified Preparations[a]

Sample	Total Cytochrome P-450 (nmoles/mg protein)	Percent High-Spin Cytochrome P-450
Microsomes (from 3-MC treated rats)	1.2	9.8%
Step III, partially purified cytochrome P-450	4.8	5.6
Step IV, partially purified cytochrome P-450	10.4	<<1

[a]Immature male rats were injected with 3-MC (25 mg/kg /day) for 3 days.
Microsomes were prepared from the livers of 100 rats and the hemoprotein
was partially purified (Step III) by the method of Lu and Levin (1972).
The hemoprotein was further purified (Step IV) by the method of Levin
et al., 1974).

(4) 3-MC induces, on a genetic level, the formation of a
newly synthesized cytochrome P-450 in a "preferred" high-spin configuration which is independent of bound substrate, and may have
different substrate specificities and spectral properties than
P-450 from control or PB-treated animals (Goujon et al., 1972;
Nebert and Kon, 1973; Nebert et al., 1973a; Nebert et al., 1973b).

(5) 3-MC treatment results in the induction of a low-spin
ferric hemoprotein with a different primary structure than that
found in control or PB-induced cytochrome P-450, which, when substrate-bound, will also be high-spin (Fujita et al., 1973; Stern
et al., 1973; Peisach et al., 1973; Levin et al., 1973; Lu et al.,
1973b).

Hypotheses (1)-(3) originated as arguments against the suggestions of Alvares and co-workers (Alvares et al.,1967; Alvares
et al., 1968b; Alvares et al., 1971) that new protein synthesis is
an absolute requirement for the new spectral species of cytochrome
P-450 which appears after 3-MC induction. There is, in fact, no
supportive evidence for these three hypotheses although arguments
against them can be formulated. Levin et al., (1974) and Lu et
al, (1973a) working with solubilized 3-MC-induced cytochrome P-450
(Step IV) have shown that they can remove more than 99% of the
membrane phospholipid and purify the cytochrome more than 7-fold
with respect to the original microsomal protein, yet retain the substrate specificity, enzymic activity and optical spectral properties of the CO derivative as found in microsomes from 3-MC-

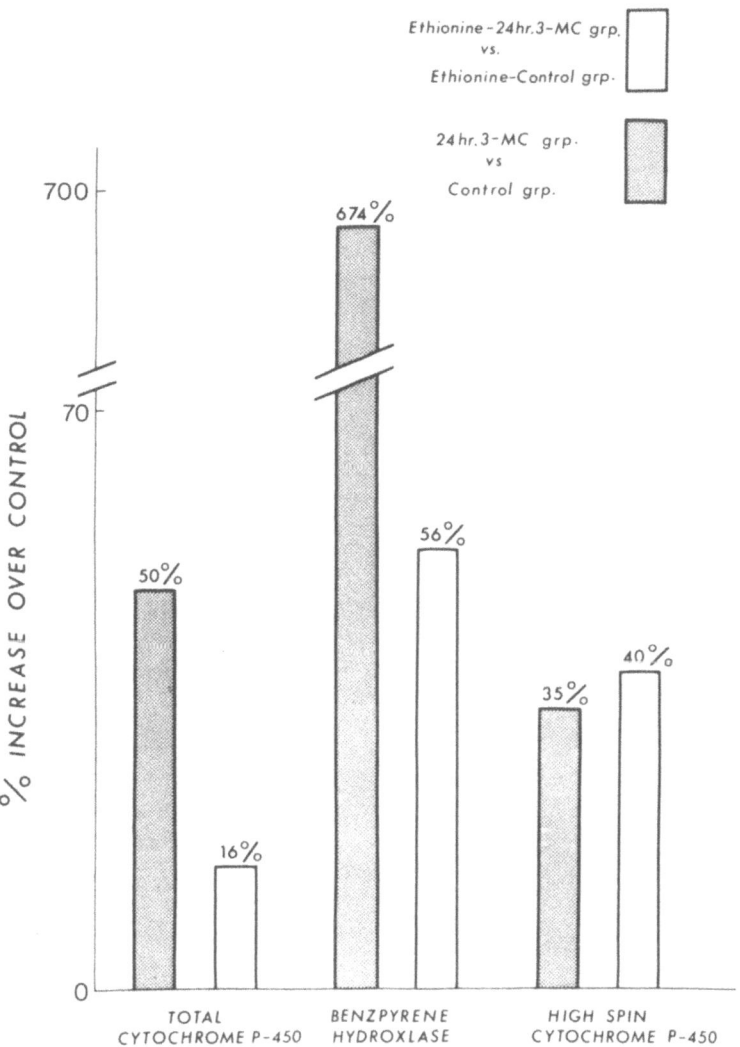

Fig. 2. A comparison of the percent increases of cytochrome P-450, benzpyrene hydroxylase and high-spin cytochrome P-450 between 24 hr, 3-MC microsomes and control microsomes (hatched areas) and between ethionine 24 hr, 3-MC microsomes and ethionine-control microsomes (clear areas).

treated animals. In addition, we have shown here that this preparation contains essentially no high-spin species (<<1%). The lack of stoichiometrically bound substrate in these partially purified preparations negates hypotheses (1) and (3) and the removal of greater than 99% of membrane lipid makes (2) untenable.

Most investigations concerning spin-state changes following 3-MC induction have come from studies with mice, rats and rabbits; attempts to correlate data among these animal species have been, at best, confusing. Nebert and co-investigators (Nebert and Kon, 1973; Goujon et al., 1972) working with two different strains of mice, one-3-MC inducible and the other non-inducible, have shown that in vivo administration of 3-MC to inducible strains results in a 3-fold increase in high-spin cytochrome P-450 as determined by EPR whereas 3-MC treatment of non-inducible strains resulted in no increase in the high-spin species. They concluded that 3-MC treatment of inducible strains resulted in the synthesis of cytochrome P-450 in a preferred high-spin configuration, and 3-MC or its metabolites, per se, do not effect this change (hypothesis 4). This is in sharp contrast to our present study on the immature rat (Table I and Figure 2) where we demonstrate that the percent increase of high-spin cytochrome P-450 in non-induced 3-MC-treated rats, i.e., ethionine-24 hr, 3-MC groups, is identical to increases seen in induced rats, i.e., 24 hr, 3-MC group. As can be seen in Figure 2, there is no correlation between 3-MC induction of cytochrome P-450 and the increase in the amount of the high-spin species in the rat. Furthermore, preliminary studies using the C57BL/6J inducible strain of mice showed qualitatively similar results to those obtained with rats. That is, 3-MC treatment of mice resulted in a greater than 500% increase over control of benzpyrene hydroxylase activity, while the increase of total cytochrome P-450 was 50%. Ethionine treatment completely blocked this induction of cytochrome P-450 as well as the benzpyrene hydroxylase activity. However, while 3-MC treatment resulted in an 80% increase of the high-spin species over control, this increase was not significantly blocked by pretreatment with ethionine (65% increase). Therefore the data presented here indicate that hypothesis (5) is the most likely.

Although the data presented in this paper only allude to the manner by which 3-MC and/or its metabolites may interact with rat and mouse liver cytochrome P-450, we can conclude that:

(1) 3-MC administration induces the synthesis of a new form of cytochrome P-450 having different optical and enzymic properties than cytochrome P-450 found in control or PB-treated rats and this induction is dependent upon protein and RNA synthesis.

(2) This new form of cytochrome P-450, in the rat and mouse, is a low-spin ferric hemoprotein although interconversion from low- to high-spin can occur in the presence of appropriate substrates.

(3) No clear correlation exists between 3-MC induction of cytochrome P-450 and benzpyrene hydroxylase activity compared to increases in the amount of high-spin cytochrome P-450 in the mouse and the rat.

Thus the presence of the g = 7.9 resonance ascribable to high-spin cytochrome P-450 is independent of induced RNA and protein synthesis in rat and mouse liver after 3-MC treatment. These results are in marked contrast to the recent suggestions of Nebert et al., (1973a) that the g = 7.9 resonance is related to the inductive process in both immature rats and mice.

FOOTNOTES

*   The portion of this investigation carried out at the Albert Einstein College of Medicine was supported in part by United States Public Health Service Research Grant HL-13399 from the Heart and Lung Institute and by National Science Foundation Grant GB-36422 to J. Peisach, and as such this is Communication 317 from the Joan and Lester Avnet Institute of Molecular Biology.

**   Predoctoral candidate of the Medical Scientist Training Program, supported by Grant 5T5-GM-1674 from the U. S. Public Health Service.

*** Recipient of Public Health Service Research Career Development Award 2-K3-GM-31,156 from the National Institute of General Medical Sciences.

1   In this communication, we will refer to the hepatic hemoproteins(s) induced after 3-methycholanthrene treatment as cytochrome P-450 rather than cytochrome P-448 (Alvares et al., 1967), P-446 (Hildebrandt et al., 1968), or P_1-450 (Jefcoate and Gaylor, 1969).

REFERENCES

Alvares, A. P., Schilling, G. R., Levin, W. and Kuntzman, R. (1967) Biochem. Biophys. Res. Commun. 29, 521.

Alvares, A. P., Schilling, G. R. and Kuntzman, R. (1968a). Biochem. Biophys. Res. Commun. 30, 588.

Alvares, A. P., Schilling, G. R., Levin, W. and Kuntzman, R. (1968b). J. Pharmacol. Exp. Ther. 163, 417.

Alvares, A. P., Schilling, G. R., Levin, W. and Kuntzman, R. (1971). J. Pharmacol. Exp. Ther. 176, 1.

Berzofsky, J. A., Peisach, J. and Blumberg, W. E. (1971). J. Biol. Chem. 246, 3367.

Comai, K. and Gaylor, J. L. (1973). J. Biol. Chem. 248, 4947.

Conney, A. H., Gillette, J. R., Inscoe, J. K., Trams, E. R. and Posner, H. S. (1959). Science 130, 1478.

Conney, A. H. (1967). Pharmacol. Rev. 19, 317.

Feher, G. (1957). Bell Syst. Tech. J. 26, 449.

Fujita, P., Shoeman, D. W. and Mannering, G. J. (1973). J. Biol. Chem. 248, 2192.

Gnosspelius, Y., Thor, H. and Orrenius, S. (1969-1970). Chem. Biol. Inter. 1, 125.

Goujon, F. M., Nebert, D. W. and Gielen, J. E. (1972). Mol. Pharmacol. 8, 667.

Hildebrandt, A., Remmer, H. and Estabrook, R. W. (1968). Biochem. Biophys. Res. Commun. 30, 607.

Imai, Y. and Sato, R. (1966). Biochem. Biophys. Res. Commun. 23, 521.

Imai, Y. and Siekevitz, P. (1971). Arch. Biochem. Biophys. 144, 143.

Jefcoate, C. R. E. and Gaylor, J. L. (1969). Biochemistry 8, 3464.

Levin, W., Ryan, D., West, S. and Lu, A. Y. H. (1973). Drug Metab. Disp. 1, 602.

Levin, W., Ryan, D., West, S. and Lu, A. Y. H. (1974). J. Biol. Chem. 249, 1747.

Levine, W. G. and Singer, R. W. (1972). J. Pharmacol. Exp. Ther. 183, 411.

Lowry, O. H., Rosebrough, N. J., Farr, A. L. and Randall, R. J. (1951). J. Biol. Chem. 193, 265.

Lu, A. Y. H. and Levin, W. (1972). Biochem. Biophys. Res. Commun. 46, 1334.

Lu, A. Y. H., Kuntzman, R., West, S., Jacobson, M. and Conney, A. H. (1972). J. Biol. Chem. 247, 1727.

Lu, A. Y. H., Levin, W., West, S., Jacobson, M., Ryan, D., Kuntzman, R. and Conney, A. H. (1973a). J. Biol. Chem. 248, 456.

Lu, A. Y. H., Levin, W., Ryan, D., West, S., Kuntzman, R., Stern, J. O. and Peisach, J. (1973b). Fed. Proc. 32, 762.

Miyake, Y., Gaylor, J. L. and Mason, H. S. (1968). J. Biol. Chem. 243, 5788.

Murakami, K. and Mason, H. S. (1967). J. Biol. Chem. 242, 1102.

Nebert, D. W. and Gelboin, H. V. (1968). J. Biol. Chem. 243, 6242.

Nebert, D. W. and Kon, H. (1973). J. Biol. Chem. 248, 169.

Nebert, D. W., Robinson, J. R. and Kon, H. (1973a). J. Biol. Chem. 248, 7637.

Nebert, D. W., Considine, N. and Kon, H. (1973b). Drug Metab. Disp.. 1, 231.

Omura, T. and Sato, R. (1964). J. Biol. Chem. 239, 2379.

Peisach, J. and Blumberg, W. E. (1970). Proc. Nat. Acad. Sci. USA. 67, 172.

Peisach, J., Stern, J. O. and Blumberg, W. E. (1973). Drug Metab. Disp. 1, 45.

Rickert, D. E. and Fouts, J. R. (1970). Biochem. Pharmacol. 19, 381.

Schenkman, J. B., Greim, H., Zange, M. and Remmer, H. (1969). Biochim. Biophys. Acta 171, 23.

Sladek, N. E. and Mannering, G. J. (1966). Biochem. Biophys. Res. Commun. 24, 668.

Stern, J. O., Peisach, J., Blumberg, W. E., Lu, A. Y. H. and Levin, W. (1973). Arch. Biochem. Biophys. 156, 404.

Winer, B. J. (1962) in "Statistical Principles in Experimental Design," McGraw-Hill Book Co., New York, p. 77.

AN ANALYSIS OF THE OPTICAL TITRATIONS OF THE 430 AND 455 NM CHROMO-

PHORES OF ETHYL ISOCYANIDE COMPLEXES OF MAMMALIAN HEPATIC CYTOCHROME

P-450*

J. Peisach**
Departments of Pharmacology and Molecular Biology
Albert Einstein College of Medicine
Yeshiva University
Bronx, New York 10461

SUMMARY

A computational method is presented from which one may calculate the pK for the spectral changes of the 430 and 455 mm
chromophores of ethyl isocyanide complexes of rabbit and rat liver
microsomal cytochrome P-450. For the rat liver protein from control or phenobarbital-treated animals, the pK for the loss of the
430 nm absorption is approximately equal to the pK for the gain of
the 455 nm absorption, confirming that the two chromophores are
in pH equilibrium with one another. For a soluble preparation in
which the chromophores exhibit the pH equilibrium property, this
equilibrium is also maintained even after the addition of ethyl
isocyanide. It is concluded that the appearance of a 430 nm absorption after the addition of ethyl isocyanide to reduced cytochrome P-450 does not necessarily represent conversion of the protein to cytochrome P-420.

— — — — — — — —

Ever since the initial observations of Imai and Sato on liver
microsomal cytochrome P-450 (1), the unusual optical spectrum in
the Soret region formed after the addition of ethyl isocyanide to
the reduced protein has been used to differentiate this type of
heme protein from others using the technique of optical difference
spectroscopy. A typical Soret maximum is observed at 430 nm and
a second peak, extraordinary for heme proteins, is seen at 455 nm.
When the pH of the microsomal suspension is raised, the addition
of ethyl isocyanide produces a lesser absorption difference near

430 nm while the 455 nm absorption difference is increased. These
authors suggested that the interconversion of the 430 and 455 nm
chromophores is pH dependent and were able to demonstrate this
phenomenon using microsomes from phenobarbital-induced rabbits. If
the cytochrome P-450 is denatured to form cytochrome P-420, only
the 430 nm absorption difference is seen after the addition of
ethyl isocyanide, regardless of the pH of the preparation (2).

In subsequent studies using rat liver microsomal cytochrome
P-450, it was observed by Sladek and Mannering (3) that the crossover of a plot of optical density versus pH for the 430 and 455 nm
absorption differences was not the same either for control or
phenobarbital-treated animals compared to that for 3-methylcholan-
threne-treated animals. These authors, however, suggested similar
pH dependent interconversions of the 430 and 455 nm chromophores in
all preparations, even though the crossovers were different.

In this communication we describe the analysis of published
data for the Soret absorption differences of ethyl isocyanide complexes of cytochrome P-450 in liver microsomes and also in soluble
preparations derived from the same tissue (4,5). In this study, we
demonstrate that in certain cases, the 430-to-455-nm interconversion
is indeed pH dependent, as the pK's for the transition involving
the loss of 430 nm absorption are essentially the same as those for
the increase of 455 nm absorption. However, in some cases this
interconversion cannot be demonstrated and the formation of another
state of the cytochrome is suggested. Finally for a purified preparation of liver microsomal cytochrome P-450 (5) which shows a pH-
dependent interconvertibility of 430 and 455 nm chromophores, a pH
jump method was employed to demonstrate that this optical transition
can take place reversibly in the presence of ethyl isocyanide and
that the 430 nm chromophore does not represent the conversion of
protein to cytochrome P-420.

MATERIALS AND METHODS

Published plots of data of Imai and Sato (1), Sladek and
Mannering (3), and Ryan et al., (5) were photographed, enlarged,
and printed on grid paper. For each set of data, coordinates of
the data points were read independently by the author and his research associate and were averaged. Least square analyses of the
data were computed using the algorithm

$$Y = A + \frac{B}{1 + 10^{(X-C)}}$$

Here, Y is the optical absorption at any pH, A is the optical
absorption at the end of the titration, B is the computed optical

absorption difference at the pH extrema of the titration (low pH value minus high pH value), X is the pH at which the spectrum is observed and C is the computed pK. Data were plotted on a Hewlett-Packard Model 7004B x-y recorder interfaced to a Honeywell Model 6000 computer with a Time Share Peripheral Plotter Controller.

For the study involving soluble cytochrome P-450, a sample of fraction PB Step III (4) was diluted to a concentration of 1 μM in 1 ml of 5 mM phosphate buffer, pH 6.5. Sodium dithionite was added, followed by ethyl isocyanide. After one minute the spectrum of the ethyl isocyanide complex versus buffer was recorded in the Soret region using a Model DW-2 Aminco spectrophotometer. Phosphate buffer, 0.1 ml of 1.0 M, pH 7.7, was added and the spectrum was once again recorded. In another experiment the protein was diluted in 5.0 mM phosphate buffer, but now at pH 7.7. The ethyl isocyanide complex was prepared, the spectrum studied and 0.1 ml of 1.0 M phosphate buffer, pH 6.5, was added. The spectrum was once again recorded.

RESULTS

The least squares fit to the algorithm for the rabbit liver microsomal cytochrome P-450 data is given in Figure 1. As can be seen, the pK for the loss of the 430 nm absorption is equal to the pK for the increase of the 455 nm absorption, confirming the original conclusion of Imai and Sato (1). The computed ratios of absorption for both high and low pH forms of each chromophore, ΔA_{430} and ΔA_{455}, are given in Table I.

Figure 2 shows similar treatment of data for rat liver microsomal cytochrome P-450 from control, phenobarbital- and 3-methylcholanthrene-treated animals. It should be noted that the pH crossover as well as the pK's for the two former preparations are essentially the same. The computed ratios of optical absorption differences at high and low pH for both chromophores, ΔA_{430} and ΔA_{455}, are not the same as for the rabbit protein and are also different from that for microsomal cytochrome P-450 from 3-methylcholanthrene treated rats (Table I.) In this last preparation, the pK's for the diminution of optical absorption for the 430 nm chromophore and the increase of optical absorption for the 455 nm chromophore differ significantly from one another.

Similar treatments of data for soluble, partially purified preparations derived from rat liver are summarized in Table I. Here, only five data points have been published for each preparation. The lack of correlation between pH crossover, pK and type of animal treatment is evident although the average of optical absorption properties of an A and B subfractional pair is roughly the same as that of the fraction (PB III or MC III) from which it was derived.

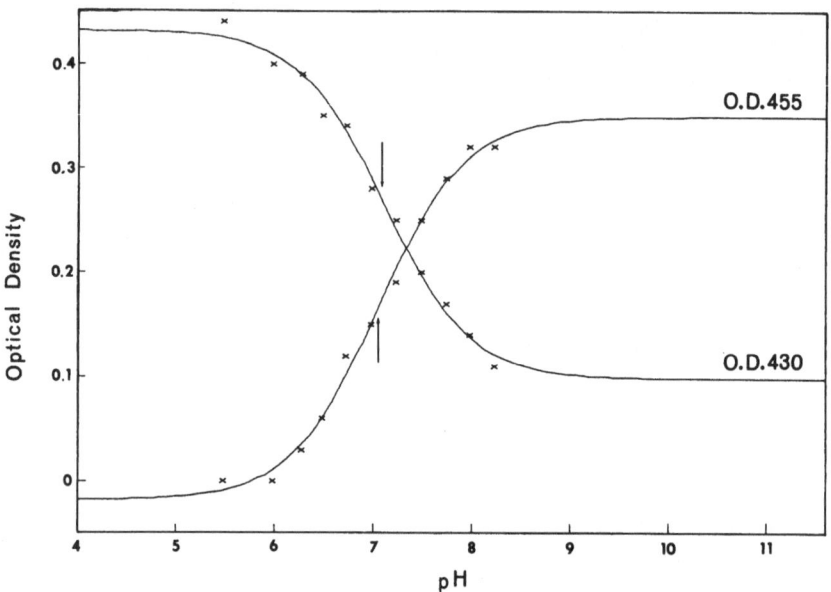

Figure 1. Computer generated titration curve based on the analysis of data from Imai and Sato (1) on the ethyl isocyanide complexes of rabbit liver microsomal cytochrome P-450 from pheno-barbital-treated animals. The data were processed and fit to the algorithm as described in the text. The crosses in the figure are based on digitalization of published data. The up-directed arrow is at the pK for the reduction of the 430 nm absorption while the down-directed arrow is at the pK for the increase of the 455 nm absorption, respectively, as a function of pH. The computed difference in optical density of a particular chromophore at high and low pH, ΔA, as well as the pK's and pH titration curve crossover, is given in Table I.

In one of the soluble fractions, PB III, derived from pheno-barbital-treated rats, the pK's for the 430 and 455 nm chromophores are equal (Table I). Using a direct photometric technique, the ratios of 430 and 455 nm absorptions at pH 6.8 are given in Table IIA. Raising the pH of the cytochrome P-450-ethyl isocyanide complex increases the 455 nm absorption, lessens the 430 nm absorption, and thus decreases the spectral ratio. For the complex prepared at pH 7.7 (Table IIB), the spectral ratio is the same as observed for the complex the pH of which was raised from 6.5 to

TABLE I

Preparation		pH Crossing	pK 430	pK 455	ΔA 430	ΔA 455	$\Delta A_{430}/\Delta A_{455}$
I&S	PB	7.3	7.1	7.1	0.336	0.367	0.91
S&M	control PB	7.4	6.9	7.1	0.132	0.061	2.2
	PB	7.4	7.0	6.8	0.422	0.217	1.9
	3-MC	6.9	6.3	6.9	0.237	0.155	1.5
L&L	PB-III	7.7	6.9	7.0	0.117	0.063	1.9
	PB A	6.9	6.3	6.8	0.218	0.075	2.9
	PB B	7.9	7.1	7.5	0.130	0.082	1.6
	3-MC III	6.9	6.3	6.2+	0.218	0.121+	1.7+
	3-MC A	7.1	6.8	7.3	0.103	0.046	2.7
	3-MC B	6.9	6.6	6.4	0.118	0.098	1.2

+ With the optical data available, the successive iterations of the fitting algorithm used to determine the pK for the 455 nm chromophore did not converge. The pK given in this table was based on a calculation which takes as the $\Delta A_{430}/\Delta A_{455}$ the average of the ratios for the two fractions 3-MC A and 3-MC B which are the components of the 3-MC III preparation. Therefore, the value for ΔA_{455} for the 3-MC III preparation is also based on the computed ratio.

Analysis of the computed titration curves for the ethyl isocyanide complexes of various preparations of cytochrome P-450. The terms I&S, S&M, and L&L refer to the preparations of Imai and Sato (1), Sladek and Mannering (3) and Ryan et al., (5) respectively. The pH crossing refers to the pH at which the titration curves expressing the loss of 430 nm absorption and the gain of 455 nm absorption as a function of pH, cross. the pK for the titration based on optical observations at 430 nm (pK_{430}) and the pK for the titration based on optical observations at 455 nm (pK_{455}) were computed as described in the text. The terms ΔA_{430} and ΔA_{455} are the computed differences in optical absorption of the individual chromophores at pH extrema in the titration curves while the term $\Delta A_{430}/\Delta A_{455}$ is the ratio of these optical absorption differences.

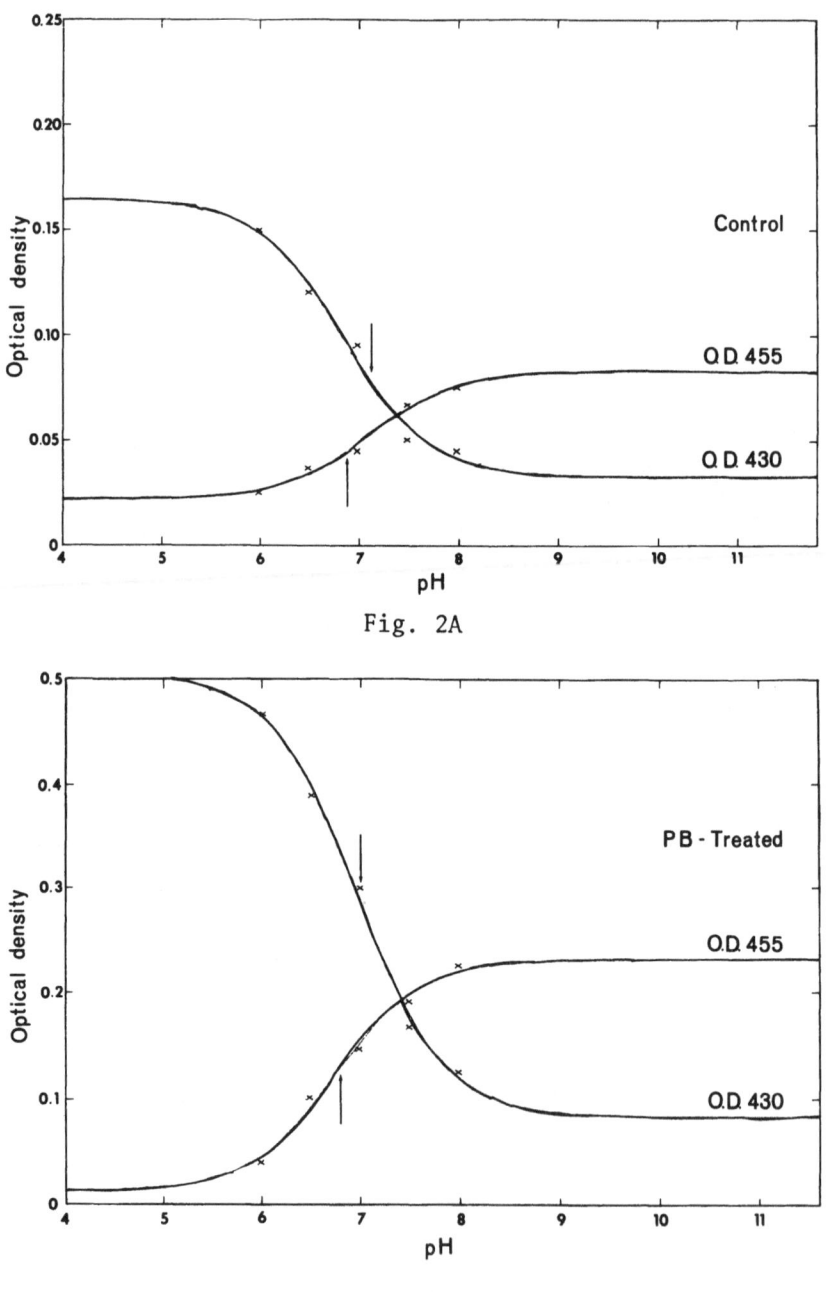

Fig. 2A

Fig. 2B

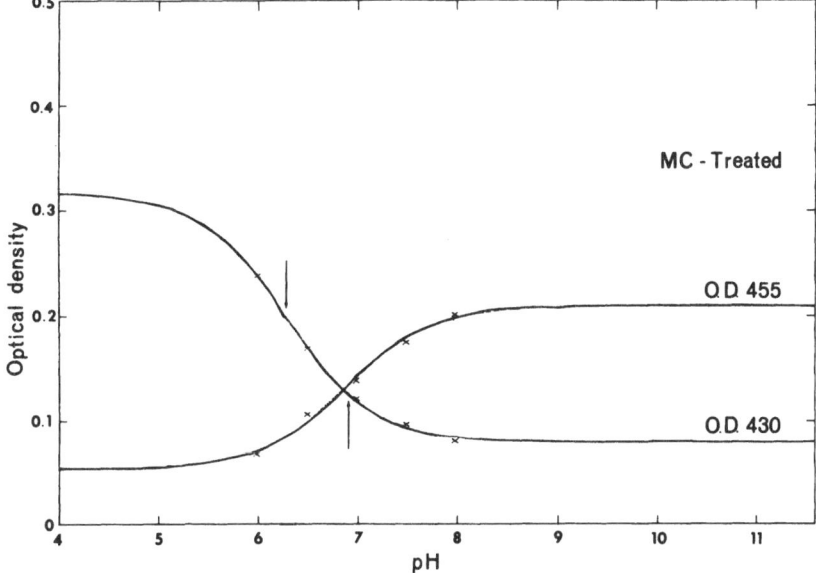

Figure 2A-B-C. Computer generated titration curves based on the analysis of data from Sladek and Mannering (3) on the ethyl isocyanide complexes from control, phenobarbital- and 3-methylcholanthrene-treated rats. The data were processed and fitted to the algorithm as described in the text. For further clarification of this Figure, see Figure 1.

TABLE II

pH	A $\Delta A_{430}/\Delta A_{455}$	B $\Delta A_{430}/\Delta A_{455}$
6.5	3.1	3.2
7.7	1.8	1.8

Ratios of optical absorption of the 430 and 455 nm chromophores, $\Delta A_{430}/\Delta A_{455}$, of the ethyl isocyanide complex of partially purified cytochrome P-450 (Fraction PB Step III (4)) obtained from phenobarbital treated rats. In column A, the soluble protein was diluted in a solution weakly buffered at pH 6.5 and the optical spectrum was recorded. Concentrated buffer was added to raise the pH to 7.7 and after baseline and dilution corrections, the ratio of optical absorption was determined. In column B, the protein was diluted in a solution buffered at pH 7.7, the spectrum was recorded and concentrated buffer was added to drop the pH to 6.5. Again, the spectral ratio was determined.

7.7 (Table IIA). Now dropping the pH to 6.5 increases the spectral ratio to that observed for the complex originally prepared at
this lower pH.

DISCUSSION

The presence of both 430 and 455 nm chromophores is ubiquitous
to ethyl isocyanide complexes of all preparations of cytochrome
P-450, whether bacterial (6,7) or mammalian (1,2) in origin. Although Griffin and Peterson (7) insist that only a single Soret
absorption near 455 nm is found using the protein isolated from
P. putida, a 430 nm absorption is clearly observable in their published absolute and difference spectra for preparations containing
no cytochrome P-420.

From stopped flow experiments, Imai and Mason (8) have demonstrated that the addition of ethyl isocyanide to reduced cytochrome
P-450 from a mammalian source leads first to the appearance of the
455 nm absorption followed by the one at 430 nm. It can be argued
that ethyl isocyanide is a chaotropic reagent which is sufficiently
disruptive to protein structure, much like urea++, as to convert
it to cytochrome P-420 which exhibits the characteristic 430 nm
absorption (2). An argument against this hypothesis is that even
though for some rat liver microsomal preparations (Figures 1 and
2A and B), the relative intensities of the 430 and 455 nm chromophores observed after the addition of ethyl isocyanide are dependent
upon the pH established before the addition of ligand, for a soluble
preparation derived from the same source (Table II) the relative
intensities of the chromophores are dependent upon the pH established after the addition of ligand. This demonstrates a pH-dependent equilibrium even in the presence of ethyl isocyanide.
Thus the appearance of the 430 nm absorption after the addition of
ethyl isocyanide is not necessarily indicative of protein denaturation.

To what structures, then, do we ascribe both the 430 and 455
nm chromophores? Much evidence has been presented to indicate that
cytochrome P-450 preparations are heterogeneous (9) insofar as that
probing the local environments of the heme shows more than one
chemical structure (10). For the ferric protein, the presence of
an alkyl thiol, such as from cysteine, has been implicated from
EPR studies (11,12). Peisach et al., (13) have suggested that the
formation of the 430 or 455 nm chromophores after the addition of
ethyl isocyanide to the ferrous protein is dependent upon the relative bond strengths of the two axial ligands of the heme. That

++ J. O. Stern and J. Peisach, unpublished observations.

is, displacement of the sulfur containing ligand would lead to the formation of a 430 nm chromophore while displacement of the <u>trans</u> axial ligand would lead to the formation of the 455 nm chromophore.

Recently, Stern and Peisach (14) have demonstrated with model compounds that the unusual Soret absorption of CO complexes of cytochrome P-450 is dependent upon the presence of a mercaptide anion as the ligand <u>trans</u> to CO. If conditions are not sufficiently basic so as to dissociate a proton from the thiol employed in these studies, the 450 nm chromophore does not appear. These experiments suggest that in the case of ethyl isocyanide ligation, the appearance of the unusual 455 nm chromophore in the reduced protein may also be dependent upon the dissociation of a proton from the sulfur-containing ligand while the 430 nm chromophore may not.have this requirement. Thus, at elevated pH where the removal of a proton is facilitated, the 455 nm absorption would predominate. The fact that this absorption appears before the one at 430 nm after ethyl isocyanide addition (8) would suggest that in the reduced protein structure, the proton is either dissociated or strongly hydrogen-bonded before the addition of ligand. After ligand addition, redistribution of bonding would take place.

What then of the preparations of cytochrome P-450 where the pK's for the disappearance of the 430 nm chromophore and the appearance of the 455 nm chromophore are different? One possibility is that the difference is within the experimental error of the analysis (Table I, L&L, S&M). For the Sladek and Mannering studies (3), only five data points are published for each preparation while for the Imai and Sato studies (1), much more data were taken. Our analysis would suggest that for cytochrome P-450 from 3-methylcholanthrene-treated rats (Table I), the 430 and 455 nm chromophores are not in pH-dependent equilibrium with one another. The lowered pK for the disappearance of 430 nm chromophores may represent the formation of an intermediate state, possibly one with a 430 nm absorption with a different absorption coefficient. Certainly the relative absorption coefficients (expressed as a function of ΔA in Table I) are quite different for the different preparations, regardless of pK or source.

The lack of correlation of pH crossover with source of cytochrome P-450 for the soluble preparations (Table I) may be grounded in the paucity of data or possibly in differences in structure which are manifest in differences in absorption coefficients of the 430 and 455 nm chromophores. Once these preparations are purified to a level where physical probes indicate homogeneous ligand binding to the heme, more can be said about these spectral differences among the various purified as well as the microsomal preparations.

FOOTNOTES

*   This investigation was supported in part by United States Public Health Service Research Grant HL-13399 from the Heart and Lung Institute and is Communication 323 from the Joan and Lester Avnet Institute of Molecular Biology.

**   Recipient of Public Health Service Research Career Development Award 2-K3-GM-31, 156 from the National Institute of General Medical Sciences.

ACKNOWLEDGEMENTS

The author would like to thank Drs. W. Levin and A. Y. H. Lu for a generous sample of partially purified cytochrome P-450 from phenobarbital treated rats, and their hospitality at Hoffman-La Roche; Dr. W. E. Blumberg, Bell Laboratories, for his aid with computations; and Mr. J. O. Stern for his aid with the experiments.

REFERENCES

1   Imai, Y. and Sato, R. (1966) Biochem. Biophys. Res. Commun. 23: 5-11.
2   Omura, T. and Sato, R. (1964) J. Biol. Chem. 239: 2370-2378.
3   Sladek, N. E. and Mannering, G. J. (1966) Biochem. Biophys. Res. Commun. 24: 668-674.
4   Levin, W., Ryan, D., West, S. and Lu, A. Y. H. (1974) J.Biol. Chem. 249: 1747-1753.
5   Ryan, D., Lu, A. Y. H., West, S. and Levin, W. Submitted for publication.
6   Appleby, C. A. (1967) Biochim. Biophys. Acta 147: 399-402.
7   Griffin, P. and Peterson, J. A. (1971) Archives Biochem. Biophys. 145: 220-229.
8   Imai, Y. and Mason, H. S. (1971) J. Biol. Chem. 246: 5970-5977.
9   Levin, W. and Kuntzman, R. (1969) Mol. Pharmacol. 5: 499-506.
10   Stern, J. O., Peisach, J., Blumberg, W. E., Lu, A. Y. H. and Levin, W. (1973) Arch. Biochem. Biophys. 156: 404-413.
11   Blumberg, W. E. and Peisach, J. (1971) in Probes of Structure and Function of Macromolecules and Membranes, Vol. II (Chance, B., Yonetani, T. and Mildvan, A. S., eds.) pp. 215-229, Academic Press Inc., New York.
12   Bayer, E., Hill, H. A. O., Röder, A. and Williams, R. J. P. (1969) Chem. Commun. 1969, 109.
13   Peisach, J., Stern, J. O. and Blumberg, W. E. (1973) Drug Metab. and Disposition 1: 44-61
14   Stern, J. O. and Peisach, J. (1974) J. Biol. Chem., in press.

IMPLICATION OF LIGAND MODIFIED SPECTRA OF CYTOCHROME P-450 ASSOCIATED WITH PREGNENOLONE SYNTHESIS IN MITOCHONDRIA FROM CORPUS LUTEUM

V. I. Uzgiris, E. N. McIntosh, P. Graves and H. A.
Salhanick, Department of Population Sciences and
Department of Obstetrics and Gynecology, Harvard
University
Boston, Massachusetts 02115

It is now well established that many reagents can modify
the cytochrome P-450 spectrum. These modifications are produced.
by the interaction of ligands with hemoprotein or the iron of the
heme prosthetic group (13). The modified cytochrome P-450 difference spectra have been classified as Type I if the induced
maximum in the Soret region is at 390 nm and the minimum at
420 nm, as reverse Type I if the positions of the maximum and
minimum are reversed, and as Type II if the maximum and minimum
are reversed and both shifted by 5 to 10 nm to the red end of the
spectrum (12).

This report examines the spectral changes produced by several ligands in relation to their inhibitory effects on the enzymatic activity of the bovine corpus luteum mitochondrial cytochrome P-450 which catalyzes cholesterol conversion to pregnenolone.
Unlike the preparations from the adrenal cortex or the liver, mitochondria from the corpus luteum appear to contain only one cytochrome P-450-linked monooxygenase system.

METHODS AND MATERIALS

Mitochondria from fresh bovine corpora lutea were prepared
in 0.25 M sucrose, 0.1 mM EDTA by differential centrifugation.
The solubilization of mitochondrial cytochrome P-450 with sodium
cholate followed the general procedure of Mitani and Horie (9),
with modifications described previously (7). For the spectral
studies, in addition to mitochondria, two other preparations of
solubilized mitochondrial cytochrome P-450 were used. In one
preparation, cytochrome P-450 was solubilized with sodium cholate

and salted out of ammonium sulfate solution at 60% saturation.
In the second preparation, the hemoprotein was in a more purified
state following Sephadex G-25 gel filtration and repeated ammonium
sulfate fractionation.

The common steroids were purchased from Steraloids. Cyanoketone (2α-cyano-17β-hydroxy-4,4,17α-trimethyl-androst-5-en-3-one)
was a gift of Dr. S. Lieberman and aminoglutethimide [2-ethyl-2
(p-aminophenyl)-glutarimide] was a gift of Dr. J. J. Chart and
Dr. B. Steinetz at Ciba-Geigy Co.

Protein content of mitochondrial suspensions was measured
by the biuret procedure (16) but protein content in solubilized
cytochrome P-450 preparations was determined according to
Sutherland et al., (15).

Assay of cholesterol conversion to pregnenolone used the
published procedure of Hochberg et al., (3) which is based on the
radioactivity counting of the side-chain fragment of 26[^{14}C]
cholesterol. The latter was purchased from New England Nuclear.
In comparison with the enzymatic assay based on the identification
and measurement of steroid products of cholesterol side-chain
cleavage reaction (18), the side-chain fragment assay (3) was
found to be more reliable.

RESULTS AND DISCUSSION

Cholesterol, the natural substrate for the cholesterol monooxygenase system can be demonstrated to yield a Type I spectral
change. Because of the large content of endogenous cholesterol
in the usual enzyme preparations, however, that spectral alteration is difficult to demonstrate. In contrast, a number of
neutral steroids readily induce reverse Type I spectral change.
Three examples of such steroids are presented in Figures 1 and 2.
The first of these, 20α-hydroxycholesterol, causes a typical reverse Type I spectral change, is a metabolite, and is a strong
inhibitor of cholesterol side-chain cleavage (11). It can also
serve as a substrate and give rise to pregnenolone and isocaproaldehyde. The spectral modifications depend on ligand concentration (Figure 1). Although the absorbance increments in the
difference spectrum of crude cytochrome P-450 are small, reliable
measurements can be carried out with solubilized or membrane-
bound corpus luteum preparations. A comparison of the spectral
affinities of 20α-hydroxycholesterol for both membrane-bound and
solubilized cytochrome P-450 is presented in Figure 3. It is
evident that there are no marked differences between these two
preparations.

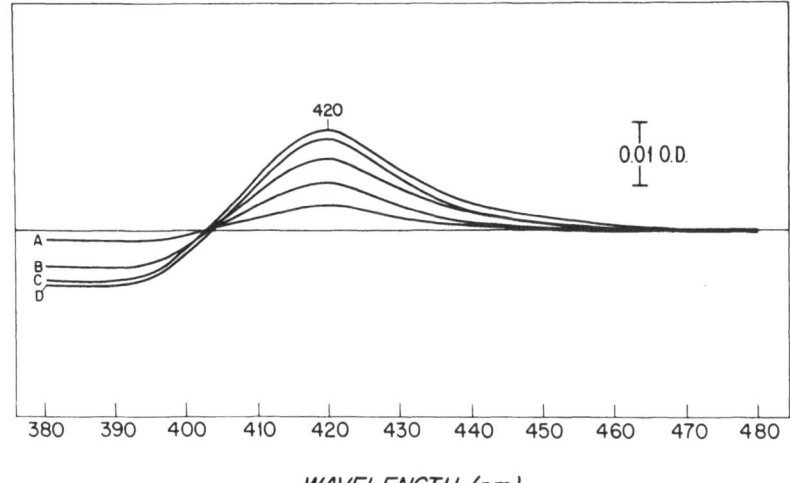

WAVELENGTH (nm)

Figure 1. Reverse Type I difference spectra of mitochondrial cytochrome P-450 induced with 20α-hydroxycholesterol. A
sample of mitochondria from bovine corpus luetum was diluted to
1.3 mg protein (cytochrome P-450, 0.27 μM) per ml of solution
consisting of 300 mM mannitol, 0.1 mM EDTA, 5 mM HEPES,* and 2.5
mM phosphate buffer, pH 7.45. To obtain the difference spectrum,
the mitochondrial suspension was divided into two cuvettes and
the base line of equal light absorbance recorded. Recordings
were repeated following successive additions of 20α-hydroxy-
cholesterol in propylene glycol to the sample cuvette and propylene glycol to the reference cuvette. Final concentration of
20α-hydroxycholesterol: (A) 1.0 μM; (B) 5.0 μM; (C) 50 μM;
(D) 100 μM.

The interpretation of the significance of induced spectral
changes in general, and particularly in the case of 20α-hydroxy-
cholesterol, is complex. Most investigators are agreed that they
represent binding to the enzyme such that the heme-apoprotein coordination is altered (13,17). Several sites for binding are
possible. In the case of most steroids, direct coordination to
the heme iron, with displacement of ligand groups of the native
protein is unlikely since such neutral compounds are of low
ligand-field strength. More likely binding sites are on the
apoprotein at the substrate site or at a site which would modify
the tertiary structure, with subsequent alterations in axial

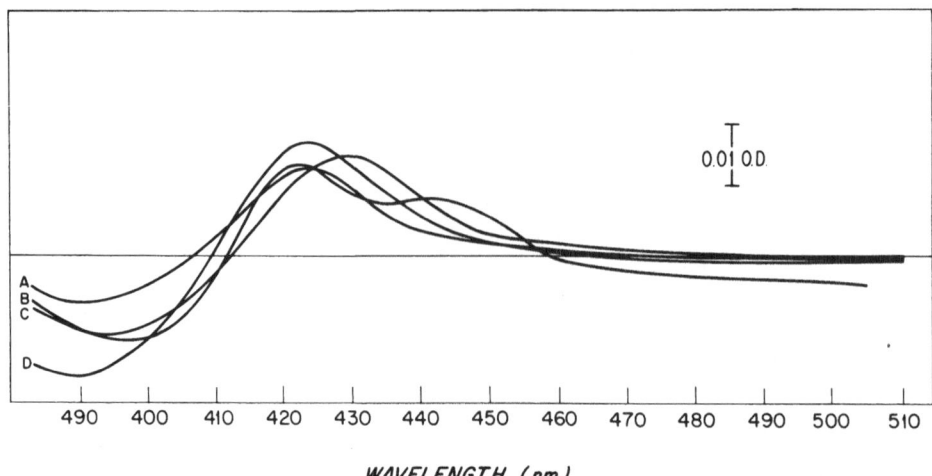

WAVELENGTH (nm)

Figure 2. Difference spectra of aerobic bovine corpus
luteum mitochondria induced by pregnenolone, cyanoketone, aminoglutethimide, and 1-benzyl-imidazole. A sample of mitochondria
was diluted to 0.26 µM cytochrome P-450 in a solution as in
Figure 1. Ligands were dissolved in propylene glycol and added
in 50 µl volume to the mitochondrial suspension. Final concentrations and ligands tested were: (A) 0.5 mM pregnonolone; (B)
0.25 mM cyanoketone; (C) 0.5 mM aminoglutethimide; (D) 0.5 mM
1-benzyl-imidazole.

ligands of the hemeporphyrin. Binding to the protein at non-
specific sites without any modifying effect on the hemoprotein
spectrum is, of course, likely but little is known about binding
at those sites.

The correlation between spectral changes and effects of
ligands on enzyme activity is not yet agreed upon. For example,
most Type I spectral ligands are substrates for the appropriate
cytochrome P-450 enzymes. With our preparations from corpus
luteum, 20α-hydroxycholesterol is an excellent substrate but it
produces a reverse Type I spectral change. Admittedly, if the
substrate site were devoid of cholesterol, it might well be that
20α-hydroxycholesterol would cause a Type I spectral alteration
and, indeed, that is what was reported by Harding et al., (2).
On the other hand, 20α-hydroxycholesterol has been shown to inhibit labeled cholesterol cleavage and to display non-competitive
kinetics as does pregnenolone (11). The possibility exists that

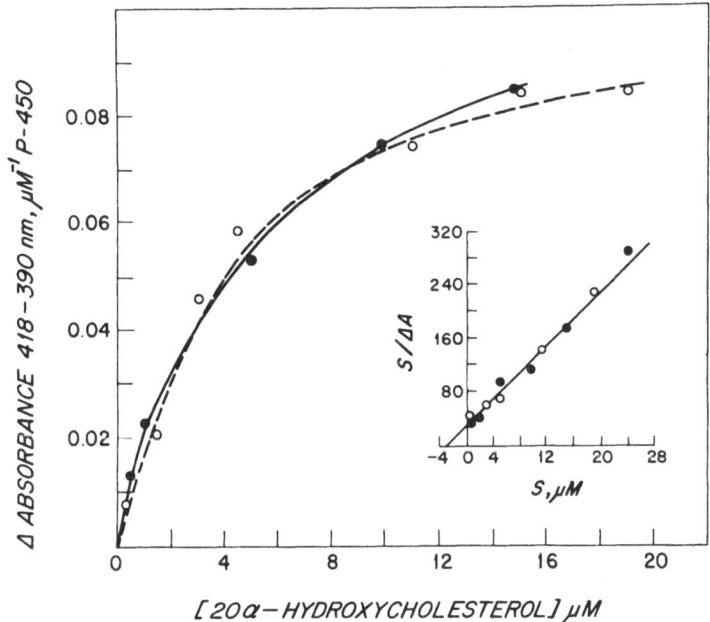

Figure 3. Affinities of 20α-hydroxycholesterol for solubilized and membrane-bound mitochondrial cytochrome P-450. A
sample of mitochondria was diluted as in Figure 1. The solubilized mitochondrial cytochrome P-450 was diluted to a 0.25 μM
concentration in 0.1 M phosphate buffer, pH 7.4. Difference
spectra were recorded following successive additions of 20α-hy-
droxycholesterol in propylene glycol to the sample cuvette and
solvent to the reference cuvette. The absorbance difference between 418 nm and 390 nm was measured and adjusted to an equivalent
absorbance difference of 1 nanomole cytochrome P-450/ml solution
and is plotted against 20α-hydroxycholesterol concentration. The
inset gives the same data in S/ΔA   vs. S plot. Titrations with
20α-hydroxycholesterol: mitochondria (——●——), and solubilized
mitochondrial cytochrome P-450 (-----0----). Apparent spectral
dissociation constant, K_s, 3 μM.

two cytochrome P-450's are involved in cholesterol side-chain
cleavage and that 20α-hydroxycholesterol is an intermediate. If
this is the case, then the concerted reaction theory proposed by
Lieberman and co-workers may not be valid (5). However, no accumu-
lation of this intermediate has been demonstrated either in vivo

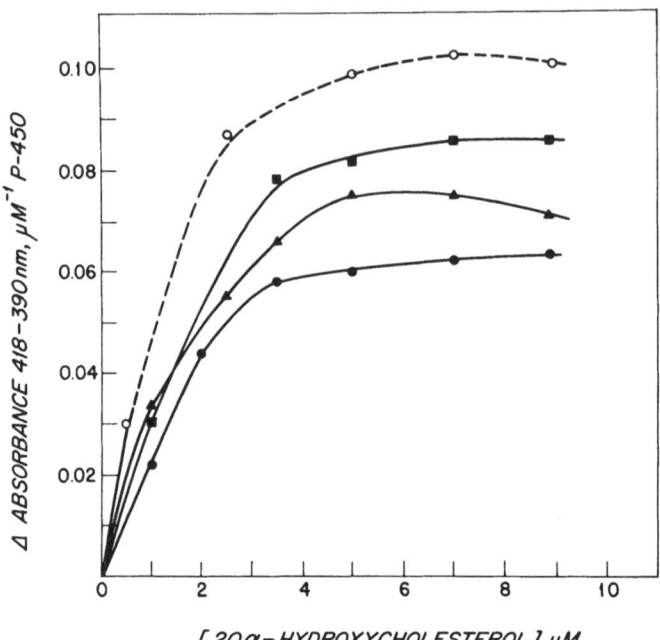

Figure 4.   Effect of added sodium cholate on the intensity of 20α-hydroxycholesterol induced spectral change of solubilized cytochrome P-450. Both sample and reference cuvettes contained solubilized cytochrome P-450 diluted to 0.24 µM cytochrome P-450 in 0.1 M phosphate buffer, pH 7.4. Difference spectra were recorded following successive additions of 20α-hydroxycholesterol in propylene glycol. Sodium cholate was neutralized and added to the protein solution prior to the titration with 20α-hydroxycholesterol. Sodium cholate was added to obtain the following concentrations: 0 mM (----0----); 1.7 mM (——■——); 3.3 mM (——▲——): 5.8 mM (——●——).

or in vitro and since the predominant cytochrome P-450 as ascertained by difference spectrometry of corpus luteum preparations must be the cholesterol monooxygenase, one must assume that low concentrations of another P-450, if it exists, would have exceedingly high turnover rates. No hint of such an enzyme has been found during purification or on reconstitution of the enzyme system and until further evidence is advanced, the existence of a separate 20α-hydroxycholesterol oxygenating enzyme is doubtful.

A more plausible interpretation supported by the inhibition

data is that cholesterol is the preferred substrate. Therefore, in its presence, 20α-hydroxycholesterol shows the reverse Type I shift and competes as a substrate, but is cleaved more slowly than cholesterol. This interpretation will be clarified when sufficient amounts of purified corpus luteum cytochrome P-450 devoid of cholesterol are obtained.

The second example concerns pregnenolone, the natural product of the cholesterol monooxygenase reaction (Figure 2, Curve A). It not only inhibits cholesterol cleavage but generates a reverse Type I spectral change with corpus luteum mitochondrial cytochrome P-450. From titration experiments with purified P-450, a binding constant (K_s) of 27 μM was obtained. This is very close to the value of approximately 20 μM for the half-maximal inhibition of cholesterol 20α-hydroxylase complex as calculated from the data of Ichii et al., (4). Pregnenolone acts as a classical feedback inhibitor of the cholesterol monooxygenase complex and the spectral changes suggest that it binds at non-substrate sites. Such binding would presumably lead to alterations in the tertiary structure of the enzyme, causing deformation of the catalytic site.

Two different binding phenomena are depicted in Figure 4. Cholate, a highly polar steroid, does not affect the spectrum at concentrations as much as 2 mM in contrast to the spectral changes induced by neutral steroids. The amphiphilic properties of cholate and deoxycholate have been used to advantage in dispersing phospholipid-protein membranes, therefore, they must affect not only the lipophilic properties of cytochrome P-450 but also change the binding of other compounds. When solubilized cytochrome P-450 was treated with neutralized cholate and subsequently titrated with 20α-hydroxycholesterol, a diminished response was obtained as the concentrations of cholate were increased. This is consistent either with minor destruction of cytochrome P-450 or with modification of 20α-hydroxycholesterol binding. Therefore, solubilized cytochrome P-450 was incubated for 15 min with neutralized cholate at several concentrations. The results, shown in Figure 5, indicate that the diminished response to 20α-hydroxycholesterol was caused by cholate denaturation of P-450, since the magnitude of spectral change is directly related to both ligand and cytochrome P-450 concentration.

The data on cyanoketone provide a third example of steroid interaction with cholesterol monooxygenase. This steroid inhibits not only 3β-hydroxysteroid dehydrogenase and steroid Δ-isomerase, but also cholesterol monooxygenase. Neville and Engle (1968) have determined the dissociation constant, K_i, to be 18 μM for the Δ-isomerase-cyanoketone complex (8). The concentration of cyanoketone which produces half-maximal inhibition of cholesterol cleavage in mitochondria from bovine corpus luteum was found to be

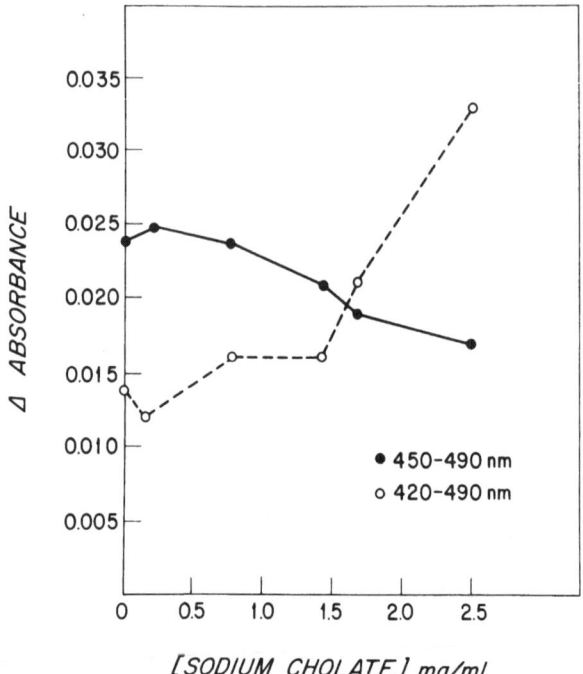

[SODIUM CHOLATE] mg/ml

Figure 5. Recovery of cytochromes P-450 and P-420 after incubation in the presence of sodium cholate. Solubilized cytochrome P-450 was diluted to 0.3 mg protein per ml 0.1 M phosphate
buffer, pH 7.4 (cytochrome P-450, 0.27 µM), and divided equally
between sample and reference cuvettes. After addition of neutralized sodium cholate to the contents of both cuvettes and a 15
minute incubation at 25°, reduced carbon monoxide difference
spectrum was recorded from 380 to 500 nm. The absorbance difference was measured between 450 or 420 nm relative to 490 nm.

higher than the value for isomerase, approximately 102 µM (P.
Graves, unpublished observations). Not only does the spectral
trace reveal the reverse Type I related spectral properties in
common with the other two steroids, but also an additional band
at around 445 nm (Figure 2, Curve B).

An even more complicated interpretation is required for the
spectral data on cyanoketone. The 2α-cyano group of the keto-
steroid resembles HCN in that it prevents reduction of cytochrome
a_3. The formation of the nitrile-cytochrome oxidase complex is
more rapid than liganding with the cytochrome P-450 (Figure 6,

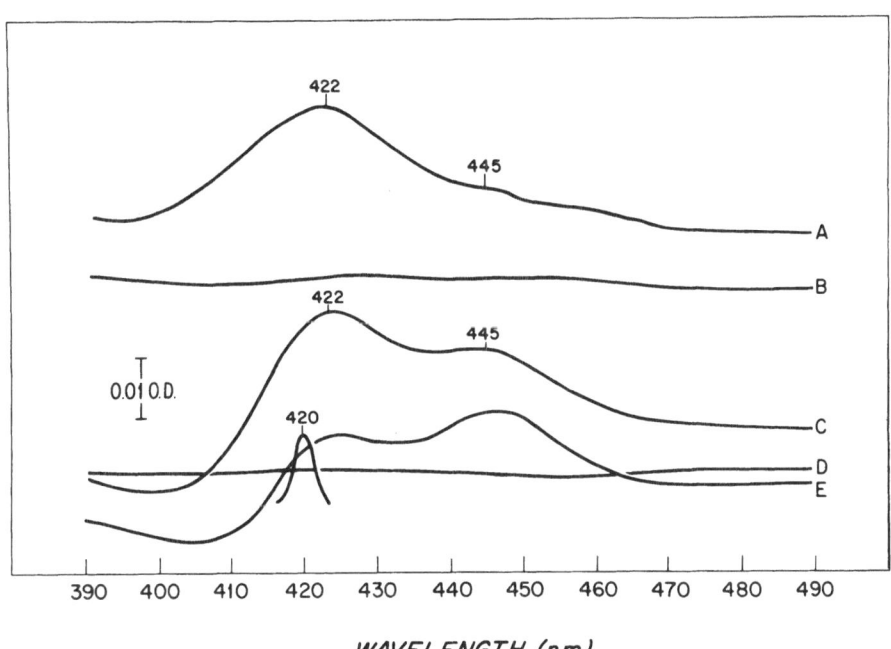

WAVELENGTH (nm)

Figure 6. Effect of cyanoketone on the difference spectra
of bovine corpus luteum mitochondrial cytochrome P-450 in oxidized
and reduced states. The mitochondrial suspension was diluted to
a concentration equivalent to 0.4 μM cytochrome P-450 in buffered
mannitol solution and after dividing equally between sample and
reference cuvettes an equal absorption base line was recorded
(Curve D). Cyanoketone in ethanol was added to the sample cuvette
and an equal volume of ethanol to the reference cuvette. Difference spectra were recorded with the following final concentrations
(μM) of cyanoketone: 50 (Curve E) and 250 (Curve C). A fresh
sample of mitochondrial suspension was incubated at 25° in presence
of 1 mM KCN and 10 mM succinate. After a 20 minute incubation
and recording of a base line (Curve B) cyanoketone was added to
the sample cuvette and the cyanoketone-reduced minus reduced difference spectrum recorded (Curve A). Final concentration of cyanoketone was 250 μM.

Curve E.). The band of reduced cytochrome a appears with cyanoketone in the reduced-oxidized difference spectrum as a consequence of small amounts of endogenous citric acid cycle substrates present in our freshly isolated mitochondria from corpus
luteum. When mitochondria are pre-incubated with succinate in
the presence of 1 mM KCN prior to the addition of the steroid
nitrile, only the cytochrome P-450-related band at around 422 nm
is visible (Figure 6, Curve A). The P-450-related band is a reverse Type I spectral change, but not a typical hemochromogen
band seen with certain nitrogenous ligands. Additional evidence
for the cytochrome oxidase-cyanoketone complex was obtained from
experiments with rabbit heart mitochondria which are devoid of
cytochrome P-450 (Figure 7). Mitochondria were prepared according
to the method of A. L. Smith (14). The reduced-CO difference
spectrum demonstrates the expected CO-cytochrome oxidase complex
(Figure 7, Curve B) with the γ band at around 430 nm and a slight
shoulder in the 440 nm region indicative of CO-cytochrome a_3(19).
Of greater interest is the similarity of dithionite-reduced minus
oxidized difference spectrum and the cyanoketone induced spectrum
(Figure 7, Curves D and C). The similarity of band positions indicates that cyanoketone prevents the re-oxidation of the respiratory chain enzymes from another tissue as well as from the ovary.
That cyanoketone binds to cytochrome a_3 very tightly is evident
from an experiment in which mitochondria were pre-incubated for
15 min with the steroid prior to reduction with dithionite. The
cyanoketone-reduced minus reduced spectrum shows a trough at
445 nm (Figure 7, Curve A). This spectral change indicates that
the steroid prevented cytochrome a_3 from being reduced.

Thus, the terminal oxidases of both electron transport chains
are inhibited by cyanoketone, but by very different mechanisms.
Cyanide at a concentration of 4 µM reduces cytochrome oxidase
activity by half (20) but cholesterol cleavage is not inhibited
at concentrations as high as 2 mM. Cyanoketone, however, demonstrates spectral changes with both enzyme systems at concentrations as low as 50 µM, and at 500 µM, blocks the corpus luteum
cholesterol cleavage activity almost entirely.

Cyanoketone has been used both in vitro and in vivo as an
"irreversible" inhibitor of steroidogenesis. The fact that other
steroid nitriles in our study had no similar effects on cytochrome
oxidase indicates a need for further studies in the pharmacology
of this compound.

The typical Type II spectral changes have been extensively
studied in microsomes from liver and adrenal cortex. In Figure 2,
two examples are shown of Type II spectral changes induced in
corpus luteum cytochrome P-450 by aminoglutethimide and 1-benzyl-
imidazole. These two compounds differ in that the activity of the

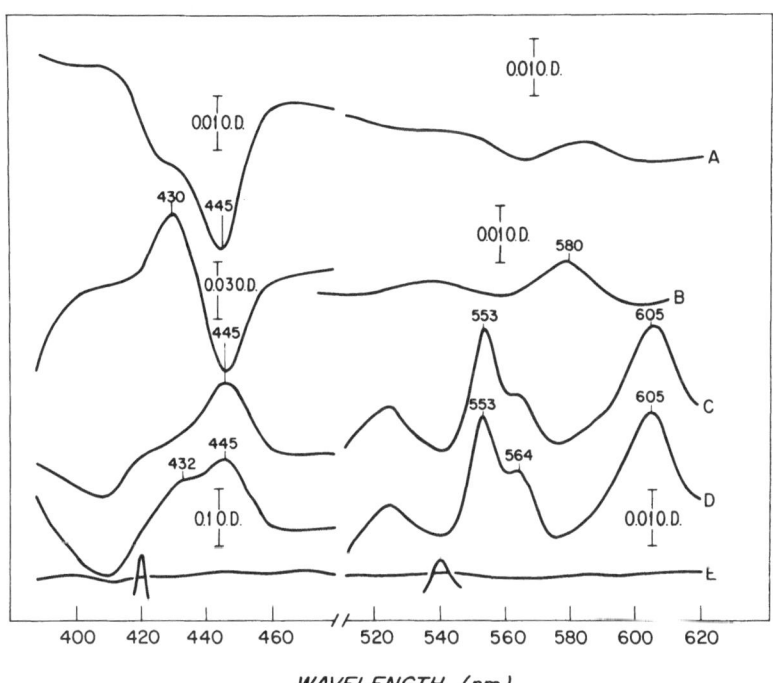

Figure 7. Effect of cyanoketone and CO on the difference spectra of cytochrome oxidase in rabbit heart mitochondria. The mitochondria suspension was diluted to contain 1.2 mg protein/ml of solution as in Figure 1, divided equally into two cuvettes, and the base line of equal absorbance recorded (Curve E). After adding 2 mg of $Na_2S_2O_4$ to the sample cuvette a reduced minus oxidized difference spectrum was recorded (Curve D). The cyanoketone induced difference spectrum was recorded (Curve C), after 250 µM was added to a fresh sample of heart mitochondria. Curve B represents the reduced-CO minus reduced difference spectrum of another sample of mitochondria after adding 2 mg of $Na_2S_2O_4$ to each cuvette and bubbling CO for 30 seconds through the contents of the sample cuvette. The cyanoketone effect on the reduced-CO minus reduced difference spectrum (Curve A) was determined by repeating the experimental conditions for Curve B except that cyanoketone was present in both cuvettes at 250 µM concentration before reduction with $Na_2S_2O_4$.

former is determined by the amino group in the para position of
the phenyl ring (since glutethimide, its analogue without amino
group in the phenyl ring is essentially devoid of effect), while
the latter is effective because of its imidazole group. The differences between these structural groups are not critical and
both compounds yield almost identical spectra presumably because
of binding at different sites from those associated with steroids.
However, they do not yield hemochromogen spectra with the reduced
mitochondria.

The relationship between spectral changes and the effects
on enzyme activity is complex for the reverse Type I ligands,
but not for the Type II compounds. Aminoglutethimide has been
used to block steroid synthesis in both the laboratory and in
human experiments. The mechanism of its action has been studied
with adrenal (6) and corpus luteum (7) cytochrome P-450 preparations. It has been concluded that in both instances aminoglutethimide decreases the reduction rate of cytochrome P-450
and displaces the substrate from the active site. The relationship between the Type II induced spectral change and the inhibition of mixed-function oxidation, derived from parallel experiments, is presented in Table I.

When the magnitude of the induced Type II spectral change was
compared with the inhibition of initial reaction rates of cholesterol monooxygenase at identical ligand concentrations, almost perfect agreement was obtained. The implication is that
Type II spectral change represents binding events directly related to the inhibition of enzyme function, by formation of an
equilibrium enzyme-inhibitor complex. This is contrary to observations on soluble adreno-cortical preparations (13,1) or drug
induced liver microsomes (10).

Several factors may account for this difference. Studies
on solubilized preparations use detergents which may modify
ligand effects. In addition, enzyme coupling in reconstituted
systems may not be as favorable as in the native membrane, resulting in a modified turnover number. Above all, adrenocortical and liver microsomal preparations, in most instances,
contain more than one species of cytochrome P-450, which would
interfere in the attempts to demonstrate a direct relationship
between spectral changes and enzymatic activity.

SUMMARY

The implications of ligand modified spectra of cytochrome
P-450 in mitochondria from corpus luteum are considered. Mito-
chondria from bovine corpus luteum contain a single cytochrome
P-450 which oxidizes cholesterol to pregnenolone and

TABLE I

Effects of Aminoglutethimide on Induced Spectral Change and Inhibition of Cholesterol Monooxygenase Activity

Aminoglutethimide, µM	Spectral Change[a] % of max.	Inhibition of Control[b] %
5	8.8	13
25	31	33
50	44	47
250	83	80
500	97	91
750	100	93

[a]   Spectral Change % of max. - per cent of the maximal spectral change was calculated by dividing absorbance difference between induced maximum at 426 nm and minimum at 393 nm at each amino- glutethimide concentration by the maximal spectral change ($90x\ 10^{-3}\ cm^{-1}$, $µM^{-1}$), obtained at 750 µM concentration and mul- tiplying by 100. Mitochondria from bovine corpus luteum were diluted to contain 0.3 nmoles cytochrome P-450/ml (0.9 mg protein/ ml) solution as in Figure 1. Difference spectra were recorded after adding aminoglutethimide in propylene glycol to the sample cuvette and solvent to the reference cuvette.

[b]   Inhibition of Control % - The control rate was 18 pmoles [26-14C] cholesterol cleaved per mg mitochondrial protein per min (see Methods and Materials). The effect of aminoglutethimide on the reaction is reported as the ratio of the aminoglutethimide modified rate and the control rate converted to per cent. Media and additions were identical to the spectroscopy conditions.

isocaproaldehyde. These mitochondria and the cytochrome P-450 purified from these mitochondria yield Type I spectral changes with substrates, reverse Type I spectral changes with certain steroid substrates, steroid products and unrelated steroid ligands. Nitrogenous ligands yield Type II spectral changes. Mitochondrial

and purified cytochrome P-450 preparations are equivalent in this
study. The inhibitory effects on the cholesterol monooxygenase
are directly related to the spectral changes induced by Type II
ligands. Lastly, it is suggested that a similar relationship exists with reverse Type I ligands.

FOOTNOTE

*   HEPES = N-2-hydroxyethylpiperazine-N'-2-ethane-sulfonic
acid.

ACKNOWLEDGEMENTS

    This work was supported by United States Public Health
Service Grant AM-10081 from the National Institute of Allergy
and Metabolic Diseases, National Institutes of Health, Bethesda,
Maryland, and Research Contract NIH-70-2319 from the National
Institute of Child Health and Human Development, National
Institutes of Health, Bethesda, Maryland.

REFERENCES

1   Burstein, S., Co., N., Gut, M., Schleyer, H., Cooper, D. Y.
         and Rosenthal, O. 1972. Substrate-induced difference
         spectra and cholesterol to pregnenolone conversion with
         adrenal heme protein P-450. Biochemistry 11: 573-577.
2   Harding, B. W., Whysner, J. A., Cheng, S. C. and Ramseyer,
         J. 1970. Substrate induced changes in adrenal corti-
         cal cytochrome P-450, pp. 294-301. Hormonal Steroids,
         Excerpta Medica Int. Congr. Series #219, Proc. IIIrd.
         Int. Congr., Hamburg, Sept. 1970.
3   Hochberg, R. B., van der Hoeven, T. A., Welch, M. and
         Lieberman, S. 1974. A simple and precise assay of the
         enzymatic conversion of cholesterol into pregnenolone.
         Biochemistry 13: 603-609.
4   Ichii, S., Omata, S., Kobayashi, S. 1967. Purification and
         some properties of cholesterol 20α-hydroxylase from hog
         adrenal mitochondria. Biochim. Biophys. Acta. 139:308-
         318.
5   Luttrell, B., Hochberg, R. B., Dikon, W. R., McDonald, P.D.
         and Lieberman, S. 1972. Studies on the biosynthetic con-
         version of cholesterol into pregnenolone. J. Biol.
         Chem. 247:1462-1472.
6   McIntosh, E. N., and Salhanick, H. A. 1969. The effect of
         steroid hydroxylase inhibition on the rate of reduction
         of adrenal mitochondrial cytochrome P-450. Biochem.
         Biophys. Res. Commun. 37:552-558.
7   McIntosh, E. N., Mitani, F., Uzgiris, V. I., Alonso, C. and
         Salhanick, H. A. 1973. Comparative studies on

mitochondrial and partially purified bovine corpus luteum cytochrome P-450. Ann. N.Y. Acad. Sci. 212:392-405.

8   Neville, A.M., and Engel, L.L. 1968.   Inhibition of steroid Δ-isomerase of the bovine adrenal gland by substrate analogues. Endocrinology 83:873-876.

9   Mitani, F. and Horie, S. 1969.   Studies on P-450.   V. On the substrate-induced spectral change of P-450 solubilized from bovine adrenocortical mitochondria.   J. Biochem. 65:269-280.

10 Orrenius, S., Wilson, B.J., von Bahr, C. and Schenkman, J.B. 1972. On the significance of drug-induced spectral changes in liver microsomes, pp. 55-77. In G.S. Boyd and R.M.S. Smellie (ed.). Biological Hydroxylation Mechanisms. Academic Press, N.Y.

11 Raygatt, P.R. and Whitehouse, M.W. 1966. Substrates and in- hibitor specificity of the cholesterol oxidase in bovine adrenal cortex. Biochem. J. 101, S19:819-830.

12 Schenkman, J.B., Cinti, D.L., Orrenius, S., Moldeus, P. and Kraschnitz, R. 1972.   The nature of the reverse Type I (modified Type II) spectral change in liver microsomes. Biochemistry 11:4243-4250.

13 Schleyer, H., Cooper, D.Y., Levin, S.S. and Rosenthal, O. 1972. Heme protein P-450 from the adrenal cortex: interaction with steroids and the hydroxylation reaction, pp. 187-206. In G.S. Boyd and R.M.S. Smellie (ed.). Biological Hydroxy- lation Mechanisms. Academic Press, N.Y.

14 Smith, A.L. 1967.   Preparation, properties and conditions for assay of mitochondria: slaughter-house material, small scale. Methods Enzymol. 10:81-86.

15 Sutherland, E.W., Cori, C.F., Haynes, R. and Olsen, N.S. 1949. Purification of the hyperglycemic-glycogenolytic factor from insulin and from gastric mucosa. J. Biol. Chem. 180: 825-837.

16 Szarkowska, L. and Klingenberg, M. 1963.   On the role of ubi- quinone in mitochondria. Biochem. Z. 338:674-697.

17 Tsai, R., Yu, C.A., Gunsalus, I.C., Peisach, J., Blumberg, W., Orme-Johnson, W.H. and Beinert, H. 1970.   Spin-state changes in cytochrome P-450$_{cam}$ on binding of specific sub- strates. Proc. Nat. Acad. Sci. 66:1157-1163.

18 Uzgiris, V.I., McIntosh, E.N., Alonso, C. and Salhanick, H.A. 1971. Role of reversed electron transport in bovine corpus luteum mitochondrial steroid synthesis. Biochemistry 10: 2916-2923.

19 Yonetani, T. 1960. Studies on cytochrome oxidase. I. Abso- lute and difference absorption spectra. J. Biol. Chem. 235:845-852.

20 Yonetani, T. and Ray, G.S. 1965. Studies on cytochrome oxi- dase. VI. Kinetics of the aerobic oxidation of ferrocyto- chrome c by cytochrome oxidase. J. Biol. Chem. 240:3392- 3398.

GENERAL DISCUSSION

Session II - Part 1

MANNERING: I would like to mention three observations which
may have some bearing on the interpretation of the inverse Type I
spectrum observed by Dr. Witmer: 1) of the five or six Type I
compounds we have employed, only hexobarbital gives an inverse
Type I spectrum with microsomes from 3-methylcholanthrene-treated
rats; 2) solubilized and purified (about 4-fold) cytochrome
P_1-450 (P-448) obtained by Dr. Fujita (Fujita, Shoeman and
Mannering, J. Biol. Chem., 248, 2192, 1973) in my laboratory gave
neither a Type I nor an inverse Type I spectrum with hexobarbital;
3) Dr. Lu and associates observed that purified cytochromes P-450
and P_1-450 give about the same amount of high spin signal and
that this represents only a small amount of the P-450 hemoprotein--
about 5%, if I remember correctly. This suggests that the high
spin signal observed with membrane bound P-450 hemoproteins may be
imparted by membrane components associated with the hemoproteins.
During their purification, the P-450 hemoproteins lost their
high spin characteristic because they are no longer associated
with the membrane components responsible for the high spin signal.
Membrane bound cytochrome P_1-450 conceivably has a higher affinity
than cytochrome P-450 for the membrane components that elicit the
high spin characteristic. If the membrane components responsible
for the high spin signal also give a Type I spectrum, one might
explain the hexobarbital induced inverse type spectrum as being
due to a displacement of the hypothetical Type I membrane component
by hexobarbital. The inverse Type I spectrum would be seen with
membrane bound cytochrome P_1-450, but not with membrane bound
cytochrome P-450, because the former is highly bound to Type I
membrane components, while the latter is not.

LU: The relative amount of high spin species in partially
purified rat cytochrome P-450 and P-448 preparations based on low
temperature EPR is approximately 2% and 7%, respectively (Stern
et al., Arch. Biochem. Biophys. 156, 404, 1973). Upon further
purification, both the rat cytochrome P-450 and P-448 preparations
contain less than 1% high-spin species. On the other hand, partially purified rabbit cytochrome P-448 isolated by Dr. J. Kawalek
in our laboratory is mostly a high spin species. In addition,
cytochrome P-448 preparations isolated from rat and rabbit have
different substrate specificities; i.e., rat P-448 is very good
in catalyzing benzpyrene hydroxylation, but rabbit P-448 is very
poor.

WITMER: In answer to Dr. Mannering's comments, I wouldn't
want to contest any different interpretation of our results right
now. We have done some other experiments in which we see the
shift to the low spin spectrum which corresponds to the formation

229

of a reverse Type I (R-I) spectrum. We have raised the pH, which causes the same shift and we may be removing something besides the hydrogen ion. If one must remove membrane to allow the R-I spectrum formation, and the presence of the membrane component is presumed to cause the Type I spectrum, then it appears that compounds which form both types of spectra must be attached to the same site, or compounds forming Type I spectra have an allosteric effect on R-I compounds. This has been postulated but does not really fit in with our hypothesis. We did isolate our P-448 just as Lu described the method and we do agree that there is a great difference in benzo(a)pyrene activity in the rabbit and rat. We have not measured spin content of any rat preparations, but everyone who has done this seems to agree that P-448 from the rat does not have much high spin content. Our rabbit P-448 preparations were consistently largely high spin. We ran at least 20 of these spectra. It would be hard to postulate how much membrane material was carried through with these isolated materials; the group at Tübingen had done some speculating before I joined them about the effects of phospholipids on the spin state. Perhaps Drs. Remmer or Snyder would like to comment.

SNYDER: The other point that Dr. Mannering made was of other compounds not binding to P-448 to give an R-I spectrum. We did look at some other compounds which behaved like hexobarbital - phenobarbital and benzene, for example.

MANNERING: Was that inverse Type I binding?

SNYDER: They all caused the shift from high to low spin with 3-methylcholanthrene microsomes, thus the spectra were R-I type after the additions.

MANNERING: All of the compounds?

SNYDER: Yes.

WITMER: These studies were carried out using absolute spectral techniques. We read the partially purified samples against buffer. We did most of the work with hexobarbital and phenobarbital, but as benzene metabolism and toxicity are our big interests, we also studied the effects of benzene extensively in regard to spectral changes.

LEVIN: May I make a comment? These are the same sort of discrepancies that started several years ago when comparing the rabbit and the rat with regard to the spin state of the P-448 hemeprotein. The two investigators have conflicting results, but one used rats and the other used rabbits. P-448 in the rat is a low-spin ferric hemeprotein whether measured by EPR (Archiv. Biochem. Biophys. 156,

404, 1973 and Fed. Proc. 33, 1387, 1974), n-octylamine binding
(Drug Metab. & Disposition 1, 602, 1973) or absolute spectra of
the partially purified cytochrome P-448 (Archiv. Biochem. Biophys.
153, 543, 1972). In the rabbit after 3-MC treatment, the P-448
hemeprotein appears to be mostly in the high-spin form when measured by these same criteria as Dr. Witmer has shown. This has
also been demonstrated using n-octylamine (Molec. Pharmacol. 6,
391, 1970) and EPR (Proc. Nat. Acad. Sci. 67, 172, 1970). Therefore, the differences seen in the reverse Type I spectrum between
the 3-MC treated rat and rabbit may be a reflection of these
differences. If the high-spin signal obtained using rabbit cytochrome P-448 is due to a substrate of some type bound to the
hemeprotein, then subsequent binding spectra may be altered by
what is already bound to the hemeprotein.

WITMER: Yes, however we can say that the high spin state
of P-448 in the 3-methylcholanthrene induced rabbit is not a result of the 3-MC being bound to the heme protein in a 1:1 ratio.

NEBERT: I agree with Dr. Levin that there are species differences in the spin state of P-450 iron between the rat and the
rabbit, when one compares the control with the 3-MC-treated animal:
the 1-2% high spin iron in the control rat is increased to 4-5%
by 3-MC treatment, and the more than 80% high spin iron in the
control rabbit is increased to more than 95% by 3-MC treatment
(Nebert and Kon, J. Biol. Chem. 248, 169-178, 1973). The 60% high
spin iron in the genetically "responsive" control mouse increases
to more than 80% after 3-MC treatment, whereas no change in the
amount of high spin iron occurs in the 3-MC-treated "nonresponsive"
mouse--although 3-MC is bound to P-450 and is metabolized at
similar rates as those in control mice (Nebert et al., J. Biol.
Chem. 248, 7637-7647, 1973). This finding indicates that config-
urational changes in the cytochromes P-450--influenced by the
membrane environment and/or as a result of required new RNA and
protein synthesis--may influence the spin state of the hemoprotein
iron. However, since fatty acids and steroids including cholesterol
are not only components of the membrane but also substrates for
these monooxygenases, our hypothesis becomes semantic. The increase in high spin iron associated with genetically mediated increases in P_1-450 might just as easily be explained on the basis
that an increased amount of any Type I substrate is now more
readily bound to the hemoprotein in aromatic hydrocarbon-treated
"responsive" animals than that in control or aromatic hydrocarbon-
treated "nonresponsive" mice. The aromatic hydrocarbon inducer such
as 3-MC, however, does not appear to be the Type I substrate causing
this effect (cf. refs. J. Biol. Chem. 248, 169-178 and 7637-7647,
1973) for further discussion).

HOLTZMAN: Another point that should be considered, which is
something that has bothered me a great deal, is that when you add

hexobarbital to any solution at 1.5-2.0 mM you find a marked increase in the pH of these solutions. Now you mentioned that you
used benzene and different pH's and observed effects. The problem
with this is that even using high concentrations of buffer, as
high as 300 mM phosphate, one can still alter the pH with the barbiturates.

WITMER: It is true that if hexobarbital is not sufficiently
buffered the pH is raised with high concentrations. We carried
out some experiments with Dr. Sies in Munich in which we perfused
whole livers and observed the spectral changes in the liver lobe
with his special equipment. In some experiments we found that the
pH had increased, so we have always been very careful to prevent
this change by adequate buffering of the solution. We do observe
Type R-I spectral changes when the pH does not change.

HOLTZMAN: In that case, how do you add the hexobarbital?

WITMER: We add it in a solution which has been highly
buffered.

HOLTZMAN: I ask because I have used 300 mM and still see
some effect.

WITMER: The amount of hexobarbital one can add is limited
because of its limited solubility in buffer. That's the whole
problem.

HOLTZMAN: Right. A second point I would like to raise is
the problem of using animals pretreated with phenobarbital. Recently, we observed that there actually is a substantial amount of
phenobarbital that remains bound to the microsomes and which can
be pulled off with albumin or solvents. These microsomes also
produce the so-called reverse Type I spectra. Other people have
postulated that there is some endogenous substrate. However, it
may well just be that the phenobarbital is binding in the region
of the heme, altering the pH of the heme to give some of these
odd effects which you observed here.

WITMER: In the isolated P-450s, the preparations have been
dialyzed several times so that any such components not tightly
bound should no longer be present. Covalently bound endogenous
substrates have not been sufficiently investigated.

HOLTZMAN: The affinity constant for phenobarbital is about
100 μM.

WITMER: Our figures show that it is in that region.

REMMER: I want to make clear that Drs. Witmer, Snyder and I had the unique opportunity, by working together with Dr. Sies in Munich, to show binding spectra in the intact liver. Dr. Sies has the apparatus to determine, directly, spectrophotometrically the spectral changes caused by perfusing the liver with hexobarbital and other substances. The changes are detected on the edge of the liver lobe and are very small but the results are beautiful. This shows that the Type I binding spectra are real and do occur in vivo following perfusion with hexobarbital. We also checked the red absorbance band - that is the 650 nm region. On addition of hexobarbital there was an increase in absorbance in the 650 nm region and when the hexobarbital was washed out with buffer there was a decrease in this 650 nm absorbance. The other problem which we wanted to study was what happens if we use 3-methylcholanthrene-pretreated rats. We also got a Type I spectrum with low concentrations of hexobarbital; however, we did not finish these experiments with higher concentrations of hexobarbital as we did not have adequate buffer capacity. I hope that the next time we see you we can tell you more about such studies.

VORE: I've taken adult Long-Evans rats; lyophilized the microsomes and extracted with butanol and in the extracted preparation hexobarbital no longer causes the reverse Type I spectra. Since I can prevent this by organic solvent extraction, it seems that maybe something has been removed from the microsomes by organic solvent extraction.

WITMER: We have also found that acetone extraction inhibits the formation of an R-I spectrum, so some fat soluble substance is required for this type of binding.

NEBERT: Actually, as we've pointed out before with mouse, rat, or rabbit liver microsomes (Nebert & Kon, J. Biol. Chem. 248, 169-178, 1973), the addition of only 10 µl of acetone to a 1-ml solution containing 30 to 50 mg of microsomal protein will completely destroy the high spin signal. I therefore am not surprised if Type I or Type II spectra are altered after solvent treatment such as acetone. Disruption of the membrane lipids may change the configuration of the P-450 active-site(s). Solvents also may preferentially displace certain substrates bound to P-450. Changes in spin state and room temperature difference spectra are therefore bound to occur. Of additional interest is that aryl hydrocarbon hydroxylase activity remains unchanged in these preparations in which all high spin P-450 iron has been converted to low spin iron by acetone in vitro (Nebert & Kon, J. Biol. Chem. 248, 169-178, 1973).

NARASIMHULU: I agree that disruption of membrane-lipids may alter cytochrome P-450 and that addition of acetone to aqueous suspension of microsomes may very well disrupt the membrane-lipids which are phospholipids. But treatment of lyophilized microsomes under essentially anhydrous conditions(essentially, because water which is more intimately associated with membranes may still be present in the lyophilized preparations) with acetone is entirely a different matter.

NEBERT: The addition of acetone to an aqueous or mostly anhydrous system should make little difference. Changes in spin state or in room temperature difference spectra are going to occur.

NARASIMHULU: I think it is well known that phospholipids are essentially insoluble in dry acetone. Therefore, it is highly unlikely that one can disrupt membrane-lipids with acetone under essentially anhydrous conditions. However, aqueous acetone can very efficiently dissolve phospholipids and extract them from membranes. Therefore, one would expect that it will make a world of difference if acetone is added to the microsomes in the presence of sufficient amount of water, especially in regard to those enzymes whose integrity may be dependent upon their association with phospholipids.

I should like to ask Dr. Witmer a question. Have you added cholate to the microsomes to see if any spectral changes occur?

WITMER: No, we didn't do that, but when we add substrates to microsomes to study these spectral effects we get the same effects that we do with the purified preparations, so the effects are not caused by cholate. I haven't actually tried the addition of cholate alone to the microsomes to compare the spectral results.

NEBERT: With mouse liver microsomes in vitro, we found that 1 mg of sodium deoxycholate per mg of protein decreases the high spin signal about 25-50% (Nebert & Kon, J. Biol. Chem. 248, 169-178, 1973) and that 1.5 mg of sodium cholate per mg of microsomal protein decreases the high spin signal about 20%. However, we did not look at rat liver, and the response--of what little high spin there is to start with--to detergents may differ from the response in mouse liver microsomes.

STERN: With regards to that paper you were mentioning, the Lu preparation normally uses 1 mg cholate/mg protein. I vividly remember that table in your latest JBC paper and you had four concentrations of cholate. You measured high-spin cytochrome P-450 by the electron paramagnetic resonance at $g = 8$, with zero cholate added. You then went up to 1.5 mg cholate/mg protein and there was no decrease in the $g = 8$ absorption, with 4 mg cholate/mg protein you saw a small decrease and at a final concentration of 8 mg cholate/mg protein the $g = 8$,signal was lost. So with the concentration of cholate used in the Lu preparation, by your own data, one wouldn't expect to see any loss of the high-spin form of cytochrome P-450.

HOLTZMAN: Isn't there cholate in the buffer?

STERN: It's all less than 0.1 mg per mg protein.

WITMER: If you dialyze it against buffer containing cholate, the final cholate concentration is known.

KAMIN: I'd like to change the subject and ask Dr. Stern: I'm still puzzled by the alternate possibilities of interpreting your data on the increase in P-450 as against the increase in benz-pyrene hydroxylase activity in the ethionine experiments. It seems to me that, formally, the most simple explanation is that in the benzpyrene hydroxylase assay, the cytochrome P-450 concentration may not be rate limiting. Protein synthesis may represent the synthesis of some other factor, non-heme, which then increases the activity, disproportional to the increase in heme. This, of course, relates to the question of whether the P-450 has become itself more active, or whether some other component has been synthesized which makes the total apparent enzyme activity higher. A critical question here is: what is the rate limiting factor in the benzpyrene hydroxylase assay and how does this relate to your data? I'd appreciate some expansion on that.

STERN: As far as the rate limiting step in the benz-pyrene hydroxylation, I think perhaps Dr. Levin could help me out. I'm not really that familiar with that.

LEVIN: I wish that I knew.

PEISACH: It's pretty important.

KAMIN: I agree.

LEVIN: Perhaps I can give a possible explanation for the large increase in BP hydroxylase after 3-MC treatment even though the total hemeprotein content only increases 50-100%. Dr. Lu and associates (J. Biol. Chem. 247, 1727, 1972; J. Biol. Chem. 248, 456, 1973) have shown that the 3-MC induced hemeprotein (cyto-chrome P-448) is much more active in the metabolism of BP than the hemeprotein found in control or PB treated animals on a per nmole basis. In the presence of a fixed amount of reductase and lipid and various amounts of hemeprotein obtained from either control, PB or 3-MC treated animals, the BP hydroxylase activity of the system containing cytochrome P-448 was approximately 5 fold greater per nmole of hemeprotein than when control or PB hemeprotein was sub-stituted in the system. That is, the cytochrome P-448 per se was

much more efficient in hydroxylating BP. Thus, a qualitative
change in the hemeprotein after 3-MC treatment had occurred.
Therefore, although the total hemeprotein in microsomes only increases 50-100% after 3-MC treatment, this hemeprotein metabolizes
BP at a 5 fold increased rate per nmole than the control hemeprotein.

KAMIN: Does this, then, have the identical spectrum of the
448-CO heme? Is it the same heme, or how close or how different?

LEVIN: Yes, the CO-spectrum of the isolated hemeprotein
from 3-MC treated rats is still at 447-448 nm. The ethylisocyanide
spectra of the isolated hemeprotein is also similar to that observed
using microsomes. Spectrally it is very similar to what is present
in microsomes.

STERN: May I add something, Dr. Levin. Also by EPR studies,
looking at the 3-MC microsomes and then going to the next step, we
are up to Step 4, now, as far as heme is concerned, the ligands to
the heme at least seem to be identical in the Step 3 and 4, partially purified preparations, as compared to the microsomes which
they were derived from.

KAMIN: Are we yet approaching the state where we could say
whether there is a heme-bearing peptide and a second sub-unit which
may be different? Are we anywhere near doing physical studies so
that we can say we have a new protein?

LEVIN: Alvares and coworkers (Biochem. Biophys. Res. Comm. 29,
521, 1967 and J. Pharmacol. Expt. Ther. 163, 417, 1968) have shown
that protein and RNA synthesis are both required for the induction
of cytochrome P-448.

KAMIN: I would like to know if you are doing work on gels to
see whether these two materials band differently. Do you get a
common band plus a different band on SDS?

NEBERT: The published studies of Drs. Alvares and Siekevitz
(Biochem. Biophys. Res. Comm. 54, 923-929, 1973) and Welton and
Aust (Biochem. Biophys. Res. Comm. 56, 898-906, 1974)--and the
elegant data presented this morning by Dr. Lu and by Dr. Coon--
all show that several cytochromes of the "P-450-type" are in fact
distinguishable by differing molecular weights when microsomes or
microsomal subfractions are electrophoresed.

SCHENKMAN: Just a point I'd like to make concerning Dr.
Kamin's question. In agreement with some of the statements of
Dr. Levin, there are differences in the hemoprotein induced by
3-methylcholanthrene. For example, Dr. Nebert knows very well
from his own studies, as well as from my studies, that with the

methylcholanthrene-induced hemoprotein, one <u>can</u> see a Type I binding
spectrum with benzpyrene. This, one does not see with the regular
cytochrome P-450. So, in addition to a higher turnover, the substrate specificity may be a very important contribution.

HILDEBRANDT: With regard to the lack of correlation between
benzpyrene hydroxylation and cytochrome P-450 spin state or
activity, I still have some doubts concerning the suitability of
benzpyrene hydroxylase activity as an indicator for cytochrome
P-450 content as well as activity. I wonder if anybody knows about
the demonstration of a photochemical action spectrum for the carbon monoxide inhibition of benzpyrene hydroxylation.

NEBERT: For the past 2 years, Dr. Cooper and I have been
talking about doing just that experiment.

COON: I believe we need to admit what we don't know. The
discussion this morning was concerned with attempts to correlate
carcinogenicity with P-450 levels as judged by CO difference
spectra and people working in that area assume that those of us
doing enzymatic assays know precisely what we are doing. In all
honesty, no one yet knows what the rate limiting factor is in
hydroxylation. We can, by solubilizing and reconstituting, make
a particular component rate limiting, but when we calculate turnover numbers, that is, moles of some compound hydroxylated per
mole of P-450, we are not yet taking into account the other electron acceptors. We do not know whether Factor C is rate limiting
in microsomes from certain tissues or even in liver microsomal
P-448 preparations.

MANNERING: A further response to Dr. Kamin's questions.
The hemes of cytochromes P-450 and P_1-450 are identical, the
hemoproteins are not. Maines and Anders (Arch. Biochem. Biophys.
159, 201, 1973) isolated crystalline hemes from cytochrome P-450
and P_1-450 and showed them to be identical by identifying their
fragments using coupled gas liquid chromatography and mass
spectroscopy. Dr. Shoeman (Shoeman, Vane and Mannering, Mol.
Pharmacol. 9, 372, 1973), working in my laboratory, converted
cytochrome P-450 and P_1-450 to soluble cytochromes P-420 and
P_1-420, respectively, and showed that they migrated at different
rates on acrylamide gel.

STUDIES ON THE INTERACTION OF WATER WITH MICROSOMAL CYTOCHROME P-450

Jordan L. Holtzman

Clinical Pharmacology Section, Veterans Administration
Hospital, Minneapolis, Minnesota 55417; and
Departments of Pharmacology and Medicine
University of Minnesota
Minneapolis, Minnesota 55455

There are many possible approaches to the study of such complex enzyme systems as the hepatic microsomal mixed-function oxidases. Among those which have been actively investigated have been
the development of model systems such as Fenton's reagent, the
purification of bacterial systems which are readily solubilized
like the camphor demethylase system of P. putida, and the more difficult detergent solubilization of the microsomes followed by
partial purification by standard techniques. Although these have
proven to be fruitful routes of investigation, since each system
appears to have its own idiosyncrasies, it is still important to
verify the observed biochemical phenomena in intact microsomes.
Over the past several years I, and a number of collaborators, have
used a variety of techniques to investigate the kinetic and
physico-chemical properties of the microsomal membranes in order
to better understand the control mechanisms and chemical processes
involved in the mixed-function oxidase reactions. I wish to summarize some of these studies which indicate that, although the
microsomal membranes are lipoprotein complexes, a portion of the
oxidases are in an aqueous medium and that this aqueous shell can
have profound effects on the mixed-function oxidase and associated
activities.

It is generally assumed that the microsomal mixed-function
oxidases are buried in the lipid membrane and that the activity
for a variety of substrates is related in part to their lipid solubility. Yet a variety of studies would indicate that at least a
portion of the components of some of the electron chains in the

239

microsomes and other organelles are actually in an aqueous phase.
Probably some of the earliest evidence for this was suggested by
the original solubilization and purification procedure of
Strittmatter for cytochrome b_5 and the reductases (Strittmatter
and Velick, 1956, 1957). In this procedure these components were
removed from the membrane by proteolytic digestion. Since such
proteolytic enzymes, as trypsin, are not highly lipid soluble, it
would suggest that these enzymes are attacking susceptible peptide
linkages in the aqueous phase. More recently Spatz and Strittmatter
(1971) have confirmed this concept utilizing the detergent preparations of Ito and Sato (1968) for cytochrome b_5. They have demonstrated that with this preparation the cytochrome has both hydrophobic and hydrophilic regions.

An alternative approach to the question of the importance of
an hydration shell around the microsomal membrane has been to study
the various biochemical phenomena in deuterated water and compare
the activities to those in protonated water. Such a substitution
can inhibit enzymatic activity by one of two mechanisms. The first
is by a primary isotope effect which occurs when a proton is directly involved in a rate limiting step. In such a case the difference in the masses of the proton and deuteron leads to a lower
zero point energy for the latter so that it takes more energy to put
the molecules into the state of the activated complex. By the
Arrhenius equation this difference in energy would represent a decrease in the rate of the reaction. Such an effect may be observed
in hydrolytic reactions. Similarly, this inhibition can be seen
when a deuterium-carbon bond is substituted for a hydrogen-carbon
bond and the rate limiting process in the reaction is the fission
of that bond. In either case, the rate for the protonated species
is two to five fold greater than that for the deuterated species.

A secondary isotope effect can occur when the deuterated
species binds to one of the reacting species, such as an enzyme,
and changes its reactivity without actually participating in the
reaction. Since deuterated water has a stronger hydrogen bond
than does protonated water, it may increase the hydration shell
around an enzyme. This can alter its tertiary structure and catalytic activity.

Tyler and Estabrook (1966) examined the effect of deuterated
water on mitochondrial activity. They found that it inhibited both
the basal oxygen uptake and the increase normally seen with the
addition of ADP. They concluded that this was a primary isotope
effect due to an inhibition of a "cryptic protonolysis" step in
the mitochondrial chain. Moreover, they stated that the inhibition
was primarily due to a reduced concentration of protonated water.
They confirmed this by showing a similar inhibition with such solvents as glycerol which reduced the concentration of water without
destroying the mitochondria.

When we examined the effect of the substitution of deuterated water for protonated water on the mixed-function oxidase activity, we found that it was a noncompetitive inhibitor of ethylmorphine N-demethylase (Holtzman and Carr, 1970, 1972). Since this reaction has no formal requirement for water, one could postulate some "cryptic protonolysis" step; yet, with a 20-50% inhibition, the reduction in activities is more in line with a secondary isotope effect. Further, the inhibition was quite variable within this range, suggesting a more indirect mechanism. Another alternative is that the inhibition could be due to changes in the pKa of the substrate leading to changes in the reactive species. That is, if the reactive species were the unprotonated form then a decrease in the pKa would lead to a lower concentration of this species. This could be particularly important with morphine and its congeners with have pKa's of 8.0 - 8.1 so that both the protonated and unprotonated forms would be markedly altered by small changes in the pKa when the incubations are run at pH 7.4 - 7.6. If the pKa of ethylmorphine were different in deuterated water with a change in the concentration of the active species, this should have led to a change in the Km for the ethylmorphine, but none was observed (Holtzman and Carr, 1972).

On the other hand, when we examined some of the effects of deuterated water on the known intermediate reactions, we found some interesting results. These effects are quite nicely illustrated by the NADPH-cytochrome P-450 reductase (Table I). Clearly, deuterated water has little or no effect on the basal reductase, that is the reductase activity in the absence of ethylmorphine, but significantly inhibits the stimulation which occurs with the addition of ethylmorphine. Further, in line with the results of Gigon et al. (1968, 1969), the stimulation of the reductase in the presence of protonated or deuterated water had a 1:1 stoichiometry with the formation of formaldehyde in the same buffer. That is, deuterated water inhibited the N-demethylase and the stimulation of the reductase to the same extent, confirming the importance of this stimulation in controlling the state of N-demethylation.   We concluded that the stimulation of the reductase probably resulted from some change in the hydration of the microsomal protein and that the deuterated water inhibited this change.

Interestingly, aniline hydroxylation was similarly inhibited by deuterated water, yet aniline does not stimulate the reductase, but actually inhibits it. Hence, it would seem that the inhibition of aniline hydroxylase could not go through the same mechanism as that for the ethylmorphine N-demethylase, unless of course aniline does stimulate the reductase but this stimulation is masked by the binding of aniline to the heme of the cytochrome P-450. The net effect on the reductase would then depend on the relative effects

TABLE I

EFFECT OF DEUTERATED WATER ON NADPH-CYTOCHROME P-450 REDUCTASE AND
ETHYLMORPHINE N-DEMETHYLASE ACTIVITIES
OF MALE RAT HEPATIC MICROSOMES

Assay	Buffer	2mM ethylmorphine	Rate	Δ[a]
			nmoles/min/mg protein	
NADPH-cytochrome P-450 reductase	H_2O	–	21.6 ± 0.3[b]	
	H_2O	+	33.5 ± 0.8	11.9
	D_2O	–	19.9 ± 0.2	
	D_2O	+	27.5 ± 0.5	7.6
Ethylmorphine N-demethylase	H_2O	+	11.0 ± 0.10	
	D_2O	+	7.5 ± 0.05	

[a]Δ is the rate of NADPH-cytochrome P-450 reductase in the presence minus that in the

absence of 2 mM ethylmorphine

[b]Averages $\pm$ standard errors of triplicate incubations

of aniline in stimulating the reductase and in inhibiting it by
removal of the substrate of the reduction, i.e., cytochrome P-450.

Not all of the microsomal mixed-function oxidases are inhibited by deuterated water. Björkhem (1972) has found that the
hydroxylation of some positions of the steroid nucleus are markedly
affected, while others are not. Further, those positions which
are affected show marked inhibition by CO but no primary isotope
effect if deuterium is substituted for the hydrogen at that
position. On the other hand, the hydroxylation of those positions
which are not affected by deuterated water are similarly only
slightly inhibited by CO, but show a significant primary isotope
effect. These studies would suggest that there are two broad
classes of hydroxylases, one of which is on the surface of the
membrane and the other buried in the membrane.

More recently I have been examining the effect of deuterated
water on the binding spectra of various substrates to microsomes.
We found that the Type I spectrum of ethylmorphine was essentially
unaffected by deuterated water, but on the other hand deuterated

water significantly enhanced the apparent absorptivity of the
Type II binding spectrum of aniline without altering either the
λ min., λ max., or the isosbestic point (Figure 1). There are
three points of some interest. The first is that since the binding difference spectra are due to shifts in the absolute spectrum
to the blue for the Type I and to the red for the Type II, the
fact that the various peaks and troughs remain at the same wavelengths suggests that the change in absorptivity is due to an increase in absorptivity of heme when aniline is bound without any
effect on the shape in the absolute spectrum of the Soret peak.
Secondly, the bound water which is enhancing the spectrum cannot
be bound to the sixth ligand of the iron since that is where the
aniline is bound. This would, therefore, suggest that water is
bound near but not directly to the heme. Thirdly, this may lend
a different interpretation to the inhibition of the mixed-function
oxidases by the deuterated water. It may well be that the stimulated reduction of the heme may require changes in the hydration
shell about the heme itself before or during reduction. These
changes may be more difficult in deuterated water than in protonated water and, hence, lead to the inhibition of the overall
activity.

Before these data can be totally accepted it is necessary to
demonstrate that the changes are not simply due to changes in the
pKa of the aniline. Since it is the unprotonated form which is
presumed to bind to the heme, small changes in the pKa could
theoretically have big effects on this form. Actually this is
not the case since in both protonated and deuterated water the
pKa is 4.58. The value for the pKa in deuterated water may be
0.4 pH units off due to an isotope effect on the voltage of the
glass electrode(Glascoe and Long, 1960); but at a pH or pD of
7.4, the per cent of unprotonated aniline in water is 99.9, so
that small shifts would have little effect.

Secondly, if we look at a double reciprocal plot of the
binding, we find again that the increase in absorbance is completely noncompetitive (Figure 2). If there was a significant
change in the actual binding species, that is if it were really
the protonated form which bound so that miniscule changes in
the pKa would markedly alter the free concentration of the binding
species, then there should be a significant change in the Ks.
Since no such change occurred, I believe that we are secure in
assuming that the change in the Type II binding is due to changes
in the hemoprotein and not in the aniline.

This titration is also of interest because it illustrates
a phenomenon which I have consistently observed, but which I
have not seen previously reported; that is, there appear to be
two Type II binding sites for aniline. Both are similarly

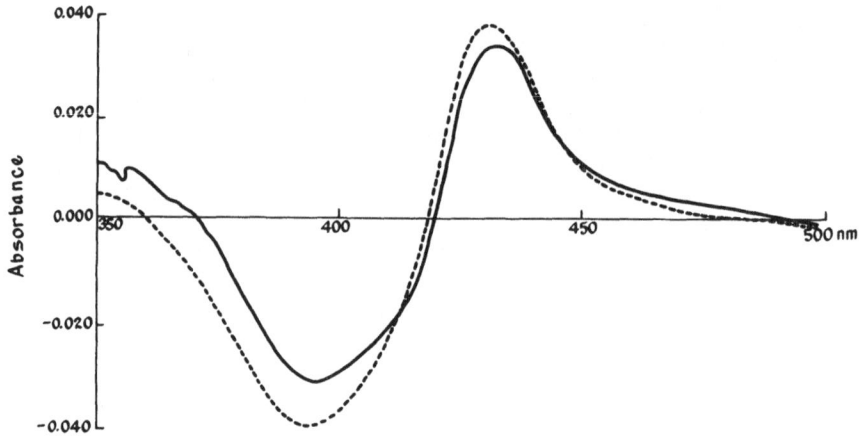

Fig. 1. The difference binding spectrum of aniline to hepatic microsomes from a male rat in protonated water (———) and deuterated water (- - -).

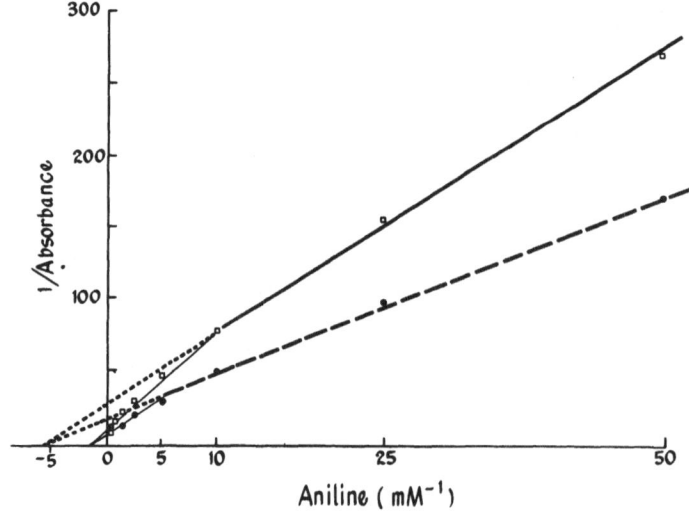

Fig. 2. Titration of the binding spectrum of aniline to hepatic microsomes (Δ430-390) from male rats in protonated (———) and deuterated water (- - -).

affected by deuterated water, making it unlikely that they could represent the two different types of cytochrome P-450 suggested by Björkem's study (1972). What is even more puzzling is that the low affinity site is no longer evident when the spectrum is obtained at pH 7.0. It should be noted that this second site is probably not cytochrome b_5 since it has six ligands and cannot bind other ligands under these conditions, although it is possible to displace one of these ligands but only at very low pH's (Ikeda et al., 1974).

When I examined the effect of deuterated water on the apparent absorptivity of the CO and ethylisocyanide binding spectra, I found no changes (Table II). Since both these spectra are obtained with the reduced species, it would indicate, as suggested above, that reduction of the heme leads to a decrease in the hydration shell. That is, there is no longer water in a position near the heme to enhance the spectrum of ligands bound to the reduced form of the heme.

In contrast to Tyler and Estabrook's suggestion (1966) that the effect of deuterated water is due solely to a decrease in the concentration of protonated water, other agents which decrease the water concentration have an effect on the Type II binding spectrum opposite to that of deuterated water (Table III). Clearly, increasing the concentration of glycerol, methanol, or dimethylsulfoxide, only decreases the intensity of the spectrum. This would support the concept that enhancement of this spectrum seen with deuterated water is due to an increase rather than a decrease in the hydration shell of the hemoprotein.

TABLE II

Effect of D_2O on the Absorbance of Cytochromes P-450 and b_5

	Absorbance In	
	H_2O	D_2O
Cytochrome P-450	43.9	43.7
CO complex	±0.4	±1.4
Cytochrome b_5	61.7	60.0
	±0.7	±0.0
Equivalence pH for ethylisocyanide complex	7.6	7.6

TABLE III

Effect of Organic Solvents on the Type II Binding Spectrum of

Aniline[a]

Solvent	%[*] (v/v)	$\Delta A_{430-387}$
Glycerol	0	0.0340
	1	0.0317
	5	0.0298
	10	0.0230
	20	0.0251
Methanol	0	0.0344
	0.3	0.0314
	1.0	0.0290
	2.5	0.0260
	5.0	0.0200
Dimethylsulfoxide	0	0.0340
	1	0.0273
	5	0.0215
	10	0.0145
	20	0.0073

[a] Aniline concentration in sample cuvette was 0.28 mM

[*] The numbers indicate the percentage of the organic solvent (v/v) in a mixture of buffer and the solvent.

A final line of evidence which suggests that there is a significant hydration shell about the hemoproteins is that there are very definite pH effects on this system. I wish to mention but two. The first is that lowering pH has marked effects on the stimulation of the NADPH-oxidase by ethylmorphine but only a slight effect on the basal oxidase (Figure 3). The difference between the stimulated oxidase and the basal oxidase clearly represents the electrons going to the N-demethylation of the ethylmorphine since it has a 1:1 stoichiometry to the formation of formaldehyde, and is similarly inhibited by deuterated water and CO (Holtzman and Carr, 1970, 1972). The significance of the basal oxidase is unclear. It could represent simply the auto-oxidation of the reduced flavoproteins. On the other hand, it is inhibited by CO to the same extent as are

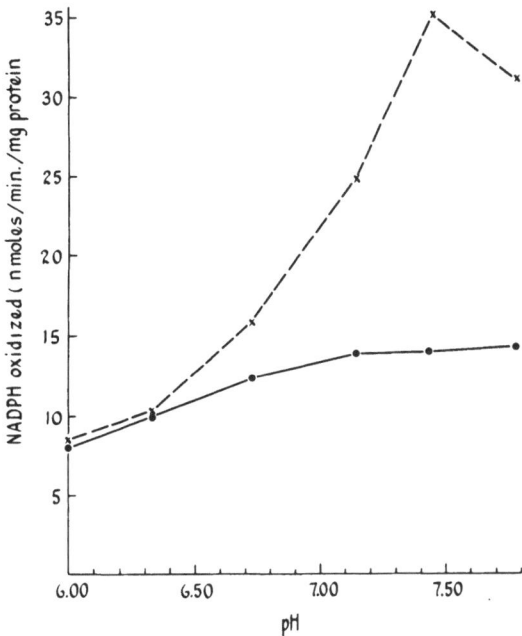

Fig. 3. The effect of pH on the NADPH oxidase of hepatic microsomes from male rats. Activity was determined kinetically from the Δ340-390 in an Aminco DW-2 spectrophotometer without (———) and with ethylmorphine (2 mM) (- - -).

the "CO-insensitive" hydroxylations described by Björkem (1972). The basal NADPH-oxidase may, therefore, represent deeply buried cytochrome P-450's. Supporting this concept is the fact that it is very insensitive to low pH's. This basal oxidase may be important for other hydroxylases in the microsomes, such as the O-deethylase of ethylmorphine. In our hands this enzyme shows about 20% of the activity of the N-demethylase in male rats (Thompson and Holtzman, 1974). The electrons for this activity must come from somewhere. Since both N-demethylation and O-deethylation are going on simultaneously, the only place they could come through is the basal NADPH oxidase activity.

The final activity which I wish to mention is the effect of pH on the Type I binding of ethylmorphine. We have observed a decrease in the ΔA_{max} with a lowering of the pH from 7.6 to 7.2, but this inhibition is associated with a concomitant decrease in the Ks for the ethylmorphine (Table IV). This decrease could occur for either of two reasons. On the one hand, the binding site may be sensitive to pH, while on the other, the protonated form may be the actual binding species. Since the pKa for ethylmorphine is about 8.1, at the Ks the concentration of the

TABLE IV

Effect of pH on Type I Binding of Ethylmorphine

pH	Ks	ΔA_{Max}
7.20	0.0286	0.0087
7.60	0.0426	0.0127

protonated form is 33 μM at pH 7.6 and 26 μM at pH 7.2. These values may well be within the usual experimental error. The decrease in ΔA_{max} is again probably related to the changes in the hydration shell of the hemoprotein.

In conclusion, these data would suggest that there are two categories of cytochrome P-450, one which is on the surface of the microsomal membrane, has a hydration shell near the heme, is sensitive to deuterated water and CO, and is known to be stimulated by the presence of substrates. The other is buried in the membrane and lacks a hydration shell as well as sensitivity to deuterated water and CO. The stimulation of the reductase associated with the first type of hydroxylase appears to involve some alteration in the hydration shell about the hemoprotein with a loss of part of the water shell.

REFERENCES

Björkem, J. 1972. On the rate-limiting step in microsomal hydroxylation of steroids. Eur. J. Biochem. 27:354-363.
Gigon, P. L., Gram, T. E. and Gillette, J. R. 1968. Effect of drug substrates on the reduction of hepatic microsomal cytochrome P-450 by NADPH. Biochem. Biophys. Res. Commun. 31:558-562.
Gigon, P. L., Gram, T. E., and Gillette, J. R. 1969. Studies on the rate of reduction of hepatic microsomal cytochrome P-450 by reduced nicotinamide adenine dinucleotide phosphate: Effect of drug substrates. Molec. Pharmacol. 5:109-122.
Glascoe, P. K. and Long, F. A. 1960. Use of glass electrodes to measure acidities in deuterium oxide. J. Phys. Chem. 64:188-190.
Holtzman, J. L. and Carr, M. L. 1970. Inhibition of hepatic microsomal mixed function oxidases by D_2O. Life Sciences 9:1033-1038.
Holtzman, J. L. and Carr, M. L. 1972. Inhibition by deuterated water of the mixed-function oxidases of hepatic microsomes of the male rat. Molec. Pharmacol. 8:481-489.
Ikeda, M., Iizuka, T., Takao, H. and Hagihara, B. 1974. Studies on the heme environment of oxidized cytochrome b_5. Biochim.

Biophys. Acta. 336: 15-24.

Strittmatter, P. and Velick, S. F. 1956. The isolation and prop-
     erties of microsomal cytochrome. J. Biol. Chem. 221: 253-264.

Strittmatter, P. and Velick, S. F. 1957. Purification and prop-
     erties of microsomal cytochrome reductase. J. Biol. Chem.
     228: 785-799.

Thompson, J. A. and Holtzman, J. L. 1974. Kinetics of N- and O-
     dealkylation of ethylmorphine by hepatic microsomes from
     male rats. Pharmacologist 16: 407.

Tyler, D. D. and Estabrook, R. W. 1966. The influence of deuterium
     oxide and organic solvents on the interaction of respiratory
     chain components. J. Biol. Chem. 241: 1672-1680.

DRUG METABOLISM IN ISOLATED RAT LIVER CELLS

Robert Grundin, Peter Moldéus, Helena Vadi and Sten
Orrenius, Department of Forensic Medicine, Karolinska
Institutet and Christer von Bahr, Department of Medicine,
Huddinge University Hospital and Dan Bäckström and
Anders Ehrenberg, Department of Biophysics, University
of Stockholm, Sweden

Although considerable knowledge has been gathered on the functional aspects of microsomal monooxygenation, comparatively little
has so far been known about the intracellular regulation of this
process. For such studies, we have found the isolated rat liver
cell system to be a very useful model, combining the convenience
of an in vitro system with the access to the complex mechanisms of
the intact in vivo system. This model has the advantage over the
perfused liver that it readily lends itself to the study of rapid
reaction sequences and makes quantitation of short-term drug metabolic reactions easier. It is also superior to liver slices which
often show considerable leakage of adenine and pyridine nucleotides and where substrate penetration and oxygen diffusion may
present problems depending on the relative thickness of the slice.

Although isolated liver cells have been extensively used in
the investigation of intermediary metabolism (cf. Krebs et al.,1974),
only few studies have been concerned with cytochrome P-450 linked
processes (Henderson & Dewaide, 1969; Holtzman et al., 1972).
Moreover, the need for the addition of exogenous NADPH to support
the reaction studied (Henderson & Dewaide, 1969) and difficulties
in quantitation of the reaction products formed (Holtzman et al.,
1972) make the results of these studies questionable. This paper
summarizes studies on drug metabolism performed in our laboratories
during the last year using an isolated liver cell system with retained viability criteria which readily catalyzes cytochrome P-450
linked drug hydroxylation in the absence of any added cofactors.

Isolation procedure. Isolated rat liver cells were prepared
by a method modified after Quistorff et al., (1973) and Seglen

(1972). Portal vein cannulation and total hepatectomy were performed under ether anaesthesia. Perfusion on the isolated liver was then carried out using a perfusion apparatus slightly modified after Hems (Hems et al., 1966), following the scheme given in Table I. All media contained 2% bovine serum albumin (cf. Moldéus et al., 1974 and von Bahr et al., 1974).

The viability of the isolated cell preparation was estimated by measurement of basal oxygen uptake and gluconeogenesis from lactate. Intactness of the cell membrane was established by the following criteria: trypan blue exclusion frequency more than 90-95%, absence of stimulation of oxygen uptake upon addition of ADP and no or minimal stimulation of oxygen uptake upon addition of succinate (Moldéus et al., 1974).

State of cytochrome P-450. Differential absorption spectra could be obtained in the isolated liver cell suspension using essentially the same technique as in isolated liver microsomes. Introduction of carbon monoxide into the sample cuvette - both cuvettes containing a suspension of liver cells isolated from control rats - produced a rapid increase in light absorption with a maximum at about 453 nm (Figure 1A) (Moldéus et al., 1973). Subsequent addition of alprenolol[1-(2-allyl-phenoxy)-3-isopropylaminopropanol] (Figure 1A) or hexobarbital (loc. cit.) caused a further increase in absorption with a slight shift in peak position to 450 nm. Addition of sodium dithionite, finally, produced a maximal light absorption in this region. When the same experiment was repeated with liver cells isolated from phenobarbital treated rats (Figure 1B) similar absorption spectra were obtained, however the maximal absorption peaks (after dithionite addition) were of much greater amplitude, showing that the cellular concentration of cytochrome P-450 had been markedly increased by this treatment. Only a minor portion of the cytochrome P-450 present was endogenously reducible in the absence of added substrate and this fraction might be used as an indirect estimate of the amount of cytochrome P-450 already bound to endogenous substrates, since it is most probable that only the substrate-bound fraction of cytochrome P-450 can accept endogenous reducing equivalents and thus interact with carbon monoxide. This would in turn imply that the major fraction of cytochrome P-450 in the isolated liver cell is present in the oxidized, non-substrate-bound state.

Substrate binding. Binding of substrate to oxidized cytochrome P-450 is considered to be an initial step in the monooxygenation process and such substrate interaction often produces a so-called Type I spectral change (λ max at about 385 nm and λ min at about 420 nm) (Schenkman et al., 1967). When alprenolol, hexobarbital or lidocaine was added to a suspension of isolated liver cells, a Type I spectral change could be recorded. The maximal magnitude of

TABLE I

Preparation of Isolated Rat Liver Cells.

1. Perfusion with Locke's medium + EGTA* and erythrocytes (3 min)

2. Perfusion with Hank's medium + Collagenase + Ca^{2+}+erythrocytes (5 min)

3. Disruption of liver

4. Incubation with Hank's medium + Collagenase without Ca^{2+} (5 min)

5. Filtration 100 mesh

6. Centrifugation. 80 x g for 2 min + suspension in Krebs-Henseleit's buffer (Repeated twice)

* Ethyleneglycol-bis (β-aminoethylester)-N,N´-tetraacetic acid.

the spectral change obtained with these drugs increased after pheno-barbital pretreatment of the rats in a similar way in isolated liver cells as in liver microsomes (Table II). In this respect, the induction phenomenon followed the same pattern in the two preparations. On the other hand, the maximal magnitude of the Type I spectral change increased about fivefold for hexobarbital and lidocaine (when measured per mg of microsomal protein or per 10^6 cells) but only about twofold for alprenolol. Thus, it seems that hexobarbital and lidocaine bind differently to cytochrome P-450 or bind to different forms of cytochrome P-450 as compared to alprenolol.

In isolated liver cells, a higher concentration of drug was needed per nmole of cytochrome P-450 than in microsomes to produce half maximal spectral change, probably since the drug was also partly bound to other intracellular fractions. Furthermore, when correlating simultaneous spectral titrations and studies on the concentration dependent cell/medium distribution of unchanged alprenolol in a nonmetabolizing cell system, we found that there was an inflexion in the intracellular/extracellular drug distribution curve at a concentration near to that giving a maximal Type I spectral change (Figure 2). This finding indicates that the "high affinity binding site" represented by cytochrome P-450 (Grundin et al., 1974) contributes to the threshold phenomenon in the cell/medium distribution of alprenolol. By the method of Ullrich (1969) using iodine to denature the cytochrome, we found that, in control rats, about 30% of cytochrome P-450 could form an enzyme-alprenolol complex. Since

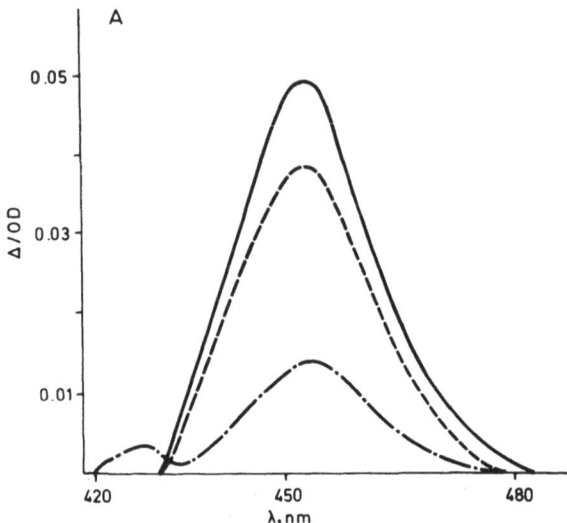

Figure 1A-B.   Carbon monoxide difference spectra of isolated
rat liver cells.   Each cuvette contained 3.6×10^6 cells per ml of
Krebs-Henseleit buffer, pH 7.5 (2% albumin).   CO was bubbled for 20
sec into the sample cuvette and the difference spectrum recorded
in an Aminco DW-2 spectrophotometer (-.-).   Alprenolol was then
added to the same cuvette to a final concentration of 0.3 mM (---).
Finally a few crystals of $Na_2S_2O_4$ were added to the sample cuvette
(——).   (A) represents cells from control rat and (B) cells from
phenobarbital treated rat (80 mg/kg body weight i.p. daily for 3
days).   Temperature was 37°C.

the concentration of cytochrome P-450 per gram of rat liver was
about 30 nmoles, this would give a value of about 10 nmoles per
gram liver for the capacity of the cytochromal "high affinity
binding site" for aprenolol.   Taken together, these findings suggest that binding of alprenolol to cytochrome P-450 may be involved
in the threshold phenomenon in the cell/medium distribution of
alprenolol, and consequently to some extent in its dose dependent
liver extraction in the perfused liver (Grundin et al., 1974).

Drug interaction with liver microsomes or partially purified
cytochrome P-450 is often associated with alterations of the electron paramagnetic resonance (EPR) characteristics of the cytochrome
(Cammer et al., 1966 and Peisach et al., 1973).   The typical EPR
spectra of cytochrome P-450 were also observed in isolated liver
cells (Figure 3).   The g_z = 2.42 and g_y = 2.25 signals of low spin
P-450 were as well characterized as in microsomes (Figure 3A),

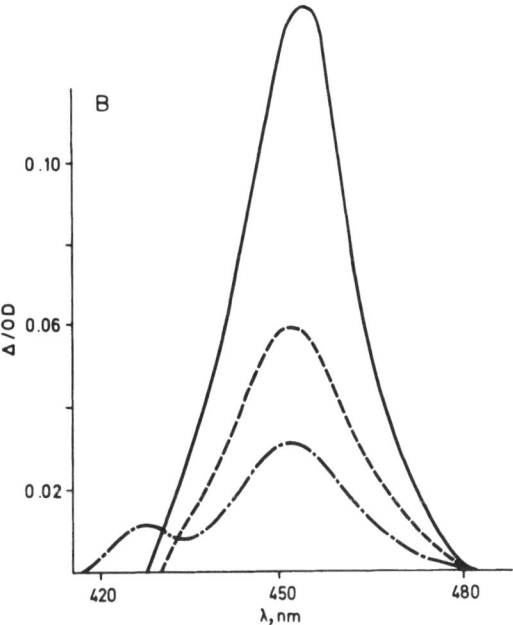

Figure 1B.

whereas g_x in the 1.9 regions was overlapped by mitochondrial iron sulfur signals. Upon addition of lidocaine or alprenolol to a cell suspension there was a decrease of the low spin signal and an appearance of a high spin signal at g = 7.9 (Figure 3B). For lidocaine these changes were markedly more pronounced. Furthermore, alprenolol induced a partial shift of the low spin signal, easily observed at g_z = 2.39.

Rate of drug uptake. The binding of substrate to cytochrome P-450 in microsomes is extremely rapid and occurs within milliseconds. The time course for uptake of drug into isolated liver cells was studied by recording the appearance of the Type I spectral change after drug addition, which was found to arise very rapidly and was fully developed within a few seconds (Figure 4). By relating concentration of intracellular alprenolol to rate of change in absorbance it could be estimated that the velocity for distribution of alprenolol to the cytochrome was more than 500 times faster than the rate of metabolism of alprenolol (see below). Thus cell entry of alprenolol seems not to be rate limiting for the overall liver disposition of this drug.

The same experimental technique was further used to study factors influencing drug uptake by the liver cell. The initial

TABLE II

Effect of Phenobarbital Treatment on the Maximal Magnitude of Drug Induced Type I Spectral Change

	10^3 x ΔA max (385-420)	
	Alprenolol	Lidocaine
Microsomes		
Control	3.3/mg protein	3.7/mg protein
PB	6.0/mg protein	17.1/mg protein
$\frac{PB}{Control}$	1.8	4.6
Cells		
Control	$1.2/10^6$ cells	$1.3/10^6$ cells
PB	$2.4/10^6$ cells	$6.0/10^6$ cells
$\frac{PB}{Control}$	2.0	4.6

PB = microsomes or cells isolated from phenobarbital treated animals (80 mg/kg i.p. daily for three days).

rate of appearance of the spectral change induced by addition of hexobarbital was dependent on drug concentration and showed no evidence of saturation in the concentration range studied. Further, the rate of uptake increased by a factor of about 1.3 upon a temperature increase of 10°C (Q_{10}), which is in accordance with what could be expected for passive diffusion. Finally, preincubation of the liver cells with rotenone to decrease the ATP level caused no marked decrease in the rate of formation of the spectral change (von Bahr et al., 1974). These results support the view that drugs enter the liver cell by a nonenergy-dependent, passive diffusion process. Although the existence of a nonenergy-requiring carrier involved in the uptake of hexobarbital cannot be completely excluded on the basis of the present data, the finding that no saturation of the rate of formation of the Type I spectral change seems to occur upon increasing hexobarbital concentration argues against the involvement of a carrier mediated transport process.

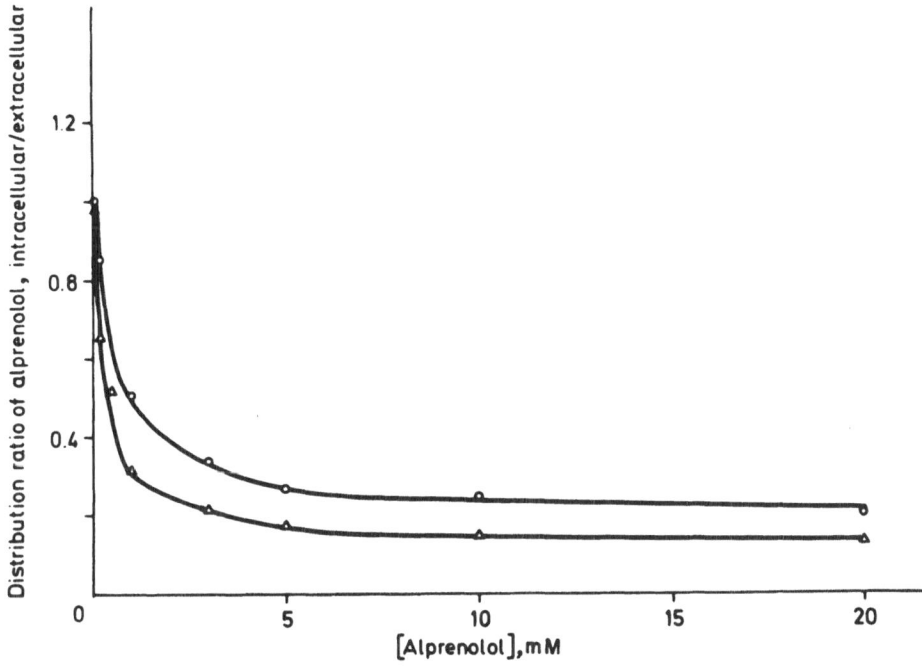

Figure 2. Distribution of alprenolol to isolated rat liver cells. Increasing concentrations of H[3] labelled alprenolol were added to suspensions of 4.2 x 10[6] cells per ml Krebs-Henseleit buffer (2% albumin) at 10°C. Cells and suspension medium were rapidly separated by centrifugation (80 x g) and radioactivity measured in the supernatant by LSC.   Δ-Δ cells from control animals.   o——o cells from phenobarbital treated animals (80 mg/kg body weight i.p. daily for 3 days). The concentration of alprenolol giving maximal Type I spectrum was 1 mM. One experiment typical of four.

Drug metabolism. Upon addition of alprenolol or nortriptyline to a liver cell suspension, a Type I spectral change was observed (Figure 5). At a temperature compatible with metabolism (37º), after the initial appearance of such a spectrum an absorption peak at 437 nm could often be seen to appear and disappear rapidly during the initial minutes. This spectral change was followed by another transient increase in absorption at about 446 nm (Moldéus et al., 1973). The nature of these absorption peaks has not yet been established.

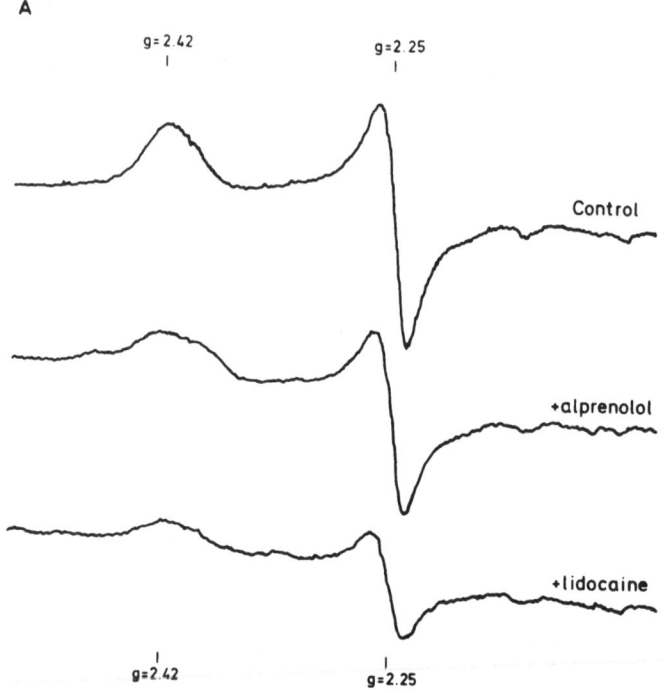

Figure 3 A-B. EPR spectra in liver cells obtained from phenobarbital treated unstarved male rats. Phenobarbital treatment:
80 mg/kg body weight i.p. daily for three days   Cell concentration
was 40,000 cells/μl. Concentration of alprenolol and lidocaine when
added was 33 mM. EPR conditions: A: Temperature 77K, microwave
power 30 mW, scanning rate 200 Gauss/min, modulation amplitude
12 Gauss and microwave frequency 9.13 GHz. B: Temperature 4.2K,
microwave power 3 mW, scanning rate 100 Gauss/min, modulation
amplitude 11 Gauss and microwave frequency 9.21 GHz. Calibrated
and matched EPR quartz tubes were used.

Their occurrence during the early phase of drug metabolism
and short duration could of course suggest that they may reflect
different states of cytochrome P-450 during the monooxygenation
reaction, but the fact that the 437 nm peak has as yet only been
demonstrated with substrate molecules containing substituted amino
groups could also suggest that this peak appears as a result of the
interaction of an amino group formed during metabolism with the
heme moiety of cytochrome P-450.

The study of drug metabolism in the isolated liver cell system is complicated by several methodological problems. Formation

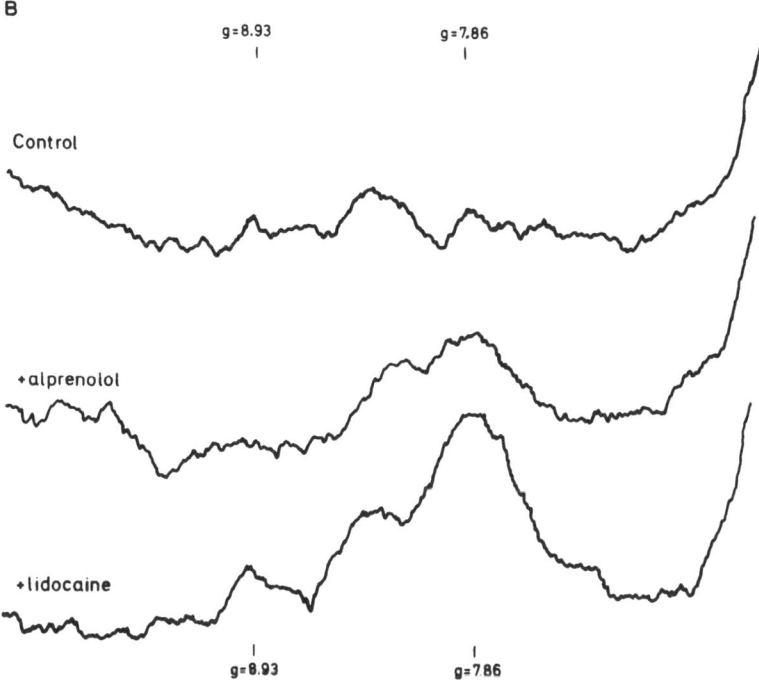

Figure 3B.

of hydroxylated products can not easily be measured since hydroxylation most often occurs in several positions of the drug molecule and since the hydroxylated products are often subject to rapid further metabolism. Therefore a substrate disappearance technique had to be used in the present work (Moldéus et al., 1974). We selected for our metabolic studies as a substrate the β-receptor blocking drug alprenolol, the metabolism of which results in the formation of the p-hydroxylated (major metabolite) and dealkylated products which are then further metabolized by conjugation (Bodin, 1974). This drug was chosen, since it is rather rapidly metabolized and shows a high affinity for interaction with cytochrome P-450, having a K_s-value of about 0.3 μM in microsomes. These properties permitted the use of rather low substrate concentrations in the metabolic experiments, which is of great advantage when a substrate disappearance method is used.

The metabolism of alprenolol without addition of cofactors was linear with time up to 60 min (Figure 6) and with cell concentration up to 3×10^6 cells per ml. The apparent Michaelis

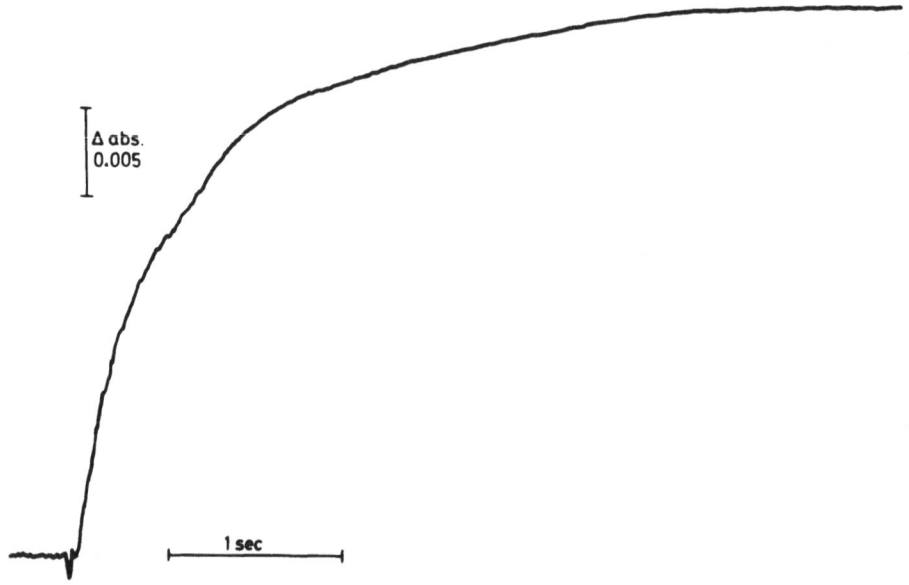

Figure 4.   Rate of formation of Type I spectral change upon
addition of alprenolol to liver cells isolated from control rats.
25 µl of 0.1 M alprenolol was rapidly mixed, using an Aminco type
plunger, with a suspension of liver cells in Krebs-Henseleit
buffer, pH 7.5 (2% albumin) containing 2.9×10^6 cells per ml.
ΔA (385-420 nm) was recorded in an Aminco DW-2 spectrophotometer.
Temperature was 37°C.   One experiment typical of five.

constant was low, about 9.5 µM, and of the same order of magnitude
as that found in isolated microsomes.   However, V_{max} for alprenolol was 1.0 nmole per 10^6 cells per min, which is about double the
activity found in microsomes.   Pretreatment of the rats with phenobarbital, which increases the cytochrome P-450 content 3-4 times,
resulted in about 100% stimulation of alprenolol metabolism in
both liver microsomes and isolated liver cells.

The wellknown inhibitors of drug metabolism SKF 525-A and
metyrapone as well as a competitive substrate, imipramine, markedly
inhibited alprenolol metabolism in isolated liver cells, SKF 525-A
being the most potent inhibitor.   The relative inhibitory effects
were the same in isolated liver cells as in microsomes, 5 µM SKF
525-A giving more than 55% inhibition and 500 µM metyrapone pro-
ducing an inhibition of about 75% in both systems, indicating similar
inhibitory mechanisms in microsomes and in isolated liver
cells (Figure 7).

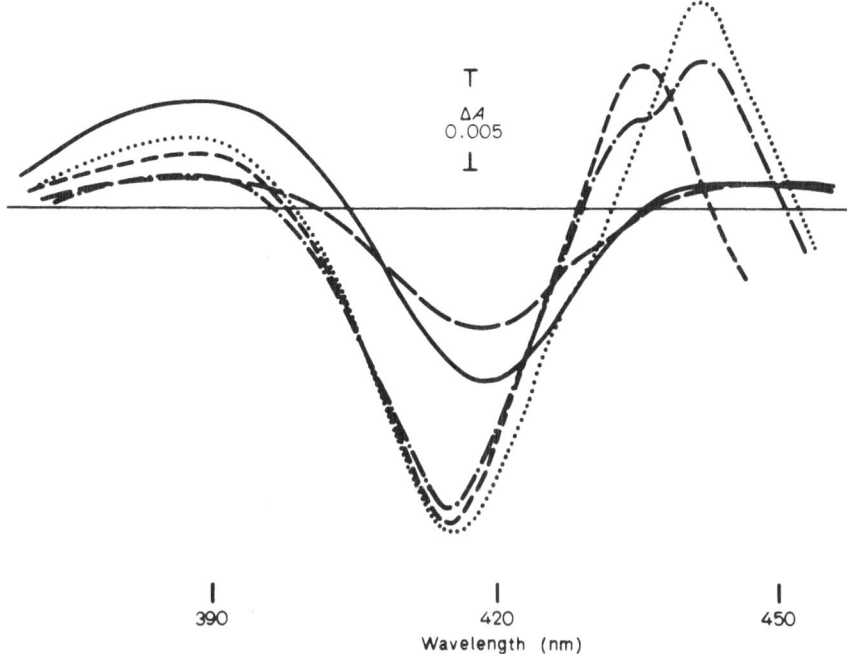

T
ΔA
0.005
⊥

390	420	450

Wavelength (nm)

Figure 5. Spectral changes produced by addition of alprenolol to isolated liver cells from fasted control rats. Each cuvette
contained a suspension of 2.2×10^6 cells per ml. Alprenolol was
added to the sample cuvette to a final concentration of 67 µM and
repetitive scanning was performed in an Aminco DW-2 spectrophotometer. (——) 15 s, (- - -) 90 s, (-.-) 120 s, (···) 180 s,
(— —) 240 s. Temperature was kept at 37°C. (Data from Moldéus
et al., 1974).

Effect of NADPH level. Drug metabolism in liver cells isolated from fasted control rats was unaffected for more than 40 min
by the addition of glucose or lactate, substrates of glycolysis
and gluconeogenesis, respectively, and known to increase the
NADPH/$NADP^+$ ratio (Figure 8). However, in liver cells isolated
from fasted, phenobarbital treated rats, where the rate of alprenolol metabolism was approximately doubled, both lactate and glucose
appreciably stimulated the metabolic activity, stimulation being
pronounced already after 5 min. This finding suggests that a constant high level of reducing equivalents is necessary to obtain
maximal drug metabolism and that the generation of cytoplasmic
NADPH apparently becomes rate limiting in the liver cell when drug
metabolism is increased by phenobarbital induction (Figure 8).

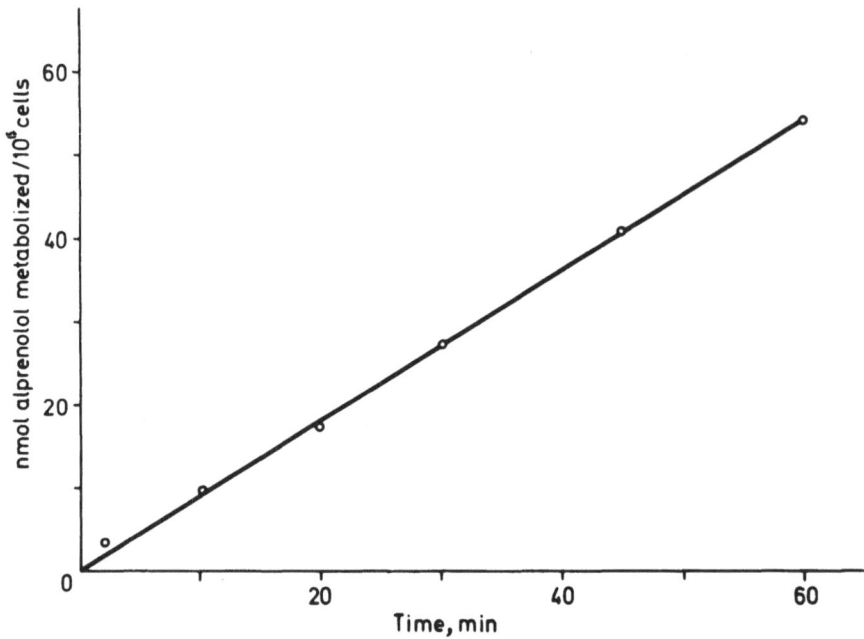

Figure 6.   Metabolism of alprenolol in isolated liver cells isolated from control rats.   Incubations were performed at 37°C with cells isolated from control rats at a cell concentration of 1.5 x 10^6 cells per ml incubation medium and a substrate concentration of 0.2 mM alprenolol.   Isolation and quantitation of unchanged alprenolol as described by Moldéus et al., (1974).

After incubation of liver cells, isolated from fasted control rats, for 15 min at 37°C almost 70% of the total cellular NADPH was found to be in the reduced state.   Lactate and glucose further increased the NADPH/NADP$^+$ ratio, maintaining about 80% in the reduced form.   When alprenolol was incubated with these liver cells, no change in the NADPH/NADP$^+$ ratio was obtained, either in the absence or presence of lactate or glucose.

In liver cells isolated from fasted, phenobarbital treated rats less NADPH was present in the reduced form after 15 min of incubation, but again both lactate and glucose increased the level to about 80 and 90% reduced, respectively.   However, contrary to the finding with control cells, when alprenolol was present the level of NADPH decreased from 51% to 37%, and when alprenolol was added together with lactate, the NADPH/NADP$^+$ ratio was lower than with lactate alone (Table III).   Thus, in liver cells isolated from phenobarbital treated rats the increased NADPH utilization due to drug metabolism could not be compensated for.   It should also be

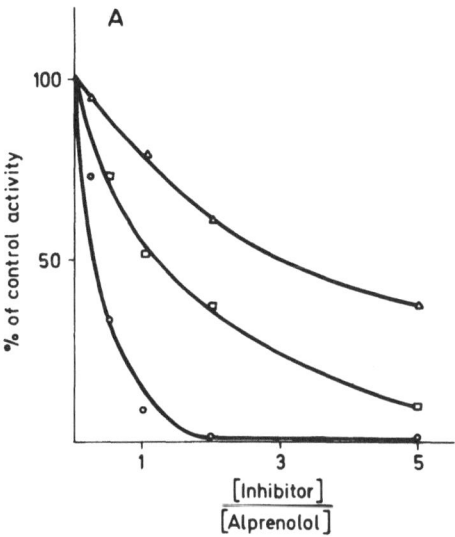

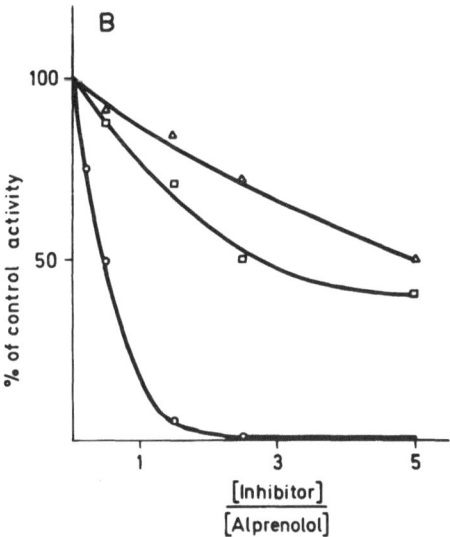

Figure 7A-B. Effect of inhibitors on the metabolism of alprenolol in microsomes (A) and isolated liver cells (B). Inhibition of metabolism of alprenolol (20 μM) by addition of increasing concentrations of metyrapone (Δ), imipramine (□) and SKF 525-A
(o). The incubation medium contained 1.5×10^6 cells per ml or
1 mg of microsomal protein per ml. Incubations for 10 min at 37°C.
100% metabolism = 0.8 nmole alprenolol metabolized/nmole P-450 per
min. Values are means of three experiments (Data from Grundin et
al., 1974).

TABLE III

Effect of Different Agents on the Reduction State of NADPH in
Liver Cells Isolated from Starved Control and Phenobarbital Pretreated Rats

Additions	NADPH reduced	
	Control	Phenobarbital pretreatment
	%	%
None	69	51
Alprenolol, 75 μM	72	37
Lactate, 10 mM	80	83
Lactate + alprenolol	79	74
Glucose, 10 mM	79	67
Glucose + alprenolol	82	68
Rotenone, 25 μM	41	21

Isolated liver cells were suspended to a concentration of 8.8×10^6
cells per ml in Krebs-Henseleit buffer, pH 7.5, containing 2%
albumin. Incubations were for 10 min at 37°C. Values are means of
5 experiments. Assay of NADPH according to Klingenberg (1970).
Data from Moldéus et al., (1974).

noted that the values represent total cellular concentrations and that the effects on the cytoplasmic NADPH/NADP$^+$ ratio might be even more pronounced.

Inhibitors of the mitochondrial respiratory chain had little effect on drug metabolism when added to isolated microsomes in concentrations sufficient to cause maximal effects on the mito-chondria. However, the inhibitors rotenone and antimycin A caused a very marked inhibition of alprenolol metabolism in liver cells isolated from fasted control rats (Table IV), antimycin A being the more potent inhibitor rendering up to 80% inhibition. Since these inhibitors do not affect the cytochrome P-450 system directly, the inhibition was most likely due to inhibition of the generation of cytoplasmic NADPH in these cells, addition of rotenone indeed causing a decrease in the cellular NADPH concentration (Table III). In liver cells isolated from fed rats, alprenolol metabolism was only inhibited about 20% by the addition of these inhibitors, so obviously inhibition of the mitochondrial respiratory chain did not decrease the cytoplasmic NADPH generation to the same extent in these cells as in liver cells from fasted rats.

Inhibition of the mitochondrial respiratory chain caused a decrease in the cellular ATP concentration; we found only 5-15% ATP remaining in the isolated liver cells after 5 min of incubation with rotenone and there is little doubt that the effect on alpreno-lol metabolism seen with inhibitors of the mitochondrial respiratory chain is due to the decrease in cellular ATP concentration.

In the fed state cytoplasmic NADPH is generated mainly by en-zymes of the pentose-phosphate pathway following glycogen break-down, a pathway requiring little energy (Thurman & Scholz, 1969). A decreased ATP concentration would consequently not affect cyto-plasmic NADPH generation appreciably when glycogen is present in the cell.

In the fasted state, however, reducing equivalents are de-rived from mitochondrial oxidations, mainly of fatty acids, and transported to the cytosol either via an α-ketoglutarate-isocitrate shuttle involving mitochondrial and cytoplasmic isocitrate dehy-drogenase (Quagliariello et al., 1968), or via a shuttle mechanism involving malic enzyme (Thurman & Scholz, 1969). A decrease in the cellular concentration of ATP would cause an inhibition of both these shuttles, the mitochondrial transhydrogenase involved in the α-ketoglutarate-isocitrate shuttle and the pyruvate carboxy-late in the shuttle involving malic enzyme both being energy re-quiring enzymes. Consequently, in liver cells isolated from fasted rats the cytoplasmic NADPH generation and thereby drug metabolism would be inhibited by inhibition of the mitochondrial respiratory chain.

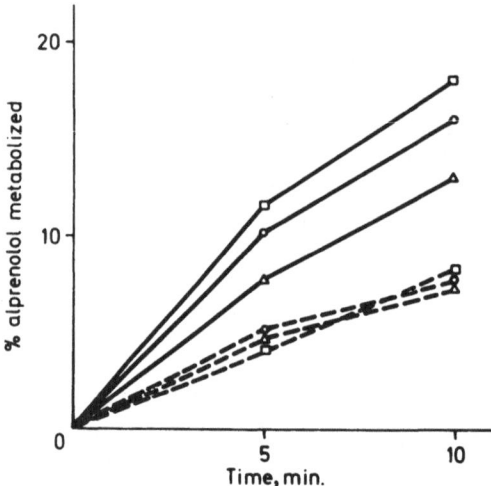

Figure 8. Effect of glucose and lactate on metabolism of
alprenolol in liver cells isolated from starved control and phenobarbital treated rats. Incubations performed at 37°C contained
3×10^6 cells per ml and 37.5 µM alprenolol. Cells from control
(---) and phenobarbital treated animals (———). (△———△) no addition, (o———o) addition of 10 mM lactate, (□———□) addition of
10 mM glucose. One experiment typical of five.

TABLE IV

Effect of Respiratory Inhibitors on Metabolism of Alprenolol in
Liver Cells Isolated from Fasted and Fed Rats.

Additions	Metabolism of alprenolol	
	fasted	fed
	nmole/10^6 cells per min	
None	0.64	0.70
Rotenone, 10 µM	0.24	0.56
Antimycin A, 1.6 µM	0.12	0.51

Isolated liver cells from control rats were suspended in Krebs-
Henseleit buffer, pH 7.5, containing 2% albumin to a concentration
of 2.3×10^6 cells per ml. Alprenolol concentration was 20 µM.
Incubations were for 5 min at 37°C. Values are means of three
experiments.

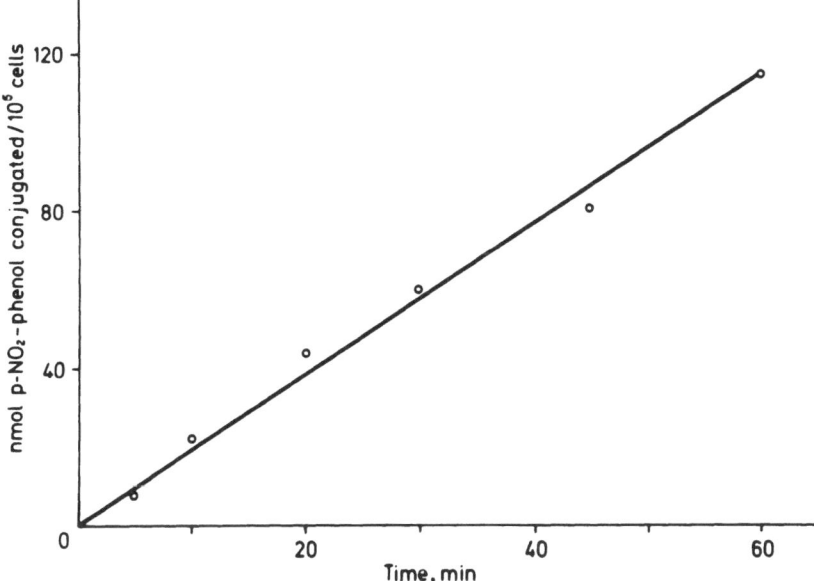

Figure 9.   Conjugation of p-nitrophenol in isolated rat
liver cells.   0.5 mM p-nitrophenol was incubated with liver cells
from control animals suspended to 3 x 10^6 cells per ml in Krebs-
Henseleit buffer, pH 7.5 (2% albumin).   Disappearance of p-nitrophenol was measured according to Hänninen (1968).   Incubation at
37^oC.

Conjugation with glucuronic acid.   After a substrate has become hydroxylated by the cytochrome P-450 system it is often conjugated prior to excretion, most frequently with glucuronic acid
by the action of glucuronyl transferase.   We studied the rate of
conjugation of p-nitrophenol in the isolated liver cells and
found that it was linear with time for up to 60 min. (Figure 9).
By incubation of the conjugation products with β-glucuronidase,
90% of the conjugated p-nitrophenol could be recovered, but less
than 5% by incubation with sulphatase.   Interestingly, the rate
of glucuronidation, without addition of cofactors in the isolated
liver cell was much faster than in microsomes in the presence of
saturating amount of UDPGA and was almost equal to that found in
liver microsomes activated by treatment with digitonine or
Triton X-100.

CONCLUSIONS

This study has shown that drugs rapidly enter the liver cell most probably via a simple passive diffusion process. The drug molecules are then bound to cytochrome P-450, which in the cell is present mainly in the oxidized form free of substrate. The binding which produces a Type I spectral change can also be illustrated by changes in the EPR spectrum with a decrease in the low spin signal and an appearance of a high spin signal. The isolated liver cells catalyze the cytochrome P-450 linked drug metabolism at a higher rate than do liver microsomes. NADPH generation may become rate limiting under certain conditions, e.g., in cells isolated from fasted, phenobarbital treated rats. The cytoplasmic NADPH generation is probably mediated by different pathways in the fed and fasted state, and glucuronidation seems to be considerably faster in isolated liver cells than in liver microsomes.

ACKNOWLEDGMENTS

This study was supported by the Swedish Medical Research Council (Grant no. 03X-2471 and 13X-542), the Swedish Natural Sciences Research Council (Grant no. FO321-026) and by Funds from Karolinska Institutet (Sten Orrenius, Christer von Bahr, Robert Grundin).

REFERENCES

von Bahr, C., Vadi, H., Grundin, R., Moldéus, P. and Orrenius, S. (1974). Spectral studies on the rapid uptake and subsequent binding of drugs to cytochrome P-450 in isolated rat liver cells. Biochem. Biophys. Res. Commun. 59: 334-339

Bodin, N. O. (1974) Identification of the major urinary metabolite of alprenolol in man, dog and rat. Life Sciences 14: 685-692

Cammer, W., Schenkman, J. B. and Estabrook, R. W. (1966) EPR measurements of substrate interaction with cytochrome P-450. Biochem. Biophys. Res. Commun. 23: 264-268

Grundin, R., Moldéus, P., Orrenius, S., Borg, K. O. Skånberg, I. and von Bahr, C. (1974) The possible role of cytochrome P-450 in the liver "first pass elimination" of a β-receptor blocking drug. Acta Toxicol. Pharmacol. 35: 242-260

Hänninen, O. (1968) On the metabolic regulation in the glucuronic acid pathway in the rat tissues. Ann. Acad. Sci. Fenn. A, II, 142: 43-45

Hems, R., Ross, B. P., Berry, M N. and Krebs, H. A. (1966) Gluconeogenesis in the perfused rat liver. Biochem. J., 101: 284-292.

Henderson, P. Th. and Dewaide, J. H. (1969) Metabolism of drugs in isolated rat hepatocytes. Biochem. Pharmacol. 18: 2087-2094

Holtzman, J. L., Rothman, V. and Margolis, S. (1972) Metabolism of drugs by isolated hepatocytes. Biochem. Pharmacol. 21: 581-584

Klingenberg, M. (1970) In Bergmeyer, H. V. (ed.) pp. 1975-1990 Methoden der Enzymatischen Analyse,Verlag Chemie, Weinheim

Krebs, H. A., Cornell, N. W., Lund, P. and Hems, R. (1974) Isolated liver cells as experimental material, pp. 726-753. In F. Lundqvist and N. Tygstrup (eds.) Regulation of hepatic metabolism, Munksgaard, Copenhagen.

Moldéus, P., Grundin, R., von Bahr, C. and Orrenius, S. (1973) Spectral studies on drug cytochrome P-450 interaction in isolated rat liver cells. Biochem. Biophys. Res. Commun. 55: 937-938.

Moldéus, P., Grundin, R., Vadi, H. and Orrenius, S. (1974) A study of drug metabolism linked to cytochrome P-450 in isolated rat liver cells. Eur. J. Biochem. 46: 351-360.

Peisach, J., Stern, J. O. and Blumberg, W. E. (1973) Optical and magnetic probes of the structure of cytochrome P-450's. Drug Metabolism and Disposition 1: 45-61

Quagliariello, E., Papa, S., Meijer, A. J. and Tager, J. M. (1968) Substrate transport in mitochondria and control of metabolism. In L. Ernster and Z. Drahota (eds.) pp. 335-346. Mitochondria, Structure and Function, FEBS Sumposium, vol. 17., Academic Press, New York.

Quistorff, B., Bondesen, S. and Grunnet, N. (1973) Preparation and biochemical characterization of parenchymal cells from rat liver. Biochem. Biophys. Acta, 320: 503-516

Schenkman, J. B., Remmer, H. and Estabrook, R. W. (1967) Spectral studies of drug interaction with hepatic microsomal cytochrome. Molec. Pharmacol. 3: 113-123

Seglen, P. O. (1972) Preparation of rat liver cells. I. Effect of Ca^{2+} on enzymatic dispersion of isolated, perfused liver. Experimental Cell Research 74: 450-454

Thurman, R. G. and Scholz, R. (1969) Mixed function oxidation in perfused rat liver. The effect of aminopyrine on oxygen uptake. Eur. J. Biochem. 10: 459-467

Ullrich, V. (1969) On the hydroxylation of cyclohexane in rat liver microsomes. Hoppe-Seylers Z. Physiol. Chem. 350: 357 365.

ROLE OF PHOSPHOLIPIDS IN ADRENOCORTICAL MICROSOMAL HYDROXYLATION

REACTIONS: ACTIVATION OF LIPID-DEPLETED MICROSOMAL PREPARATIONS

BY NON-IONIC DETERGENTS

Shakunthala Narasimhulu
Harrison Department of Surgical Research
School of Medicine
The University of Pennsylvania
Philadelphia, Pennsylvania 19174

INTRODUCTION

Phospholipids are believed to play at least two distinct roles in biological membranes. As integral parts of membrane structure, phospholipids are believed to impart much of the uniqueness to membranes and to provide a matrix in which membrane proteins are imbedded. A second role is their requirement for the function of certain membrane-associated enzymes.

Cytochrome P-450, the oxygen-activating component of several mixed-function oxidase systems, is membrane-bound in the mammalian species. Indirect evidence such as inactivation of the cytochrome by agents which attack phospholipids leads to the thought that the enzymatic activity of the cytochrome depends upon its association with phospholipids in the membrane (1). Lu et al., (2) reported the first direct evidence for the requirement of phospholipids for the electron transport to cytochrome P-450 in deoxycholate-solubilized, purified and reconstituted mixed function oxidase systems of hepatic endoplasmic reticulum. Chaplin and Mannering (3) showed that the treatment of hepatic microsomes with phospholipase C resulted in release of 70% of the phospholipids, 40% loss in the metabolism of Type I compounds and only 15% loss in the metabolism of Type II compounds. But the enzyme-treatment apparently completely eliminated Type I binding and enhanced Type II binding. They further indicated that phospholipase-treated microsomes failed to respond to the addition of phospholipids. Liebman et al., (4) and Tagg and Mitoma (5) made similar observations upon lipid-

271

depletion of liver microsomes by extraction with isooctane. In
the case of bovine adrenocortical microsomes (7), extraction with
n-butanol resulted in 80 to 85% removal of phospholipids and 90 to
95% loss in hydroxylation activity. However, in contrast to liver
microsomes, the hydroxylation activity lost upon removal of phospholipids from the adrenal microsomes could be considerably (40 to
60%) restored by replenishing the depleted microsomes with phospholipids and completely restored by the addition of the non-ionic
detergent Triton X-114.

The present report deals with the effects of non-ionic detergents and phospholipids on the steroid 21-hydroxylase system in I,
the lipid-depleted microsomes and II, the partially resolved hydroxylase system. In addition, data on the phospholipid composition and micrographs of these preparations will be presented.

LIPID-DEPLETED MICROSOMES

a) Extraction of Phospholipids from the Microsomes.
Choosing the appropriate solvent and conditions for the extraction
of membrane lipids is difficult because successful lipid-depletion
seems to be inevitably accompanied by some irreversible inactivation of membrane-bound enzymes.

n-Butanol was introduced by Morton (6) in 1950 for removal
of lipids from biological membranes in order to solubilize membrane-associated proteins. When applied to bovine adrenocortical
microsomes according to the procedure previously described in
detail (7, 8) n-butanol removes most of the phospholipids with
preservation of the components of the hydroxylase system in a re-
activatable state. The procedure is essentially as follows:
Aqueous suspensions of the microsomal fraction are lyophilized.
The lyophilized powder is homogenized in dry n-butanol (1.0 ml per
10 mg of microsomal protein) and centrifuged at $4^{\circ}C$. The extraction is repeated a second time. The sediment is washed twice with
dry acetone (0.5 ml per 10 mg of microsomal protein) to remove
residual butanol. The acetone layer is removed after centrifugation. Residual acetone in the sediment is removed by blowing dry
nitrogen over it. The dry lipid-depleted microsome powder is
stored at $-20^{\circ}C$ in the presence of a dessicant.

b) Phospholipid Composition and Morphology of Microsomal
Preparations. Table I shows the phospholipid composition of the
original microsomes and of the phospholipid-depleted (PLD) microsomes. Phosphatidylcholine and phosphatidylethanolamine are the
major phosphatides which account for for 50 and 29% respectively of
the total phospholipids of the cortical microsomes. This observation is similar to those reported for hepatic endoplasmic reticulum (12, 13). The phospholipid composition and content of

TABLE I

Phosphatides of Bovine Adrenocortical Microsomes

PHOSPHATIDES	Phosphorus nmoles per mg of protein		PLD/Control %
	'Control' Microsomes	'PLD' Microsomes	
Total	1014	200	19.5
Phosphatidyl(p)- Choline	502	47	9.4
(p)-Ethanolamine	297	48	16.2
(p)-Serine plus (p)-Inositol	150	104	69.5
Sphingomyeline	54	10	18.5

Phospholipids were extracted from microsomes with chloroform/
methanol 2:1 (v/v) according to Folch et al., (9); separated by
thin layer chromatography (10); and phosphorus content determined
according to the procedure of Bartlett (11).

the cortical microsomes changed markedly after extraction of the
microsomes with n-butanol according to the procedure described
earlier. By this procedure, about 85% of the phospholipids were
removed. Analysis of the remaining phospholipids in the depleted
microsomes indicated that 91% of the phosphatidylcholine, 84% of the
phosphatidylethanolamire, 81% of sphingomyelin, and 30% of phosphatidylserine plus phosphatidylinositol were extracted. In the
lipid-depleted microsomes, phosphatidylserine plus phosphatidylinositol account for 52% of the total phospholipids as opposed to
14.8% in the original microsomes. On the other hand, phosphatidylcholine accounts for 24% of the total phospholipids as opposed
to 60% in the original microsomes. The results indicate that n-
butanol-extraction of the microsomes according to the procedure described results in a preferential removal of basic phosphatides (14).

The electron micrographs of the original microsomes and
of the phospholipid-depleted microsomes revealed that the morphology
of the membranous components has been distinctly altered after
lipid-depletion. The form of the microsomal membranes in the

original microsome preparation is largely vesicular (Figure 1a)
with many profiles showing the attached ribosomes. In the depleted
microsomes (Figure 1b), the vesicular nature of the membrane is
not as pronounced and the membrane is amorphous in appearance.
There is, however, some residual material present which probably
corresponds to the vesicular membranous material of the original
microsomes. This material appears to form largely linear as
opposed to vesicular structures. The hydroxylase components are
presumably associated with these structures. The possibility that
activation of the hydroxylase by phospholipids and detergents is
associated with the restoration of the vesicular structures remains to be investigated.

c) <u>Steroid Hydroxylation.</u> Since lipid-depletion often causes
irreversible inactivation of membrane-bound enzymes, reactivation
of an inactivated enzyme by replenishing it with lipids is an
important criterion to establish lipid-dependency of enzyme function.

The steroid hydroxylation activity lost upon depletion of the
adrenal microsomes of most of their lipids could be considerably
restored by the addition of phospholipids. Under the conditions
of the experiments illustrated in Figure 2, the addition of
Asolectin (soy bean phosphatides mixture) micelles to the lipid-
depleted microsomes restored 50% of the hydroxylation activity.
When added alone, purified egg lecithin (curve II, Figure 2) was
not as effective and purified lysolecithin was completely without
effect (not shown). However, lecithin plus lysolecithin mixed
micelles restored the activity of the lipid-depleted microsomes to
the same extent as Asolectin but at a much lower concentration
(curve III, Figure 2) while at higher concentrations they inhibited the reaction. Asolectin, though failing to restore the
activity completely, did not inhibit the reaction.

The reactivation of the lipid-depleted microsomes was not a
unique property of phospholipids. Non-ionic detergents such as
Triton X-114 or Triton N-101 could also restore hydroxylation
activity. Most effective was Triton X-114, which completely restored the activity to the level of the original microsomes. On
the other hand, Triton N-101, although nearly as effective as
X-114 at lower concentrations, inhibited the reaction at higher
concentrations. In addition, Triton N-101 inhibited the X-114 re-
activated PLD microsomes. The nature of the inhibition by Triton
N-101 is not clear because both Tritons inhibit the substrate-
binding reaction (Figures 4 and 5) apparently by competing with
the steroid substrate. If the detergents had no other effects,
one would not expect any inhibition of the hydroxylation reaction
because the assay system contains saturating concentrations of
the substrate 17-hydroxyprogesterone which should overcome the
inhibition by the detergents. This is, in fact, the case with

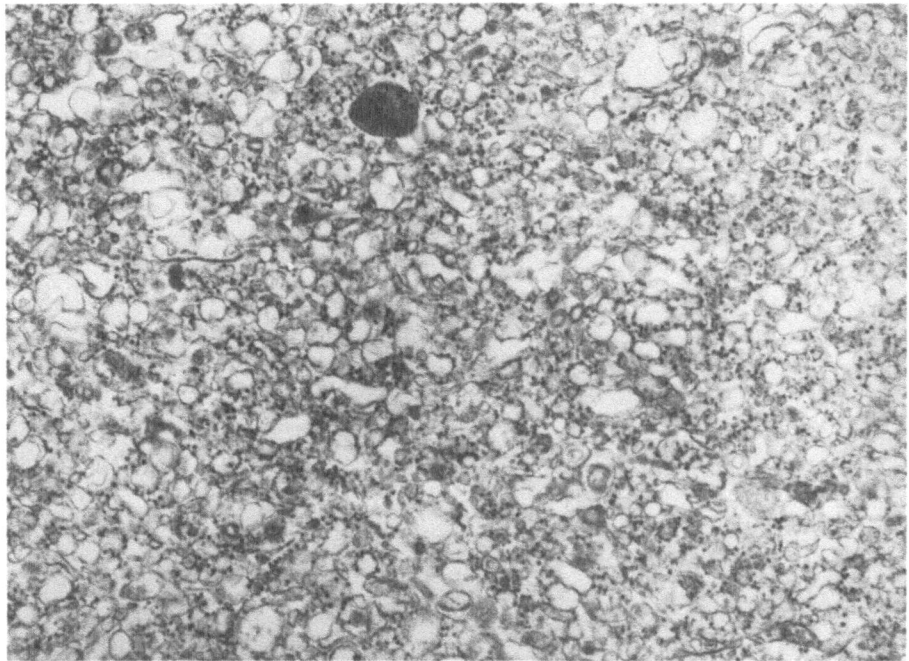

Figure 1a.   An electron micrograph of Control Microsomes.

Triton X-114. Since Triton N-101 still inhibits the hydroxylation reaction at high substrate concentrations, this detergent probably has additional effects on the components of the hydroxylase system.

In view of the report of Cater et al., (15) that phosphatidyl-choline restored the activity of deoxycholate-inhibited hepatic microsomes, attempts were made to reactivate the Triton N-101 inhibited microsomes by addition of phospholipids but the activity could not be restored.

d) Reduction of Cytochrome P-450. In addition to the loss of steroid hydroxylation activity, phospholipid-depletion of the microsomes was accompanied by a marked inhibition of cytochrome P-450 reduction by NADPH. Figure 3 shows the time course of the reduction of the cytochrome. The slow reduction after lipid de-pletion is shown by curve C. Similar to the hydroxylation activity, the addition of phospholipid micelles or a suitable detergent re-stored the rate of reduction of the cytochrome. The initial rate of reduction in the presence of Asolectin micelles (curve B) was approximately 5 times the initial rate of the control (curve C).

Figure 1b. An electron micrograph of Lipid-Depleted
Microsomes.

In the presence of Triton X-114 (curve A), the initial rate was
about 13 times the control rate. The slow rate of reduction in
the lipid-depleted microsomes was not altered when either the
substrate or the activating agent was added alone. The simultaneous presence of the substrate and one of the activating agents
was required for the restoration of P-450 reduction in the lipid-
depleted microsomes. The increase in the rate of reduction of
the cytochrome caused by the addition of either Asolectin or
Triton X-114 in the presence of the substrate, was paralleled by
similar increase in the rate of substrate hydroxylation.

The results indicate that phospholipids are in some way required for the electron transport to cytochrome P-450 in the
lipid-depleted adrenal microsomes. Since detergents can replace
phospholipids, the latter are probably not directly involved in
the electron transport but rather agents that facilitate the reaction between hydrophylic and hydrophobic functional groups of
the electron transport system, as suggested by Fleischer et al.,
(16) on the basis of extensive studies of the mitochondrial
electron transport system.

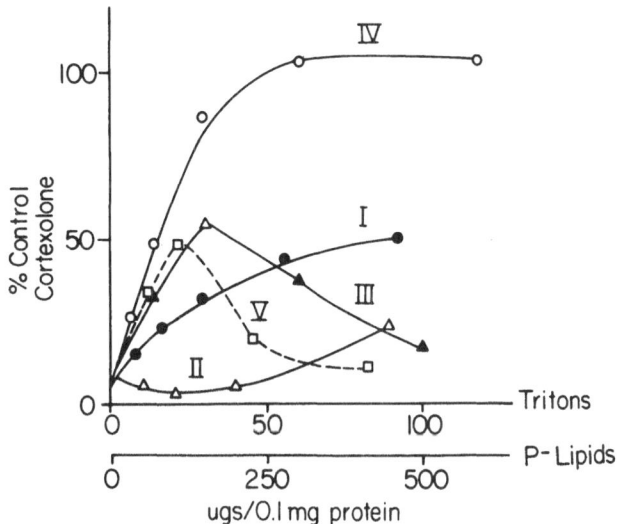

Fig. 2.   Effect of Phospholipids and Non-ionic Detergents on Steroid C-21 Hydroxylation in Lipid-Depleted Microsomes.   The reaction medium and the procedure for the assay of C-21 hydroxylated product (cortexolone) of 17-hydroxyprogesterone have been previously described (13).   The incubation mixture containing 1.0 mg of the microsomal protein was incubated for 30 min at 37°C.   The activity of the control microsomes (taken as 100%) in these experiments ranged between 6 and 7 nanomoles product formed per mg of protein per min.   Phospholipids were added as micelles prepared by sonication.

Curve I.....Asolectin (1 gm in 20 ml of 0.05M Na-PO$_4$ buffer pH 7.4 sonicated for 20 min and centrifuged at 8000 x g for 20 min.

Curve II....Purified egg lecithin (20 mg in 1.0 ml of the buffer sonicated until clear under N$_2$).

Curve III...Purified egg lecithin plus purified egg lysolecithin (1.7:1.0) 20 mg in 1.0 ml of the buffer and sonicated until clear under N$_2$).

Curve IV....Triton X-114.

Curve V.....Triton N-101

e) <u>Substrate-Cytochrome P-450-Binding Reaction</u>. As indicated earlier, steroid hydroxylation and P-450 reduction were nearly completely inhibited upon removal of the bulk of phospholipids from the microsomes. The substrate-cytochrome P-450 binding reaction, however, as measured by the criterion of substrate-induced Type I spectral change was qualitatively identical to that observed in the original microsomes (17). Likewise, spectrophotometric titration of PLD microsomes with the Type I substrate 17-hydroxy-progesterone indicated that the apparent substrate-dissociation constant (K_s) was not altered. Furthermore, replenishing of the depleted microsomes with phospholipid (Asolectin) micelles also did not alter the dissociation constant (Figure 4). In contrast, the addition of the detergents, Triton X-114 (Figure 4) or Triton N-101 (Figure 5), to the depleted microsomes caused a three fold increase in the dissociation constant. However, at the concentrations tested, neither phospholipids nor the detergents had any effect on the maximum Type I spectral change as indicated by the lack of effect on the slopes of the linear plots (Figures 4 and 5).

The results indicate that the detergents inhibit the binding of the substrate to cytochrome P-450 apparently competitively while those phospholipids which were tested have no effect on the binding under the present experimental conditions. However, the lack of effect of phospholipids does not rule out their involvement in the substrate-cytochrome P-450 binding reaction. This is because the 15 to 20% of the microsomal phospholipids still remaining in the depleted microsomes and, perhaps, intimately associated with cytochrome P-450 may be required for lipid soluble substrates such as steroids to gain access to the cytochrome.

THE PARTIALLY RESOLVED SYSTEM

a) <u>Resolution of the Hydroxylase System without the Use of Detergents</u>. Detergent-dependent stimulation of the activity of lipid-depleted or lipid-freed enzyme preparations has been reported in the case of several membrane enzymes (18, 19). In addition, it has been suggested that a detergent can play a role in a purified membrane-enzyme equivalent to the role of lipid in the native membrane (19). On the other hand, complex lipid requirements for detergent-solubilized enzymes have also been encountered (20), which may or may not imply development of an artificial requirement for phospholipids upon treatment of an enzyme system with detergents.

In view of these reports and of the difficulties involved in removing bound detergent from the detergent-solubilized enzymes, it is highly desirable to achieve dissociation of a multicomponent enzyme system, such as steroid hydroxylase, without the use of detergents.

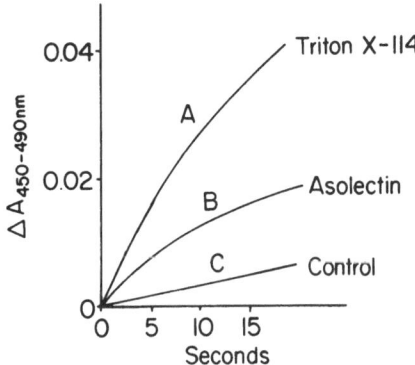

Fig. 3. The Effect of Asolectin and Triton X-114 on the Reduction of Cytochrome P-450. 2.0 ml of the reaction medium which includes albumin-containing glycylglycine buffer prepared as previously described (13) were placed in an Aminco anaerobic cuvette (equipped with a plunger assembly to make additions under anaerobic conditions). To this medium, the following were added: 0.3ml (6.0 mg protein) of lipid-depleted microsomes suspended in 0.25 M sucrose; 0.4 ml of 0.05 M $NaPO_4$ buffer, pH 7.4, containing either 4 mg of Triton X-114 or approximately 18 mg of Asolectin. An equivalent volume of buffer was added to the control. 17-Hydroxyprogesterone (about 60 µg) dissolved in 25 µl of methanol was placed on a stirring rod and the methanol evaporated. The steroid was stirred into the reaction mixture in the cuvette. This mixture was bubbled for 7 min with CO which had been passed through Fieser's deoxygenizing mixture (a solution of sodium dithionite and sodium anthroquinone-2-sulfonate in 0.1 M NaOH). Catalase, glucose and glucose oxidase were then added to remove the residual oxygen. The reaction was started by the addition of NADPH-generating system by means of the plunger.

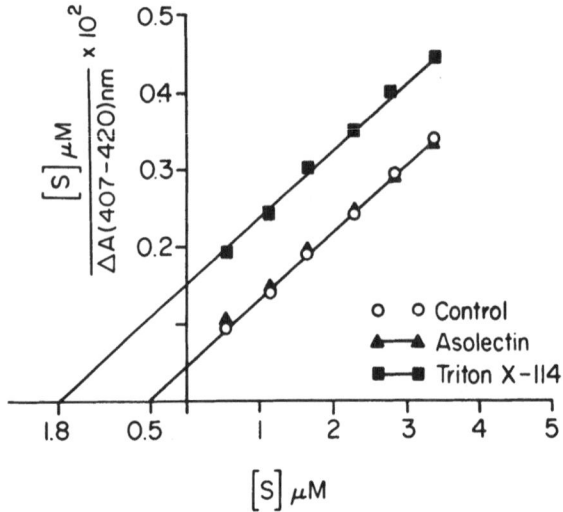

Fig. 4.   The Effect of Asolectin and Triton X-114 on the
Apparent K_S of 17-Hydroxyprogesterone.   3.0 ml of the enzyme
buffer mixture containing 3.0 mg of microsomal protein were
placed in a spectrophotometric cuvette.   The steroid was added
as methanolic solution in not more than 2 µl volume per addition.
The decrease in absorption at 420 nm with respect to 407 nm was
recorded with an Aminco-Chance dual wavelength spectrophotometer.
The concentrations of Triton X-114 and Asolectin were 1.4 mg and
3.0 mg per ml.   The following algebraic transformation of the
Michaelis-Menten equation was used for the plot:

$$\frac{S}{\Delta A} = \frac{K_S}{\Delta A \; max} + \frac{1}{\Delta A \; max} . S.$$

   Procedure for Resolution.   The phospholipid-depleted microsomes
are suspended in 0.25 M sucrose (10 mg of protein per ml) and subjected to sucrose density gradient centrifugation; the gradients
are: 0.9 M sucrose (18.0 ml) and 1.2 M sucrose (7.0 ml) both containing 0.01 M Tris buffer, pH 7.4.   About 5.0 ml of the lipid-
depleted microsomal suspension in 0.25 M sucrose are layered over
the gradient and centrifuged at 40,693 x $g_{av.}$ for one hour.   This
procedure results in a partial (65%) resolution of the hydroxylation system into a 'Particulate Fraction', recovered as sediment,
and a soluble 'Reductase Fraction' which remains in the 0.25 M
sucrose layer.   The active principle in this fraction is considered soluble by the criterion that it is not sedimentable by
centrifugation at 100,000 x g for one hour.

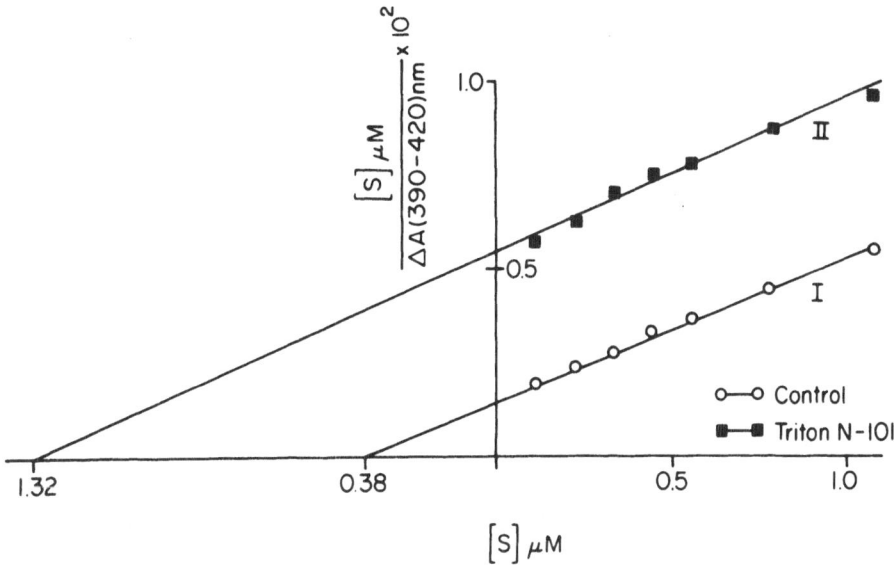

Fig. 5.   The Effect of Triton N-101 on the Apparent K_s of 17-Hydroxyprogesterone. The experimental conditions were the same as in Figure 4. But the entire spectrum was scanned, (500 to 360 nm) after each addition of the steroid, with an Aminco-Chance spectrophotometer in a split beam mode.

Analysis of the two fractions indicated that cytochrome P-450 remains bound to the 'Particulate Fraction' while a significant portion of the NADPH-cytochrome c reductase activity is associated with the 'Reductase Fraction.' There was no detectable amount of cytochrome P-450 present in the 'Reductase Fraction.' The specific activity of the NADPH-cytochrome c reductase in this fraction was found to be 27.2 nmoles per mg of protein per min at 25°C as opposed to 10.7 nmoles in the particulate fraction. The total NADPH-cytochrome c reductase activity recovered in the 'Reductase Fraction' and in the 'Particulate Fraction' were 22% and 37% respectively. These values represent the total activities recovered from the original microsomes (21). These values do not represent true recoveries because for reasons not clear, there appears to be considerable enhancement of the NADPH-cytochrome c reductase activity upon depletion of the microsomes of their phospholipids.

Immunochemical studies of Masters et al., (22) indicate the enzyme of bovine adrenocortical microsomes involved in the reduction of cytochrome P-450 can also catalyze the reduction of

cytochrome c by NADPH. Since the 'Reductase Fraction' contains
NADPH-cytochrome c reductase activity, it can be assumed that the
active component in this fraction is the reductase for cytochrome
P-450. Preliminary experiments have indicated that the 'Reductase
Fraction' enhances the reduction of cytochrome P-450. However,
in view of reports on the requirement of a soluble activating
factor for hepatic microsomal hydroxylation (23) and on the possible involvement of a lipoprotein factor in the binding of substrates to hepatic microsomal cytochrome P-450 (4), the presence
of such unknown factors in the 'Reductase Fraction' must also be
considered because of the similarity of the microsomal electron
transport system of adrenal cortex and of liver (22).

　　　b) Reconstitution of the Hydroxylation Activity. Although
detergents are not required either to deplete the microsomes of
their phospholipids or for the partial resolution of the hydroxylase
system therefrom, the presence of either a detergent or phospholipids was required for the expression of the hydroxylation activity by the lipid-depleted microsomal preparations.

　　　Since Triton X-114 restored the activity in the unresolved
PLD microsomes completely (Figure 2),this detergent rather than
phospholipids was used as the activating agent in experiments on
reconstitution of the hydroxylation activity. Table II shows the
activity at various steps of resolution and reconstitution. Neither
the 'Particulate Fraction' nor the 'Reductase Fraction' was found
to be significantly active when tested either individually or in
combination. In the presence of Triton X-114, however, the 'Particulate Fraction' exhibited 36% of the hydroxylase activity of the unresolved PLD microsomes while the 'Reductase Fraction' was found
to be inactive (not shown in the table). When 'Reductase Fraction'
was added to 'Particulate Fraction' in the presence of Triton X-114,
hydroxylase activity increased with increasing "Reductase" concentrations (Figure 6). At the highest concentration tested, the activity was about 40% higher than that of unresolved lipid-depleted
microsomes. Furthermore, the activity of the unresolved system
could also be increased by supplementing it with the 'Reductase
Fraction' (not shown in the table). These results indicate that
the concentration of the reductase in the unresolved lipid-depleted
microsomes is rate limiting. It is also possible that in the unresolved systems the particle-associated reductase is under some
type of control.

　　　The curve in Figure 6 showing the activity of the 'Particulate
Fraction' as function of the concentration of the 'Reductase Fraction' indicates a biphasic response. The change in slope
occurs when the activity level of 25 nmoles per min has been
reached, which is nearly equal to the activity of the unresolved
lipid-depleted system in the presence of Triton X-114 (Table III).
In the second phase the increase in hydroxylase activity with

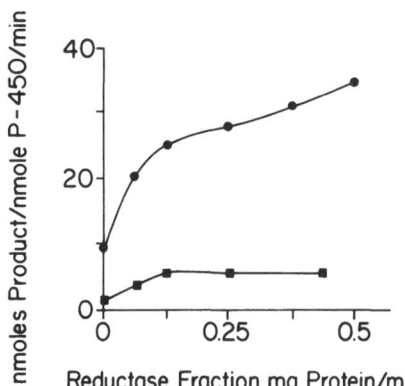

Fig. 6. Activity of the 'Particulate Fraction' as a Function of the 'Reductase Fraction'. Experimental conditions were the same as those described in the legend to Table II.

Circles......In the presence of Triton X-114
Squares......In the presence of phospholipids.

increasing 'Reductase Fraction' continues at a slower rate. The significance of the biphasic curve is not known. It is possible that cytochrome P-450 in these preparations is heterogeneous, at least with respect to its ability to react with the reductase.

If lecithin plus lysolecithin micelles were used as the activating agent instead of Triton X-114 for the recombined system, the maximum activity restored was only 15 to 20% as compared to 48% restoration by these phospholipids in the unresolved lipid-depleted microsomes.

COMMENTS

In view of the high concentration of the phospholipids required for the reactivation of the lipid-depleted microsomes, their role as coenzyme, cofactor, or reaction partner, is highly improbable. At least such a specific role cannot be associated with the bulk of the microsomal phospholipids, which were removed by extracting the microsomes with n-butanol. Furthermore, the specificity is too low for any such catalytic function. Indeed it is so low that non-ionic detergents can serve as substitutes for phospholipids. In fact, under the present experimental conditions, the detergent Triton X-114 was found to be considerably more effective than phospholipids in restoring the activity of the lipid-depleted microsomes as well as of the resolved-recombined system.

TABLE II

Reconstitution of Steroid C-21 Hydroxylation Activity

Hydroxylase Activity In "PLD" Microsomes (MCS) nmoles Of Product/nmole P-450/min			
Unresolved		Resolved	
Preparation	Activity	Preparation	Activity
"PLD" MCS	0.8	P-Fr	1.0
"PLD" MCS + Triton X-114	25.0	R-Fr	0
"PLD" MCS + P-Lipids	12.0	P-Fr + R-Fr	<1
MCS	23.0	P-Fr + Triton X-114	9.0
		P-Fr + R-Fr + Triton X-114	35.0
		P-Fr + R-Fr + P-Lipids	5.9

Experimental conditions for the assay of cortexolone are
described under methods. Concentrations of the various enzyme
preparations in the reaction mixture were: 'PLD' microsomes = 0.5
mg protein per ml; P-Fr (Particulate Fraction) = 0.3 mg per ml;
R-Fr (Reductase Fraction) = 0.5 mg protein per ml. Triton X-114 =
0.8 mg per ml and lecithin plus lysolecithin (1.7:1.0) = 1.3 mg
per ml. Incubated for 30 min at $36^{\circ}C$.

The concentrations of the detergent and of the phospholipids
used in these experiments were those required for maximum activation of 'PLD' microsomes.

Activity expressed per nmole P-450 may not represent a true
turnover number because one cannot yet say what fraction of the
total P-450 present in the enzyme preparation can function in the
hydroxylation reaction.

Reactivation of the lipid-depleted enzymes by detergents is
not surprising. As has been suggested (19), the physiologic function of phospholipids may be related to their physical properties
(ability to form micelles) which are shared by detergents. In
addition, stabilization of purified enzymes has been suggested
as a possible mechanism by which detergents as well as phospholipids
stimulate a lipid-depleted or lipid-free enzyme system (19). It

is possible that detergents and phospholipids stimulate the steroid 21-hydroxylase in the lipid-depleted enzyme preparations by a similar mechanism. In the lipid-depleted adrenocortical microsomal preparations, the possibility that the detergents act by potentiating the effect of phospholipids which are remaining in the depleted microsomes must also be considered. The ability of detergents to stimulate a lipid-depleted enzyme has been correlated with the hydrophobic lipophilic balance (HLB) number which is a measure of the relative hydrophobicity of the detergent (19). Among the detergents tested, Triton X-114 was found to be most effective in reactivating the lipid-depleted microsomal preparations. Therefore, it is possible that this detergent has the right HLB number to provide the environment required by the enzyme system to maintain conformational integrity.

ACKNOWLEDGEMENTS

This work was partly supported by ONR Contract No. N 00014-67-A-0216.0034; partly accomplished during the tenure of a dozent grant from the Alexander Humboldt Foundation, West Germany and scholarly leave from the University of Pennsylvania.

The author is greatly indebted to the following persons: Professor Hj. Staudinger for the use of the facilities of his institute and providing for technical assistance; to Dr. H. U. Schultze for phospholipid analysis and to Dr. Caroline Damsky of the University of Pennsylvania for electron micrographs. In addition, the use of the facilities of Johnson Research Foundation is gratefully acknowledged.

REFERENCES

1. Omura, T. and Sato, R., J. Biol. Chem. 239, 2370 (1964).
2. Lu, A. Y. H., Junk, K. W. and Coon, M. J., J. Biol. Chem. 244, 3714 (1969).
3. Chaplin, M. D. and Mannering, G. J., Mol. Pharmacol. 6,631(1970).
4. Leibman, K. C. and Estabrook, R. W., Mol. Pharmacol.7, 26 (1971).
5. Tagg, J. and Mitoma, C., Biochem. Pharmacol. 17, 2471 (1968).
6. Morton, R. K., Nature, 166, 1092 (1950).
7. Narasimhulu, S., 9th Intnl. Congr. of Biochemistry, Stockholm, Abst. 7c 6, 1973).
8. Narasimhulu, S., (see text) in Cooper, D. Y., Rosenthal, O. and Schleyer, H.) Hoppe-Seyler's Z. Physiol. Chem. 349, 1592 (1968).
9. Folch, J., Lees, M., Sloan Stanley, G. H., J. Biol. Chem. 226 497-509 (1957).
10. Parker, F. and Peterson, N. F., J. Lipid. Res., 6, 455-460 (1965).
11. Bartlett, G. R., J. Biol. Chem. 234, 466-468 (1959).

12. Schwartz, H. P., Dresbach, L., Polis, B. D. and Soffer, E.,
    Arch. Biochem. and Biophys. 111, 422 (1965).
13. Schultze, H. U. and Staudinger, Hj., Hoppe-Seyler's Z. Physiol.
    Chem. 352, 302 (1971).
14. Narasimhulu, S., Schultze, H. U. and Staudinger, Hj. In
    Preparation.
15. Cater, B. R., Walkden, V. and Hallinan, T., Bioch. J. 127,
    37p (1972).
16. Fleischer, S., Brierley, G., Klouwen, H. and Slauterback,
    B. D., J. Biol. Chem. 237, 3246 (1962).
17. Narasimhulu, S., Arch. Biochem. Biophys. 147, 391 (1971).
18. Fleischer, S., Klouwen, H. and Brierley, G., J. Biol. Chem.
    236, 2936 (1961).
19. Umbreit, J. N. and Strominger, J. L., Proc. Nat. Acad. Sci.
    70, No. 10, 2997 (1973).
20. Umbreit, J. N. and Strominger, J. L., Proc. Nat. Acad. Sci.
    69, 1972 (1972).
21. Narasimhulu, S., Drug. Met. and Disposition 2, 573, (1974) in
    press.
22. Masters, B. S. S., Baron, J., Tyler, W. E., Isaacson, E. L.
    Lospolluto, J., J. Biol. Chem. 246, 4143 (1971).
23. Nelson, D. O., Lorusso, D. J. and Mannering, G. J. Biochem.
    Biophys. Res. Commun. 53, 995 (1973).

ON THE STRUCTURE OF PUTIDAREDOXIN AND CYTOCHROME P-450$_{cam}$ AND

THEIR MODE OF INTERACTION*

Karl Dus

Department of Biochemistry
University of Illinois
Urbana, Illinois 61801

CAMPHOR METHYLENE HYDROXYLASE OF PSEUDOMONAS PUTIDA

The 5-exo-methylene hydroxylase of D(+)-camphor of Pseudomonas putida requires three components, a flavoprotein called putidare-doxin reductase, an iron-sulfur protein, putidaredoxin, and a variant b-type cytochrome, P-450$_{cam}$, all of which show good solubility in aqueous buffers and have been obtained in homogeneous form (Tsai et al., 1971). Cytochrome P-450$_{cam}$, the substrate and oxygen reactive component, is reduced by putidaredoxin which also acts as an effector forming very tight complexes with the hemeprotein in the presence of substrate (Tyson et al., 1972). Since these complexes can be readily formed in vitro, and since the reductase is not very specific and can be effectively replaced by other flavoproteins, the putidaredoxin-cytochrome P-450$_{cam}$-substrate complexes are of great interest to the study of the mechanism of electron transfer in this multienzyme system.

Due to its excellent solubility, cytochrome P-450$_{cam}$ is the only member of the group of P-450 hemeproteins which has been carefully characterized by chemical and physico-chemical techniques (Dus et al., 1970); it also has been crystallized in the form of its camphor complex (Yu and Gunsalus, 1970). Thus, much of our present knowledge concerning the properties of P-450 cytochromes has been obtained from studies of this bacterial protein, and its structural similarity to those P-450s which are tightly associated with membranous structures and therefore not readily available is of great interest (Dus et al., 1974). Very recently, the complete amino acid sequence of putidaredoxin, and its homology to that of adrenodoxin have been established (Tanaka et al., 1974). It is

287

the aim of this report to show how details of structural and immuno-chemical information on both putidaredoxin and cytochrome P-450$_{cam}$ can be used to gain better insight into the mode of interaction be-tween these two proteins and possibly also the mechanism of hy-droxylation.

PUTIDAREDOXIN

Structure and Homology to Adrenodoxin. The striking simi-larity in amino acid composition between putidaredoxin and adreno-doxin (Table I) and the many similarities observed in a great num-ber of their physico-chemical properties are now substantially underscored by the extensive sequence homology (Table II) found between these iron-sulfur proteins. It should be noted that the alignment of sequences shown in this table differs somewhat from that of Tanaka et al., (1974) because preference is given to the concept of maximum coincidence and inserted sequences in both chains, between residues 14 and 15 in adrenodoxin and between resi-dues 25 and 26 in putidaredoxin, making these extended chains more suitable for comparison to other ferredoxins. Clearly, the con-cept of homology is best supported in the area around the four cysteine residues believed to attach the (Fe-S)$_2$ active site struc-ture to the respective polypeptide chain because they are in iden-tical positions in both proteins, namely, 39 and 46, 45 and 52, 48 and 55, and 86 and 92, respectively. Without giving weight to the gaps or deletions which were invoked to permit favorable alignment of the two sequences, the average base change per codon was cal-culated to be 0.64. These findings are of great significance for the recognition of general structural features related to the func-tion of redox proteins designed to transfer electrons to P-450 cytochromes as well as for tracing evolutionary vestiges of early developments in the diversion of species.

But the observations of similarity have also created a serious dilemma because the two iron-sulfur proteins are not able to sub-stitute for one another in hydroxylase assays using the reconsti-tuted systems. Perhaps the discrepancy can be resolved by focusing on the difference in the number of sulfhydryls associated with each protein. Adrenodoxin contains only 5 cysteines, four of which are tied up to the (Fe-S)$_2$-prosthetic group while the fifth has a freely available sulfhydryl group. Putidaredoxin, on the other hand, has 6 half-cystines which we believe to be divided into four cysteines linked to the prosthetic group and one disul-fide bridge. A disulfide loop involving 13 residues (73-85) at or very close to the binding site of putidaredoxin to cytochrome P-450$_{cam}$ would certainly be a sufficiently distinctive structural feature to explain complete lack of interchangeability for the otherwise so closely related proteins. Unfortunately, it is not easy to secure conclusive evidence for the presence of this

TABLE I

Camphor-exo-5 hydroxylase		Steroid-11-β hydroxylase
Amino Acids	Putidaredoxin	Adrenodoxin
Asp	9	11
Asn	4	7
Thr	5	10
Ser	7	7
Glu	6	7
Gln	4	4
Pro	4	1
Gly	8	8
Ala	9	7
Val	14	7
Met	3	3
Ile	6	8
Leu	6	12
Tyr	3	1
Phe	1	4
His	2	3
Lys	3	5
Trp	1	0
Arg	5	4
CyS/2	6	5
Total	106	114
Free SH	4	5
N-terminus	Ser	Ser
C-terminus	Trp	Ala
Mol. Wt.	11,600	12,600
Prosthetic Group	(FeS)$_2$	(FeS)$_2$

disulfide bridge. Most of the evidence we have today has been obtained indirectly. Let me say here that no free sulfhydryls can be titrated in putidaredoxin without immediate destruction of the (FeS)$_2$ prosthetic group, and that the reaction comes to completion when four sulfhydryls have been titrated (Tsai et al., 1971).

Immunochemical Experiments. Since putidaredoxin is a labile iron-sulfur protein we were unable to find any evidence for antibody production in rabbits injected with an emulsion of the native form of this protein with complete Freunds adjuvant. However, when we enhanced its stability by cross linkage with glutaraldehyde (Figure 1) using a reagent to protein ratio of 40:1 we obtained

TABLE II

PROPOSED SEQUENCE HOMOLOGY

```
                        1                                          10
Putidaredoxin   H-Ser  -    -    Lys  Val  Val  Tyr  Val  Ser  His  Asn  Gly  Thr  Arg  Arg  Gln  Leu  Asp  Val  Ala  Asp
Adrenodoxin     H-Ser  Ser  Ser  Glu  Asp  Lys  Ile   -   Thr  Val  His  Phe  Ile  Asn  Arg   -    -    -    -    -   Asp
                1                                          10

                20                                                      30
Pd              Gly  Val  Ser  Leu  Met  Gln   -    -    -    -   Ala  Ala  Val   -   Ser  Asn  Gly
Ad              Gly  Glu  Thr  Leu  Leu  Thr  Lys  Gly  Ile  Gly  Asp  Ser  Leu  Leu  Asp  Gln  Asn  Asn   -
                                                            30

                     40                                  50
Pd              Ile  Tyr   -   Asp  Ile  Val  Gly  Cys  Gly  Ser  Ala  Ser  Cys  Ala  Thr   -   Val  Tyr  Val
Ad              Leu  Asp  Ile  Asp  Gly  Phe  Gly  Ala  Cys  Gly  Thr  Leu  Ala  Ser  Ser  Cys  His  Leu  Ile  Phe  Glu
                          40                                  50                                           60

                                   60                              70                              80
Pd              Asn  Glu  Ala  Phe  Thr  Asp  Lys  Val  Pro  Ala   -   Ala  Asn  Glu  Arg  Glu  Ile  Gly  Met  Leu  Glu  Cys  Val
Ad              Gln  His  Ile  Phe   -   Glu   -   Leu  Glu  Ala  Ile  Thr  Asn  Glu   -   Glu  Ile  Asn  Met  Leu  Asp  Leu
                                   70                                                        80

                80                           90                            100
Pd              Thr  Ala  Glu  Leu  Lys  Pro  Asn  Ser  Arg  Leu  Cys  Cys  Gln  Ile  Ile  Met  Thr  Pro  Gln  Leu  Asp  Gly  Ile
Ad              Ala  Tyr  Gly  Leu  Thr  Asp  Arg  Ser  Arg  Leu  Gly  Ile  Gln  Ile  Leu  Leu  Thr  Lys  Ala  Met  Asp   -   Asn
                                        90                            100

                100               106
Pd              Val   -   Val  Asp  Val  Pro  Asp   -    -   Arg  Gln  Trp-OH
Ad              Met  Thr  Val  Arg  Val  Pro  Asp  Ala  Val  Ser  Asp  Ala-OH
                                    110                        114
```

Without considering gaps or deletions (–) the average base change per codon is 0.64.

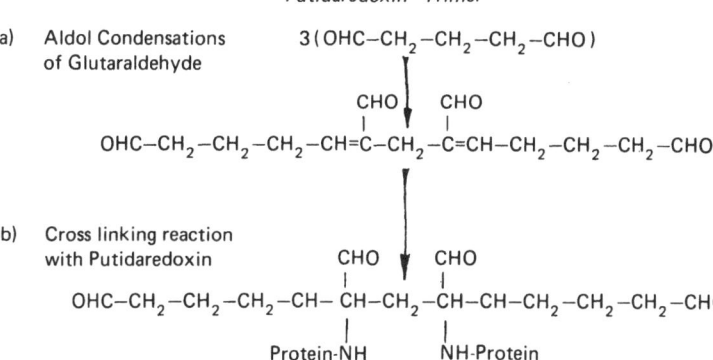

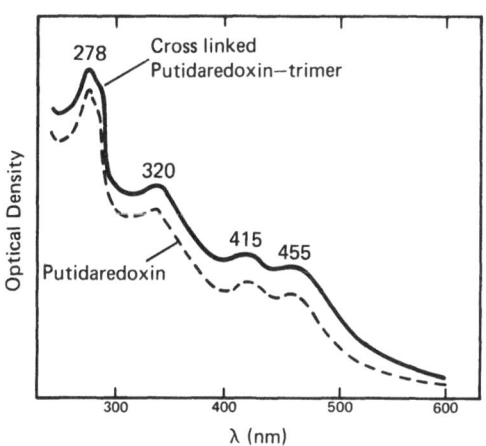

Figure 1.   Cross Linkage and Absorption Spectra of Putidare-
doxin-Trimer.   The optical absorption spectra of putidaredoxin
monomer and trimer refer to the oxidized forms in Tris-HCl buffer,
pH 7.4, 50 mM.   From Litchfield et al., (1974).

about 10% of the protein in form of an enzymatically active trimer
with unchanged spectral properties, and this trimer, as isolated
by gel filtration, elicited in good titers antibodies which gave
complete precipitin reactions with native putidaredoxin.   Thus,
no major antigenic site of the iron-sulfur protein was obstructed
in the trimer, and the ability to interact with cytochrome P-450$_{cam}$
in the hydroxylase reaction was also unimpaired.   If putidaredoxin had free sulfhydryl groups, they would have readily reacted

with the cross linking reagent, and the resulting polymer most
likely would have lost its capacity for binding of the heme protein.
But no evidence for reaction of glutaraldehyde with cysteines was
found in the trimer. Instead, two of the three lysines present
in putidaredoxin which are probably exposed on the periphery of
the protein, perhaps those in positions 2 and 59, were found to be
responsible for the formation of the trimer. Again, this may be
taken as evidence that probably no free sulfhydryls are available
in putidaredoxin, and that the active site including the proposed
disulfide loop is still fully exposed in the trimer.

In reference to a discussion between Drs. Peterson and Baron
this morning in which Dr. Baron stated that he was unable to observe cross reaction between putidaredoxin and anti-adrenodoxin
antibodies I would like to mention here that we have carried out
the complementary experiment with essentially the same negative
result. Our anti-putidaredoxin trimer antibodies apparently were
unable to cross react with adrenodoxin.

Interaction with Cytochrome P-450$_{cam}$. Both physical and
kinetic measurements indicated that putidaredoxin, cytochrome
P-450$_{cam}$ and camphor in a 1:1:1 ratio form an active complex; but
ratios of putidaredoxin to cytochrome P-450$_{cam}$ as high as 6:1 with
higher catalytic activity are also indicated by circular dichroism,
isoelectric focusing, and turnover number for the hemeprotein.
Perhaps the most instructive demonstration of this phenomenon is
given by crossing of focused protein bands in electrofocusing
(Figure 2). All of the cytochrome (pI=4.5) present as a prefocused band is converted into complexes during the crossing of
excess putidaredoxin (pI<< 4.5) and forms new bands with intermediate isoelectric points dependent on the ratio of putidaredoxin
to P-450$_{cam}$. These complexes were isolated and the ratio of components was determined by quantitative amino acid analysis.

CYTOCHROME P-450$_{cam}$

Inhibition of Camphor Hydroxylation by Anti-P-450$_{cam}$ Antibodies. In contrast to putidaredoxin, cytochrome P-450$_{cam}$,
especially in form of its substrate complex, is quite stable and
it elicited antibodies in good titers. These antibodies were
partially purified by ammonium sulfate precipitation and then
digested by pepsin to yield the corresponding univalent Fab fragments. This was deemed necessary to obtain soluble antigen-anti-
body complexes for inhibition studies with the reconstituted camphor hydroxylase. As shown in Figure 3, complete inhibition of
camphor hydroxylation can be accomplished with these fragments although a rather high antibody concentration is required. This
experiment proves, however, that inhibition is not due to general
precipitation of proteins but results from specific antigen-anti-
body complex formation. After this point was clearly established

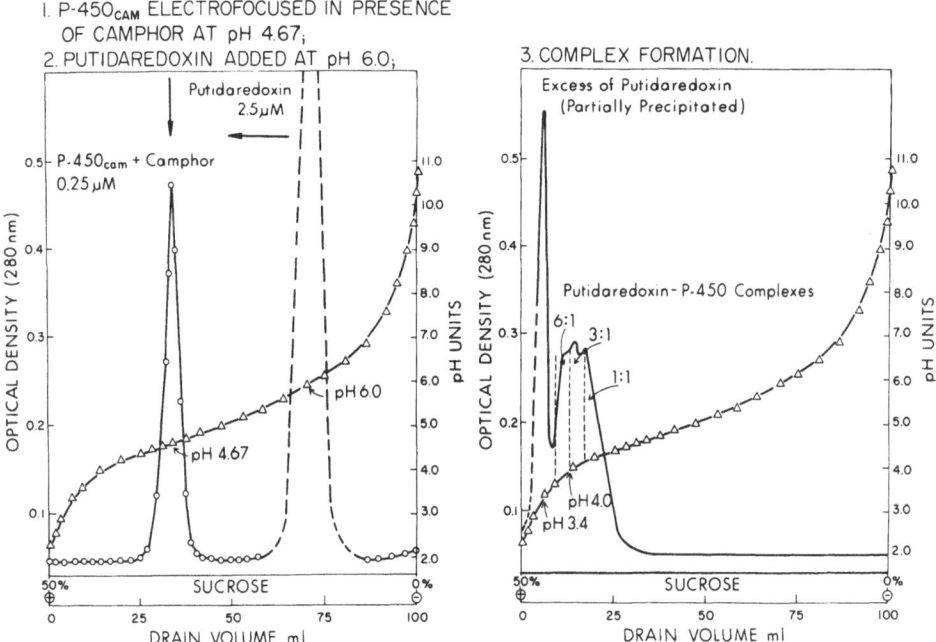

Figure 2. Formation of Putidaredoxin-Cytochrome P-450 cam Complexes by Crossing of Bands During Electrofocusing. After focusing P-450 cam in form of its camphor complex a 10-fold excess of putidaredoxin was injected via a plastic needle at a position roughly corresponding to pH 6.0. Continuation of the electrofocusing experiment for another 8 hr period caused putidaredoxin to cross the P-450 cam band due to its much lower pI. In the process the P-450 cam band disappeared and a group of closely spaced black-brown bands appeared around pH 4 together with partially precipitated excess of putidaredoxin at pH 3.4. These bands were withdrawn and the ratio of components in each band was determined by amino acid analysis. From Dus et al., (1971).

we investigated inhibition of camphor hydroxylation by undegraded purified antibodies against cytochrome P-450 cam; putidaredoxin trimer, and the complex formed between cytochrome P-450 cam, camphor and putidaredoxin as shown in Figure 4. Much smaller antibody concentrations were required in these experiments and nearly identical inhibition curves were obtained with anti-putidaredoxin-P-450 cam complex antibodies, irrespective whether putidaredoxin or P-450 cam was chosen as the limiting component in the hydroxylase assay, indicating that the complex obtained by simply mixing and

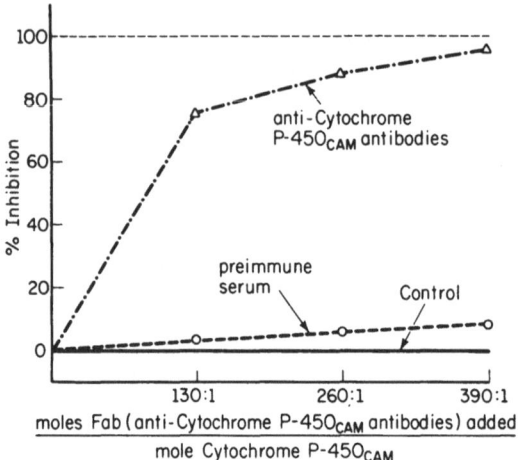

Figure 3.   Inhibition of Camphor Hydroxylation by Anti-Cytochrome P-450$_{cam}$ Fab Fragments. In the hydroxylase assay, NADH oxidation monitored at 340 nm is inhibited progressively by addition
of monovalent Fab fragments of Anti-P-450$_{cam}$ IgG. P-450$_{cam}$ was preincubated with these fragments for 10 min at 5° before adding the
other components. From Litchfield et al., (1974).

incubating putidaredoxin and P-450$_{cam}$ in a ratio of 2:1 at room
temperature probably contained the two proteins in 1:1 stoichiometry.
Excess of putidaredoxin seems necessary to guarantee that all
P-450$_{cam}$ is present in form of the complex while free putidaredoxin,
as demonstrated before, cannot elicit antibodies and therefore
does not interfere. It is of great interest that this complex is
sufficiently stable, soluble, and antigenic to permit good antibody
production in rabbits because the antibody population directed
against the complex should be devoid of antibodies specific for the
mutual binding sites of putidaredoxin and P-450$_{cam}$.   Provided
these binding sites contain major antigenic determinants of these
proteins it will then be relatively easy to gain access to site
specific antibodies by precipitating the complex with the respective
antibody populations made against the individual proteins. In each
case one may obtain antibodies specific for the site involved in
the interaction between these two proteins (Litchfield et al., 1974).
Further experiments designed to obtain more information on this
point will be discussed at a later point in this report.

    Competitive Binding Assays and the Antigenic Determinants of
Cytochrome P-450$_{cam}$. Another line of investigation of cytochrome

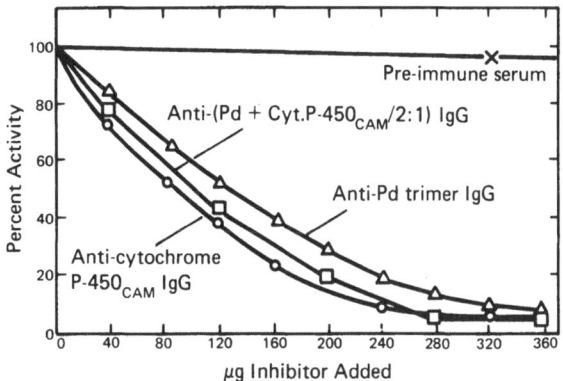

Figure 4. Inhibition of Camphor Hydroxylation by Anti-Cytochrome P-450$_{cam}$ Antibodies (IgG), Anti-Putidaredoxin-Trimer Antibodies (IgG), and Anti-Putidaredoxin-P-450$_{cam}$ complex Antibodies
(IgG). NADPH oxidation was monitored at 340 nm. The hydroxylase
assay and preincubation period were the same as given in Figure 3.
From Litchfield et al., (1974).

P-450$_{cam}$ which we have recently pursued with great interest is
application of immunochemical techniques to the search for characteristic structural features expressed on the surface of this
protein. Figure 5 shows an attempt to determine the approximate
number of major antigenic sites of this hemeprotein by extrapo-
lating backwards from the maximum of the precipitin curve to zero
antigen concentration. The number of major sites projected by
this approach is about 5. In order to locate these 5 antibody
binding sites of P-450$_{cam}$ we tested for cross reacting material
(CRM) remaining after partial degradation of the P-450$_{cam}$-camphor
complex by enzymes and selective chemical reagents. The purpose
of this study was twofold. First, we explored the stability of the
protein under various conditions, in particular its conformational
stability around the heme attachment and substrate binding sites
(Dus et al., 1973,A). Second, we tried to determine the nature
of the groups involved in forming the major antigenic sites, a
necessary step toward resolving the total anti-P-450$_{cam}$ antibody
population into discrete fractions, each specific for a single
site. Our experimental test was based on a sensitive radioimmunoassay measuring inhibition of binding between ^{125}I-labeled P-450$_{cam}$
and anti-P-450$_{cam}$ antibodies by competing CRM. The maximum inhibition by unlabeled P-450$_{cam}$ was normalized to 100% (Dus et al.,
1973,A). As shown in Figure 6, following the iodination procedure
of Landon et al., (1967) which is specific for iodination of

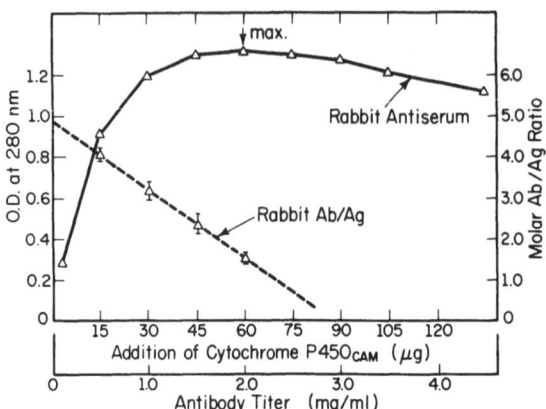

Figure 5. Major Antigenic Sites of Cytochrome P-450$_{cam}$. The
molar antigen to antibody ratio assumed to be 1:1 at the maximum
of the precipitin curve has been projected backwards to zero
antigen concentration. The precipitin reaction was carried out in
50 mM phosphate-150 mM NaCl, pH 7.0, with 48 hr incubation at 5°.
From Dus et al., (1973).

tyrosyl residues, no other amino acid residues of P-450$_{cam}$ were
iodinated. Quantitative amino acid analysis indicated that 50% of
the tyrosyl residues were converted to their monoiodo derivatives.
The iodinated hemeprotein retained the characteristic spectral
properties and practically all of the enzymatic activity of the
unmodified protein (Dus et al., 1973,B). Employing this assay for
competitive binding experiments with labeled P-450$_{cam}$, unlabeled
competing antigens, and anti-P-450$_{cam}$ antibodies, we tested for
antibody binding sites remaining after derivatization, denaturation,
and both chemical and proteolytic degradation(Figure 7). It is of
great interest that digestion with proteolytic enzymes leads to
rapid loss of almost all antigenic determinants. The tryptic digest
shown in this figure (Figure 7A) is a good example. Essentially the
same results were obtained with chymotrypsin and thermolysin.
Apparently no undigestable "core"-peptide is formed from P-450$_{cam}$.
We have a special interest in the tryptic digest because one peptide
is released almost instantaneously leading to a loss of about 20%
of the original antibody binding capacity. It is easy to isolate
and purify this peptide. It is about 15 residues long and contains
the only Trp residue of the protein together with a cluster of 3
Arg (Dus et al., 1973,A). This explains both its rapid release by
trypsin and its apparently high antigenicity. It must be located
at the very periphery of the molecule, and probably accounts for
one of the 5 major antigenic sites determined from the precipitin

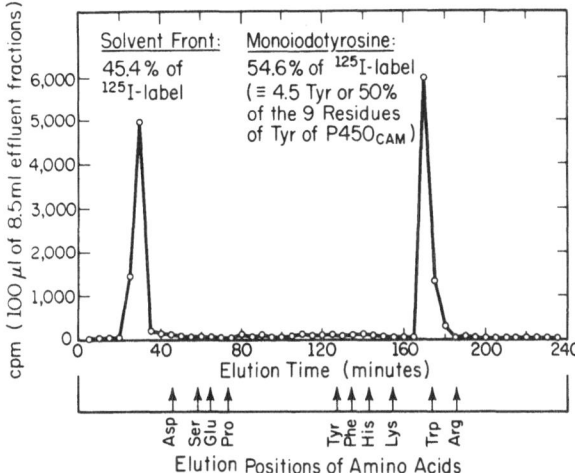

Figure 6. Separation of Enzymic Digest of ^{125}I-labeled Cytochrome P-450 cam. Iodination was carried out with ^{125}I, and chloramine T as the oxidizing agent, in 50 mM phosphate buffer, pH 7.0 , by stirring the protein solution (1 mg/ml) on ice in the cold room and stopping the reaction by addition of $NaHSO_3$. The reaction mixture was dialyzed and lyophilized. Complete digestion of the labeled P-450 cam to the level of free amino acids was then accomplished with an enzyme cocktail consisting of pronase and leucine aminopeptidase. The resulting mixture of amino acids was resolved and recorded on an automatic amino acid analyzer, Beckman, Model 120. The effluent was collected in 8.5 ml fractions, and 100 µl of each fraction were transferred to scintillation vials and counted with the ^{14}C window of a Beckman Model LS-30 scintillation counter.

curve. By exhaustive digestion with a mixture of carboxypeptidases A and B we have demonstrated that Trp and several Arg are released from the C-terminus of the protein. Thus, part of the carboxyl end of the protein is located on the surface of P-450 cam which also explains why this protein is so readily digested by carboxypeptidase. A very different picture is presented by cleavage of P-450 cam by BrCN under mild conditions (Figure 7B). Even after digestion for 72 hr, more than 75% of the antibody binding capacity is retained. As you will see in a later part of this report this finding was of great significance and led to the isolation of a small hemepeptide which retains at least 2 but possibly even 3 of the major antigenic sites of the native hemeprotein, indicating that the heme attachment site is close to the surface of the molecule and consists of a tightly structured segment of linear sequence rather than a combination of small portions of different parts of the polypeptide chain brought together at the heme site by folding into the 3-dimensional shape of the molecule

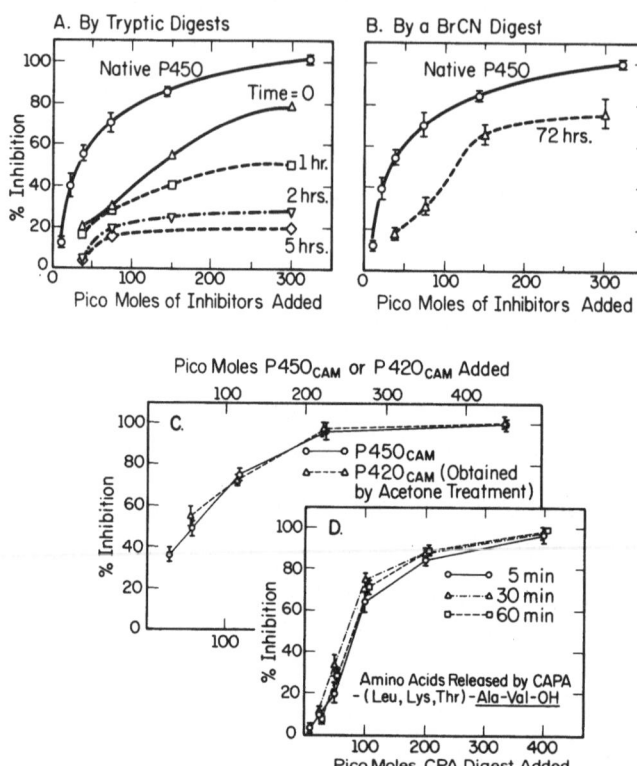

Figure 7. Competitive Binding Assay as a Probe of Structural
Integrity of the P-450_cam Degradation Products. To unlabeled
antigen in 0.15 M NaCl–0.05 M sodium phosphate buffer, pH 7.5, a
constant volume of antibody was added to give a total volume of
0.30 ml. After a 2 hr incubation at room temperature one equivalent of ^{125}I-labeled P-450_cam, monoiodinated at five of its nine
tyrosine residues, was added. After an additional 48 hr incubation
at 10° the precipitates were collected by centrifugation, washed
once with cold buffer, and dissolved in 0.2 ml of 0.2 N NaOH for
counting with the ^{14}C window in a Beckman liquid scintillation
counter, Model LS-30. The inhibition of binding by CRM is based
on maximal P-450_cam-anti-P-450_cam antibody binding normalized
to 100%.

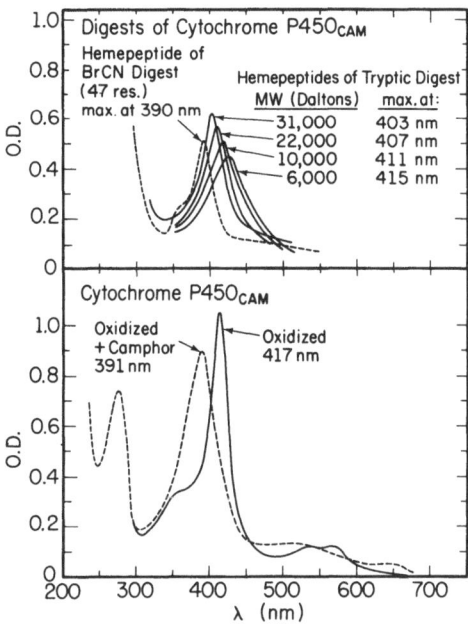

Figure 8.   Spectral Properties of Cytochrome P-450$_{cam}$ and its Smallest BrCN Hemepeptide Compared to Those of Trypsin Derived Hemepeptides.   All spectra were recorded with a Cary 14 spectrophotometer in 50 mM phosphate buffer, pH 7.0.

(Dus et al., 1973,B).   Since BrCN is specific for cleavage of Met residues it is quite conceivable that at least 2 of these residues are readily accessible to the reagent such that cleavage precedes unfolding at the heme attachment site.   In addition to this immunochemical evidence, this concept that the BrCN hemepeptides retain most of their conformational structure is further supported by their absorption spectra (Figure 8), having a Soret maximum at 390 nm, and by the stoichiometric amount of heme which travels with these peptides throughout purification.   The smallest of these hemepeptides is the one most interesting for sequencing and also happens to be the one most amenable to complete purification under mild conditions.   In Figure 7C, I have demonstrated that P-420 obtained by acetone treatment is immunochemically indistinguishable from P-450, probably because treatment with complete Freunds adjuvant may lead to conversion of P-450 to P-420.   Finally, in Figure 7D, you can see that short time incubation of P-450$_{cam}$ with carboxypeptidase alone, at low enzyme to substrate ratios (1:200), completely releases the ultimate and penultimate residues, Val and Ala, respectively, without measurable loss of antibody binding capacity.   This observation was paralleled by

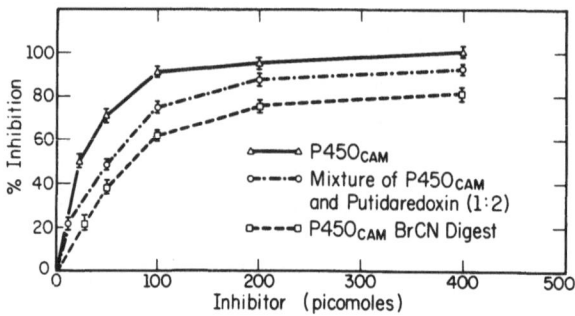

Figure 9. Inhibition of Binding Between ^{125}I-Labeled Cytochrome P-450$_{cam}$ and its Antibodies by Unlabeled P-450$_{cam}$ and CRM.
To unlabeled antigen in 0.15 M NaCl-0.05 M sodium phosphate buffer,
pH 7.5, a constant volume of antibody was added to give a total
volume of 0.30 ml. After a 2 hr incubation at room temperature
one equivalent of ^{125}I-labeled P-450$_{cam}$, monoiodinated at five of
its nine tyrosine residues, was added. After an additional 48 hr
incubation at 10° the precipitates were collected by centrifugation, washed once with cold buffer, and dissolved in 0.2 ml of
0.2 N NaOH for counting with the ^{14}C window in a Beckman liquid
scintillation counter, Model LS-30. The inhibition of binding by
CRM is based on maximal P-450$_{cam}$-anti P-450$_{cam}$ antibody binding
normalized to 100%.

retention of spectral properties and catalytic activity in des-
Val-Ala-P-450$_{cam}$. Thus, it seems that these two residues form an
extended arm at the surface and are to some extent expendable.

Figure 8 shows the characteristic differences in spectral
properties between degradation products of P-450$_{cam}$ obtained by
proteolytic digestion and BrCN cleavage. Heme-containing peptides
of the tryptic digest, for instance, clearly indicate the progressive shift with decreasing molecular weight from the Soret
maximum of the cytochrome-substrate complex to that of the free
hemeprotein while even the smallest BrCN hemepeptide still retains
the spectral characteristics of the P-450-camphor complex and
gives proper reduced CO spectra as well. Whether the substrate
is still present in this hemepeptide, and in stoichiometric
amounts, is a matter of speculation at this time. It would seem
logical to assume that the presence of the substrate is responsible
for the fact that the heme is still locked in place although no
covalent link is present in this b-type cytochrome. On the other
hand, conformational changes close to the heme attachment site
induced by the BrCN cleavage could conceivably be responsible for

such a spectral shift. Since the hemepeptide is no longer active catalytically we intend to use radioactively labeled camphor to investigate this point.

Figure 9 extends the competitive binding assays I have shown before to the P-450-putidaredoxin complex (Litchfield et al.,1974). Again, the apparent number of anti-p-450$_{cam}$ antibody binding sites has decreased by roughly 20% as compared to the uncomplexed P-450$_{cam}$. It is of interest to note that putidaredoxin as part of this complex is obviously protected against denaturation during emulsification and injection into the rabbit. Unfortunately, we have not yet been able to accomplish iodination of putidaredoxin under conditions sufficiently mild to retain all of its spectral and immunochemical properties. Thus, we cannot say whether the drop in antibody binding sites observed for P-450$_{cam}$ as part of the complex also applies to putidaredoxin.

I should mention that one sulfhydryl group of P-450$_{cam}$ readily engages in dimer formation via intermolecular disulfide bridges. These dimers form at high protein concentrations in the absence of mercaptoethanol during chromatography and electrophoresis and can be separated from the monomer by gel filtration on Sephadex or by acrylamide gel electrophoresis. They show an apparent molecular weight of 90,000 daltons by ultracentrifugal analysis (Lipscomb et al., 1974) and are readily reconverted to monomer by treatment with mercaptoethanol. Interestingly, the dimer has the same spectral and catalytic properties as the monomer but its antibody binding capacity is 20% lower than that of the monomer, and in the competitive binding assay gives a curve which is practically identical to that produced by the putidaredoxin-P-450$_{cam}$ complex. Again, we seem to have identified another binding site of P-450$_{cam}$ because this site is different from either the Trp-Arg or the putidaredoxin binding site. The 2-3 binding sites associated with the hemepeptide, however, may well include the putidaredoxin binding site. In summary, we obtain a detailed picture of the projected major antibody binding sites of P-450$_{cam}$ (see Table III). It is too early to decide whether the sites associated with the hemepeptide are due to the heme, the substrate, or other groups.

Sulfhydryl Groups and S-Alkyl Derivatives of P-450$_{cam}$. As shown in Table IV, the presence of 6 half-cysteines in P-450$_{cam}$ is documented by amino acid analyses of the alkylated or performate oxidized protein. All six residues are present as free sulfhydryls and can be titrated in the denatured protein by PBM (Yu and Gunsalus, 1974) or other sulfhydryl reagents as well as amperometrically (Figure 10). But only 4 of these 6 SH groups can be titrated in the native P-450$_{cam}$. A surprising phenomenon was observed when we titrated these 4 sulfhydryl groups with N-ethyl maleimide (NEM) or Azobenzene sulfenyl bromide (ABS-Br) (Lipscomb et al., 1974).

TABLE III

Major Antigenic Determinants of Cytochrome P-450$_{cam}$

Hemepeptide	2-3
Trp-Arg Site at C-terminus	1
SH group (+ surrounding area) involved in dimer formation	1
Binding site to Putidaredoxin*	1
	5-6

*May be included in hemepeptide area

TABLE IV

Cytochrome P-450$_{cam}$: Total Number of Sulfydryl Groups

Aminoethyl Cysteine	6
Cysteic Acid	6
Boyer's Method (PMB)	6
Amperometric Titration (PMA in SDS)	5.7-6.5

One sulfhydryl reacted extremely fast and apparently had no influence on spectral or catalytic properties, which is in agreement
with the fact that the dimer did not contain this group. Therefore, it is the one which is located close to the surface of the
protein and involved in dimerization.

The other 3 accessible sulfhydryl groups are apparently
brought close together by the three-dimensional structure of the
protein and are probably located in the vicinity of the heme
group. Titration of these groups causes conformational changes
which by using suitable reagents (e.g. NEM or ASB-Br) can be revealed as sizable spectral shifts (Figure 11, Table V) as well as
changes in redox potential and EPR signals and loss of DPNH reactivity (Lipscomb et al., 1974). The properties of these S-alkyl
derivatives are similar to those of the substrate free low-spin
form of ferri-cytochrome P-450$_{cam}$ yet neither the capacity to
bind camphor nor the ability to form reduced CO complexes with
maxima at 446 nm (Figure 12) has been lost, and the ability to
form product has hardly been altered. By implication this means
that the interaction with putidaredoxin as an effector has not
been altered significantly. That several of the sulfhydryl groups
of P-450$_{cam}$ are in close proximity to each other is also suggested
by an incidental observation during exhaustive iodination of this
protein which produces a similar shift of the Soret maximum from
391 to 417 nm accompanied by formation of an intramolecular disulfide bridge.

In the camphor free P-450$_{cam}$ a fifth sulfhydryl is available
for slow titration while it is apparently protected in the enzyme-
substrate complex, perhaps by forming a hydrogen bond to the camphor carbonyl group. It is interesting that the EPR spectra (see
Figure 13) of the P-450$_{cam}$ S-alkyl derivatives containing up to 4,
or even 5 reacted SH groups closely resemble those of the low-spin
ferric protein with signals at g = 2.4, 2.2, 1.9 while derivatization of the last (6th) SH group suddenly produces a signal at g=6
indicating conversion to pure axial symmetry at the heme (Lipscomb
et al., 1974). Thus, the last sulfhydryl is strongly implicated in
chelation to the heme iron. This interpretation is further supported by the presence of a single sulfhydryl in the small BrCN
derived hemepeptide which I mentioned to you this morning (Dus et
al., 1974). Derivatization of the last two SH groups causes a
spectral change from predominantly P-450$_{cam}$ character to exclusively
P-420$_{cam}$ type. A simplistic interpretation of these results is
given in Figure 14. It still needs to be demonstrated by differential labeling of SH groups and isolation of pure, S-alkylated
peptides, that indeed one sulfhydryl is chelating to the heme iron,
and that three readily available sulfhydryls are close to each
other and in the vicinity of the heme and substrate attachment
sites. Perhaps one of the longer BrCN hemepeptides may still
contain the essential components of all these features.

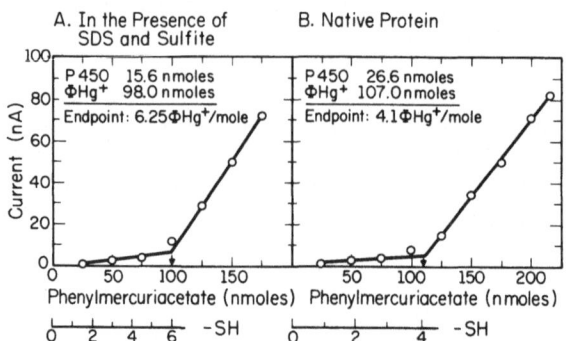

Figure 10. Amperometric Titration of Sulfhydryl Groups of Cytochrome P-450$_{cam}$.

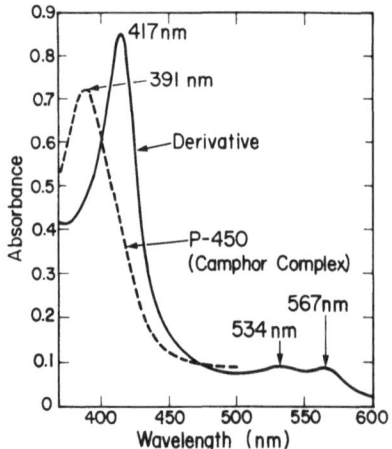

Figure 11. Derivatization of Ferric-Cytochrome P-450 Camphor Complex with N-Ethyl Maleimide (NEM$_4$P-450): Spectral Shift.

TABLE V

Spectral Properties of N-Ethyl Maleimide Derivative

	γ	β	α
		Oxidized	
P-450 (No Cam)	417	538	569
Derivative*	416	536	568
		Reduced	
P-450 (No Cam)	411	541	
Derivative	410	543	
		Reduced + CO	
P-450	446	550	
Derivative	446	550	

*Derivative's Soret not Shifted by Camphor

SUMMARY

Structural and immunochemical experiments with putidaredoxin, cytochrome P-450$_{cam}$, and their 1:1 complex have led us to the following conclusions: Despite the remarkable sequence homology between putidaredoxin and adrenodoxin which permits a tentative assignment of cysteines binding to the (Fe-S)$_2$ prosthetic group, these redox proteins cannot replace each other in reconstitution experiments because putidaredoxin contains a disulfide loop close to its P-450$_{cam}$ binding site. This feature may also be responsible for the complete lack of immunochemical cross reactivity between these proteins. The stability of putidaredoxin can be enhanced significantly by cross linkage with glutaraldehyde without change in spectral, catalytic, or immunochemical properties. Putidaredoxin also gains stability by binding to the P-450-camphor complex in a 1:1 ratio. Precipitation of this complex with anti-P-450$_{cam}$ antibodies gives access to site specific antibodies directed against the putidaredoxin binding site of P-450$_{cam}$.

A series of putidaredoxin-cytochrome P-450$_{cam}$-substrate complexes with ratios of 1 to 6 molecules of redoxin per molecule of cytochrome have been obtained by migration of excess redoxin across prefocused P-450$_{cam}$ in electrofocusing.

Complete inhibition of camphor hydroxylation was achieved by

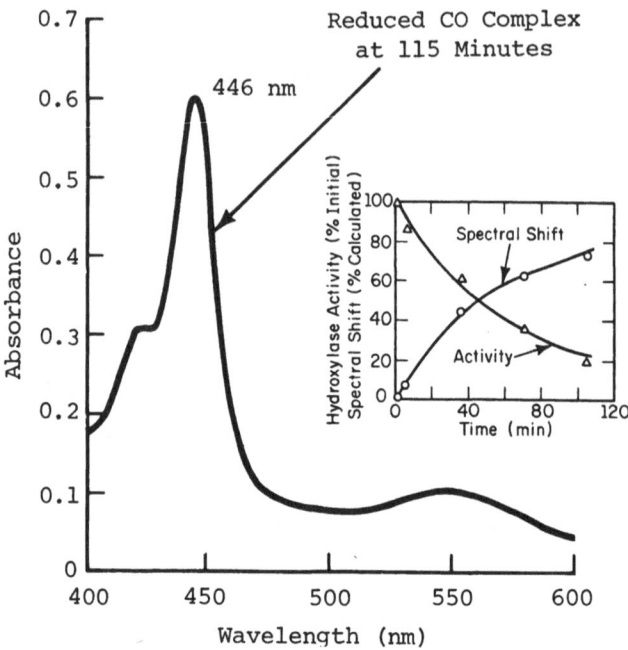

Figure 12.   Reduced CO-complex of S-alkylated Cytochrome
P-450$_{cam}$ (NEM$_4$P-450). The decrease in hydroxylase activity with
time, expressed in percent of initial activity of the reconstitued
hydroxylase prior to alkylation, and the corresponding increases
in Soret absorption at 417 nm are given by the inserted figure
while the main figure refers to the spectrum of the reduced CO-
complex of the derivative obtained after 115 minutes of reaction.

anti-P-450$_{cam}$ antibodies, their Fab fragments, anti-putidaredoxin-
trimer antibodies, and antibodies directed against the putidare-
doxin-P-450$_{cam}$ complex. Five major antigenic sites were tentatively established for P-450$_{cam}$, two of which seem to be associated
with the BrCN hemepeptide while one each relates to the putidaredoxin binding site, the Trp-Arg site close to the C-terminus, and
the site surrounding the most reactive SH group which gives rise to
dimer formation. Iodination of P-450$_{cam}$ at tyrosyl residues only
permitted use of a sensitive radioimmunoassay procedure for testing
of cross reacting material (CRM) remaining after degradation of
P-450$_{cam}$ with BrCN and enzymes, denaturation with acetone, and
complex formation with the redoxin. The BrCN hemepeptide still has
a Soret maximum at 390 nm and reacts with CO yielding a P-420
spectrum.

All 6 half-cystines of P-450$_{cam}$ are present as free sulfhydryls

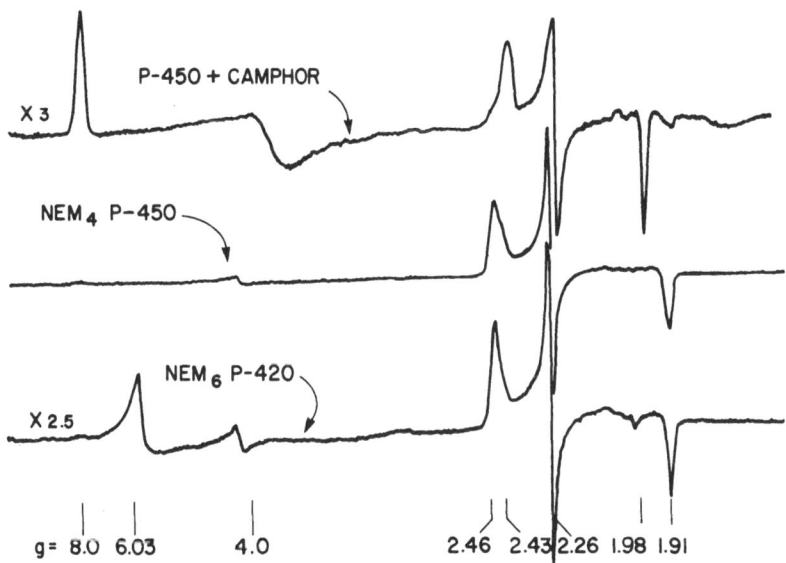

Figure 13.   EPR Spectra of Cytochrome P-450$_{cam}$ and its NEM-
Derivatives.

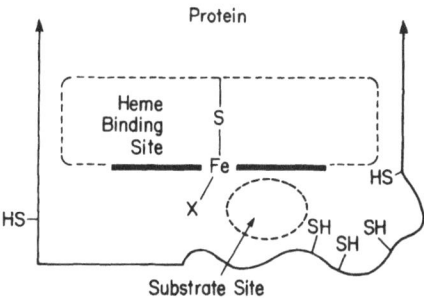

Figure 14.   Schematic View of Cytochrome P-450$_{cam}$ Sulfhydryl
Groups.

and can be titrated after denaturation but only 4 of them are
available in the P-450-camphor complex. Three of these are close
to each other and the heme, and work in concert; their alkylation
with N-ethyl maleimide (NEM) leads to shifts of the Soret from
391 to 417 nm and concomitant changes in redox potential, EPR-
signals and DPNH - reactivity. The fifth SH group is protected
by camphor while the 6th SH group, still present in the BrCN hemepeptide, is implicated in chelation to the heme iron by a drastic
change in EPR spectra, reflecting pure axial symmetry at the heme
after complete alkylation by NEM.

FOOTNOTES

*   Supported by Grant GM-18902 from the National Institutes of
Health.

REFERENCES

Dus, K., Katagiri, M., Yu, C.A., Erbes, D.L., and Gunsalus, I. C.
      1970. Chemical characterization of cytochrome P-450$_{cam}$.
      Biochem. Biophys. Res. Commun. 40: 1423-1430.
Dus, K., Lipscomb, J. D. and Gunsalus, I. C. 1971. Structure-
      function parameters of the pure soluble camphor methylene
      hydroxylase components and systems. 162nd ACS National
      Meeting, Washington, D. C., Abstracts, Biol. No. 44.
Dus, K., Litchfield, W. J., Miguel, A. G. and Harrison, J. E.
      1973. A. Structural stability of cytochrome P-450$_{cam}$.
      FASEB Meeting, Atlantic City, Abstracts No. 1583. Fed. Proc.
      32: 502.
Dus, K., Miguel, A. G., Smith, P. C., Litchfield, W. J. and
      Harrison, J. E. 1973. B. "Hemepeptides" derived from cyto-
      chrome P-450$_{cam}$. 9th International Congress of Biochemistry,
      Stockholm, Abstracts 7 d 4.
Dus, K., Litchfield, W. J., Miguel, A. G., van der Hoeven, T. A.,
      Haugen, D. A., Dean, W. L. and Coon, M. J. 1974. Immuno-
      chemical and Compositional Comparison of Cytochromes P-450$_{cam}$
      of Pseudomonas putida and P-450$_{LM}$ of Phenobarbital-Induced
      Rabbit Liver Microsomes. This volume, p. 48-70.
Lipscomb, J. D., Dus, K., Harrison, J. E. and Gunsalus, I. C.
      1974. Cytochrome P-450$_{cam}$: Cysteinyl mercapto reactivities
      and derivatives, Biochem. Biophys. Res. Commun, in press.
Litchfield, W. J., Emptage, M. and Dus, K. 1974. Immunochemical
      studies with antibodies against cytochrome P-450$_{cam}$ and
      putidaredoxin. Biochemistry-Biophysics Meeting, Minneapolis,
      Abstracts No. 823, Fed. Proc. 32: 1369.
Landon, J., Livanou, T. and Greenwood, F. C. 1967. The prepara-
      tion and immunological properties of ^{131}I-labeled

adrenocorticotrophin. Biochem. J. 105: 1075-1083.

Tanaka, M., Haniu, M., Yasunobu, K. T., Dus, K. and Gunsalus, I.C. 1974. The Amino acid sequence of Putidaredoxin, an iron-sulfur protein from Pseudomonas putida. J. Biol. Chem. 249: 3689-3701.

Tsai, R. L., Gunsalus, I. C. and Dus, K. 1971. Composition and structure of camphor hydroxylase components and homology between putidaredoxin and adrenodoxin. Biochem. Biophys. Res. Commun. 45: 1300-1306.

Tyson, C. A., Lipscomb, J. D. and Gunsalus, I. C. 1972. The Roles of putidaredoxin and P-450 complexes in methylene hydroxylation. J. Biol. Chem. 247: 5777-5781.

Yu, C.-A., Katagiri, M. and Gunsalus, I. C. 1974. Cytochrome P-450$_{cam}$: I. crystallization and properties. J. Biol. Chem. 249: 94-101.

METABOLIC CONTROL OF CYTOCHROME P-450$_{cam}$

Julian A. Peterson and Donald M. Mock
Department of Biochemistry, The University of Texas
Health Science Center, Southwestern Medical School
5323 Hines Boulevard
Dallas, Texas 75235

INTRODUCTION

Long before cytochrome P-450 was discovered, it was known that the metabolism of steroids was inhibited by CO (Ryan and Engel, 1957). It was a study of carbon monoxide inhibition of steroid metabolism by adrenal cortex which led to the discovery of the involvement of cytochrome P-450 in this reaction (Estabrook et al., 1963). Following this original observation, carbon monoxide inhibition of various monooxygenase reactions has been used in a number of laboratories to study the metabolism of these compounds. Philadelphia has been the site of a number of memorable experiments in the elucidation of the role of this enzyme in drug and steroid metabolism. Therefore, it is quite fitting that the subject of this paper be the control of cytochrome P-450 reactions and the role which carbon monoxide inhibition of cytochrome P-450 has played in obtaining an understanding of this control.

The metabolism of camphor by Pseudomonas putida which has been studied in both our laboratory and that of Dr. Gunsalus has taught us many things about mammalian monooxygenase reactions (Peterson et al., 1973; Gunsalus et al., 1973). The use of this model system for the study of monooxygenase reactions has not been without its associated problems. One of our initial experiments involved an attempt to measure the carbon monoxide (CO) inhibition of camphor hydroxylation which was catalyzed by the soluble cytochrome P-450 of this microorganism. The initial experiments were unsuccessful in demonstrating the expected inhibition. The lack of CO inhibition of camphor hydroxylation left us in a very

311

embarrassing position in our attempts to explain how this soluble
enzyme could be a good model for the mammalian microsomal enzyme
if one of the most basic properties of the mammalian system, CO
inhibition, couldn't be demonstrated.

In an extended study of the kinetics of the reactions of
cytochrome P-450$_{cam}$, an explanation for the initial observation
of no CO inhibition was discovered. The greater understanding of
the metabolic control which came about as a result of this kinetic
study has permitted us to predict conditions under which the
bacterial hydroxylation would be sensitive to CO.

REACTION CYCLE

In a beautiful series of experiments Gunsalus' group (Katagiri
et al., 1968) showed that the electron transport pathway for cytochrome P-450$_{cam}$ (Figure 1) was very similar to the previously elucidated adrenal mitochondrial cytochrome P-450 (Omura et al., 1965).
Several major differences are immediately apparent. The components
of this pathway are all soluble, readily purified proteins and the
electron donor is NADH rather than NADPH. The electrons are
transferred from NADH via a flavoprotein dehydrogenase to putidaredoxin, an iron sulfur protein, and finally to cytochrome P-450
the terminal oxidase.

The reaction cycle which is the starting point for this investigation is shown in Figure 2. This cycle was originally proposed in a slightly different form by Estabrook (1968). The point
on the cycle which is best thought of as the origin is the substrate
free ferric form of the enzyme. The initial reaction is the binding
of substrate by the ferric low spin form of the enzyme. The substrate binding reaction is extremely rapid as is indicated by the
second order reaction rate constant of $4.8 \times 10^{6} M^{-1} sec^{-1}$ (Griffin
and Peterson, 1972). In the usual buffered reaction medium which
is 6 mM in camphor, the $t_{\frac{1}{2}}$ of the binding reaction is calculated
to be 24 μsec. For all practical purposes this reaction is instan-
taneous. The magnitude of the equilibrium constant also indicates
that in the presence of camphor-saturated buffer there would be
essentially no camphor-free enzyme in the solution.

The second reaction in this cycle has some very interesting
properties which permit it to play a significant role in the control of the overall reaction cycle. The rate constant for the reaction is practically independent of the putidaredoxin concentration. As an interesting aside, it has been found that for all
practical purposes the substrate free form of the enzyme is not
reducible with NADH (Peterson, 1971; Gunsalus et al., 1974).

The next form of cytochrome P-450 on this cycle (Figure 2)

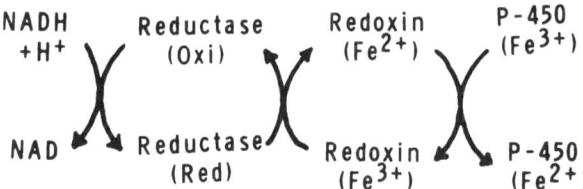

Figure 1.   Electron Transfer System in the Camphor Hydroxylation Reaction.   The electron transfer system is after Katagiri et al., (1968).   The reductase is a flavoprotein and the redoxin is an iron sulfur protein, putidaredoxin.

represents an extremely interesting form of the enzyme. It is the form which can interact with either CO or oxygen.   These molecules can compete with each other for a single binding site on the enzyme, the ferrous iron (Peterson and Griffin, 1972).   The reaction of CO with ferrous cytochrome P-450 is second order and the presence of the substrate, camphor, results in an inhibition of the CO binding reaction.   The reaction of oxygen with ferrous cytochrome P-450 is also second order but the reaction is approximately 20 times faster than the comparable reaction with CO.   The magnitude of the equilibrium binding constant of ferrous cytochrome P-450 for oxygen indicates that in normal air saturated buffer the enzyme will be effectively instantly converted to the oxy-form.

The next form of cytochrome P-450 which is seen on this reaction cycle is the oxy-cytochrome which in the case of the bacterial enzyme is very stable with a half life in the absence of putidaredoxin of 45 min at 4 degrees.   The oxy-form reacts with a second molecule of reduced putidaredoxin in what is   probably an essentially irreversible manner to give the hydroxylated product, hydroxy-camphor.   An interesting feature of this step in the cycle is the kinetics of the reaction.   When these experiments were first tried, we attempted to fit the data using a typical pseudo first order analysis of a second order reaction.   The reaction was done in the presence of a large excess of reduced putidaredoxin and the disappearance of the 418 nm band of oxy-cycochrome P-450 was followed in the dual wavelength stopped flow apparatus (Peterson et al., 1973).   A pseudo first order plot of the data was linear with respect to the disappearance of oxy-cytochrome P-450, but when we plotted the pseudo first order rate constant against the concentration of

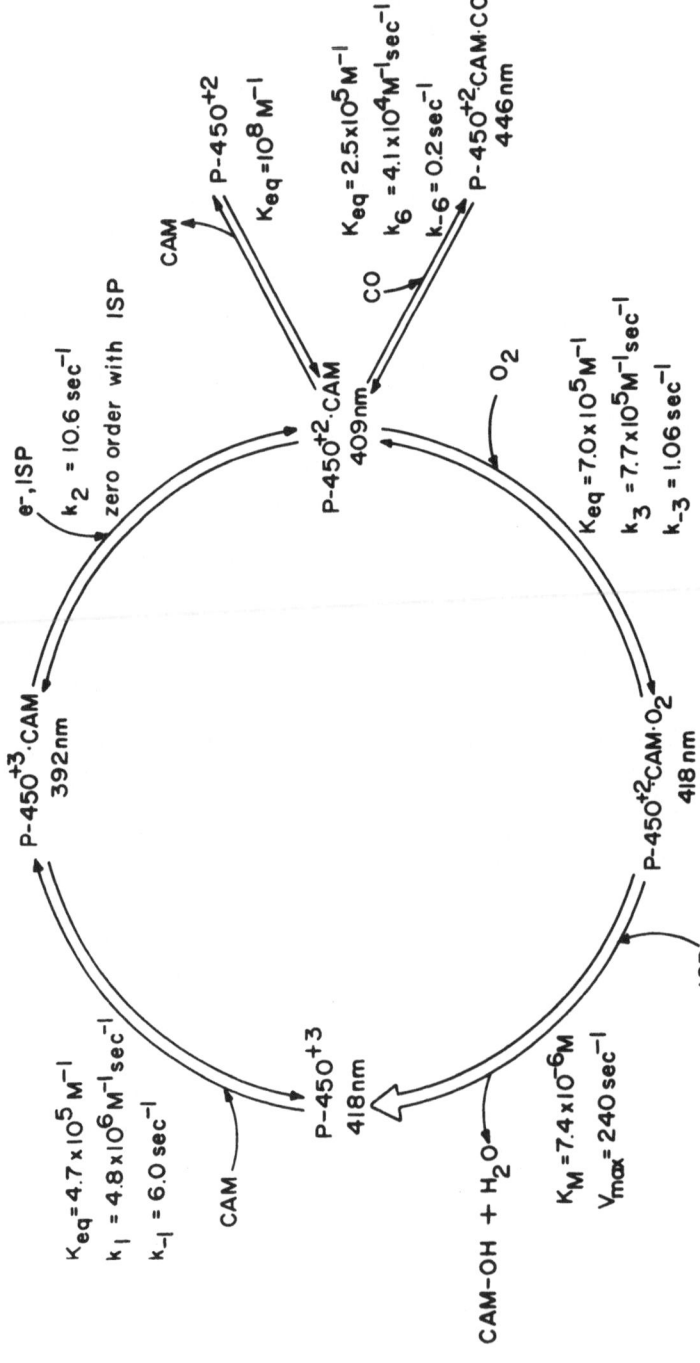

Figure 2. Reaction Cycle of Cytochrome P-450. The abbreviations used in this figure are: P-450, cytochrome P-450; CAM, camphor; ISP, putidaredoxin; CAM-OH, 5-hydroxy camphor, the hydroxylated product.

putidaredoxin used, a straight line was not obtained. If the reaction had been a typical second order reaction, the plot would
have given a straight line. Several other types of analysis were
attempted and finally a Lineweaver-Burk plot was attempted and the
results were linear. These results are interpreted to mean that
putidaredoxin serves as a "substrate" of oxy-cytochrome P-450.
The binding reaction is extremely rapid and an equilibrium is
established between free and bound forms of the enzyme. The enzyme complex then undergoes a decay to products with the V_{max}
shown in Figure 2. The K_m and V_{max} shown in this figure were obtained from this type of kinetic analysis.

From these results several interesting features of the reaction became apparent. a) Since the rate constant for the introduction of the first electron was the same as the rate of the overall hydroxylation reaction this step in the cycle might serve to
control the maximum velocity of the complete reaction cycle. b)
The introduction of the second electron was a function of the concentration of reduced putidaredoxin and in the presence of low
ratios of putidaredoxin to cytochrome P-450 this second step could
become rate limiting. c) If the addition of the first electron is
rate limiting the predominant form of the enzyme in the steady state
of hydroxylation should be the substrate bound high spin form of the
enzyme. Under these conditions the amount of ferrous-cytochrome
P-450 should be very small and oxygen and CO should compete for it.
The rate and equilibrium constants for oxygen binding are larger
than for CO binding. Also, any oxy-cytochrome P-450 which is formed
is immediately reacted with the excess putidaredoxin thus preventing
the establishment of the equilibrium conditions necessary for the
expression of CO inhibition. A special case exists in which the
ratio of CO to oxygen is very large (e.g., 100 to 1) at which point
CO will begin to inhibit the reaction. In retrospect an examination of the data which we obtained initially shows that as the oxygen to CO ratio was approaching zero there was a slight but repro-
ducible inhibition of oxygen consumption. d) if the concentration
of reduced putidaredoxin is decreased significantly then the second
reduction reaction becomes rate limiting and the amount of oxy-
cytochrome P-450 present in the steady state should be larger.
Under these conditions oxygen and CO reach an equilibrium propor-
tioning of the available ferrous cytochrome P-450 between the oxy-
and carbon monoxy-forms. These conditions would be expected to produce significant inhibition of oxygen uptake by CO. The most
favorable ratio of putidaredoxin to cytochrome P-450 which will give
these results is 1 to 1.

CARBON MONOXIDE INHIBITION

The validity of the kinetic constants which were determined
in stopped flow experiments was amenable to testing by analysis

of the concentration of the components present during the steady
state of camphor metabolism. It was felt that one of the most
sensitive tests would be the demonstration of carbon monoxide inhibition under very closely controlled conditions and that the
inhibition should be reversible by monochromatic light of 450 nm.
In addition the rate of oxygen consumption for any CO to oxygen
mixture should be predictable. To test our results we constructed
a special oxygen electrode vessel which has two of the sides parallel
and fitted with optical windows so that the sample in the vessel
could be irradiated with selected wavelengths of light. The vessel
was equipped with a Clark type oxygen electrode and the oxygen concentration was determined polarographically as has been described
previously. To begin an experiment, the vessel was filled with a
solution of buffer which had been equilibrated with selected mixtures of oxygen, carbon monoxide, and nitrogen. The appropriate
concentrations of enzymes were added and the reaction was begun by
the addition of NADH or putidaredoxin. The type of data which can
be obtained with such a vessel is shown in Figure 3 where the
dashed line curve represents an experiment which was performed in
the absence of carbon monoxide to show that the rate of oxygen consumption was linear with time. There was no effect of the light
on the rate of oxygen consumption in the absence of carbon
monoxide.

A second part of this experiment is shown by the solid line
curve. The original buffer contained both carbon monoxide and
the oxygen-nitrogen mixture at a fixed ratio but as the oxygen was
consumed during the experiment the ratio of carbon monoxide to
oxygen increased. As can be seen in this curve, as the ratio of
carbon monoxide to oxygen increased the rate of oxygen consumption
decreased as is predicted. In the middle of the curve the light
reversal of CO inhibition by 450 nm light is shown. The rate of
the reaction increases almost to the rate in the absence of carbon
monoxide. When the light is turned off the rate of oxygen consumption again decreases to a rate which would have been observed
if the light hadn't been turned on.

STEADY STATE ANALYSIS

The following analysis of the reaction cycle of cytochrome
P-450$_{cam}$ is based on several simplifying assumptions: a) the reaction sequence and rate constants previously shown in Figure 2
are correct; b) the experimental conditions were chosen such that
the introduction of the second electron is rate limiting. Hence,
the concentrations of the remaining complexes of cytochrome P-450$_{cam}$
approach values determined by the appropriate equilibrium constants
during the steady state; c) equation 1 describes the irreversible
rate limiting step.

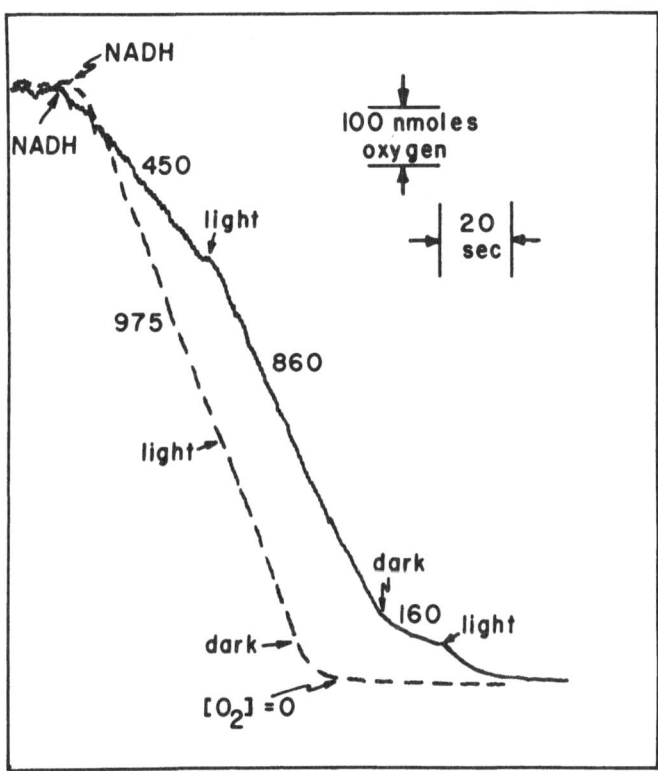

Figure 3. Oxygen Uptake by the Cytochrome P-450 Monooxygenase System. The reaction was performed in the oxygen electrode vessel which had been modified for photochemical reactions as described in the text. The reaction mixture contained in a final volume of 8.7 ml of 0.05 M potassium phosphate buffer, pH 7.4; cytochrome P-450, 6.8 nmoles; putidaredoxin, 12 nmoles; putidaredoxin reductase, an amount sufficient to maintain the putidaredoxin in the reduced state; camphor, 5 mM. The reaction was initiated by the addition of NADH (250 µM) to the mixture. Oxygen consumption was followed in the absence of CO (- - - -) and in the presence of CO (1 mM,————). The numbers beside the curve give the rate of oxygen consumption in nmoles per min. At the points indicated by "LIGHT" and "DARK," light of 450 nm was switched on or off respectively.

$$1) \qquad E \cdot O_2 + ISP_R \overset{k_7}{\underset{k_{-7}}{\rightleftarrows}} E \cdot O_2 \cdot ISP \overset{k_8}{\rightarrow} E + P$$

This equation is of the type which is used to account for the Michaelis-Menten type behavior of this step. An analysis of equation 1 reveals that the overall reaction rate is determined by equation 2.

$$2) \qquad -\frac{dO_2}{dt} = \frac{k_8 \left[E \cdot O_2 \right]}{\left(1 + \dfrac{k_7}{[ISP]} \right)}$$

The concentration of $E \cdot O_2$ is very difficult to determine even in the absence of CO; however, certain equations containing readily measurable quantities can be substituted for $E \cdot O_2$. In the presence of CO the simplifying assumption is made that essentially all of the cytochrome P-450 is present either as the oxy- or the carbon monoxy-complex. This is expressed by the relationship:

$$3) \qquad E_t = E \cdot O_2 + E \cdot CO$$

Using the equilibrium constants shown in Figure 2 and our experimental conditions it can be shown that the concentrations of all other forms of the enzyme are at least 3 orders of magnitude less than the concentrations of the oxy- or CO-complexes.

The definitions of the equilibrium constants for CO association (eq. 4) and oxygen association (eq. 5) can be rearranged and combined to give equation 6 where E_t is the total cytochrome P-450 in the reaction mixture and K_{O_2} and K_{CO} are the equilibrium binding constants for oxygen and CO respectively.

$$4) \quad K_{CO} \frac{[E \cdot CO]}{[E][CO]} \; ; \quad 5) \quad K_{O_2} = \frac{[E \cdot O_2]}{[E][O_2]}$$

$$6) \quad [E \cdot O_2] = [E_t] \left\{ 1 + \frac{[CO] K_{CO}}{[O_2] K_{O_2}} \right\}^{-1}$$

The resultant quantity can be substituted in the equation for the rate of the reaction (eq. 2) to give equation 7. The quantities shown in equation 7 are all readily measurable and the equation is a variables-separable first order differential equation which can be integrated by inspection.

$$7) \quad -\frac{dO_2}{dt} = \frac{k_8[E_t]}{\left(1 + \frac{k_7}{[ISP]}\right)} \left\{ 1 + \frac{CO}{O_2} \frac{K_{CO}}{K_{O_2}} \right\}^{-1}$$

From this equation it can be seen that if the concentration of CO is zero the terms in CO drop out and we are left with the typical Michaelis-Menten expression. If, however, the reaction mixture contains CO the rate of oxygen consumption is decreased by the term in equation 7 which contains CO. The decrease in oxygen consumption is a function of: a) the ratio of the association constants of oxygen and CO with cytochrome P-450; and b) the ratio of the concentrations of CO and oxygen. Since irradiation of cytochrome P-450 with light of 450 nm decreases the magnitude of K_{CO}, it can be readily seen why the rate of oxygen consumption approaches that of the uninhibited reaction.

We have taken the data shown in Figure 4 by the solid line curve and using the assumptions which we have described above, we have determined the values of k_8, k_7 and the ratio of the equilibrium constants for CO and oxygen binding which best fit the data. The estimation procedure used to obtain the value of these constants was developed by R. Feldman of The National Institutes of Health. The program uses a nonlinear estimation procedure which is part of a laboratory analysis program called "MLAB" written for a Digital Equipment Corporation PDP10. The calculated values are superimposed on the plot in Figure 4 and are indicated by the triangles. For comparison the dashed line indicates the rate of oxygen consumption in the same reaction mixture except with irradiation by 450 nm light. In the corner of this figure are given the calculated values of k_8, K_m and the ratio of the equilibrium constants (cf. equation (1) and the text). As can be seen the only value which differs appreciably from previously published values is the ratio of equilibrium constants. We cannot as yet identify which of our previously determined equilibrium constants was in error. Since the value of the binding constant for oxygen was determined by an indirect procedure, we presume that the value for this constant in Figure 2 is inaccurate. We are in the process of redetermining this value.

APPLICATION TO MAMMALIAN SYSTEMS

The above analysis and application of our kinetic data was fun and its results gave us a lot of satisfaction. The main question which still remains for us is "is there any application of these data to the much more complex mammalian enzyme?" We

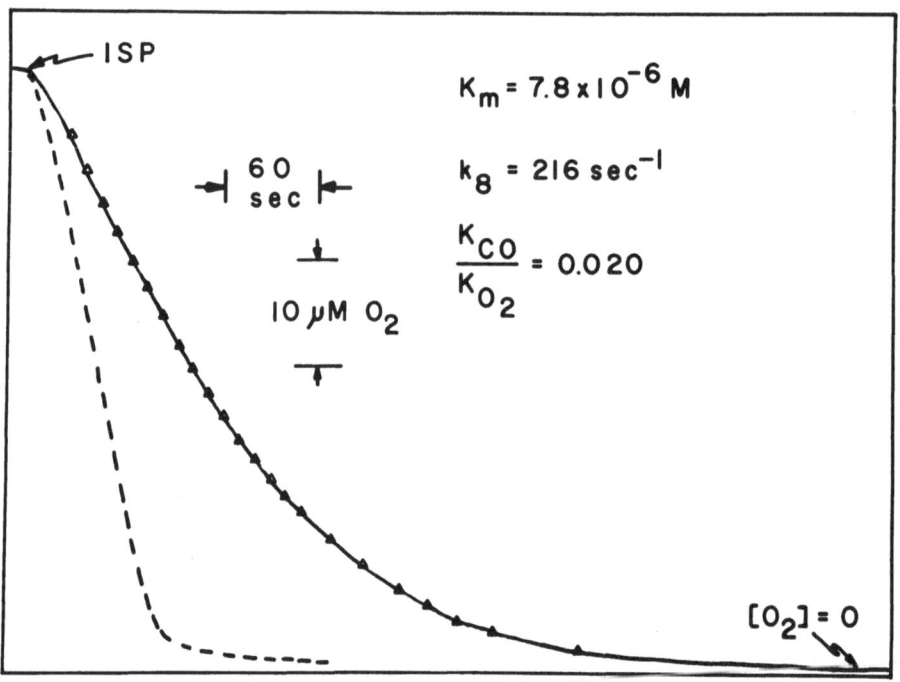

Figure 4. Carbon Monoxide Inhibition of Cytochrome P-450$_{cam}$.
The reaction was performed in the special oxygen electrode vessel
described in the text. The buffer was 0.05 M potassium phosphate
(pH 7.4) and it initially contained: KCl, 0.1 M; camphor, 5 mM;
carbon monoxide 670 μM; NADH, 0.46 mM; putidaredoxin, 0.13 μM;
cytochrome P-450, 0.15 μM; putidaredoxin reductase, an amount
sufficient to keep essentially all of the putidaredoxin in the reduced form. The solid line curve is the experimental curve for
oxygen consumption in the dark in the presence of CO while the
dashed curve represents the same reaction performed in the light
(450 nm). The triangles (Δ) represent the computed time course
for oxygen consumption as indicated in the text.

think so and in this section we will try to convince the readers
of this fact and in addition illustrate some areas which need much
further work.

    The initial work of Gillette's group on the kinetics of cyto-
chrome P-450 reduction in control microsomes tended to indicate
that NADPH cytochrome P-450 reductase was the rate limiting enzyme
in the overall hydroxylation of some drugs by control rat liver

microsomes (Gigon et al., 1969). Later Estabrook's group (Estabrook et al., 1971) demonstrated the presence of a new spectral form of cytochrome P-450 in rat liver microsomes prepared from phenobarbital pretreated animals. This new spectral form was observed during the oxidative metabolism of some drugs such as ethylmorphine. Estabrook's group was able to identify, on the basis of its spectral properties, this new intermediate as oxy-cytochrome P-450. However, if the introduction of the first electron into cytochrome P-450 is rate limiting, as was observed by Gillette, then there should be no appreciable quantity of the oxy-form present in the steady state.

Several groups have made the observation that the effectiveness of carbon monoxide to serve as an inhibitor of cytochrome P-450 catalyzed reactions frequently was a function of the pretreatment of the animal. These data have been used to support the contention that there are several forms of cytochrome P-450 with differing sensitivities toward carbon monoxide inhibition. We hope to convince the readers that these data could be used to support the contention that the rate limiting step in hydroxylation had been changed by pretreatment of the animals with inducing agents. If the rate limiting step in the monooxygenation reaction is changed as we have demonstrated above in the case of the bacterial cytochrome P-450 then it is not unreasonable to assume that the sensitivity to carbon monoxide will be altered. It goes without saying that the only alteration which we made to the bacterial cytochrome P-450 used in our analysis described above was in the nature of the rate limiting step. The cytochrome itself was unaltered. However, by the criteria which some people have used in their studies of carbon monoxide inhibition of microsomal cytochrome P-450, we would have to say that there is a new form of cytochrome P-450 present in our reaction mixture with a different specificity and sensitivity to carbon monoxide.

The data of Table I show that there is a very definite effect of pretreatment of the animals on the NADPH-cytochrome P-450 reductase activity. The rate of cytochrome P-450 reduction in microsomes from phenobarbital pretreated animals is faster than the rate of ethylmorphine demethylation. If these rates are calculated on the basis of the amount of cytochrome P-450 reduced in the fast phase of the reaction, the rate constant for cytochrome P-450 reduction in the presence of ethylmorphine using microsomes from phenobarbital pretreated animals is 100 nmoles per min. This presents conditions which are similar to those described above for the bacterial cytochrome P-450. The rate of introduction of the first electron is faster than the rate of overall metabolism thereby permitting oxy-cytochrome P-450 to accumulate in the steady state. The best conditions for observing the new spectral species of cytochrome P-450 is in microsomes prepared from animals which had been pretreated with phenobarbital (Estabrook et al.,1971).

TABLE I

NADPH-Cytochrome P-450 Reductase Activity

Treatment of Animals	Substrate	Rate Constant (sec^{-1})		Fast % of total
		Fast phase	Slow phase	
Control	---	0.51	0.10	45
	Hexobarbital	0.78	0.15	48
	Ethylmorphine	1.08	0.21	47
PCN	---	0.73	0.16	45
	Hexobarbital	0.95	0.21	60
	Ethylmorphine	1.66	0.37	64
3-MC	--	0.20	0.07	80
	Hexobarbital	0.34	0.09	56
	Ethylmorphine	0.74	0.19	22
PB	---	0.64	0.22	53
	Hexobarbital	1.06	0.39	58
	Ethylmorphine	1.74	0.17	46

Microsomes were prepared from control, 3-methylcholanthrene (3-MC), pregnenolone-16α-carbonitrile (PCN), or phenobarbital (PB) pretreated animals by standard procedures. The microsomes were suspended in 50 mM Tris-HCl buffer (pH 7.4) containing: KCl, 150 mM; $MgCl_2$, 10 mM; beef heart electron transport particles, 0.2 mg/ml; and succinate, 5 mM. After 5 min preincubation at room temperature, CO saturated buffer was added to the reaction mixture to bring the final concentration of microsomes to 2 mg per ml. This mixture was put in one of the driving syringes of an Aminco Morrow stopped flow apparatus. The other driving syringe contained the buffer described above with 0.6 mM NADPH. The reaction was initiated by mixing the contents of the two syringes in the stopped flow aparatus. The time course of the reduction was followed in the dual wavelength mode using the wavelength pair 450-490 nm. The final concentrations of hexobarbital and ethylmorphine in the reaction mixtures were 2mM and 8 mM, respectively.

Thus, the objection which has been raised against the new spectral species being an oxy-form is invalid because the rate of reduction of cytochrome P-450 is faster than the overall rate of hydroxylation. We are currently testing the sensitivity of CO inhibition of various cytochrome P-450-catalyzed monooxygenation reactions. Preliminary results indicate that our original hypothesis described above will be upheld even in the mammalian system.

CONCLUSION

We have shown that the kinetic data obtained by stopped flow spectrophotometry can be used in a steady state analysis of the metabolic control of the reaction cycle of cytochrome P-450. These data were used successfully to predict conditions under which cytochrome P-450cam would be sensitive to CO inhibition. The analytical procedure was extended to include the mammalian cytochrome P-450 to explain why oxy-cytochrome P-450 can be observed in some preparations and not others.

REFERENCES

Estabrook, R. W., Cooper, D. Y. and Rosenthal, O. 1963. The light reversible carbon monoxide inhibition of the steroid C21-hydroxylase system of the adrenal cortex. Biochem. Z. 338: 741-755.

Estabrook, R. W., Hildebrandt, A. G., Remmer, H., Schenkman, J. B., Rosenthal, O. and Cooper, D. Y. 1968. The role of cytochrome P-450 in microsomal mixed function oxidation reactions, pp. 142-177. In B. Hess and Hj. Staudinger, (eds.). Biochemie des Sauerstoffs, 19. Colloquium der Gesellschaft für Biologische Chemie, Springer-Verlag, Berlin.

Estabrook, R. W., Hildebrandt, A. G., Baron, J., Netter, K. J. and Liebman, K. 1971. A new spectral intermediate associated with cytochrome P-450 function in liver microsomes. Biochem. Biophys. Res. Comm. 42: 132-139.

Gigon, P. L., Gram, T. E. and Gillette, J. R. 1969. Studies on the rate of reduction of hepatic microsomal cytochrome P-450 by reduced nicotinamide adenine dinucleotide phosphate: Effect of drug substrates. Mol. Pharmacol. 5:109-122.

Griffin, B. W. and Peterson, J. A. 1972. Camphor binding by Pseudomonas putida cytochrome P-450: Kinetics and Thermodynamics of the reaction. Biochemistry 11: 4740-4746.

Gunsalus, I. C., Meeks, J. R., Lipscomb, J. D., Debrunner, P. and Munck, E. 1974. Bacterial monooxygenases - the P-450 cytochrome system, pp. 561-613. In O. Hayaishi (ed.). Molecular mechanisms of oxygen activation. Academic Press, Inc., New York.

Gunsalus, I. C., Tyson, C. A. and Lipscomb, J. D. 1973. Cytochrome P-450 reduction and oxygenation systems, pp. 583-602. In

T. E. King, H. S. Mason and M. Morrison (eds.). Oxidases and related redox systems. Vol. 2, University Park Press, Baltimore.

Katagiri, M., Ganguli, B. N. and Gunsalus, I. C. 1968. A soluble cytochrome P-450 functional in methylene hydroxylation. J. Biol. Chem. 243: 3543-3546.

Omura, T., Sanders, E., Cooper, D. Y., Rosenthal, O. and Estabrook, R. W., 1965. Isolation of a non-heme iron protein of adrenal cortex functional as a TPNH-flavoprotein-cytochrome P-450 reductase for hydroxylation reactions, pp. 401-423. In A. San Pietro (ed.). Non-heme iron proteins: role in energy conversion, The Antioch Press, Yellow Springs, Ohio.

Peterson, J. A. 1971. Camphor binding by Pseudomonas putida cytochrome P-450. Arch. Biochem. Biophys. 144: 678-693.

Peterson, J. A. and Griffin, B. W. 1972. Carbon monoxide binding by Pseudomonas putida cytochrome P-450. Arch. Biochem. Biophys. 151: 427-433.

Peterson, J. A., Ishimura, Y., Baron, J. and Estabrook, R. W. 1973. Cytochrome P-450: it's function and chemistry during substrate hydroxylation, pp. 565-577. In T. E. King, H. S. Mason and M. Morrison (eds.). Oxidases and related redox systems, Vol. 2, University Park Press, Baltimore.

Ryan, K. J. and Engel, L. L. 1957. Hydroxylation of steroids at carbon 21. J. Biol. Chem. 225: 103-114.

GENERAL DISCUSSION

Session II - Part 2

STERN: I have a comment to Dr. Dus. I was very pleased to
see his results where he was able to titrate five of the six sulfhydryl groups and still retain the integrity of the P-450$_{cam}$·CO
spectrum. We have been very interested in the unusual absorption
spectrum of P-450·CO complexes for the past few years and have been
trying to duplicate the spectrum in model systems (J. O. Stern and
J. Peisach, J. Biol. Chem. 249, 7495-7498, 1974). Other investigators (Y. Imai and R. Sato J. Biochem. 64, 147-159, 1968) have
shown that aggregated heme model systems, in the presence of isocyanides, can lead to a Soret absorption near 450 nm. Since it
has been unequivocally demonstrated that cytochromes P-450 are
mononuclear in heme, these studies reveal little about the unique
heme-ligation found in these hemoproteins. I would like to report
here on the studies done by Dr. Peisach and myself where we have
been able to simulate the P-450·CO spectrum in an organic solvent
keeping the heme mononuclear (Figure 1). Here, we see two Soret
bands, one at 412 nm which is characteristic of a typical heme·CO
complex, and another at 448 nm similar to that seen in cytochrome
P-450·CO complexes. The requirements for the formation of the
448 peak are the presence of mononuclear ferrous heme, mercaptide,
a very strong base, dithionite, and CO, as shown in Table I. If
we leave out the thiol, we only form the typical heme·CO absorption
at 412 nm. That's also true if we omit the strong base. Therefore, this P-450-like·CO spectrum requires the presence of both
thiol and strong base, which leads us to believe that RSH is present
as a mercaptide anion (RS$^-$). This ligand is thought to be trans
to the CO in all cytochrome P-450·CO complexes. (When we removed
dithionite from the reaction mixture we still saw this 448 peak
and we repeated the experiments in the absence of CO. We observed
a peak at 421 nm typical of ferrous heme, demonstrating that excess
thiol is capable of reducing heme in this solvent system.)

If we change the environment of the P-450-like·CO model compound by adding a more non-polar solvent (in essence this would
be equivalent to creating a more hydrophobic heme pocket in the
protein) we see a shift in this 448 peak down to 446 nm, thereby
giving us a clue that the heme in cytochrome P-450·CO from 3-
methylcholanthrene-treated animals (λ_{max} = 448 nm) or cytochrome
P-450$_{cam}$·CO (λ_{max} = 446 nm) may be in a more hydrophobic environment than the cytochrome P-450·CO complexes which absorb at precisely 450 nm.

UNIDENTIFIED SPEAKER: May I comment on this, please. This
is free hemin in solution, in organic solvents?

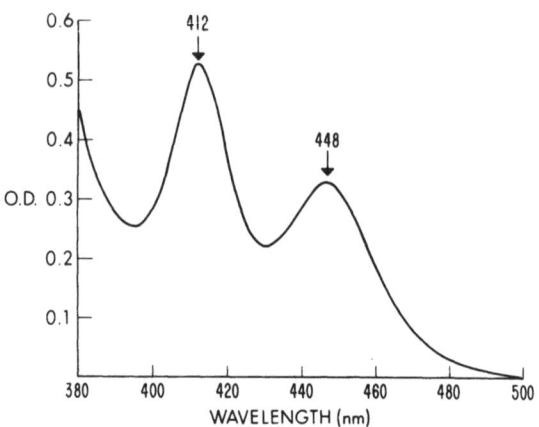

Figure 1. Optical absorption spectrum in the Soret region of hemin chloride in dimethylsulfoxide-ethanol, after reduction by thiol in the presence of CO and a strong base (tetramethyl-ammonium hydroxide, 1M). (For further details, see Stern and Peisach, J. Biol. Chem. 249, 7495, 1974).

TABLE I

Model Studies of Cytochrome P-450·CO
Soret Absorption Maxima

Components	λ_{max}(nm)
Heme, RSH, OH$^-$, $S_2O_4^{-2}$, CO	412,448
$-S_2O_4^{-2}$	412,448
$-S_2O_4^{-2}$ and CO	421
-RSH	412
-OH$^-$	412

STERN: Right.

UNIDENTIFIED SPEAKER: Well, whenever you have a free heme porphyrin of any sort with strong base present, you get μ-oxo heme.

STERN: Yes, that's true for <u>ferric</u> hemes and we have demonstrated that, but in the presence of excess thiol, the reduction of the heme and binding to CO occurs before μ-oxo formation. μ-Oxo heme has its Soret absorption at 408 nm (E. B. Fleischer and T. S. Srivastava, J. Am. Chem. Soc. 91, 2403-2405, 1969) and not at 412 or 448 nm.

PETERSON: We've also done mercurial titrations with cytochrome P-450 which we reported two years ago at the Stanford Conference, and our cytochrome P-450 preparation, for some reason, is slightly different from Dr. Gunsalus' and Dr. Dus'. We have only 5 sulf-hydryls in our fully denatured protein and we also find that if you titrate with mercurials, you only titrate 4 readily accessible ones in our protein but you still have camphor binding, still high spin and it still gives you the carbon monoxide complex at 446.

STERN: And, if you remove the 5th, it goes from a 446 peak to a 420 peak?

PETERSON: Yes, in our case if you remove the 5th, yes, it goes to 420. To do that you have to add very high concentrations of urea of guanidine; you have to denature the protein to make the 5th one, the last one, accessible.

DUS: Dr. Stern, I very much appreciate your comments concerning the heme-iron ligands in cytochrome P-450$_{cam}$. I would like to add that we have some experiments going, in which we are looking at the infrared spectra of the cytochrome P-450·CO adduct, in collaboration with Dr. James Alben at Columbus, Ohio. For P-450·CO Dr. Alben found a peak at 1940 cm^{-1}, that means at a lower frequency than that for Mb·CO or Hb·CO (1944 and 1960 cm^{-1}, respectively). Again this points to stronger π bonding of the CO to Fe which, in view of the broader absorption ($\Delta\gamma 1/2$ of P-450·CO is about 13.5 cm^{-1} as compared to $\Delta\gamma 1/2$ of Mb·CO or Hb·CO which are 8-10 cm^{-1}), cannot mean distortion because of lack of space. The observation would be, however, compatible with more σ contribution from the remaining ligand, being a more anionic ligand. Compared to Mb and Hb this would favor cysteine over histidine as the remaining ligand in P-450·CO.

PEISACH: I think we've pretty much clinched the fact that for the CO complex of the ferro protein, the axial ligands are CO and RS⁻? What are the ligands in the ferric protein and, especially, in the high spin ferric protein? What is the difference between

high spin and low spin ferric cytochrome P-450? At the last
meeting on cytochrome P-450 organized by Dr. Cooper, we speculated
that the difference between high spin and low spin ferric cytochrome P-450 was grounded in a difference in a particular axial
ligand. At that time, we suggested that for the low spin case,
an obligate ligand to the heme is negatively charged mercaptide
anion. The crystal field of any ferric heme complex, we believe,
is sufficiently large that all heme compounds with this ligand
will of necessity be low spin. On the other hand, neutralizing
the charge on the sulfur, such as by adding a proton for example,
decreases the crystal field contribution of that particular ligand
so that in some cases, the complex is now high spin. (The unusual
EPR of high spin ferric cytochrome P-450 more than likely speaks
for asymmetric overlap of the pi-electrons of the sulfur and d-
electrons of the iron.) After listening to Dr. Holtzman and to
others here today, I'm convinced that the heme site of ferric
cytochrome P-450 is very sensitive to chaotropic effects such as
the nature of solvent and the presence of substrate. These reagents will affect the structure of the protein and indirectly will
affect the state of charge (e.g., possible protonation or depro-
tonation) at this particular site.

BRESNICK: Question and comment for Dr. Orrenius. About
three years ago, Dr. Cantrell, when he was in my laboratory, and I
published a paper in the Journal of Cell Biology in which we
described a method for the isolation of hepatocytes as well as
reticulo-endothelial cells from liver. We were interested at that
time in determining what the contribution of each of these cell
types was to the overall aryl hydrocarbon hydroxylase activity
that one would find in liver. In that study, if cells came from
pretreated animals, i.e., 3-MC stimulated animals, aryl hydrocarbon
hydroxylase activity was seen in the reticulo-endothelial cell
types. Accordingly, I would urge you to take at look at Kupffer
cell metabolism of some of your pharmacologic agents. The second
aspect of that study is that we were interested in seeing whether
we could culture these hepatocytes and reticulo-endethelial elements for short terms to see a little bit more about the induction
phenomenon. We found that the cells did not do very well in culture in terms of aryl hydrocarbon hydroxylase activity. Activity
began to fall off very, very rapidly, starting with something like
two hours, and was virtually down by 90% by 4 to 6 hours of culture.
At the same time, Trypan Blue exclusion tests did not reveal anything unusual. Oxygen consumption also did not reveal anything
unusual, but at least that particular enzyme, the aryl hydroxylase,
really went to pot very shortly while in culture.

ORRENIUS: In regard to your first comment, there are various
isolation techniques which give you various ratios of Kupffer cells
to hepatocytes. Our technique gives us less than 10% Kupffer cells.

BRESNICK: That wasn't my comment.

ORRENIUS: You suggested that we should take a look at the contribution by the Kupffer cells.

BRESNICK: No, I wasn't claiming that your preparation was contaminated with Kupffer cells, but I was asking that you might look at the metabolism by these cells. We have the feeling, especially in stimulated animals, that the reticulo-endothelial cells might actually contribute a fair amount to metabolism in liver.

ORRENIUS: Yes, I agree with you. That is very possible, and it is a very good idea to check the activity. With regard to the second comment, we haven't tried that. We know it's difficult.

HOLTZMAN: We've also looked at metabolism in the hepatocytes and have found that for the metabolism of ethylmorphine in an in vitro system with isolated hepatocytes, there is a very good cor-relation between the percentage of cells which will continue to exclude Trypan Blue and the activity of the ethylmorphine N-demethy-lase.

The second point which I thought was rather interesting, in contra-distinction to what Dr. Orrenius has found, is that we find a tremendous difference in the K_m of ethylmorphine in the isolated microsomes which is normally about 250 µM and in that found for the isolated hepatocytes, which in our hands turns out to be about 45 µM. This would suggest that there is an active uptake of ethyl-morphine by the hepatocytes so that the isolated hepatocytes can metabolize a lower concentration.

ORRENIUS: Doesn't this test rely upon the determination of formaldehyde?

HOLTZMAN: No.

ORRENIUS: How was the assay done?

HOLTZMAN: We tried to determine the formaldehyde colorimetri-cally and found that it is totally destroyed by the intact hepato-cyte. We do it by a radioactive technique which we recently pub-lished in the Journal of Pharmacology and Experimental Therapeutics. In this assay we could separate the tritiated water from the ethylmorphine by the use of a column procedure where we run the incubate through a XAD-2 column and only the product, tritiated water, comes through. Alternatively we have used a lyophilization technique.

REMMER: Since one year we did isolation studies of liver cells, but in our hands the Trypan Blue method was not sufficient for estimating the viability of the cells. We found that it was not sensitive to alterations of the potassium content of the cells and the membrane potential. We had a lot of cells that did not contain Trypan Blue yet did not have a normal potassium content and therefore we are worried. My question is how can we improve the Trypan Blue method so that it shows whether the membrane potential is normal.

ORRENIUS: I think we improved the method already, and we rely on a collection of viability criteria, such as Trypan Blue exclusion, oxygen consumption, leakage of enzymes, and leakage of adenine and pyridine nucleotides.

CINTI: Dr. Orrenius, with reference to your oxygen consumption studies, you observed a large increase in oxygen consumption when succinate was present. Why didn't you see similar results with added β-hydroxybutyrate? Is it a transport problem?

ORRENIUS: I think, in a way, it is a transport problem because β-hydroxybutyrate enters the cell much better than succinate. When succinate gets in, however, it is so much better a substrate.

THURMAN: I have a question for Dr. Orrenius. You show beautiful spectra for cytochrome P-450 in liver cells. However, in view of the fact that you have 85% inhibition of respiration with antimycin A, why do you fail to see cytochrome oxidase-CO in your spectra?

NEBERT: Did you add CO to both reference and sample cuvettes?

ORRENIUS: No. CO was added only to the sample cuvette and we used only endogenous reducing equivalents. But the system was not anaerobic.

CINTI: In reference to that point, we are looking here at cells from phenobarbital-treated rats. In this case we have at least 5 to 10 times more P-450 hemoprotein than any other mitochondrial hemoproteins, and that could explain why one doesn't see the cytochrome oxidase interference.

NEBERT: This might explain it because, for example, in hematoma cultures, the cytochrome oxidase is overwhelming, and you have to use other techniques.

NARASIMHULU: Another possible reason for the absence of a cytochrome oxidase CO spectrum in the liver cells could be that the partition constant, between CO and O_2, for cytochrome a_3 is

10 to 12 times higher than for cytochrome P-450. Thus the ratio
CO/O_2 might not have been high enough to form appreciable amounts
of cytochrome $a_3 \cdot CO$ complex.

COON: A question about the hepatocytes. Is the effect of
hexobarbital on the reduction of P-450 one of extent or of rate?
I'm curious about this because in microsomal suspensions, drugs
are commonly thought to increase the rate of P-450 reduction, but
this effect has not been observed in the reconstituted system.

ORRENIUS: It's difficult to tell. We haven't done these
measurements yet. It is probably, however, a matter of extent.

COON: So it's due to a shift in equilibrium?

ORRENIUS: I would guess so.

LU: A question for Dr. Narasimhulu. Perhaps I misunderstood
you. Did you say that we have suggested that lipid is required
for substrate binding?

NARASIMHULU: I thought so. I might have misunderstood.

LU: Other people have suggested that phospholipids may be
involved in substrate binding, but neither our laboratory nor
Dr. Coon's laboratory has suggested that lipid is required for
substrate binding. I was very happy to see your results because
we have reached the conclusion that phospholipids probably are not
involved in substrate binding based on two observations. One
is that our partially purified preparation has less than 1 nmole
of phospholipid per nmole P-450, yet the P-450 can bind both
Type I and Type II substrates. Secondly, Dr. Vore in our laboratory has shown that the removal of 80-90% of the phospholipids
from lyophilized microsomes by organic solvents does not affect
the binding of several Type I substrates. Indeed, the extracted
microsomes show even better Type I binding than the unextracted
microsomes. We believe that at least one function of lipid is its
involvement in the reduction of cytochrome P-450 by TPNH and reductase as first reported by Dr. Strobel in Dr. Coon's laboratory.

NARASIMHULU: I am in complete agreement with you that phospholipids are indeed required for the expression of the enzymatic
activity by certain hydroxylase preparations. However, at the
present time, I cannot rule out the requirement of phospholipids
for the substrate-cytochrome P-450 binding reaction in the adrenocortical microsomes, although removal of 85% of the phospholipids
from the microsomes failed to alter either the Type I spectral
change or the concentration of the substrate required to produce
half maximum change. This is because the possibility exists of

the residual phospholipids (15%) in the depleted microsomes being
involved in one way or other in the binding of the substrate to
the cytochrome. In this respect, it is possible that steroid
hydroxylases, due to their relatively high substrate specificity
are different from the hepatic system.

LU: I was also very happy to see your results showing phospholipid is required for the adrenal steroid system because this
is one question we always get. Dr. Schenkman mentioned that lipid
could simply relieve the inhibition caused by detergents present
in our preparations, but the studies by you, Dr. Vore, and Dr.
Coon's laboratory really indicate that this is not the case.

JERINA: I'd like to turn to the question of the hydrophobicity
of the active site. Dr. Holtzman, you've shown us a clear decrease in rate of about 30% on going to D_2O for reduction of
P-450 in the presence of ethylmorphine, which had been attributed
to hydration effects of the heme site, if I understood this correctly. This corresponds to a K_{D_2O}/K_{H_2O} of about 0.7, which is
a relatively small primary isotope effect. Also you have indicated
that there is a rather sharp pH maximum for the reaction, as a
function of pH. I wonder what correction has been made for hydrogen ion activity in passing from H_2O to D_2O. If no correction has
been made, your data simply reflect the difference in acidity in
the medium and may have nothing to do with hydration whatsoever.
In addition, the strong isotope effect you observe could be due
entirely to the isotope effect for proton transfer during reduction,
and thus again could be unrelated to hydration. Could you comment
on the correction made for changing from H_2O to D_2O and how you can
obviate proton transfer for this very small kinetic isotope effect.

HOLTZMAN: That clearly is a very important point which we did
consider. Many of these phenomena are significantly affected by
pH and since pH is 0.4 units lower than pD for the same potential
difference with either a standard hydrogen or calomel electrode,
we decided to determine the activity of both ethylmorphine N-
demethylase and aniline hydroxylase for various pH's and "nominal"
pD's, the "nominal" pD being the value observed on the pH meter
using a Ag/AgCl//AgCl/Ag glass electrode calibrated with standard
commercial buffers. Fortunately, the two curves paralled each
other for each pH, so that maximal activity was observed for each
solution when they were adjusted to give the same reading on the
pH meter. All studies were performed at the peak of activity.

PETERSON: Just a short quick comment on the hydrophobicity
or hydrophilicity of the active site, or substrate-binding site
of cytochrome P-450. Again, two years ago, at Stanford, we presented our data on proton relaxation with the Pseudomonas putida
cytochrome P-450, showing that in the absence of camphor as a

substrate a water molecule or a proton can very rapidly penetrate
the substrate-binding site. In the presence of camphor or any of
the classical Type II inhibitors of this particular cytochrome
P-450, water or the proton is excluded. So water can get into
the active site, at least in the bacterial enzyme.

RELATIONSHIP BETWEEN MICROSOMAL HYDROXYLASE AND GLUCURONYLTRANS-FERASE

Herbert Remmer, Karl W. Bock, Bernhard Rexer
Institut Für Toxikologie
der Universität Tübingen
74 Tübingen, Wilhelmstrasse 56

Two types of inducibility of the cytochrome P-450-system in the endoplasmic reticulum of liver cells are well known: cytochrome P-448 induced by polycyclic hydrocarbons and cytochrome P-450 induced by numerous drugs. The increase of the latter is connected with proliferation of endoplasmic membranes and consequently an enlargement of the liver.

Electron spin resonance studies (1) presented evidence that both cytochrome P-448 and cytochrome P-450 can exist in two different spin states and that the binding of drugs alters the ratios of these two spin states in opposite directions for the two cytochromes. This indicates that the two hydroxylating cytochromes possess different conformations in the environment of the heme group.

Closely associated with the unspecific mixed function oxidase in the membrane is the enzymic step which uses UDP-glucuronic acid to conjugate drugs and endogenous compounds. The enzyme responsible for this reaction, termed UDP-glucuronyl transferase, EC 2.4.1.17, has not yet been isolated. Thus, it is an open question whether several enzymes are involved in this conjugation reaction, or only one or two, the specificity of which may be derived from the lipid environment which in turn determines their conformation.

The glucuronyl transferases differ in many respects from the enzymes of the hydroxylating system, but they also have several properties in common. A systematic investigation of the similarities as well as of the differences of these classes of enzymes might be advantageous for explaining their arrangement and

association in the endoplasmic membranes. Therefore, we studied
the following: 1. solubilization of the enzymes; 2. induction by
phenobarbital (PB) and methylcholanthrene (MC); 3. sex specificity;
and 4. the inhibitory action of drugs. In studying the properties
of the glucuronyltransferases, we limited the substrates used to
those for which the enzymatic activities of the microsomal hydroxylases were well studied - namely, bilirubin, chloramphenicol,
1-naphthol and p-nitrophenol.

Before dealing with the similarities between the hydroxylases
and the glucuronyl transferases I would like to describe one
important difference: Whereas cytochrome P-450 loses activity when
unprotected microsomes are solubilized by sonication or detergents,
the glucuronyl transferases become much more active. The degree
of activation differed according to the substrate used (2). The
rates of glucuronidation of bilirubin and chloramphenicol increased
by factors of 3.5 and 5, respectively, while 1-naphthol and p-
nitrophenol were conjugated at 11- and 16-fold higher rates after
solubilization of the microsomes with deoxycholate. Solubilization
by deoxycholate treatment was more effective in this respect than
solubilization with Triton X-100 or by sonication.

Both the glucuronyltransferases and the hydroxylating systems
are inducible. Pretreatment of rats with phenobarbital enhances
the glucuronidation rate of all four substrates used, but the
increase varies greatly as shown in Table I. The limited inducing
action of methylcholanthrene as well as its selectivity for the
hydroxylation reaction which are well known, were also found when
we investigated the increase of glucuronyltransferase activity
(Table I). Methylcholanthrene is seen to have the greatest effect
on p-nitrophenol conjugation but has no enhancing effect on the
glucuronidation of bilirubin and chloramphenicol. Indeed there
seems to be some inhibition of chloramphenicol conjugation in the
deoxycholate treated samples following MC treatment.

Induction by phenobarbital (PB) produced the opposite results
in comparison with MC, i.e., PB had its maximum stimulatory effect
on chloramphenicol glucuronidation. The time course of the rise
and fall of this activity following PB treatment corresponds to
that of cytochrome P-450 and seems to be an indication that cytochrome P-450 and the enzyme which conjugates chloramphenicol are
coupled insofar as the induction process is concerned.

The widely differing extent of the induction due to PB- or
MC-treatment remains practically unaltered whether microsomes in
native or solubilized state are used for the test (Table I). The
amount of these enzymes does not seem to be changed by treating
microsomes with detergents. That we are dealing not with a conformation change elicited by PB or MC, but with a real induction,

TABLE I

UDP Glucuronyltransferase Activity

Rat Liver Microsomes

Substrate		No pretreatment			Induced with PB			Induced with MC		
		Nat.	X-100	DOC	Nat.	X-100	DOC	Nat.	X-100	DOC
p-Nitrophenol	(a)	3±2	38±5	57±10	5±2	61±9	79±14	12	167±20	174±20
	(b)	(1.0)	(1.0)	(1.0)	(1.7)	(1.58)	(1.39)	(4.0)	(4.39)	(3.05)
	(c)	[1.0]	[12.7]	[19]	[1.0]	[12.2]	[15.8]	[1.0]	[13.9]	[14.5]
1-Naphthol	(a)	—	—	76±15	11±3	86±18	118±19	—	—	131.5
	(b)	—	—	1.0	—	—	1.55	—	—	1.73
Bilirubin	(a)	—	—	1.2±0.1	0.7±0.8	0.6	2.50	—	—	1.2
	(b)	—	—	1.0	—	—	2.08	—	—	1.0
Chloramphenicol	(a)	0.2±0.1	0.5±0.1	1.1±0.1	0.8±0.2	2.6±0.4	4.0±0.5	—	0.5	0.7
	(b)	(1.0)	(1.0)	(1.0)	(4.0)	(5.2)	(3.6)	—	(1.0)	(0.64)
	(c)	[1.0]	[2.5]	[5.5]	[1.0]	[3.25]	[5.0]			

The measured UDP glucuronyltransferase activities (mean ± S.D.) for each substrate are given as nmoles (mg protein)$^{-1}$ min^{-1} (Line a). The relative magnitude of the induction effects for each substrate compared with untreated control animals is expressed in (Line b). Similarly, the effects of solubilization treatment are expressed as relative values, compared with the native microsomes in each case (values in [ ], Line c).

The results were obtained with male Sprague-Dawley rats (150-180 g) used as control, or after induction with phenobarbital (PB) (3 daily injections, i.p., 100 mg/kg) or 3-methylcholanthrene (MC) (one injection, i.p., 80 mg/kg in olive oil). The activities of native (nat.) microsomes were compared with the same microsomal preparations after solubilization with Triton X-100 (X-100; final concentration 0.05% w/v) or with deoxycholate (DOC; final concentration 0.23% w/v, followed by passage through Sephadex G-25).

is indicated by the fact that the apparent K_m values of the enzyme for p-nitrophenol (0.28-0.40 mM) or UDPGA (0.1-0.14 mM) are almost identical, regardless of whether normal or induced microsomes are tested, or whether they are solubilized with Triton X-100 or with deoxycholate (2).

As with the hydroxylation rate of compounds exhibiting Type I binding spectra, the conjugation of chloramphenicol in microsomes of rat liver is sex dependent (Table II). However, the sex specificity is limited and could not be observed when p-nitrophenol or bilirubin was used as substrate for conjugation.

Bilirubin had in common with chloramphenicol that its conjugation rate did not increase strikingly after solubilization. The influence of PB or MC on the conjugation rates of these two substrates was also similar and could be easily distinguished from the inducing action of PB or MC on p-nitrophenol and 1-naphthol.

The p-nitrophenol- and bilirubin-conjugation steps seem to be catalyzed by two distinct enzymes, since the Gunn rat, which is well known for its deficiency for conjugating bilirubin, has not lost its ability to form the glucuronide of p-nitrophenol (3). Surprisingly, it also retains its capacity to conjugate chloramphenicol, indicating that the conjugating enzyme for chloramphenicol, in spite of several similarities with that of bilirubin, is genetically controlled in a different manner.

Substances known to act as inhibitors of the microsomal hydroxylase system also inhibit glucuronyltransferases, indicating another relationship between these two enzyme systems (Table III). Their inhibitory effect is dependent on the substrate used for the glucuronidation. For example, SKF-525A and hexobarbital are most active in inhibiting the conjugation of bilirubin, whereas metyrapone inhibits chloramphenicol conjugation. Those remarkable specificities which have been found might also indicate that there are different types of transferases. The compounds we selected for inhibiting glucuronyltransferase activity are little or not at all conjugated as judged from in vivo studies. In another study we used p-nitroanisol which cannot be conjugated before it is oxidatively demethylated. This compound caused 50 percent inhibition of the chloramphenicol conjugation at around 1 mM. Performing the opposite experiment by measuring the inhibitory activity of chloramphenicol on the demethylation rate of p-nitroanisol, we found a K_i value around 0.2 mM. These few examples demonstrate that a mutual inhibition occurs if drugs which are predominantly or exclusively either conjugated or hydroxylated are used to inhibit the reaction in which they are presumably not involved. Further studies have to be performed before it becomes possible to arrive at a better insight into the close association of

TABLE II

Sex Specificity of UDP Glucuronyl Conjugate Activity

	Substrate	N	Male	Female	P
Cytochrome P-450					
nmoles/mg Protein	——	13	0.96±0.28	0.88±0.15	——
Transferase Activity					
nmoles/mg Protein per min.	p-Nitrophenol	1C	51±9	52±10	——
	Bilirubin	14	0.17±0.03	0.20±0.04	0.015
	Chloramphenicol	14	0.86±0.13	0.54±0.16	<0.001
Hydroxylase Activity					
nmoles/mg Protein per min.	Aminopyrine	11	8.6±2.4	6.2±1.6	<0.01

Microsomes were solubilized with Triton X-100. Details of the methods are published elsewhere (2).

TABLE III

Inhibition of Microsomal Glucuronyltransferase Activity

Substrates	Inhibitors (Conc., M)					
	SKF-525 A		Hexobarbital		Metyrapone	
	10^{-3}	10^{-4}	10^{-3}	10^{-4}	10^{-3}	10^{-4}
Bilirubin	69	30	62	19	14	11
Chloramphenicol	18	0	16	8	82	62
p-Nitrophenol	17	7	21	3	0	0
Demethylation of Aminopyrine (for comparison)	76	38	29	11	56	17

Microsomes from uninduced male Wistar rats were solubilized with Triton X-100 and tested for transferase activity and, as a comparison, also for aminopyrine demethylation, by the methods described elsewhere (2).

Values are given as percent inhibition of the control rates which were found to be (in nmoles/(mg protein x min)): bilirubin 0.23; chloramphenicol 1.10; p-nitrophenol 59; aminopyrine demethylation 10.0.

hydroxylating and conjugating enzymes in the endoplasmic membranes of liver cells.

SUMMARY

These experiments did not answer the question of whether one or several UDG-glucuronyltransferases are present in endoplasmic membranes of the liver. However, they present results which indicate that the glucuronyltransferase(s) have several properties in common with the hydroxylating cytochrome P-450 dependent enzyme system: the inducibility (which differs considerably after pretreatment with phenobarbital or 3-methylcholanthrene), the sex specificity, and the inhibition by the same compounds. The most obvious difference between the systems is the alteration of the enzyme activities after solubilization of the membranes by sonication or use of detergents. On solubilization, the activity of the glucuronyltransferase(s) increases, whereas the opposite is true for the hydroxylating system which may lose one or several components essential for its activity (such as the NADPH dependent reductase). Our experiments can best be interpreted by assuming a common micro-environment around the enzymes produced by lipids and proteins which modulate both the rate of hydroxylation and that of glucuronyl-conjugation of drugs.

REFERENCES

1.   Witmer, C., Nehls, P., Krauss, P., Remmer, H. and Snyder, R.
        This symposium.
2.   Bock, K. W., Frohling, W., Remmer, H. and Rexer, B. Biochim.
        Biophys. Acta 327: 46-56 (1973).
3.   Jansen, P. L. M., Henderson, P. Th. Biochem. Pharmacol. 21:
        2457-2462 (1972).

A POSSIBLE ROLE OF COPPER IN THE REGULATION OF HEME BIOSYNTHESIS

THROUGH FERROCHELATASE

G. S. Wagner and T. R. Tephly

The Toxicology Center, Department of Pharmacology
The University of Iowa
Iowa City, Iowa 52242

Drugs and heavy metals may alter the synthesis of hemoproteins by either inducing or inhibiting various enzymes in the
heme biosynthetic pathway (Fig. 1) (Tephly et al., 1971; Tephly et
al.,1973). Porphyrogenic drugs and inducers of the hepatic hemoprotein, cytochrome P-450, such as phenobarbital, induce hepatic
δ-aminolevulinic acid synthetase (ALAS), the initial and proposed
rate-limiting enzyme of this pathway (Granick and Urata, 1963;
Baron and Tephly, 1969; Tephly et al., 1973). Ferrochelatase
(heme synthetase, protoheme ferro-lyase, EC 4.99.1.1), the terminal enzyme in the heme pathway, is located on the inner mitochondrial membrane (Jones and Jones, 1968; McKay et al., 1969) and
catalyzes the formation of heme from protoporphyrin IX and ferrous
iron (Labbe and Hubbard, 1960). This enzyme, like ALAS, is inducible (Hasegawa et al., 1970; Tephly et al., 1971), and may be
subject to product inhibition by heme (Jones and Jones, 1970).
Heme which is formed through ferrochelatase is then available for
synthesis of cytochrome P-450 and other hemoproteins. Tephly and
co-workers (Tephly and Hibbeln, 1971; Tephly et al., 1972) have
reported that cobalt treatment produces a substantial decrease in
the level of hepatic cytochrome P-450. The existing evidence indicates that cobalt produces this effect through an inhibition of
heme biosynthesis and that its effects on heme biosynthesis may
be explained by its inhibition of ferrochelatase. This report
will describe the inhibition of ferrochelatase by cobalt and other
divalent cations and a possible role for copper in ferrochelatase
activity.

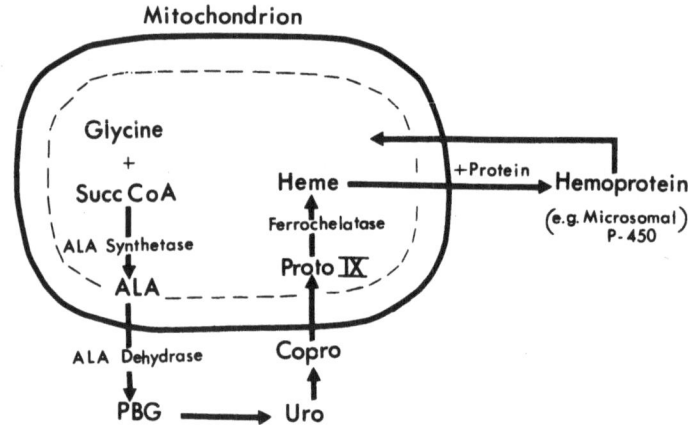

Fig. 1. The relationship between the heme biosynthetic pathway and cytochrome P-450.

METHODS

Male, Sprague-Dawley rats, 150-200 g, were used. Metal salts were administered subcutaneously in distilled H_2O (5 ml/kg), at 48 and 24 hr prior to sacrifice, and the animals were fasted 24 hr prior to sacrifice.

Mitochondria were prepared from the pooled livers of three or more rats. Animals were decapitated and livers were removed, rinsed with 0.9% NaCl, weighed and homogenized with a glass hand homogenizer in nine volumes of a solution containing 0.25 M sucrose, 0.05 M Tris-HCl buffer (pH 8.2 at 5°C) and one mM EDTA. The homogenate was then centrifuged at 1000 x g for 10 min (Sorvall RC-3 centrifuge, HG-4L rotor) to remove the nuclei and cell debris. The 1000 x g supernatant was then centrifuged at 9000 x g for 10 min (Sorvall RC2-B centrifuge, GSA rotor) to sediment the mitochondria. The mitochondria were then washed twice in the same mixture without EDTA, resuspended in 0.25 M sucrose and immediately assayed for ferrochelatase activity. For preparation of solubilized, dialyzed mitochondria, the method of Porra et al. (1967) was followed, with the exception that the final centrifugation was carried out at 105,000 x g for 60 min instead of 80,000 x g.

For preparation of microsomes livers were perfused with ice cold 0.9% NaCl prior to excision and a 25% homogenate was prepared with 1.15% KCI. The post mitochondrial supernatant was centrifuged for 36 min at 226,000 x g to sediment microsomes. Microsomal

cytochrome P-450 content was determined as described by Baron and
Tephly (1969). Protein was measured by the method of Lowry et al.,
(1951).

A sensitive radioisotopic assay based on the incorporation of
[59]Fe into heme under anaerobic conditions was used to study ferrochelatase activity. Unless otherwise noted, reaction mixtures contained 50 μM protoporphyrin IX, Tween 80 (1% v/v), 4mM GSH, 50 mM
Tris-HCl (pH 8.0 at 37°C), 1.5-2.5 mg mitochondrial protein/ml and
4000 dpm of ^{59}Fe per atom of Fe in a total volume of four ml.
Control reactions were carried out using heat-treated mitochondria
(100°C for 10 min) or aerobic conditions with protoporphyrin IX
added at the end of the incubation period.

Protoporphyrin IX solubilized in Tween 80 was placed in the
sidearm of a Thunberg tube, while all other components were added
to the main chamber. The Thunberg tube was submerged in ice until
the reaction was started and evacuated gently for 20 min. The reaction was started by tipping the protoporphyrin IX into the
tube and incubations were carried out in vacuo at 37°C for 30 min.
The reaction was stopped by placing the tubes in ice and releasing
the vacuum. All incubations were performed at least in duplicate
with a control for each pair of incubations. These conditions produced rates of iron incorporation into heme which were optimal.

Heme was extracted in acid-acetone as described by Falk
(1964). Two ml aliquots of the reaction mixture were added to
10 ml of ice cold, 5% (w/v) HCl-acetone, shaken for five minutes,
and then centrifuged to remove the precipitated protein. The
heme was extracted into ether by shaking of the HCl-acetone extract for five minutes in a mixture of 10 ml of saturated sodium
acetate solution and 10 ml of peroxide-free ether. The ether
phase was washed with 20 ml of saturated sodium acetate solution
to remove any free ^{59}Fe in the ether phase, and then evaporated to
dryness. Radioactivity was measured in a Packard model 3001 auto
gamma counter.

RESULTS AND DISCUSSION

Fig. 2 describes the inhibition of ferrochelatase activity
by Co^{++} at concentrations greater than 100 μM. The I_{50} value obtained under these conditions was 30 μM. Therefore, one explanation for the inhibition of heme biosynthesis and cytochrome P-450
synthesis by cobalt is the inhibition of ferrochelatase activity.
Studies were designed to determine whether other divalent cations
might produce a similar inhibition of ferrochelatase activity, and
if they could produce a depletion of hepatic cytochrome P-450 such
as that seen with $CoCl_2$ treatment. Table 1 shows the effects of

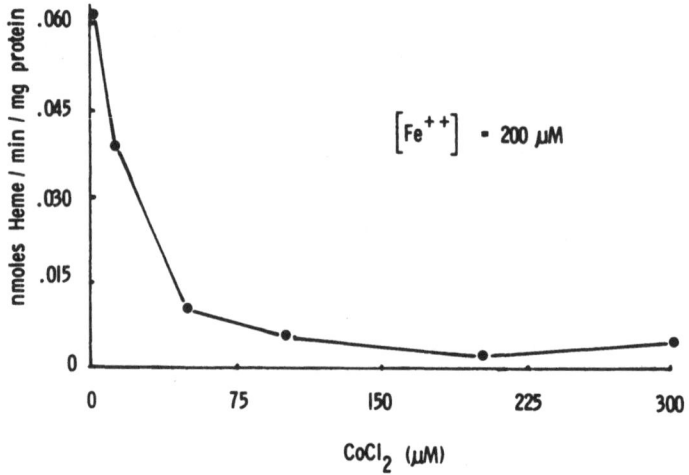

Fig. 2. The inhibition of ferrochelatase activity by cobaltous chloride.

other divalent cations on hepatic ferrochelatase activity. The alkaline earth elements, Ca^{++} and Mg^{++}, produced no appreciable inhibition. The heavy metals Cd^{++} and Pb^{++} did produce a significant inhibition, but not to the extent that was observed for the transition metals, Zn^{++}, Mn^{++}, and Co^{++}. The effect of copper differed from the other metals tested in that copper did not inhibit, but instead, produced a significant stimulation of ferrochelatase activity. It should be noted that the reaction mixture contained reduced glutathione, which reduces copper as well as iron, and that the copper was probably in the univalent state under the conditions of this experiment. The other transition metals are potent inhibitors of ferrochelatase activity, and of this series, Mn^{++} is known to be actively translocated into mitochondria (Lehninger, et al., 1967). Also, Maynard and Cotzias (1955) have shown that after treatment of rats with Mn^{++} the metal accumulates within hepatic mitochondria. Therefore, the ability of Mn^{++} to reduce rat liver cytochrome P-450 content in vivo was investigated.

Fig. 3 shows the effect of increasing doses of $MnCl_2$ and $CoCl_2$ on hepatic microsomal cytochrome P-450 content. Mn^{++}, like Co^{++} reduced the content of hepatic cytochrome P-450 and, at low doses, Mn^{++} caused a greater decrease in the cytochrome P-450 level than did Co^{++}. However, Co^{++} at higher doses was more effective than Mn^{++}. Differences in the dose response curves between Mn^{++} and Co^{++} may reflect differences in the disposition of these two metals. Greenberg et al., (1943) have shown that

TABLE I

The Effect of Various Divalent Cations on Ferrochelatase Activity

Cation Added (100 μM)	% Control Activity
Mg^{++}	80
Ca^{++}	66
Pb^{++}	42[a]
Cd^{++}	34[a]
Zn^{++}	15[a]
Mn^{++}	8[a]
Co^{++}	4[a]
Cu^{++}	122[a]

Control activity = 0.127 nmol/heme/min/mg protein.

$[FeSO_4]$ = 50 μM

[a]Denotes a statistically significant difference (p < .05) when analyzed by Student's paired t test. (N = 3)

Mn^{++} is excreted mainly via the bile and gastrointestinal tract while Co^{++} is excreted via the urine. These data suggest further that an inhibition of heme synthesis, such as that produced by Co^{++} or Mn^{++} at ferrochelatase, can effect a reduction in the level of hepatic microsomal cytochrome P-450.

Of particular interest was the finding that unlike the other transition metals, copper did not inhibit, but rather stimulated ferrochelatase activity. Interference with copper metabolism has previously been shown to reduce cytochrome P-450 content and drug metabolism. Moffit and Murphy (1973) recently reported decreases in aniline hydroxylase activity and in hexobarbital oxidase activity during copper deficiency. Zemaitis and coworkers (1973) have reported that treatment of rats with the copper chelator, diethyldithiocarbamate, produces a decrease in hepatic cytochrome P-450 content. If copper is required for ferrochelatase activity, these effects on cytochrome P-450 content and drug metabolism can be explained by an impairment in heme synthesis at ferro-chelatase with a subsequent decrease in hepatic microsomal cyto-chrome P-450 synthesis. Therefore, the stimulatory effect of copper on ferrochelatase activity was studied further.

Table II shows that as the copper concentration was increased

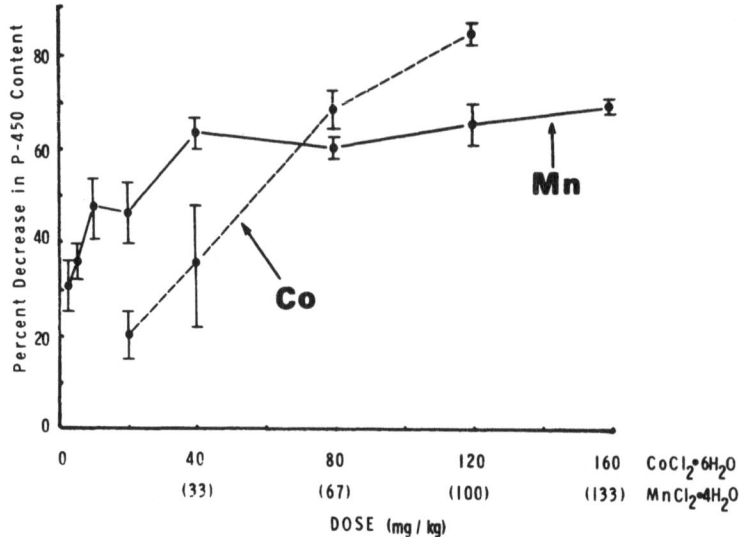

Fig. 3. The effect of $MnCl_2$ and $CoCl_2$ treatments on rat
hepatic microsomal cytochrome P-450 content.

there was a progressive reduction in the apparent K_m for iron with
no appreciable effect on the V_{max}. No effect on the apparent K_m
for protoporphyrin IX was observed. In an attempt to define the
mechanism by which copper stimulated ferrochelatase activity,
Tween 20-solubilized preparations of hepatic mitochondria were
employed. Fig. 4 shows the effect of copper on ferrochelatase
activity in a solubilized dialyzed preparation of mitochondria.
In the absence of added copper, no appreciable ferrochelatase
activity could be detected. The addition of copper to the incubation mixture progressively restored the ferrochelatase activity
and the "half-maximal activation" concentration was 0.2 μM.
Fig. 5 shows the effect of copper on the apparent K_m of iron in a
solubilized, dialyzed preparation of hepatic mitochondria. The K_m
of iron is decreased as the copper concentration is increased and,
as was observed in mitochondrial preparations, the V_{max} is not
altered. Certain other metals like mercury and nickel do not
substitute for copper as an activator of the enzyme, and Co^{++},
Mn^{++}, and Pb^{++} are inhibitory (Table 1).

Of the divalent cations tested, the transition metals were
the most potent inhibitors of ferrochelatase activity. Such an
inhibition may be due to competition with either the substrate,
iron, or the proposed activator, copper. Fig. 6 shows the

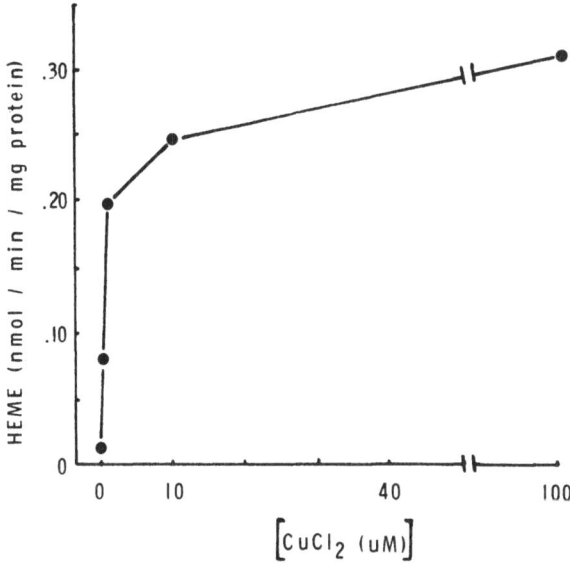

Fig. 4.   The activation of ferrochelatase activity by copper.   A solubilized preparation from rat liver mitochondria was employed (Porra et al., 1967) which had been dialyzed for 12 hr against 0.02 M Tris HCl buffer (pH 8.2) at 5°C. Reactions were carried out as described in Methods except that 0.65 mg of protein per ml of reaction mixture was used.   Copper chloride was added to the main chamber of the Thunberg tube.

reversal by copper of Co^{++} inhibition of ferrochelatase activity. In these experiments, control activity represents the activity in the absence of Co^{++}. Fig. 7 shows the reversal by copper of Pb^{++} inhibition of ferrochelatase activity. As with Co^{++}, a return toward control activity is seen as the copper concentration was increased. Next, the inhibition of ferrochelatase activity by Co^{++} and Pb^{++} in the presence of increasing concentrations of iron was examined. When the iron concentration was varied from 6.25 to 50 μM, there was essentially no change in the inhibition produced by Co^{++} or Pb^{++} (Fig. 8.) These results indicate the Co^{++} and Pb^{++} may be acting by antagonizing the activation of ferrochelatase by copper.

TABLE II

The Effect of Copper on the Apparent K_m for Iron

Cu^{++} Added (μM)	K_m (μM)	V_{max} (nmol/heme/min/mg protein)
0	59	0.14
5	32	0.19
25	20	0.16
100	17	0.15

[Protoporphyrin IX] = 50 μM

SUMMARY

Experiments have been performed which show that it is possible to regulate heme biosynthesis by regulating ferrochelatase activity. Treatment of rats with the transition metals, Co^{++} and Mn^{++}, has been shown to inhibit ferrochelatase activity and to produce a dose-dependent decrease in hepatic cytochrome P-450 content. Unlike other transition metals, copper stimulated ferrochelatase activity and there was an interaction beween Cu^{++} and Fe^{++} in the system such that the K_m of Fe^{++} was dependent on the concentration of copper. After solubilizing and dialyzing hepatic mitochondrial preparations, ferrochelatase activity was lost and could be restored by the addition of copper. In addition, copper, but not iron, reversed the inhibition of ferrochelatase activity produced by Pb^{++} or Co^{++}. This study suggests that cytochrome P-450 content may be decreased during copper deficiency due to a decrease in ferrochelatase activity and a subsequent defect in heme biosynthesis. The synthesis of other hemoproteins may also be affected by copper deficiency. For example, it is well known that copper deficiency can lead to anemias (Lee et al., 1968) which might be explained by an intracellular defect in heme biosynthesis at ferrochelatase.

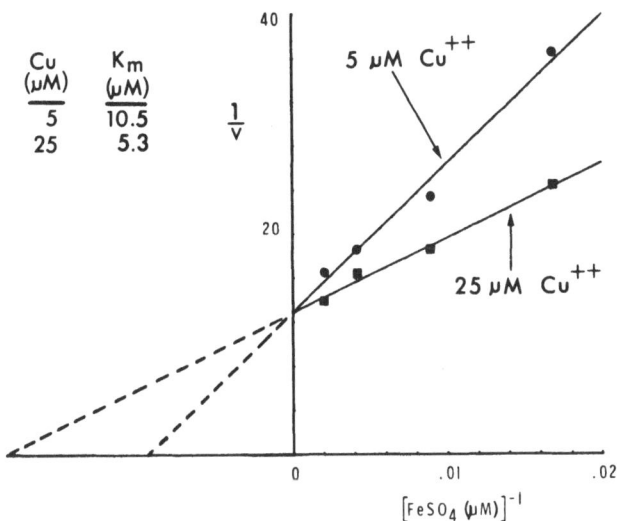

Fig. 5. The effect of copper on the Michaelis constant of iron for ferrochelatase in a solubilized, dialyzed preparation of rat liver mitochondria. The enzyme preparation is described under Figure 4. The analysis of ferrochelatase activity is described in Methods.

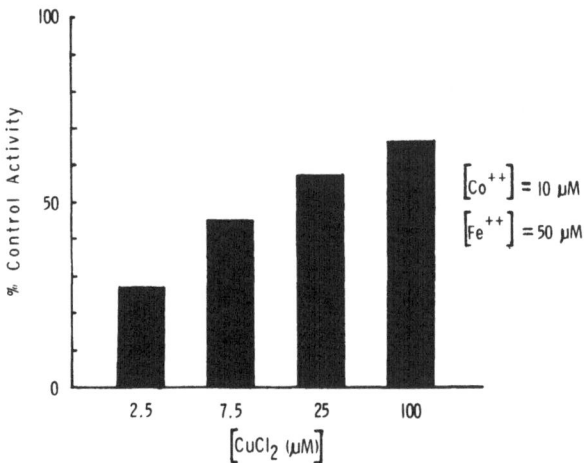

Fig. 6. The reversal of cobalt inhibition of ferrochelatase activity by copper. A solubilized, dialyzed preparation of rat liver mitochondria described under Figure 4 was employed. Control activity represents the activity obtained in the absence of cobalt.

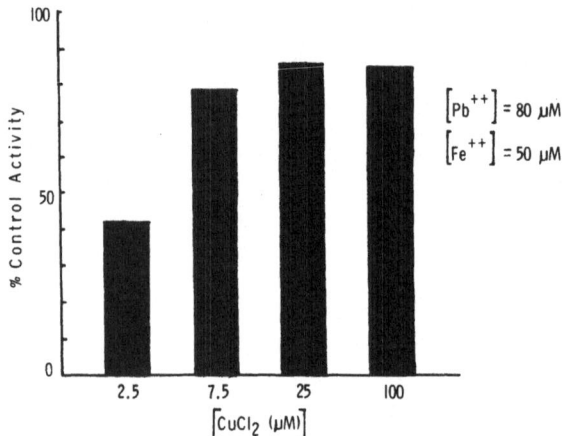

Fig. 7. The reversal of lead inhibition of ferrochelatase activity by copper. Conditions were as described under Figure 6. Control activity represents the activity obtained in the absence of lead.

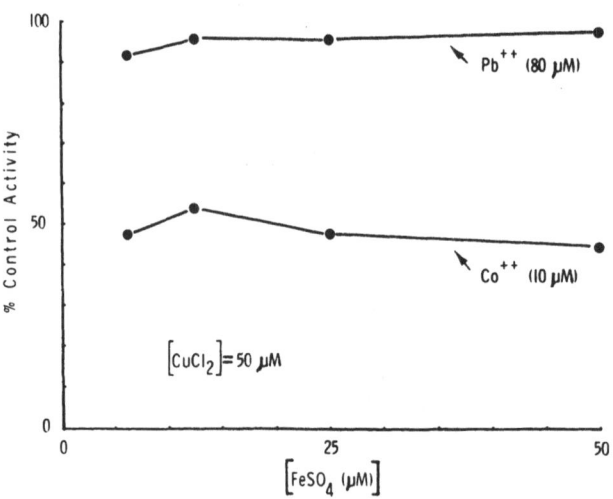

Fig. 8. The effect of iron on the inhibition of ferro-chelatase activity by cobalt and lead. Conditions were as described under Figure 6 except that iron concentrations were varied in the presence of 10 μM Co^{++} and 80 μM Pb^{++}. Control activity represents that activity obtained in the absence of inhibitor.

REFERENCES

Baron, J. and Tephly, T. R. 1969. Effect of 3-amino-1,2,4-triazole on the stimulation of hepatic microsomal heme synthesis and induction of hepatic microsomal oxidases produced by phenobarbital. Mol. Pharmacol. 5:10-20.

Falk, J. E. 1964. Porphyrins and Metalloporphyrins, B. B. A. Library 2:183, Elsevier Publishing Company, New York.

Granick, S. and Urata, C. 1963. Increase in activity of δ-aminolevulinic acid synthetase in liver mitochondria induced by feeding of 3,5-dicarbethoxy-1,4-dihydrocollidine. J. Biol. Chem. 238:821-827.

Greenberg, D. M., Copp, D. H. and Cuthbertson, E. M. 1943. Studies in mineral metabolism with the aid of artificial radioactive isotopes. VII. The distribution and excretion, particularly by way of the bile of iron, cobalt, and manganese. J. Biol. Chem. 147:749-756.

Hasegawa, E., Smith, C. and Tephly, T. R. 1970. Induction of hepatic mitochondrial ferrochelatase by phenobarbital. Biochem. Biophys. Res. Commun. 40:517-523.

Jones, M. S. and Jones, O. T. G. 1968. Evidence for the location of ferrochelatase on the inner membrane of rat liver mitochondria. Biochem. Biophys. Res. Commun. 31:977-982.

Jones, M. S. and Jones, O. T. G. 1969. The structural organization of heme synthesis in rat liver mitochondria. Biochem. J. 113:507-514.

Jones, M. S. and Jones, O. T. G. 1970. Permeability properties of mitochondrial membranes and the regulation of heme biosynthesis. Biochem. Biophys. Res. Commun. 41:1072-1079.

Labbe, R. F. and Hubbard, N. 1960. Preparation and properties of the iron-protoporphyrin chelating enzyme. Biochim. Biophys. Acta 41:185:191.

Lee, R. G., Nacht, S., Lukens, J. N. and Cartwright, G. E. 1968. Iron metabolism in copper-deficient swine. J. Clin. Invest. 47:2058-2069.

Lehninger, A. L., Carafoli, E. and Rossi, C. S. 1967. Energy-linked ion movements in mitochondrial systems. Advan. Enzymol. 29:259.

Lowry, C. H., Rosebrough, N. I., Farr, A. L. and Randall, R. J. 1951. Protein measurements with the Folin reagent. J. Biol. Chem. 193:265-275.

Maynard, L. S. and Cotzias, G. C. 1955. The partition of manganese among organs and intracellular organelles of the rat. J. Biol. Chem. 214:489-495.

McKay, R., Druyan, R., Getz, G. S. and Rabinowitz, M. 1969. Intra-mitochondrial localization of δ-aminolevulinate synthetase and ferrochelatase in rat liver. Biochem. J. 114:455-461.

Moffitt, A. E., Jr., and Murphy, S. D. 1973. Effect of excess and deficient copper intake on rat liver microsomal enzyme activity. Biochem. Pharmacol. 22:1463-1467.

Porra, R. J., Vitols, K. S., Labbe, R. F. and Newton, N. A.
    1967. Studies on ferrochelatase. The effects of thiols
    and other factors on the determination of activity. Biochem.
    J. 104:321-327.

Tephly, T. R., Hasegawa, E. and Baron, J. 1971. Effect of drugs
    on heme synthesis in the liver. Metabolism 20:200-214.

Tephly, T. R. and Hibbeln, P. 1971. The effect of cobalt
    chloride administration on the synthesis of hepatic micro-
    somal cytochrome P-450. Biochem. Biophys. Res. Commun.
    42:589-595.

Tephly, T. R., Webb, C., Trussler, P., Kniffen, T., Hasegawa, E.
    and Piper, W. 1973. The regulation of heme synthesis re-
    lated to drug metabolism. Drug. Met. Disp. 1:259-266.

Zemaitis, M. A., Blackburn, W. R. and Green, F. E. 1973. Some
    effects of disulfram on drug metabolism and hepatic cell
    ultrastructure in the rat. The Pharmacologist 15:191.

MIXED FUNCTION OXIDATION AND INTERMEDIARY METABOLISM: METABOLIC INTERDEPENDENCIES IN THE LIVER*

Ronald G. Thurman and Donald P. Marazzo+
Johnson Research Foundation
University of Pennsylvania
Philadelphia, Pennsylvania 19174

and

Roland Scholz
Institut für Physiologische Chemie und Physikalische
Biochemie
Universität München
Munich, Germany

INTRODUCTION

The mixed function oxidation system located in the endoplasmic reticulum consists of a membrane-bound electron transport system with cytochrome P-450 as the terminal oxidase. The function of this system, as can be seen from a cursory inspection of other contributions to this volume, is to convert a wide variety of structurally dissimilar lipophilic compounds, such as drugs, into more polar products primarily by hydroxylation (1,2). The overall reaction requires equimolar amounts of NADPH and oxygen.

In perfused livers, addition of drugs causes a marked stimulation of respiration (3,4,5). For example, the addition of aminopyrine to perfused livers of phenobarbital pretreated rats, i.e., in the presence of an induced mixed function oxidation system (6), stimulates oxygen consumption by up to 40% of the basal respiratory rate (3). Thus, reasoning from the stoichiometry of oxygen and NADPH requirements, the mixed function oxidation system is a highly active NADPH-utilizing system. NADPH is generated in the cytosol primarily by reactions catalyzed by malic enzyme and the dehydrogenases of the pentose phosphate shunt (Figure 1). Recently, evidence was presented that in livers of well-fed animals, glycogen

355

breakdown and the subsequent generation of NADPH from glucose-6-phosphate is a limiting factor for maximal mixed function oxidation (3). In the fasted state, however, reducing equivalents from mitochondrial oxidations are utilized for extramitochondrial NADPH generation using malate as substrate. They are transferred into the cytosol most probably via an energy-dependent substrate shuttle mechanism as depicted in Figure 1.

On the basis of these theoretical considerations it is predicted that multiple interdependencies between mixed function oxidation and intermediary metabolism exist in the liver with the extramitochondrial NADP-NADPH system as an interlinking factor. For example, malate and glucose-6-phosphate are intermediates in the pathway of gluconeogenesis as well as substrates for NADPH generation. Thus, drug metabolism may interfere with gluconeogenesis. Moreover, drug metabolism and synthesis of fatty acids or sterols may compete for NADPH, if the extramitochondrial NADPH supply is limiting. On the other hand, the rate of NADPH generation depends on the substrate supply. Therefore, under certain conditions the rate of drug metabolism also depends on the metabolic state, e.g., nutrition or hormonal influences.

In the past, the regulatory properties of biosynthetic and energy-yielding processes, on the one hand, and biotransformation of drugs, on the other hand, were studied more or less independently. However, both processes occur simultaneously in the hepatocyte and their regulation in vivo can only be understood if one considers possible interdependencies. The purpose of this paper, therefore, is to review recent observations from our laboratories indicating that such interdependencies exist and that an enhanced turnover of NADPH due to mixed-function oxidation of drugs alters the biosynthetic capacity of the liver and vice versa.

Interaction of Gluconeogenesis and Aminopyrine Metabolism. Malate and glucose-6-phosphate are intermediates common to both glucose synthesis and NADPH generation. Thus, an interaction between these metabolic functions of the liver is possible at the level of reactions catalyzed by malic enzyme and the dehydrogenases of the pentose-phosphate shunt. These theoretical considerations were borne out by observations from our laboratories that substrates for mixed-function oxidation (e.g., aminopyrine, hexobarbital, etc.) inhibit hepatic glucose synthesis (7,8,9).

Aminopyrine (i.e., dimethylamino-phenylpyrazolone) diminishes gluconeogenesis from lactate in perfused livers from fasted rats (9). This inhibition was less than 20% in livers from untreated animals. However, in livers from animals where the mixed-function oxidation system was induced by pretreatment with phenobarbital (6), glucose synthesis was diminished up to 50% by the drug (Figure 2).

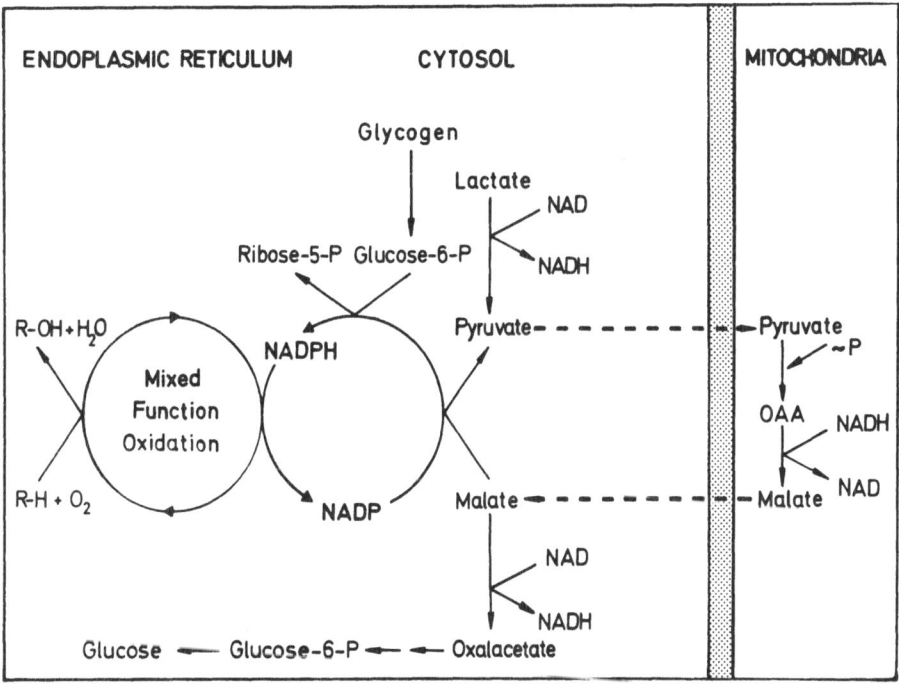

Fig. 1. Scheme depicting the site for extramitochondrial NADPH generation and its interaction with the pathway of gluconeogenesis and mixed function oxidation. R-H: substrates for mixed function oxidation; R-OH product; ∿P: energy-rich phosphate representing ATP required for carboxylation of pyruvate.

These data suggest that metabolism of aminopyrine is required for the inhibitory effect. This contention is further supported by the observations that the completely demethylated product of aminopyrine metabolism, i.e., aminophenylpyrazolone, does not effect glucose synthesis, and that the inhibitory effect of aminopyrine is prevented by metyrapone (7,8).

Further insight into the mechanism of the interaction between gluconeogenesis and drug metabolism came from experiments with dihydroxyacetone. This precursor for glucose synthesis differs from lactate since it enters the gluconeogenic sequence at the level of triose phosphates. Consequently, malate is not an intermediate in gluconeogenesis from dihydroxyacetone as it is with lactate. It was observed (9) that the inhibitory effect of aminopyrine was five-fold greater with lactate than with dihydroxyacetone as precursor indicating that interaction occurs at the level of malate rather than glucose-6-phosphate.

Moreover, the inhibition by aminopyrine was only observed at submaximal and maximal rates of glucose synthesis; no effect was observed at low rates. A hypothesis consistent with these data has been discussed in detail by Scholz, Hansen and Thurman (9). Basically, it involves the development of the idea that in the presence of active NADPH-utilizing processes a "futile cycle" in the gluconeogenic sequence involving pyruvate carboxylase, malate dehydrogenase and malic enzyme is stimulated (see Figure 1) which is compensated for by an increased flux over the pyruvate carboxylase reaction when the rate of glucose synthesis is low. On the other hand, glucose synthesis is suppressed by drugs when pyruvate carboxylase functions nearly maximally (as at high lactate concentrations), since the compensatory mechanism is limited by the capacity of pyruvate carboxylase.

Inhibition of Lipid Synthesis by Aminopyrine. Both lipid synthesis and mixed function oxidation require NADPH. Thus, competition for NADPH in the cell is theoretically possible if the NADPH supply is rate limiting or nearly rate limiting for one or both processes. Previous work from this laboratory (3) suggested that cytosolic NADPH supply is a regulating factor (see below) for mixed function oxidation in the intact cell. Moreover, it has been shown that NADPH supply is a limiting factor in fatty acid synthesis in adipose tissue (10), whereas in liver its role has been minimized (11).

To ascertain if interaction between these hepatic metabolic processes occurs, an examination of the influence of substrates for mixed function oxidation (e.g., aminopyrine) on the rate of lipid synthesis in perfused rat livers was performed. The rate of incorporation of tritium from tritiated water into the lipid fraction of tissues is employed by a number of laboratories as a measure of the rate of lipid synthesis (12,13,14,15,16,17,18) since it obviates dilution problems observed with ^{14}C-labelled glucose and acetate. In experiments with perfused liver from control and phenobarbital pretreated animals, the following was observed (Table I): (a) pretreatment with phenobarbital did not influence the control rate of lipid synthesis; (b) aminopyrine inhibited lipid synthesis 18% in livers from untreated rats but 43% in livers from phenobarbital pretreated animals; (c) aminophenylpyrazolone, i.e., a demethylated product of aminopyrine metabolism, had little effect. Similar to the inhibition of gluconeogenesis observed by aminopyrine, this effect was also inducible with phenobarbital pretreatment and was absent with a structurally similar, but non-metabolizable product (18).

The mechanism for inhibition of lipogenesis by aminopyrine may be due to the interaction of aminopyrine or one of its products directly with enzymes of the lipogenic pathway or with energy metabolism. However, since the demethylated product was without effect and since the ATP/ADP ratio in liver was unaffected by

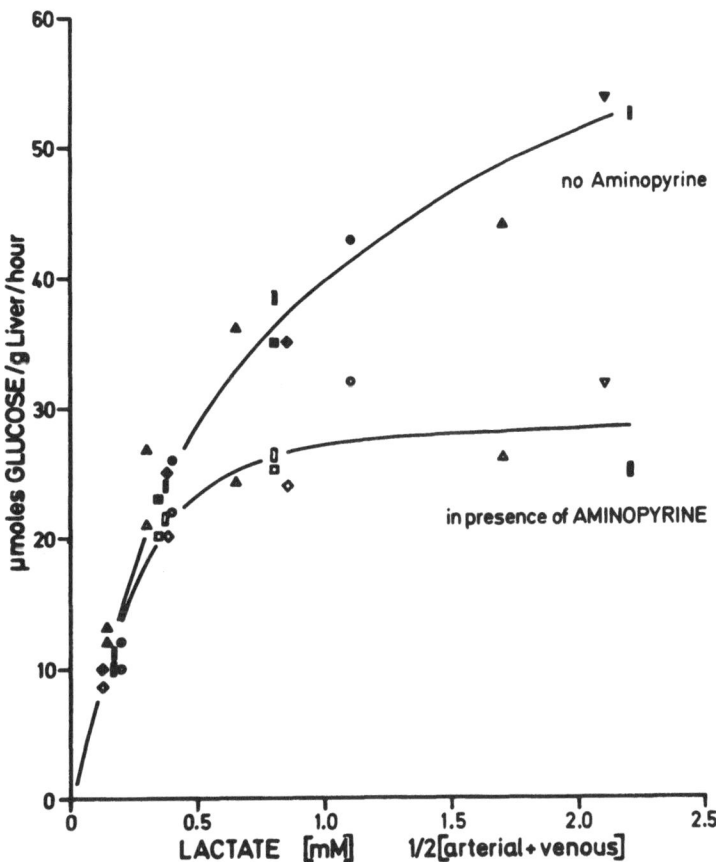

Fig. 2. Effect of aminopyrine on gluconeogenesis from lactate. Livers of phenobarbital pretreated rats (6) fasted for 24 hours were perfused in a non-circulating system as described by Scholz et al., (9). The lactate concentration of the perfusate was increased in a stepwise manner. At each step 0.2 mM aminopyrine was infused when the oxygen recording indicated that steady state conditions had been reached. Rates of glucose production (umoles per gram liver/wet weight/per hour) from 5 identical experiments (as indicated by different symbols) were plotted against the arithmetic means of lactate concentrations in the perfusate entering and leaving the liver. Closed symbols: dependency of gluconeogenic rates on the lactate concentration in the absence of aminopyrine; open symbols: in the presence of 0.2 mM aminopyrine.

TABLE I

Incorporation of Tritium from Tritiated Water into the Lipid Fraction of Perfused Rat Liver

Pretreatment	Addition	n	μg atom x g^{-1} x h^{-1}	% of Control
---	---	9	46.8 $\pm$ 9.6	100
---	Aminopyrine (0.5 mM)	6	38.6 $\pm$ 7.2	82
Phenobarbital	---	9	55.2 $\pm$ 5.4	100
Phenobarbital	Aminopyrine (0.5 mM)	9	31.6 $\pm$ 3.8	57
Phenobarbital	Aminophenyl-pyrazolone (o.5 mM)	4	48.6 $\pm$ 6.6	88

Livers from normal and phenobarbital pretreated rats were perfused with 20 mM glucose in the presence or absence of substrates or products of the mixed function oxidation system. Rates of hydrogen incorporation into newly synthesized lipids were calculated from the incorporation of tritium into the chloroform-methanol extract of liver after 2 hours of perfusion with circulating perfusate and from the specific activity of water in the perfusate. Mean $\pm$ S.E.M. (data from (18)).

aminopyrine (18), a direct interaction seems unlikely. On the
other hand, an interaction between both processes could occur due
to a competition at the level of either substrate (i.e., acetyl-
CoA) or cofactor (i.e., NADPH) supply. For example, acetylation
of products of aminopyrine metabolism could divert cytosolic
acetyl-CoA from fatty acid synthesis. However, the rates of
formation of acetyl-aminoantipyrine, i.e., the main product of
aminopyrine metabolism in vivo (19), measured in these perfusion
experiments suggest that not more than 10% of the inhibitory
effect may be accounted for by a diversion of acetyl-CoA from
fatty acid synthesis towards drug acetylation. A likely explanation of this inhibition, therefore, is a competition between mixed
function oxidation and lipid synthesis for NADPH.

This conclusion implies that in the liver NADPH supply for
lipogenesis may be rate limiting under some metabolic conditions,
in contrast to the conclusions derived from measurements of enzyme activities in tissue extracts (20). However, activities
measured in vitro rarely reflect the metabolite flux over a particular enzyme step in vivo. Moreover, in addition to fatty acid synthesis several other NADPH-utilizing processes occur simultaneously
in the intact cell. Thus, under extreme conditions, e.g., drug
metabolism following chronic drug application in a nutritional and
hormonal state favoring lipogenesis, NADPH generation may become a
limiting factor for fatty acid synthesis.

Dependence of Mixed Function Oxidation on Intermediary
Metabolism. Above we have documented the evidence which shows
that mixed function oxidation of drugs influences intermediary
metabolism, particularly the biosynthetic processes of gluconeogenesis and lipogenesis. Most likely this interaction occurs via
a competition for NADPH or NADPH-generating substrates common to
drug metabolism and intermediary metabolism. The question has been
inverted, and we now ask whether the metabolic state influences
the mixed function oxidation of drugs. This problem was studied
in rat livers perfused in different metabolic states achieved by
pretreatment of the animals or by addition of substrates, drugs,
inhibitors or hormones to the perfusate. Experimentally we have
employed two model substrates for mixed oxidation, aminopyrine
and p-nitroanisole. With aminopyrine, the stimulation of oxygen
uptake closely reflects its metabolism (3), and, with p-nitroanisole,
the excretion of its metabolic products into the perfusate has been
monitored continuously.

When aminopyrine is added to perfused livers with stepwise
increasing concentrations, oxygen consumption increases concomi-
tantly showing half-maximal effect at concentrations near 0.05 mM
independent of the metabolic state of the liver (3). This indicates
that aminopyrine metabolism is limited by substrate supply at

concentrations less than 0.1 mM. However, maximal stimulation of
oxygen consumption obtained with aminopyrine concentrations
greater than 0.5 mM varied with changes in the metabolic state.
For example, aminopyrine stimulated respiration 25% in livers from
fed animals. In the presence of antimycin A, an inhibitor of the
mitochondrial respiratory chain, a 40% stimulation of respiration
due to aminopyrine was observed (Table II) (3). However, in livers
from fasted animals aminopyrine stimulated respiration by only 16%
and this rate was totally abolished by antimycin A (for a full
treatment of these data, see (3)). The different maximal rates of
mixed function oxidation as reflected by the rates of oxygen uptake in those experiments cannot be explained by different activities of the drug metabolizing system since microsomes from fed or
fasted livers incubated in the presence or absence of antimycin A
all had turnover numbers of cytochrome P-450 between 8 and 10
(Table II). Thus, since oxygen and aminopyrine were in excess and
since the capacity of the cytochrome P-450 system appeared to be
higher than the observed rates, it was concluded that NADPH was
rate-limiting for drug hydroxylation under these conditions.

The main substrates for extramitochondrial NADPH generation
are (a) glucose-6-phosphate derived predominantly from glycogen
breakdown or from gluconeogenic precursors, and (b) malate which
may be derived via a cyclic process involving mitochondrial reactions as depicted in Figure 1. The first source for NADPH,
available only in the fed state, appears to be most efficient for
mixed function oxidation. The second source appears to be the
only possibility for generating NADPH in the fasted state in the
absence of exogenous substrates. Moreover, NADPH generation via
malic enzyme is considered to be an energy dependent process,
since the continuous supply of malate requires mitochondrial carboxylation of pyruvate (see Figure 1). Thus, inhibition of mitochondrial oxidation interrupts this cyclic process, and consequently, inhibits drug metabolism indirectly. This conclusion is
consistent with the observation that in the fasted state aminopyrine metabolism is almost completely abolished in the presence
of antimycin A (Table II).

The influence of intermediary metabolism upon mixed function
oxidation was further investigated employing a continuous readout
for the rate of product excretion into the perfusate. p-Nitroanisole, a typical substrate for the mixed function oxidation system, is 0-demethylated to p-nitrophenol (21) which is excreted
partially free and partially as the respective glucuronide (22).
Due to the yellow color of p-nitrophenolate the rate of mixed
function oxidation can be monitored continuously employing a non-
circulating perfusion system with the effluent perfusate passing
through a flow cuvette placed in a spectrophotometer (436 nm)
(23,24).

TABLE II

Stimulation of Oxygen Uptake in Perfused Rat Livers by Aminopyrine

Animals	Conditions	Oxygen Uptake in Perfused Livers			Microsomes
		Basal Rate	Stimulation by Aminopyrine	% of Basal Rate	Turnover Number Cytochrome P_{450}
		μg atom x g^{-1} x h^{-1}			
Fed	---	224 ± 29	53 ± 9	25	9.4
Fed	Antimycin A	197 ± 12	78 ± 3	40	9.9
Fasted	---	206 ± 11	32 ± 4	16	8.2
Fasted	Antimycin A	209 ± 23	< 4	< 2	7.7

Livers from fed or fasted rats pretreated with phenobarbital were perfused in a circulating system (3). Oxygen uptake was measured polarographically. After 40 min of perfusion, either 50 umoles of aminopyrine or 2 mg of antimycin A were added to 100 ml of circulating perfusate. In the latter case, aminopyrine was added 15 min after antimycin A. The increase in oxygen uptake following aminopyrine was maximal after 3 to 5 min. The maximal stimulation in each experiment was used for the calculation of the mean values in column 4 (mean ± S.E.M. with 6 experiments in each group; data from (3)). In parallel experiments, microsomes were prepared from livers and oxygen uptake following aminopyrine was measured in the presence and absence of antimycin A (30 μM). Cytochrome P-450 was determined spectrophotometrically (25), and turnover numbers for the oxidase were calculated (column 6). (Mean of data from 4 livers.)

A typical perfusion experiment with the liver of a fasted rat
is presented in Figure 3. Following a preperfusion period, the
infusion of 0.2 mM p-nitroanisole results in the rapid formation
of p-nitrophenol as reflected by the increase in the absorbance of
the perfusate at 436 nm. The subsequent addition of dihydroxyacetone, a substrate for glucose synthesis, produced a marked increase in p-nitrophenol excretion indicating a stimulation of
mixed function oxidation. Similar data (not shown) were obtained
with pyruvate and lactate as substrates for glucose synthesis (24).
Thus, in the fasted state maximal rates of mixed function oxidation
can be further increased by the addition of gluconeogenic precursors,
presumably by increasing the cellular concentration of intermediates,
such as malate or glucose -6-phosphate, which are substrates for
NADPH generation.

The data presented in this review clearly indicate that the
rate of mixed function oxidation in the presence of saturating concentrations of suitable substrates depends upon the metabolic state
of the liver. Rate-limiting steps in the sequence of the microsomal
electron transport chain, as observed in studies with isolated microsomes (25), appear to be of minor importance under the conditions of
the intact cell. Mixed function oxidation is favored in the fed
state as compared with the fasted state. This finding is in agreement with the observation of Kato and Gillette that drug metabolism
in vivo is slower following a period of starvation (26). More recently, Bock et al., reported that hexobarbital is eliminated faster
from the perfusate when livers from fed rats were perfused (27).
The conditions most favorable for mixed function oxidation in perfused liver are the following: (a) stimulation of glycogen breakdown and (b) gluconeogenesis. These processes have one characteristic in common; they increase the steady-state concentration of
important intermediates for NADPH generation such as glucose-6-
phosphate. For example, antimycin A causes a large stimulation of
glycogenolysis resulting in an elevated tissue content of glucose-
6-phosphate (28). Similar fluctuations in tissue levels of intermediates are observed when gluconeogenic substrates are added following a period of starvation. It is certainly more than coinci-
dental that the same trend in rates of mixed function oxidation is
observed, since glucose-6-phosphate is an important substrate for
NADPH generation. Thus, it now appears clear that the NADPH supply
is a rate-limiting factor for mixed function oxidation in the intact cell. Furthermore, these observations strongly support the
viewpoint that control of mixed function oxidation in vivo cannot
be studied without taking into account metabolic interdependencies
in the liver.

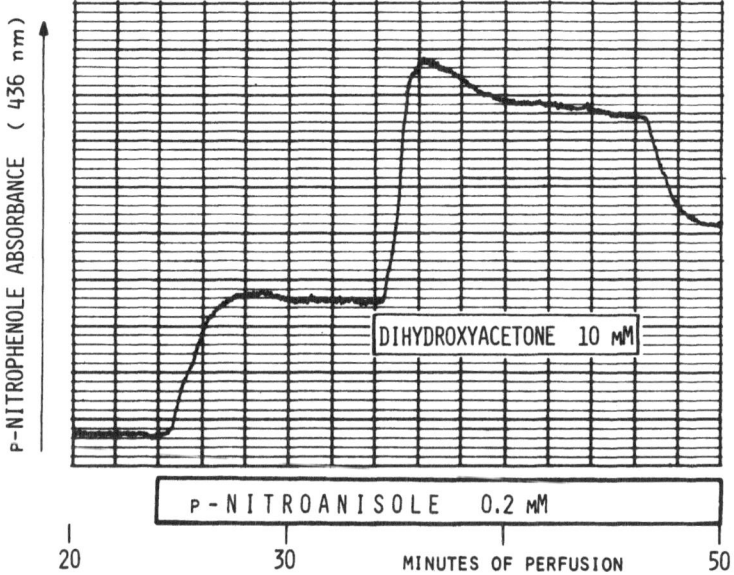

Fig. 3. Effect of dihydroxyacetone on O-demethylation of p-Nitroanisole. Livers of phenobarbital pretreated rats fasted for 24 hours were perfused with Krebs-Henseleit bicarbonate buffer (29) in a non-circulating system. The absorbance of the perfusate at 436 nm was monitored continuously employing a stream splitter and a flow cuvette attached to a spectrophotometer as reported previously (23,24). Changes in the rate of O-demethylation of p-nitroanisole are indicated by changes in the p-nitrophenolate absorbance of the perfusate. Horizontal bars represent the infusion of p-nitroanisole (0.2 mM) and dihydroxyacetone (10 mM).

SUMMARY

Both hepatic lipogenesis from glucose and gluconeogenesis from lactate are inhibited by substrates for mixed function oxidation such as aminopyrine. These effects are inducible with phenobarbital pretreatment and absent with a demethylated product of aminopyrine metabolism, aminophenylpyrazolone. The data indicate that an interaction between biosynthesis and drug metabolism occurs in the intact liver cell. In the case of lipogenesis, direct competition for cytosolic NADPH occurs, whereas in the case of gluconeogenesis, enhanced drug metabolism diverts key intermediates, e.g., malate from glucose synthesis toward NADPH generation.

Furthermore, rates of mixed function oxidation were modified following perturbations of intermediary metabolism, e.g., fasting, addition of substrates for glucose synthesis, inhibition of energy metabolism, etc. The data support the hypothesis that in the presence of sufficiently high concentrations of substrate and oxygen, the rate-limiting step for mixed function oxidation in the intact cell is the supply of NADPH.

FOOTNOTES

*   Supported, in part, by a grant from the National Institute of Alcohol Abuse and Alcoholism (AA-00288), a Career Development Award from the National Institute of Mental Health (K 2 MH 70, 155; RGT) and a grant from Deutsche Forschungsgemeinschaft, Sonderforschungsbereich 51 "Medizinische Molekularbiologie und Biochemie."

+   Present address: Georgetown University School of Medicine.

REFERENCES

1.   Mason, H. S. (1957) in Advan. Enzymol. 19, 79.
2.   Hayaishi, O. (1962) in Oxygenases (O. Hayaishi, ed.) Academic Press, New York, p. 1.
3.   Thurman, R. G. and Scholz, R. (1969) Eur. J. Biochem. 10, 459.
4.   Brauser, B., Sies, H. and Bücher, Th. (1969) FEBS Letters 2, 170.
5.   Sies, H. and Brauser, B. (1970) Eur. J. Biochem. 15, 531.
6.   Remmer, H. and Merker, H. J. (1963) Science 142, 1657.
7.   Thurman, R. G. and Scholz, R. (1970) in Regulation of Gluconeogenesis (Söling, H. D. and Willms, B., eds.) Thieme-Verlag, Stuttgart, p. 315.
8.   Thurman, R. G. and Scholz, R. (1973) Isolated Liver Perfusion and Its Applications (Bartosek, I., ed.) Raven Press, New York, p. 211.
9.   Scholz, R., Hansen, W. and Thurman, R. G. (1973) Eur. J. Biochem. 38, 64.
10.   Young, F. W., Shrago, E. and Lardy, H. A. (1964) Biochemistry 3, 1687.
11.   Wakil, S. J. (1970) in Lipid Metabolism (Wakil, S. J., ed.) Academic Press, New York, p. 1.
12.   Jungas, R. L. (1968) Biochemistry 7, 3708.
13.   Lowenstein, J. M. (1971) J. Lipid Res. 12, 179.
14.   Veech, R. L. (1974) in Alcohol and Aldehyde Metabolizing Systems (Thurman, R. G., Yonetani, T., Williamson, J. R. and Chance, B., eds.) Academic Press, New York, p. 383.

15. Scholz, R., Kaltstein, A., Schwabe, U. and Thurman, R. G.
    (1974) Ibid., p. 315.
16. Brunengraber, H., Boutry, M., Lowenstein, L. and Lowenstein,
    J. M. (1974) Ibid., p. 329.
17. Barth, Chr., Liersch, M., Hackenschmidt, J., Ullman, H. and
    Decker, K. (1972) Hoppe-Seyler's Z. Physiol. Chem. 353,
    1085.
18. Thurman, R. G. and Scholz, R. (1973) Eur. J. Biochem. 38,73.
19. Schuppel, R. and Soehring, K. (1965) Pharm. Acta Helv. 40,105.
20. Tepperman, J. and Tepperman, H. M. (1970) Fed. Proc. 29, 1284.
21. Netter, K. J. and Seidel, G., J. Pharmacol. Exp. Ther. 146:
    61, 1964.
22. Pogell, B. M. and Leloir, L. F., J. Biol. Chem. 236: 293, 1961.
23. Thurman, R. G., McKenna, W. R., Brentzel, H. J. and Hesse, S.,
    Fed. Proc., in press.
24. Thurman, R. G., McKenna, W. R. and Marazzo, D., The
    Pharmacologist, in press.
25. Omura, T. and Sato, R. (1964) J. Biol. Chem. 239, 2370.
26. Kato, R. and Gillette, J. R. (1965) J. Pharmacol. Exp. Ther.
    150, 279.
27. Bock, K. W., Fröhlich, W. and Schlotte, W. (1972) Naunyn-
    Schmiedeberg's Arch. Pharmacol. 273, 193.
28. Schwarz, F. (1968) Thesis, Medical Faculty, University of
    Munich.
29. Krebs, H. A. and Henseleit, K. (1932) Hoppe-Seyler's Z.
    Physiol. Chem. 210, 33.

DISSOCIATION OF MICROSOMAL ETHANOL OXIDATION FROM CYTOCHROME P-450

CATALYZED DRUG METABOLISM

Kostas P. Vatsis and Martin P. Schulman
Department of Pharmacology
School of Basic Medical Sciences
University of Illinois College of Medicine
Chicago, Illinois 60612

INTRODUCTION

Considerable evidence is currently available which strongly suggests that, in addition to the alcohol dehydrogenase (ADH)-catalyzed reaction, a pyrazole-insensitive pathway participates in the metabolism of ethanol in vivo (Israel et al., 1970; Lieber and DeCarli, 1972), in the isolated perfused rat liver (Papenberg et al., 1970; Thurman and McKenna, 1974), in rat and mouse liver slices (Thieden, 1971; Lieber and DeCarli, 1970; Vatsis and Schulman, 1973a; Vatsis et al., 1973), and in isolated rat liver parenchymal cells (Grunnet et al., 1973). The non-ADH pathway has two distinguishing properties, namely, it becomes increasingly important at ethanol concentrations (20-80 mM) well above the K_m of AHD (Thieden, 1971; Vatsis et al., 1973; Lieber and DeCarli, 1973), and appears to be inducible by chronic ethanol feeding (Lieber and DeCarli, 1970, 1972) as well as by treatment of animals with phenobarbital (Carter and Isselbacher, 1971; Khanna et al., 1972; Vatsis and Schulman, 1974). Though subcellular fractionation studies have indicated that the non-ADH pathway of ethanol metabolism may be localized in the microsomal fraction of rat and mouse liver (Lieber and DeCarli, 1970; Vatsis and Schulman, 1973a); Vatsis et al., 1973), its precise enzymic identity has not been elucidated unequivocally.

Based on cofactor requirements, a partial sensitivity to carbon monoxide, and certain ill-defined interactions with the hepatic mixed function oxidase system, a distinct microsomal system (MEOS) was thus described that metabolizes ethanol and drugs by a common pathway in which cytochrome P-450 is implicated as

369

the terminal oxidase for both processes (Lieber and DeCarli, 1970; Rubin et al., 1970). Much of the original basis of MEOS has been contradicted, however, since it has been demonstrated that the NADPH-dependent microsomal oxidation of ethanol can be explained by the combined activities of catalase and NADPH oxidase without an apparent need for cytochrome P-450 participation in this reaction (Roach et al., 1969; Khanna et al., 1970; Carter and Isselbacher, 1971; Nelson et al., 1973; Vatsis and Schulman, 1973b, 1974). Exhaustive studies with inhibitors have led to conflicting interpretations (Thurman et al., 1972; Lin et al., 1972; Lieber and DeCarli, 1970, 1973), and have not resolved the controversy. Results with solubilized microsomal systems have not been in agreement (Teschke et al., 1972, 1974; Mezey et al., 1973; Thurman and Scholz, 1973; Thurman et al., 1974). A recent report has shown, however, that the microsomal ethanol-oxidizing system could be effectively reconstituted by the combination of two fractions, one of them containing the partially purified NADPH-linked flavoprotein reductase, and the other being rich in catalase activity; the fraction containing catalase activity could be replaced by commercial crystalline catalase without any loss in activity. In contrast, combination of the two fractions containing partially purified reductase and cytochrome P-450, respectively, resulted in negligible ethanol oxidation rates (Barakat et al., 1974).

In the present studies, microsomal ethanol oxidation and drug metabolism have been compared in different experimental situations without the use of inhibitors. The systems studied were (1) hepatic microsomes of mutant C_sb mice which possess a thermolabile tissue catalase activity, (2) hepatic microsomes of Swiss mice of different ages, and (3) microsomes in which a portion of the cytochrome P-450 had been destroyed by prior induction of lipid peroxidation. In each instance, a lack of parallelism between ethanol oxidation and drug metabolism was noted. It is concluded that these two enzymic processes are independent of each other, that microsomal ethanol oxidation is accomplished by the concerted action of catalase and NADPH oxidase, and that cytochrome P-450 is not involved in the microsomal oxidation of ethanol.

RESULTS AND DISCUSSION

Studies with Hepatic Microsomes of C_sb Mice. It has been reported that liver slices of genetic control C_sa and mutant C_sb mice have similar catalase activities and equal rates of ethanol disappearance (Vatsis and Schulman, 1973b). Liver homogenates and microsomes from C_sb mice lost 80-90% of their catalase activity after 5 min of incubation at $37^\circ C$, whereas similar preparations from genetic control C_sa mice retained full catalase activity even after a one hour incubation.

Unlike the equal ethanol-metabolizing and catalase activities obtained with liver slices from either strain, microsomal preparations from C_sb mice lost their ability to metabolize ethanol when incubated either with NADPH or an H_2O_2-generating system (hypoxanthine and xanthine oxidase); this was accompanied by a complete loss of catalase activity. For C_sa mice, neither catalase activity nor ethanol oxidation was destroyed by incubation of liver homogenates or microsomes. Despite the absence of ethanol-oxidizing activity in C_sb microsomes, the metabolism of aniline, aminopyrine, and paranitroanisole proceeded with rates similar to those observed in microsomes from C_sa mice. Furthermore, the addition of beef liver catalase restored ethanol oxidation by microsomes from mutant C_sb mice to levels obtained with similar preparations from C_sa mice.

Pretreatment of C_sa and C_sb mice with phenobarbital (80 mg/kg once daily for three days) resulted in a two- to three-fold stimulation of ethanol oxidation by liver slices from either strain (Vatsis and Schulman, 1974). Similar increases were also obtained for microsomal NADPH oxidase, NADPH-cytochrome c reductase and drug-metabolizing activities of both strains, as well as for the NADPH-linked oxidation of ethanol by C_sa microsomes. In contrast, catalase and ethanol-oxidizing activities were again absent in C_sb microsomes. As was the case with the untreated animals, catalase addition restored ethanol oxidation by microsomes from phenobarbital-treated C_sb mice to values obtained with microsomes from phenobarbital-treated C_sa mice.

Thus, these studies clearly show that the loss of catalase activity was associated with a concomitant loss of NADPH- and H_2O_2-linked ethanol-oxidizing activities in hepatic microsomes from untreated or phenobarbital-pretreated mutant C_sb mice, but that the drug-metabolizing capacity of these same microsomal preparations bore no relationship to ethanol oxidation. It is concluded that the oxidation of ethanol by microsomes in vitro is due entirely to catalase and H_2O_2 generated from NADPH via NADPH oxidase, and that the drug-metabolizing pathway, which utilizes cytochrome P-450 as its terminal oxidase, does not represent a common detoxifying route for ethanol and drugs.

In the studies described above, the ethanol-oxidizing capacity of microsomal preparations was assayed by the disppearance of ethanol from the incubation medium using gas chromatographic techniques. When acetaldehyde formation was measured, rather than ethanol disappearance, it was observed that C_sb microsomes, incubated either with NADPH or an H_2O_2-generating system and ethanol, formed acetaldehyde with rates that were 75% of those obtained with C_sa microsomes. Nevertheless, rates of acetaldehyde formation (1.0 - 1.3) μmoles/hr/incubation mixture) were only 8 - 12% of the rates obtained

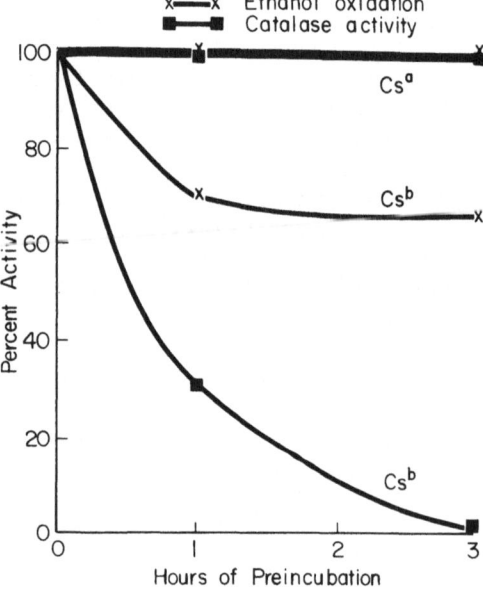

Fig. 1

Figure 1. Effect of Preincubation on Microsomal Ethanol-Oxidizing and Catalase Activities of $C_s a$ and $C_s b$ Mice. Livers from 6-10 adult male and female $C_s a$ or $C_s b$ mice were pooled and microsomes prepared from 33% homogenates in Krebs-Ringer-phosphate buffer, pH 7.4 (KRP). Once-washed microsomes were finally suspended 1:3 (w/v) in the same buffer, and portions were preincubated in air at 37°C for the time periods shown above. Aliquots were subsequently assayed for catalase and ethanol-oxidizing activities, and values were compared to those obtained with similar microsomal suspensions that had been kept in ice (zero-time or non-preincubated controls).

Catalase was assayed at 0°C by the method of Baudhuin et al., 1964. For $C_s a$ microsomes, the amount of protein used per 5.2 ml reaction mixture was the same (1.2 µg) regardless of the experimental condition. For $C_s b$ microsomes, the amount of protein in the assay mixture was varied with each condition as follows: non-preincubated microsomes, 1.2 µg (0.05 mg liver); one-hour preincubated microsomes, 3.5 µg (0.17 mg liver); and three-hour preincubated microsomes, 717 µg (33.3 mg liver).

Ethanol oxidation was assessed by acetaldehyde formation as described by Lieber and DeCarli (1970). Microsomes equivalent to 330 mg of liver (6.0 - 8.0 mg of protein) were incubated with 1.4 mM NADPH and 88 mM ethanol in a total volume of 1.5 ml of KRP. Incubations were carried out in duplicate in the main chambers of stoppered 50 ml Erlenmeyer flasks with center wells containing 1.0 ml of 0.1 M semicarbazide in 0.16 M potassium phosphate buffer (pH 7.0). Reactions were initiated by the addition of substrate, and were terminated after one hour with 0.3 ml of 50% TCA. A negligible amount of acetaldehyde (2-3%) was produced when ethanol, NADPH or microsomes were omitted from the medium. Values for blank incubations, in which TCA was added prior to starting the reaction with substrate, were subtracted from the corresponding experimental incubations. Non-preincubated control levels (100%) were (for catalase and ethanol-oxidizing activities, respectively): $C_s a$ microsomes, 19,500 µmoles H_2O_2 decomposed hr^{-1} (g liver)$^{-1}$ and 3.8 µmoles acetaldehyde formed hr^{-1} (g liver)$^{-1}$; $C_s b$ microsomes, 17,800 µmoles H_2O_2 decomposed hr^{-1} (g liver)$^{-1}$ and 3.0 µmoles acetaldehyde formed hr^{-1} (g liver)$^{-1}$.

with C_sa microsomes for ethanol disappearance (11 - 13 µmoles/hr/
incubation mixture) under similar conditions. Since these small
amounts of acetaldehyde were within the error of the gas chroma-
tograph used, they could not have been detected as ethanol disappearance by the latter method. This discrepancy between substrate disappearance and product formation has also been observed
with hepatic microsomal preparations from rats and Swiss albino
mice, and is currently under investigation. Part of this discrepancy may be explained by further metabolism of acetaldehyde to
acetate and by non-specific binding of the aldehyde to tissue constituents. As an example of the latter, it was shown recently that
acetaldehyde binds to cytochrome P-450 (Pennington et al., 1970).
By either binding or further metabolism, not all of the acetaldehyde produced from ethanol would be available for trapping by the
semicarbazide solution, the latter being the conventional method
of assessing microsomal ethanol oxidation.

Acetaldehyde formation from ethanol in C_sb microsomes might
be explained by a small amount of catalase activity previously un-
detected. Our initial studies had shown that catalase activity in
C_sb microsomes was inactivated following a 15 min incubation.
Owing to the tremendous activity of hepatic catalase, enzymic assays
were carried out in those studies with extremely small amounts of
microsomal protein (0.05 mg wet weight of liver or 0.004 mg of protein per 5.2 ml reaction volume) in order to maintain first-order
kinetics. By increasing the amount of preincubated microsomes 600-
fold, however, catalase could now be detected, but the linearity
of the reaction with respect to microsomal protein concentration
was lost. With increased protein (33 mg of liver per 5.2 ml reaction mixture) rates of H_2O_2 decomposition in 3-hr preincubated
C_sb microsomes were approximately 1% of those seen in non-preincubated preparations measured in 0.05 mg liver (Figure 1). Even
though almost all of the catalase activity was destroyed in 3-hr
preincubated C_sb microsomes, the remaining activity accounted for
the decomposition of 190 µmoles $H_2O_2hr^{-1}$ (g liver)$^{-1}$. Since this
activity is greater than that needed for the peroxidatic decomposition of ethanol by catalase in the microsomal preparation, it was
not surprising to find that C_sb microsomes, preincubated for 3 hr
and subsequently incubated with NADPH or a H_2O_2-generating system
and ethanol, exhibited rates of acetaldehyde formation that were
60% of those seen in non-preincubated controls. Thus, sufficient
catalase activity was present to account for acetaldehyde formation
in both non-preincubated and preincubated C_sb microsomes.

Studies with Hepatic Microsomes of Swiss Mice of Different Ages.
Lower rates of ethanol disappearance were observed in liver slices,
homogenates and microsomes of older mice (Table I). When microsomes from older mice were incubated with an H_2O_2-generating system
rather than with added NADPH or cytosol, rates of ethanol oxidation

were similar to those seen with young mice. It appeared that limited rates of generation of H_2O_2 in the presence of added NADPH or cytosol may have been related to, and possibly responsible for, lower ethanol oxidation in older mice. This was substantiated when it was found that the NADPH oxidase activity in microsomes from old mice was only 42% of that obtained with microsomes from young mice (Table II). In contrast, catalase and several drug-metabolizing activities were the same in the two preparations. Thus, ethanol metabolism again did not parallel drug metabolism in hepatic microsomes. Kato et al., (1970) have similarly demonstrated that the content of cytochrome P-450 as well as the activities of NADPH-cytochrome c reductase, NADPH-neotetrazolium reductase, p-nitrobenzoic acid reductase, aniline hydroxylase, aminopyrine demethylase, and hexobarbital hydroxylase were similar in young (five weeks), adult (15 weeks) and old (50 weeks) male mice (ICR strain). This is in contrast to the observed lower activities of drug-metabolizing enzymes and NADPH-linked electron transport components in hepatic microsomes of old as compared to young female rats (Kato and Takanaka, 1968); the latter findings have been confirmed in our laboratory with male rats of the Sprague-Dawley strain.

Since the reduction of cytochrome c and the oxidation of NADPH may be attributed to the same enzymic entity (Masters et al., 1965; Thurman et al., 1974), it may appear paradoxical that these two microsomal activities did not change in a parallel fashion as a function of age (compare lines two and four in Table II). That different activities of the microsomal NADPH-linked flavoprotein reductase do not necessarily have to respond in parallel as a function of a certain physiological parameter was indicated by the work of Davies et al., (1969). These investigators demonstrated that the rate of reduction of cytochrome c remained constant, whereas endogenous rates of cytochrome P-450 reduction by NADPH varied widely in different species; moreover, the substrate-stimulated (ethylmorphine) rates of cytochrome P-450 reduction were also shown to vary in rats as a function of sex, whereas, again, rates of cytochrome c reduction were unaffected by this parameter (Gignon et al., 1969). In contrast, adrenalectomy of male rats gave rise to similar decreases in rates of reduction of both cytochromes c and P-450 (Castro et al., 1970).

As shown by the lowered activity of NADPH oxidase in the older mice (Table II), the rate-limiting step in the microsomal oxidation of ethanol would appear to be the rate of H_2O_2 formation by NADPH as previously indicated by others (Thurman et al., 1972; Lin et al., 1972).

TABLE I

Ethanol Metabolism by Cellular and Subcellular Preparations from
Livers of Male Swiss Mice of Different Ages

Preparation or Subcellular Fraction	1.4 months old	12-14 months old
	μmoles ethanol oxidized hr^{-1} (g liver)$^{-1}$	
Slices	235	142
Homogenates	57	25
Post-mitochondrial Supernatant	59	24
Microsomes + Cytosol,	56	27
Microsomes + NADPH	58	27
Microsomes + H$_2$O$_2$-gen.	59	64

Incubations were carried out at 37°C with 88mM ethanol for the time
periods indicated below; each preparation was made up to a final volume of 2.5 ml with oxygenated Krebs-Ringer-phosphate buffer, pH 7.4
(KRP).   200 mg of liver slices were incubated for 30 min, while
homogenates and post-mitochondrial supernatants from 550 mg of liver
were incubated for 1 hr.   Once-washed microsomes were resuspended
1:3 (w/v) in cytosol (100,000 x g supernatant) or in KRP: microsomes in cytosol equivalent to 550 mg of liver were incubated for
1 hr.   Microsomes in buffer from 550 mg of liver were incubated for
1 hr with 1.4 mM NADPH or an H$_2$O$_2$-generating system consisting of
1.5 mM hypoxanthine and 10 mU/ml xanthine oxidase.   All incubations
were in triplicate in stoppered 25 ml Erlenmeyer flasks.   After
chilling, the contents were spun at 39,000 x g for 20 min, and ethanol determined in the supernatant fluid by gas chromatography as
previously described (Vatsis and Schulman, 1973, 1974).   Values are
the means from three separate experiments; for each experiment,
livers from 5-10 mice were pooled.   Similar results were obtained
with comparable hepatic preparations from rats and C$_s$a mice.

TABLE II

Cytochrome P-450 Content and Enzymic Activities in Hepatic Microsomes from Male Swiss Mice of Different Ages

| | Microsomal Enzymic Activities | | |
Paramater measured	1.4 months	12-14 months	% "young" Activity
Ethanol oxidation	38.1	13.8	36
NADPH oxidase	11.6	4.9	42
Catalase	47.6	45.0	95
NADPH-cyt. c reductase	284	279	98
Cytochrome P-450	1.2	1.3	108
Aniline p-hydroxylation	1.8	1.7	94
Aminopyrine N-demethylation	14.0	14.1	100
Paranitroanisole O-demethyla- tion	2.6	2.6	100

Microsomal NADPH-dependent ethanol oxidation was assayed as
described in the legend to Table I. Drug-metabolizing activities
were determined at 37°C in a medium containing microsomes from 150
mg of liver (2.5 - 3.0 mg protein), 1.2 mM NADP, 6.0 mM $MgCl_2$, 9.0
mM isocitrate, 180-200 mU of isocitrate dehydrogenase, and substrate
(16 mM aniline or aminopyrine, and 0.6 mM paranitroanisole) in a
total volume of 1.0 ml of 0.15 M Tris-HCl buffer (pH 7.5). The
incubation period was 10 min for aniline and paranitroanisole, and
3 min for aminopyrine. After termination of the reactions with
TCA, aniline hydroxylation and paranitroanisole O-demethylation
were determined by a modification of the method of Deckert and Remmer
(1972), while aminopyrine N-demethylation was measured as released
formaldehyde (Nash, 1953). Cytochrome P-450 content was estimated
by the method of Omura and Sato (1964), and reported as nmoles/mg
protein. Catalase, NADPH oxidase, and NADPH-cyt. c reductase activities were determined as previously described by Vatsis and Schulman
(1974). All enzymic activities, except for catalase, are expressed
in terms of nmoles min^{-1} (mg protein)$^{-1}$. Catalase activity is reported as μmoles H_2O_2 decomposed min^{-1} (mg protein)$^{-1}$. Values shown
are the means of two separate experiments. In each experiment,
livers were pooled from 5-10 mice, and all assays were carried out
in triplicate.

K. P. VATSIS AND M. P. SCHULMAN

TABLE III

Cytochrome P-450 Content and Enzymic Activities in Mouse Liver Microsomes Subsequent to NADPH-Induced Lipid Peroxidation

Additions to Preincubation mixture	Cytochrome P-450	Microsomal Enzyme Activities					
		Malondialdehyde formation	Aniline Hydroxylation	Aminopyrine Demethylation	Ethanol Oxidation	NADPH Oxidase	NADPH-Cyt.c Reductase
	nmoles/mg	nmoles/min/mg microsomal protein					
None	1.23	0.6	0.88	10.2	12.0	7.8	219
0.5mM NADPH	0.66	30.7	0.40	2.1	10.2	6.8	206
0.5 mM NADPH + 1.0mM EDTA	1.14	0.7	0.87	9.4	12.4	7.4	224

Livers from 15-20 young-adult male Swiss albino mice were pooled and microsomes prepared from 25% homogenates in 0.05 M Tris-0.15 M KCl (pH 7.5). Once-washed microsomes from 1.0 g of liver were preincubated in air for 30 min at 37°C in a total volume of 18.0 ml of 0.1 M K_2HPO_4-KH_2PO_4 buffer (pH 7.4) as indicated above. Similar suspensions were also kept in an ice bath (non-preincubated or zero-time controls). After preincubation, aliquots were assayed for NADPH oxidase and NADPH-cyt. c reductase activities as described in the legend to Table II. Malondialdehyde was measured according to the method of Ottolenghi (1959), and cytochrome P-450 was estimated on aliquots that were treated with 10^{-4} M menadione according to the method of Raj and Estabrook as described by Levin et al., (1973). Ethanol and drug metabolism were determined in incubation mixtures consisting of microsomes from 100 mg of liver (1.6 mg of protein), 1.4 mM NADPH, and substrate (88 mM ethanol, 8 mM aniline or aminopyrine) in a total volume of 2.0 ml of 0.1 M K_2HPO_4-KH_2PO_4 buffer (pH 7.4). Incubations were carried out for 10 min at 37°C, except for aminopyrine which was for 5 min. All incubations were performed in triplicate, with zero-time (blank) determinations in duplicate. Ethanol oxidation was assayed by acetaldehyde formation (Lieber and DeCarli, 1970). Aniline hydroxylation and aminopyrine demethylation were estimated as described in the legend to Table II. Values shown are the means of two separate experiments. Incubated control activities were 80 to 90% of those obtained for the zero-time (non-preincubated) controls.

Ethanol and Drug Metabolism in Microsomes Subsequent to Lipid Peroxidation-Induced Destruction of Cytochrome P-450. In the presence of NADPH and O_2, unsaturated fatty acids of hepatic microsomal phospholipids undergo peroxidation, a process which destroys microsomal membranes and results in malondialdehyde formation (Hochstein and Ernster, 1963; May and McCay, 1968). In addition, there is a breakdown of cytochrome P-450 heme to carbon monoxide (Schacter et al., 1972) and a parallel loss of cytochrome P-450 spectrally (Levin et al., 1973). EDTA has been reported to inhibit these degradative processes by blocking lipid peroxidation.

Preincubation of hepatic microsomes from Swiss mice with NADPH (Table III) resulted in malondialdehyde formation, destruction of cytochrome P-450, and decreased rates of aniline hydroxylation and N-demethylation of aminopyrine. No reduction was observed in rates of NADPH-linked oxidation of ethanol or in the activities of NADPH oxidase and NADPH-cytochrome c reductase. Addition of EDTA to preincubation mixtures prevented lipid peroxidation, loss of cytochrome P-450, and inactivation of the drug-metabolizing capacity of microsomes, but did not alter ethanol oxidation rates nor the activities of NADPH oxidase and NADPH-cytochrome c reductase. These findings do not support the involvement of cytochrome P-450 in the microsomal ethanol-oxidizing system, but again suggest that drug oxidations and ethanol metabolism in hepatic microsomes are separate metabolic events.

REFERENCES

Barakat, H. A., Tapscott, E., and Pennington, S. 1974. Ethanol oxidation by components of rat liver microsomes. Biochem. Biophys. Res. Commun. 60: 482-488.

Baudhuin, P., Beaufay, H., Rahman-Li, Y., Sellinger, O. F., Wattiaux, R., Jacques, P., and DeDuve, C. 1964. Tissue fractionation studies. Intracellular distribution of monoamine oxidase, aspartate aminotransferase, alanine aminotransferase, D-amino acid oxidase and catalase in rat liver tissue. Biochem. J. 92: 179-184.

Carter, E. A., and Isselbacher, K. J. 1971. The role of microsomes in the hepatic metabolism of ethanol. Ann. N. Y. Acad. Sci. 179: 282-294.

Castro, J. A., Greene, F. E., Gigon, P., Sasame, H. and Gillette, J. R. 1970. Effect of adrenalectomy and cortisone administration on components of the liver microsomal mixed function oxygenase system of male rats which catalyzes ethylmorphine metabolism. Biochem. Pharmacol. 19: 2461-2467.

Davies, D. S., Gigon, P. L., and Gillette, J. R. 1969. Species and sex differences in electron transport systems in liver microsomes and their relationship to ethylmorphine demethylation. Life Sci. 8: 85-91.

Deckert, F. W. and Remmer, H. K. 1972. In vitro inhibition of rat
    and human liver microsomal enzymes by 4-hydroxycoumarin anti-
    coagulants and related compounds. Chem-Biol. Interactions 5:
    255-263.

Gigon, P. L., Gram, T. E. and Gillette, J. R. 1969. Studies on
    the rate of reduction of hepatic microsomal cytochrome P-450
    by reduced nicotinamide adenine dinucleotide phosphate;
    Effect of drug substrates. Mol. Pharmacol. 5: 109-122.

Grunnet, N., Quistorff, B. and Thieden, H. I. D. 1973. Rate-
    limiting factors in ethanol oxidation by isolated rat-liver
    parenchymal cells. Effect of ethanol concentration, fructose,
    pyruvate and pyrazole. Eur. J. Biochem. 40: 275-282.

Hochstein, P. and Ernster, L. 1963. ADP-activated lipid peroxi-
    dation coupled to the TPNH oxidase system of microsomes.
    Biochem. Biophys. Res. Commun. 12: 388-394.

Israel, Y., Khanna, J. and Lin, R. 1970. Effect of 2,4-dinitro-
    phenol on the rate of ethanol elimination in the rat in vivo.
    Biochem. J. 120: 447-448.

Kato, R. and Takanaka, A. 1968. Metabolism of drugs in old rats.
    I. Activities of NADPH-linked electron transport and drug-
    metabolizing enzyme systems in liver microsomes of old rats.
    Jap. J. Pharmacol. 18: 381-388.

Kato, R., Takanaka, A. and Onoda, K-I. 1970. Studies on age
    difference in mice for the activity of drug-metabolizing en-
    zymes of liver microsomes. Jap. J. Pharmacol. 20: 572-576.

Khanna, J. M., Kalant, H. and Lin, G. 1970. Metabolism of ethanol
    by rat liver microsomal enzymes. Biochem. Pharmacol. 19:
    2493-2499.

Khanna, J. M., Kalant, H. and Lin, G. 1972. Significance in vivo
    of the increase in microsomal ethanol-oxidizing system after
    chronic administration of ethanol, phenobarbital and chlor-
    cyclizine. Biochem. Pharmacol. 21: 2215-2226.

Levin, W., Lu, A. Y. H., Jacobson, M., Kuntzman, R., Poyer, J. L.
    and McCay, P. B. 1973. Lipid peroxidation and the degradation
    of cytochrome P-450 heme. Arch. Biochem. Biophys. 158: 842-
    852.

Lieber, C. S. and DeCarli, L. M. 1970. Hepatic microsomal ethanol-
    oxidizing system. In vitro characteristics and adaptive
    properties in vivo. J. Biol. Chem. 245: 2505-2512.

Lieber, C. S. and DeCarli, L. M. 1972. The role of the hepatic
    microsomal ethanol oxidizing system (MEOS) for ethanol meta-
    bolism in vivo. J. Pharmacol. Exper. Therap. 181: 279-287.

Lieber, C. S. and DeCarli, L. M. 1973. The significance and char-
    acterization of hepatic microsomal ethanol oxidation in the
    liver. Drug.Metabol. Disp. 1: 428-440.

Lin, G., Kalant, H. and Khanna, J. M. 1972. Catalase involvement
    in microsomal ethanol-oxidizing system. Biochem. Pharmacol.
    21: 3305-3308.

Masters, B. S. S., Kamin, H., Gibson, Q. H. and Williams, C. H.

1965. Studies on the mechanism of microsomal triphosphopyridine nucleotide-cytochrome c reductase. J. Biol. Chem. 240: 921-931.

May, H. E., and McCay, P. B. 1968. Reduced triphosphopyridine nucleotide oxidase-catalyzed alterations of membrane phospholipids. II. Enzymic properties and stoichiometry. J. Biol. Chem. 243: 2296-2305.

Mezey, E., Potter, J. J. and Reed, W. D. 1973. Ethanol oxidation by a component of liver microsomes rich in cytochrome P-450. J. Biol. Chem. 248: 1183-1187.

Nash, T. 1953. The colorimetric estimation of formaldehyde by means of the Hantzsch reaction. Biochem. J. 55: 416-421.

Nelson, E. B., Kohl, K. B. and Masters, B. S. S. 1973. The role of the NADPH-cytochrome c reductase in the microsomal oxidation of ethanol and methanol. Drug Metabol. Disp. 1: 455-458.

Omura, T. and Sato, R. 1964. The carbon monoxide-binding pigment of liver microsomes. I. Evidence for its hemoprotein nature. J. Biol. Chem. 239: 2370-2378.

Ottolenghi, A. 1959. Interaction of ascorbic acid and mitochondrial lipides. Arch. Biochem. Biophys. 79: 355-363.

Papenberg, J., Wartburg, J. P. and Aebi, H. 1970. Metabolism of ethanol and fructose in the perfused rat liver. Enzym. Biol. Clin. 11: 237-250.

Pennington, S. N., Chattopadhyay, S. K. and Brown, H. D. 1970. A possible pathway for ethanol-induced fatty liver and modification of liver injury by antioxidants. Quart. J. Stud. Alc. 31:13-19.

Roach, M. K., Reese, W. N. and Creaven, P. J. 1969. Ethanol oxidation in the microsomal fraction of rat liver. Biochem. Biophys. Res. Commun. 36: 596-602.

Rubin, E., Gang, H., Misra, P. and Lieber, C. S. 1970. Inhibition of drug metabolism by acute ethanol intoxication. A hepatic microsomal mechanism. Am. J. Med. 49: 801-806.

Schacter, B. A., Marver, H. S. and Myer, U. A. 1972. Hemoprotein Catabolism during Stimulation of Microsomal Lipid Peroxidation. Biochim. Biophys. Acta. 279: 221-227.

Teschke, R., Hasumura, Y., Joly, J-G., Ishii, H. and Lieber, C. S. 1972. Microsomal ethanol-oxidizing system (MEOS): Purification and properties of a rat liver system free of catalase and alcohol dehydrogenase. Biochem. Biophys. Res. Commun. 49: 1187-1193.

Teschke, R., Hasumura, Y. and Lieber, C. S. 1974. Hepatic microsomal ethanol-oxidizing system: Solubilization, isolation, and characterization. Arch. Biochem. Biophys. 163: 404-415.

Thieden, H. I. D. 1971. The effect of ethanol concentration on ethanol oxidation rate in rat liver slices. Acta Chem. Scand. 25: 3421-3427.

Thurman, R. B., and McKenna, W. 1974. Activation of ethanol

utilization in perfused liver from normal and ethanol-pre-treated rats. Hoppe-Seyler's Z. Physiol. Chem. 355: 336-340.

Thurman, R. G. and Scholz, R. 1973. The role of hydrogen peroxide and catalase in hepatic microsomal ethanol oxidation. Drug Metabol. Disp. 1: 441-448.

Thurman, R. G., Ley, H. G. and Scholz, R. 1972. Hepatic microsomal ethanol oxidation. Hydrogen peroxide formation and the role of catalase. Eur. J. Biochem. 25: 420-430.

Thurman, R. G., Hesse, S. and Scholz, R. 1974. The role of NADPH-dependent hydrogen peroxide formation and catalase in hepatic microsomal ethanol oxidation, pp. 257-270. In R. G. Thurman, T. Yonetani, J. R. Williamson, and B. Chance (ed.). Alcohol and Aldehyde Metabolizing Systems. Academic Press, New York.

Vatsis, K. P. and Schulman, M. P. 1973a. Peroxidatic ethanol oxidation in homogenates of mouse liver. The Pharmacologist 15: 160.

Vatsis, K. P. and Schulman, M. P. 1973b. Absence of ethanol metabolism in 'acatalatic' hepatic microsomes that oxidize drugs. Biochem. Biophys. Res. Commun. 52: 588-594.

Vatsis, K. P. and Schulman, M. P. 1974. 'Acatalatic' hepatic microsomes metabolize drugs but not ethanol, pp. 287-298. In R. G. Thurman, T. Yonetani, J. R. Williamson, and B. Chance (ed.). Alcohol and Aldehyde Metabolizing Systems. Academic Press, New York.

Vatsis, K. P., Miller, C. and Schulman, M. P. 1973. Ethanol metabolism in vivo and by hepatic microsomes of mice of different ages. Fed. Proc. 32: 697.

GENERAL DISCUSSION

Session III

van der HOEVEN: I would like to address a question to Dr. Remmer. Your data seem to indicate that there are two forms of transferases on the basis of induction by chloramphenicol and of sex difference. The data show relative specific activity. Is there also any difference in total activity between two or three substrates?

REMMER: I did not contend that there are only two different types of transferases. What I only wanted to say is that there are two different types which are inducible. I don't know how many transferases are present, but we should not operate with the concept of different types of enzymes or of a single type of enzyme. I think this doesn't lead us further in our considerations. I don't know. It looks like there are several types of binding proteins which are involved in the hydroxylation, as well as in the conjugation reaction ; but it could also be that the environment (which is of course so important for us today in our society) is also as important for these microsystems. I have no real explanation but I wanted to say that there are a lot of similarities between the hydroxylating system and the conjugating system, pointing to the fact of a microenzyme complex.

MANNERING: In a recent study performed with Dr. Nelson we observed a decline in hexobarbital metabolism by perfused rat livers with time. After six hours of perfusion the rate of hexobarbital metabolism was only about half the initial rate. We studied a number of characteristics of the microsomes from these perfused livers--hydroxylase activities, reductase activities, substrate binding--and found that prolonged perfusion changed only one of the factors studied, namely, NADH synergism of NADPH-dependent hydroxylase, which was greatly reduced. This suggests that NADH is being used by the perfused liver when drugs are metabolized. My other comment is that the turnover of the P-450 monooxygenase system is very low when drugs are metabolized as compared to that of the metabolism of endogenous substrates by other systems. How do you explain the rather large effect of this slowly moving system on the more rapidly moving system?

THURMAN: With respect to your first comment, in fasted animals treated with antimycin A, there was no stimulation of oxygen uptake by aminopyrine despite the fact the NADH-dependent synergism requires some NADPH in microsomes. Under these conditions, addition of ethanol, which elevates the intracellular NADH, had no effect on oxygen uptake, confirming the requirement of the NADH-dependent synergism on the presence of some NADPH in the

whole organ (Thurman and Scholz, Eur. J. Biochem. 10, 459, 1969).
In response to your second question, the rates of glucose synthesis
in these livers are between 30 to 60 μmoles/[(g wet wt.)·hr]
The rates of drug metabolism are between 10 and 30 μmole/(g wet·hr)
Therefore, I don't think that the maximal rates of biosynthesis
(glucogenesis, lipogenesis) and of mixed function oxidation are all
that different.

COON: I have a question for Dr. Thurman. In your last slide
you pointed out that since catalase had no effect on the metabolism
of the usual substrates in microsomal or reconstituted systems, we
need not be concerned about hydrogen peroxide as a component of
the system. But I would like to ask you about the latest word on
the role of cytochrome P-450 in ethanol metabolism since it has
been a very controversial subject.

THURMAN: This has been a highly controversial problem. The
problem arises from the fact that microsomes indeed oxidize ethanol
and everyone agrees on this. However, there are two divergent
schools of thought concerning the mechanism of this phenomenon.
First there are those who contend that there is a unique interaction
between ethanol and cytochrome P-450 and, secondly, those who contend that hydrogen peroxide, generated by the mixed function oxidation electron transport system, reacts with catalase in microsomes to peroxidize ethanol. Lieber and his group support the
viewpoint that catalase is not involved in microsomal ethanol oxidation, while we contend the opposite. I have visited Dr. Lieber's
laboratory and examined his solubilized-reconstituted microsomal
preparation. His system oxidized ethanol. However, it also contains a protein with an Rf identical to rat liver catalase on SDS-
polyacrylamide gel, suggesting that catalase contamination might
account for activity in his preparation. This problem, then, is
not yet totally resolved. Experimental details and a summary of
five papers on this subject are now available (see Estabrook, R. W.,
Discussion summary of papers on microsomal ethanol oxidation. In:
Thurman, R. G., Yonetani, T., Williamson, J. R. and Chance, B.,
eds., Alcohol and Aldehyde Metabolizing Systems, Academic Press,
New York, 1974, p. 531).

VATSIS: I have a comment on ethanol oxidation for Dr. Coon.
With the mutant strain of mice we have at Illinois there is, I
think, absolutely no complication of the new drug metabolizing
pathway in alcohol metabolism. We have measured ethanol disappearance in liver slices and microsomes, and we have also measured the acetaldehyde formation. There seems to be some acetaldehyde formed in these mutant mice even when the microsomes are
preincubated and catalase is destroyed. We found that we have 1%
catalase left after preincubation of microsomes. As for the ratio,
even with 1% catalase left, there are about 160 μmoles peroxide
decomposed per hour under conditions when you have only 10 μmoles

of ethanol metabolized per hour. That accounted for that.

COON: Do you know whether your reaction is carbon monoxide sensitive?

VATSIS: We haven't studied that. I also have a question for Dr. Orrenius. In your slide you showed the inhibition of drug metabolism by alcohol in the isolated cells. To what do you attribute that?

ORRENIUS: I think it is a direct interaction with P-450.

SCHENKMAN: The concentrations of alcohol are quite high.

ORRENIUS: 5 mM is not too high.

KAMIN: I need my curiosity satisfied. As a clinical pharmacologist, I would like to know if there is an in vivo counterpart of your very elegant observation between NADPH and drug metabolism. Does a starving animal sleep longer on phenobarbital?

THURMAN: Kato and Gillette have reported that drug metabolism in vivo is slower following a period of starvation (J. Pharm. Exp. Ther. 150, 279, 1965).

BRESNICK: Two quick questions - one to Dr. Thurman, one to Dr. Wagner. Have you looked at glucose concentration in serum or blood of fasted animals that have been dosed with aminopyrine? This would be the counterpart to the NAD liver system?

THURMAN: That particular experiment has not been performed.

BRESNICK: In the reciprocal plots with iron ferrochelatase activity in the presence of copper, there was a suggestion, based on one point, of allosterism with low concentrations of copper. Have you looked at lower concentrations of copper to see if this is an allosteric effect of copper? You had two concentrations of copper, the lowest of which was 5 mM, if I remember -

WAGNER: 5 μM.

BRESNICK: And you drew a straight line - with my background in allosteric effects I might have drawn an allosteric line - biphasic kinetics. Have you done that set of experiments at copper concentrations lower than 5 μM?

WAGNER: We have done experiments at lower concentrations and we still see a linear plot.

HILDEBRANDT: Dr. Thurman, you don't seem to have ruled out the involvement of acetyl CoA on aminopyrine metabolism, as in humans aminopyrine is metabolized to acetylaminopyrine. How do you explain this contribution by acetyl CoA?

THURMAN: We measured the rate of formation of acetylamino-pyrine and it accounted for about 10% of the inhibition of lipid synthesis by aminopyrine (Thurman and Scholz, Eur. J. Biochem. 38, 73, 1973).

HILDEBRANDT: I see - it must be different in the rat than in the human. In the human we found that about 50% of metabolites excreted in the urine were acetylaminoantipyrine.

THURMAN: You are confusing percent inhibition with percent excreted, free or conjugated. We found three-fourths conjugated and one-fourth free. However, this still only accounted for 10% of the inhibition by aminopyrine.

ORRENIUS: One very quick question. In the cobaltous chloride inhibited system, has one looked for the presence of the apoprotein, or is the whole synthesis blocked?

WAGNER: Dr. Tephly has looked at incorporation of ^{14}C-ALA into heme and he has found ^{14}C heme formation is depressed, but as far as apoprotein reaction in presence of $CoCl_2$ is concerned, there is no decrease in microsomal protein. As for the rate of protein synthesis, I don't believe that has been looked at.

INTERACTION BETWEEN MICROSOMAL ELECTRON TRANSFER PATHWAYS

John B. Schenkman and Ingela Jansson
Department of Pharmacology
Yale University School of Medicine
New Haven, Connecticut 06510

In the past few years our interests have expanded from the mixed function oxidase in a narrow sense of the system to microsomal electron transfer systems in a broad sense. Back in 1965-66, Herbert Remmer, Ron Estabrook and I were studying the effect of substrate addition to microsomes, spectrophotometrically. We found (1) that in the presence of NADPH the hexobarbital Type I spectral trough was enlarged and shifted to longer wavelengths (Figure 1). Remmer later subtracted out the spectrum in the presence of NADPH from that in its absence, (curve c) and concluded that the shift and enlargement of the spectral change was due to a decrease in the extent of reduction of cytochrome b_5 in the sample cuvette. Cytochrome b_5, although reducible by NADPH, is reduced more rapidly and completely by NADH. For example, in Figure 2, we see that NADPH will reduce cytochrome b_5 to about 98% of the NADH reducible level. In actuality this value is variable; on occasion NADPH addition only reduced cytochrome b_5 to 75% of the NADH level. This is probably due to the presence of variable amounts of endogenous substrates of the mixed function oxidase in the microsomes, because addition of aminopyrine, a substrate of the mixed function oxidase, lowers the extent of reduction of cytochrome b_5 by NADPH to about 70% of full reduction (Figure 2). If NADH is then added, the hemoprotein becomes fully reduced (Figure 2).

Based upon observations like the above, plus earlier reports that NADH provides a synergistic effect on NADPH-supported oxidation of substrates of the microsomal mixed function oxidase (2, 3), Hildebrandt and Estabrook (4) suggested that cytochrome b_5 was a component of the mixed function oxidase, acting to transfer the second electron from NADH or from NADPH for activation of

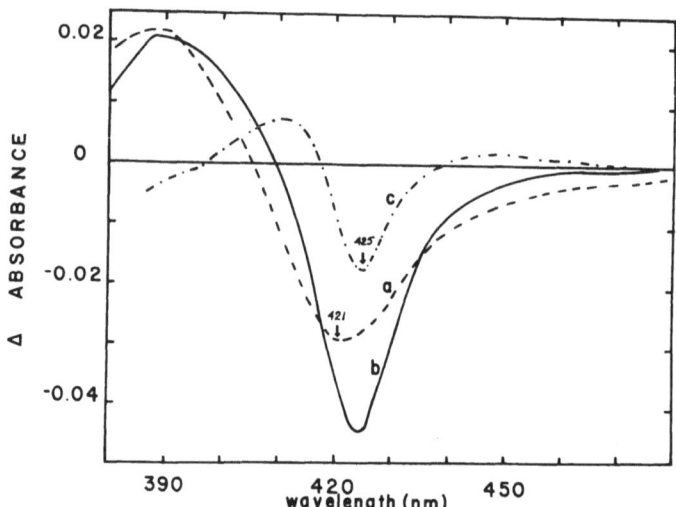

Fig. 1. The effect of NADPH on the hexobarbital induced Type I
spectral change: Rat liver microsomes (1.6 mg/ml) in 0.1M Tris,
pH 7.5 were examined by recording difference spectroscopy as reported previously (1). 0.8 mM hexobarbital was added to the sample
and the difference spectrum was recorded (curve a). NADPH (0.4 mM)
was then added to both cuvettes and spectrum b was recorded. When
curve a was subtracted from curve b, the spectral change shown as
curve c was obtained. The spectra were recorded with an Aminco
Chance recording spectrophotometer at room temperature.

molecular oxygen by cytochrome P-450. In their scheme, one electron
goes directly to substrate-bound ferri-cytochrome P-450 and a
second electron passes via cytochrome b_5 to the reduced, oxygenated
form of substrate-bound cytochrome P-450. Since NADH was without
effect on NADPH-supported reduction of cytochrome P-450, but only
stimulated NADPH-supported formation of products, it was felt (4)
that in the absence of NADH, the input of the second electron was
rate limiting in the NADPH-supported oxidation of drugs.

At this point our interest in the role of cytochrome b_5 was
aroused, since we felt that there were other possible explanations
for the effect of Type I substrates on the steady state NADPH-re-
duced level of cytochrome b_5. We favored the possibility that
cytochrome b_5 is not in the mixed function oxidase electron transfer pathway, but is in another connecting pathway, and acts as an
electron sink, serving to inhibit the mixed function oxidase by
diminishing the supply of NADPH-derived electrons; stimulation by
NADH, according to this possibility, would be by reducing

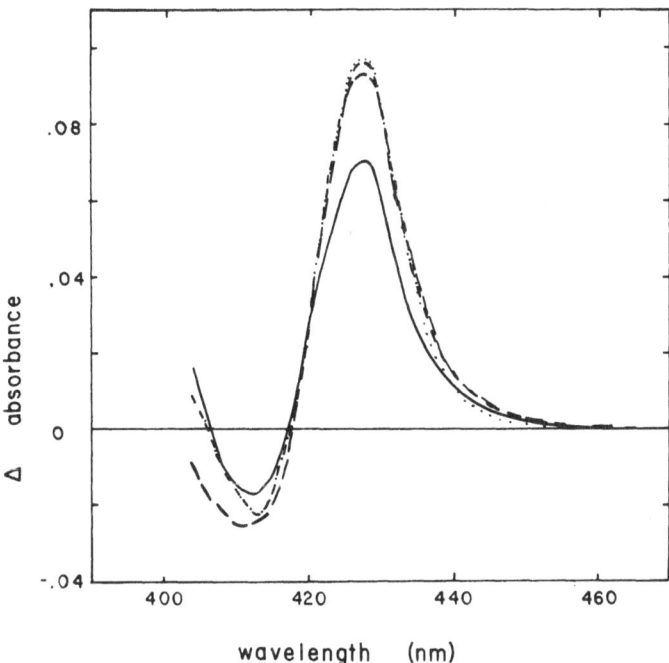

Fig. 2. Change in the steady state reduced level of cytochrome b_5 by aminopyrine. Rat liver microsomes (1.6 mg/ml) in
0.1 M Tris, pH 7.5 were examined in difference spectrum at room
temperature as described in Figure 1. Long dashes = NADPH, 1 mM;
dots = 1 mM NADH; solid line = 8 mM aminopyrine in both cuvettes,
1 mM NADPH in sample cuvette; short dashes = 1 mM NADPH + 1 mM
NADH in sample cuvette and 8 mM aminopyrine in both cuvettes.

cytochrome b_5 more rapidly, thereby eliminating its drain of
electrons from NADPH-cytochrome c (P-450) reductase (FP_2).

In considering the possible involvement of cytochrome b_5 in
drug metabolism we came to the conclusion that it was a "red
herring," distracting attention from a more important phenomenon,
the interaction between hepatic microsomal electron transfer pathways. The pathways I refer to (Figure 3) can actually all be
considered enzymes of lipid metabolism and, as such, would be expected to have some interaction. They are the mixed function oxidase, which can hydroxylate fatty acids to ω or (ω-1) hydroxy
fatty acids (5), the fatty acid desaturase, which connects the
acyl-CoA derivatives of fatty acids to their corresponding 9,10-
unsaturated derivatives (6) and also can oxidize monocyclic aromatic compounds (7), and the lipid peroxidase, which oxidizes

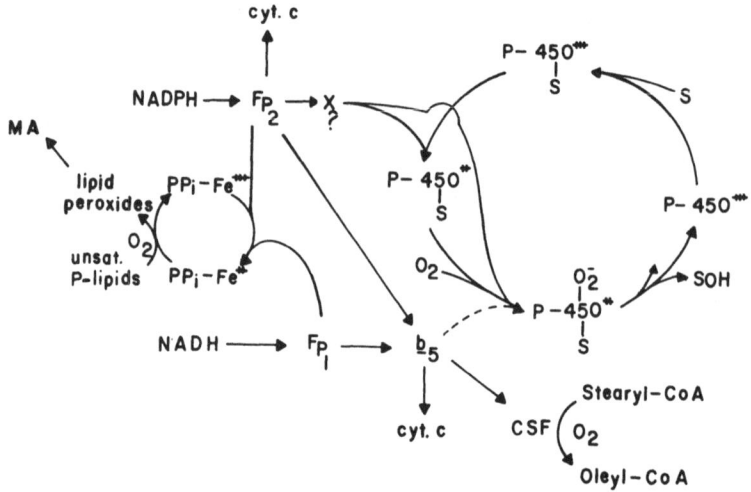

Fig. 3. Three hepatic microsomal electron transfer pathways.
MA = malondialdehyde; PPi-Fe = iron pyrophosphate; FP_1 = NADH-
cytochrome b_5 reductase; FP_2 = NADPH-cytochrome-c (P-450) reductase; b_5 = cytochrome b_5; $X^?$ = unknown postulated intermediate;
S = substrate of mixed function oxidase; CSF = cyanide sensitive
factor.

membranous unsaturated fatty acids (predominantly arachidonate)
to unspecified oxidation products and malonaldehyde (8). These
three electron transfer pathways can operate with electrons supported by NADH or NADPH.

   In an attempt to ascertain whether cytochrome b_5 is a component of the mixed function oxidase, serving to insert a rate
limiting second electron for oxygen activation, we attempted to
fortify fresh microsomes with detergent isolated cytochrome b_5.
We reasoned that if electrons from cytochrome b_5 were rate limiting, by providing the microsomes with more b_5 hemoprotein as
substrate for NADPH-cytochrome c (P-450) reductase (FP_2), we
should increase the rate of electron flow from FP_2 to cytochrome
b_5; an increase in the amount of reduced cytochrome b_5 should then
speed up electron flow to the oxygenated, substrate-bound, cytochrome P-450, and the rate of product formation. The microsomal
content of cytochrome b_5 was increased by the methods of Enamoto
and Sato (9) and Strittmatter, et al., (10), and was found to be
reducible by NADH, and to transfer electrons to exogenous cytochrome c (11). However, (Figure 4a) increasing levels of cytochrome b_5 actually inhibited NADPH-supported aminopyrine

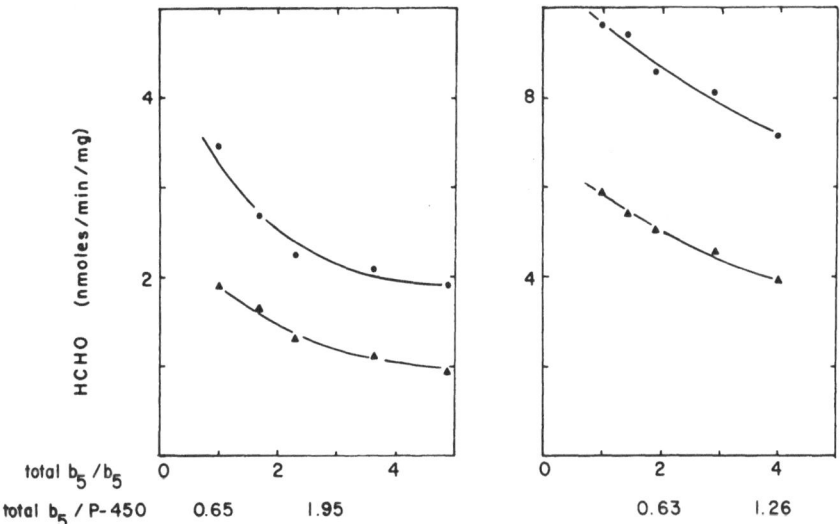

Fig. 4. Change in aminopyrine demethylase activity after fortification of rat liver microsomes with cytochrome b_5. Control rat liver microsomes (left) and phenobarbital rat liver microsomes in 50 mM Tris, pH 7.5, were used, at 1 mg protein per ml. ▲ = 0.35 mM (generated) NADPH supported reaction; ● = 0.35 mM NADPH (generated) plus 1 mM NADH supported reaction. Assay time 7 min. at 37°C, aminopyrine concentration was 8 mM.

oxidation; addition of NADH to the NADPH-supported reaction did not alleviate the inhibition. Further evidence that cytochrome b_5 is not in the mixed function oxidase electron transfer pathway is seen in Figure 4b. Phenobarbital pretreatment elevated the cytochrome P-450 content 2.5-fold, causing a similar increase in aminopyrine oxidation, but little change in cytochrome b_5 level. Thus, the ratio of cytochrome b_5 to cytochrome P-450 was lower than in control microsomes. If cytochrome b_5 is the limiting step in a second electron transfer to cytochrome P-450 we would expect a large stimulation on addition of cytochrome b_5 to these microsomes. Again (Figure 4b) there is only inhibition similar to that in control rat liver microsomes (Figure 4a).

The possibility was considered that the addition of cytochrome b_5 caused inhibition of the mixed function oxidase by altering or disrupting the microsomes. This was tested both enzymatically as well as by electron microscopy. The latter examination revealed no morphological differences even when the cytochrome b_5 content was increased 6-fold to 3 nmoles per mg microsomal protein. The possibility that contaminating detergent was present in our

cytochrome b_5, and responsible for the inhibition was obviated by
purifying the hemoprotein further to 42 nmoles per mg protein and
examining the ratio of 280 nm absorbance to 413 nm absorbance;
detergent absorption in the UV would be revealed by very high ratios
(J. Ozols, private communication). Even with highly purified
cytochrome b_5, inhibition is obtained (Figure 5) for aminopyrine,
ethylmorphine and aniline oxidation. Of interest is the observa-
ion that the pattern of inhibition of metabolism of the different
substrates is different, The added cytochrome b_5 is completely
reducible on addition of NADH, but when 4-fold higher levels of b_5
are present in the microsomes, NADPH only reduces the hemoprotein
to 70% of full reduction by 3 min in aerobic medium. NADH-cytochrome c reductase activity increased with microsomal cytochrome
b_5 (Figure 6), as the NADH-cytochrome b_5 reductase (FP_1) does not
interact with cytochrome c directly, but through mediation of cytochrome b_5 (Figure 3). NADPH-cytochrome c reductase activity
(FP_2, Figure 6) was not affected by increased content of cytochrome
b_5, since FP_2 can transfer electrons directly to cytochrome c (see
Figure 3). Although the two reductases, FP_1 and FP_2, behave normally in cytochrome b_5 fortified microsomes, proof was necessary to
show that the added hemoprotein also behaves like the native hemoprotein. Cytochrome b_5 has been shown to function as a component
of a reconstituted fatty acid desaturase system consisting of FP_1,
a cyanide sensitive factor (CSF) and cytochrome b_5 (12). This
simplified system is, of course, NADH and cytochrome b_5 dependent.
However, in liver microsomes, both NADH and NADPH can support fatty
acid desaturation (6), and proof of cytochrome b_5 function in microsomes rests more firmly on immunological evidence. Inhibition of
cytochrome b_5 reactivity with cytochrome b_5 antibody inhibited both
NADH and NADPH-supported stearyl-CoA desaturase equally well (13).
In Figure 7, it can be seen that fortification of liver microsomes
with cytochrome b_5 causes an increase in fatty acid desaturase
activity, both NADH and NADPH-supported. Thus, cytochrome b_5
bound in vitro to liver microsomes fits into a functional position
and can act in a manner similar to native cytochrome b_5. Consequently, its inhibition of mixed function oxidation must be due
to an exacerbation of an existing electron drain from this system.

The above story is but one example of an interaction between
two microsomal electron transfer pathways. Examination of the
effect of cytochrome b_5 on lipid peroxidation revealed (Figure 8)
essentially no inhibition either of NADPH-supported or NADH-
supported malonaldehyde generation. The lipid peroxidase must then
draw electrons at a point before cytochrome b_5 or, preferentially,
from the reductase.

Cytochrome b_5 is not the only component of liver microsomes
which can be isolated and used to fortify fresh microsomes. FP_1
can also be handled in this manner (14), and when microsomes are

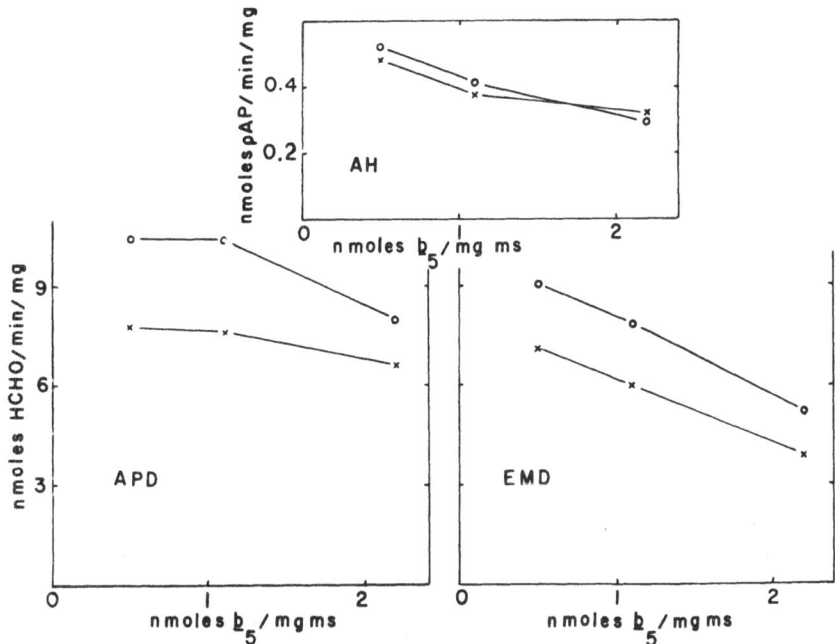

Fig. 5. The effect of fortification of rat liver microsomes with cytochrome b₅ on three mixed function oxidations. Assay medium contained 1 mg microsomal protein per ml in 50 mM Tris, pH 7.5, assay time 7 min for aminopyrine demethylation (APD) and ethylmorphine demethylation (EMD), and 15 min for aniline hydroxylation (AH) at 37°C. Aminopyrine and ethylmorphine concentrations were 8 mM and aniline was 5 mM. x = 0.35 mM NADPH (generated), 0 = 1mM NADH + 0.35 mM NADPH (generated).

fortified with this reductase (Figure 9), NADH-supported stearyl CoA desaturase activity is stimulated slightly by small increases but not further by increases up to 2.5 times endogenous FP_1. The addition of FP_1 is without effect on NADPH-supported fatty acid desaturase activity (Figure 9). These results indicate that FP_1 is not a limiting component in the fatty acid desaturase pathway of microsomes. Of interest is the finding (Figure 10) that the added FP_1 slightly elevates the NADPH-supported mixed function oxidation of ethylmorphine. This action is probably due to FP_1 binding to native microsomal cytochrome b_5, thereby decreasing the ability of the hemoprotein to drain electrons from FP_2 and the mixed function oxidase.

Oshino and Sato (15) have shown that the fatty acid desaturase

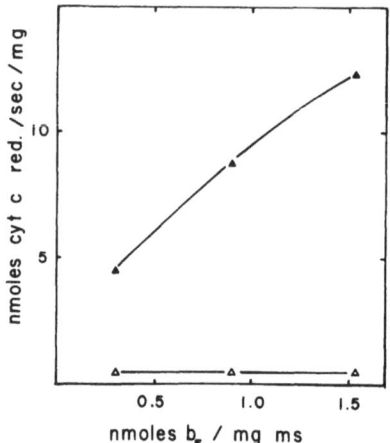

Fig. 6. The effect of fortification of microsomes with cyto-
chrome b_5 on pyridine nucleotide supported cytochrome c reduction.
The assay medium contained 60 μM horse heart cytochrome c in 0.1 M
Tris buffer, pH 7.5 and 80 mg of microsomal protein per ml.
Assay at room temperature.   Δ = 0.5 mM NADPH: ▲ = 0.5 mM NADH.

is under dietary control. Using their high carbohydrate regimen
the NADH-supported activity and NADPH-supported activity are elevated 15 and 25-fold, respectively (Figure 11). Despite the tremendous increases in the cytochrome b_5 dependent reactions, the
content of cytochrome b_5 per mg of microsomal protein was decreased
about 50% (compare Figure 11 with Figure 5). The fatty acid desaturase system evidently has a physiological connection with the
lipid peroxidase system, because induction of the former was accompanied by a complete loss of both NADPH and NADH-supported lipid
peroxidase activity (Figure 12).

Substrates of these pathways also affect the activities of
the other pathways. Several years ago, Orrenius, et al., (16)
reported that drug substrates of the mixed function oxidase cause
an inhibition of lipid peroxidation. Inhibition by aminopyrine of
lipid peroxidation can be seen in Figure 12. Some substrates of
the mixed function oxidase can also inhibit the fatty acid desaturase reaction (Figure 13). The extent of inhibition by the different substrates is variable; the Type I substrates aminopyrine
and ethylmorphine cause, if anything, slight inhibition while the
Type II compound aniline causes strong inhibition. Since the
former two compounds are oxidized by the mixed function oxidase at

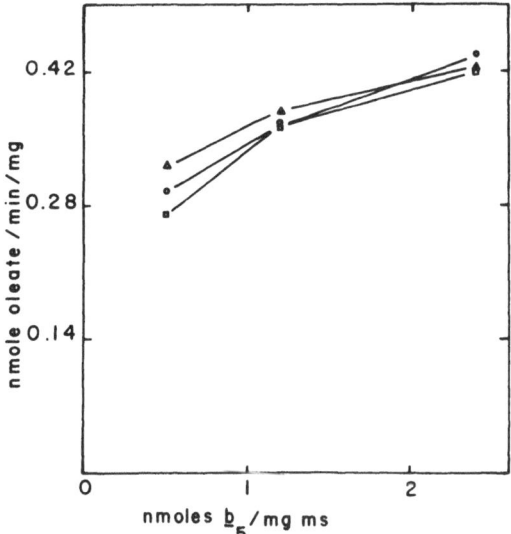

Fig. 7. Enhancement of stearyl CoA desaturase after fortification of rat liver microsomes with cytochrome b_5. Initial
rates of desaturase activity were determined from time courses
of desaturation between zero and 2.5 min by measuring ^3H_2O production from 9,10-H^3-stearyl CoA. The assay medium contained 1 mg
control rat liver microsomes fortified or unfortified with cytochrome b_5 per ml, in 0.1 M Tris, pH 7.25, and 70 μM ^3H-stearyl
CoA. Reactions were begun with addition of microsomes. 1 mM NADPH
(0), 1 mM NADH (Δ) or both pyridine nucleotides (□).

rates 10-20 times faster than aniline (19), the observed inhibition
by aniline must be caused by something other than competition
between the two systems for reducing equivalents from NADPH. Most
probably the inhibition is due to some indirect action of aniline
at the level of the desaturase system.

Stearyl CoA, a substrate of the fatty acid desaturase, causes
strong inhibition of both ethylmorphine demethylation and aniline
hydroxylation (Figure 14), as shown by Correia and Mannering (20).
However, unlike their observations, we find the addition of cyanide
does not reverse the inhibition (Figure 14). Similar effects are
obtained when NADH is present for synergism with the NADPH-supported reaction. These latter results raise questions as to the
manner of inhibition of mixed function oxidation by stearyl CoA.
Clearly, little of the inhibition is due to electron drain by
operation of the desaturase system, since the microsomes in this

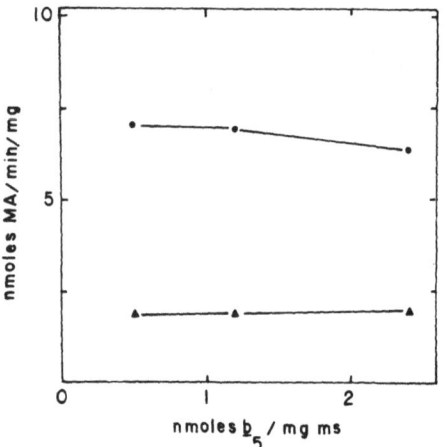

Fig. 8. The effect of fortification of rat liver microsomes with cytochrome b_5 on lipid peroxidase activity. The assay medium contained 50 µM iron pyrophosphate in 0.15 M KC1-25 mM Tris, pH 7.5, and 1 mg microsomal protein/ml; initial rates were determined in assays during 5 minutes after addition of 1 mM NADH (Δ) or NADPH (0).

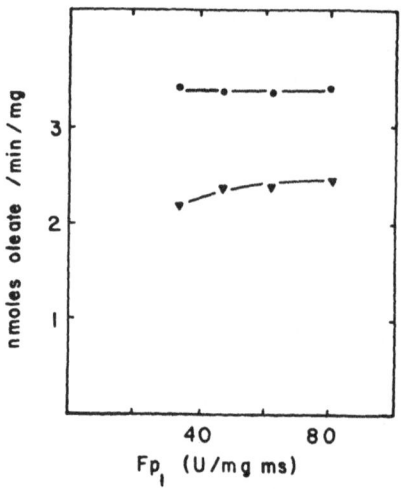

Fig. 9. The activity of stearyl CoA desaturase in FP_1 fortified liver microsomes. Assay medium contained 1 mg microsomes/ml, and 70 µM ^{3}H-stearyl CoA in 0.1 M Tris, pH 7.25, and was measured as described in Figure 7. Reactions were started with microsomes. ● = mM NADPH, ▲= 1 mM NADH.

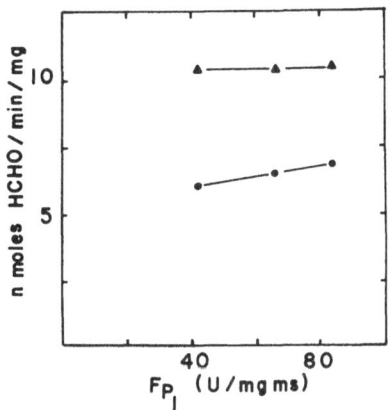

Fig. 10. The effect of fortification of microsomes with FP_1 on ethylmorphine demethylation. Assay conditions were as in Figure 5. ● = 1 mM NADPH; ▲ = 1 mM NADH.

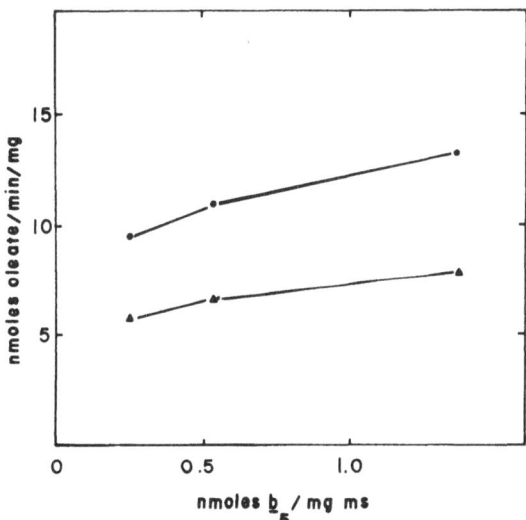

Fig. 11. Stimulation of fatty acid desaturation in b_5 fortified microsomes of desaturase-induced rats. Conditions were as described in Figure 7. ▲ = 1 mM NADH supported reaction; ● = 1 mM NADPH-supported reaction.

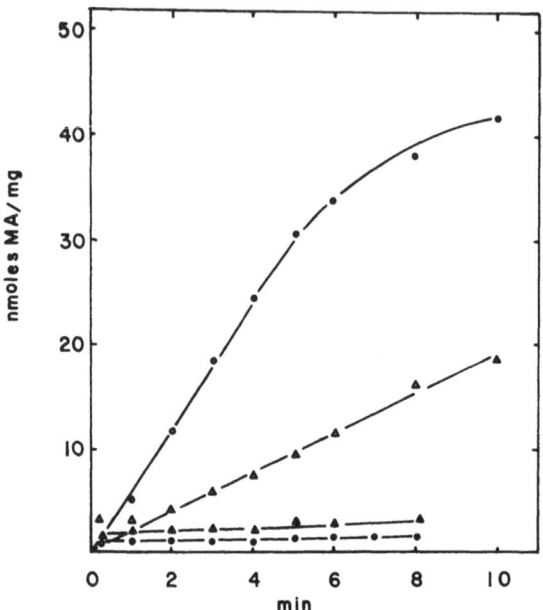

Fig. 12. Factors affecting the time course of lipid peroxidation in rat liver microsomes. The assay medium contained 1 mg
microsomes per ml in 0.15 M KCl-25 mM tris, pH 7.5, 50 µM iron pyrophosphate, and the reaction was started with 1 mM NADH (Δ) or
1 mM NADPH (0). ▲ = 8 mM aminopyrine present, 1 mM NADPH was
added to start the reaction; ● = microsomes from desaturase induced rats, 1 mM NADPH was added to start the reaction.

study were from desaturase induced animals (15), where the
latter activity was some 25-fold elevated. Similar observations
were obtained with control animals. Dr. Orrenius has suggested
to us that inhibition of drug metabolism by stearyl CoA may be
due to binding of this compound to cytochrome P-450; he has observed shorter-chain fatty acids like lauryl-CoA, I believe, to
cause a Type I spectral change. However, in experiments subsequent
to that meeting we have found this not to be the case with stearyl
CoA. Breakdown of the stearyl CoA to stearic acid also is not
responsible because the latter compound at 70 µM concentration is
not inhibitory.

As one examines the sketch in Figure 3, it is at once apparent that the three electron transfer pathways have common enzymes
of electron input, FP_1 and FP_2. The three pathways all have the
ability to utilize reducing equivalents from NADH and NADPH. However, differences exist in the relative extent to which the pyri-

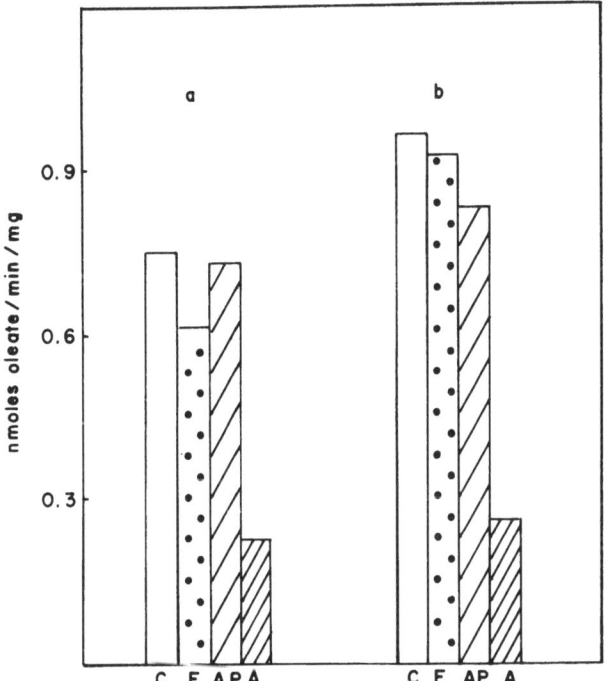

Fig. 13. Inhibition of fatty acyl CoA desaturase by substrates of the mixed function oxidase pathway. Assay conditions
were as in Figure 7. C = Control; E = ethylmorphine (8 mM) present; AP - aminopyrine (8 mM) present; A = aniline (5 mM) present;
a = 1 mM NADPH-supported; and b = 1 mM NADH-supported.

dine nucleotides can support the different pathways: the mixed
function oxidase rate with NADH alone is about 1/10 that of NADPH.
Both pyridine nucleotides together have an activity about 30%
above the rate with NADPH alone when enzyme saturating levels of
NADPH are present. As seen in Figure 7, the fatty acid desaturase
functions equally well with NADH or NADPH. However, this action
is a variable, with literature reports saying they are equally
effective (6), or that NADH is the preferred pyridine nucleotide
(21), and Figure 11, showing NADPH may be preferred at times.
The lipid peroxidase reaction also can be supported by either
NADPH or NADH. The latter pyridine nucleotide is less effective
as a source of reducing equivalents (Figure 12). Rabbit antiserum
to rat NADPH-cytochrome c (P-450) reductase (FP$_2$) strongly inhibits the NADPH-supported reduction of cytochrome c and cytochrome
P-450, as well as the oxidation of aminopyrine, ethylmorphine and

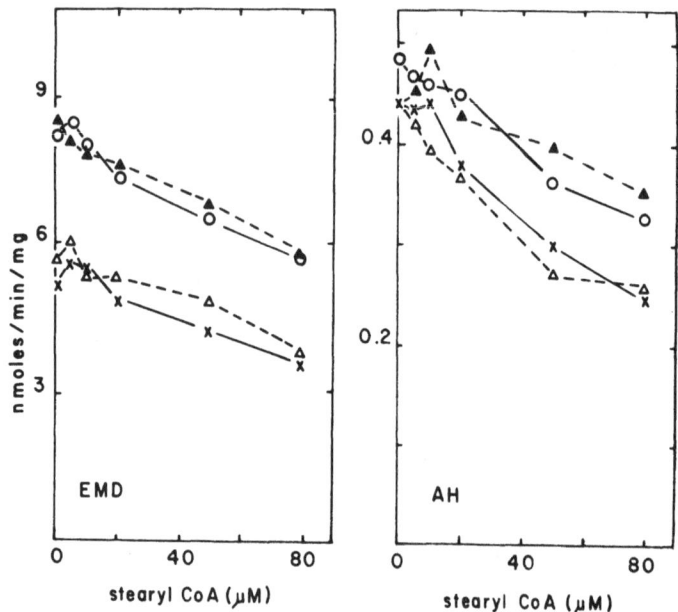

Fig. 14. The inhibition of the mixed function oxidation of ethylmorphine (EMD) and aniline (AH) by stearyl CoA, and its lack of reversal by KCN. The assay medium contained 1 mg of microsomes per ml in 50 mM Tris, pH 7.5. Conditions were as in Figure 5. X = NADPH alone (0.35 mM generated); Δ = NADPH plus 0.2 mM KCN; 0 = NADPH (0.35 mM generated) plus 1mM DACH: $\blacktriangle$ = NADPH + NADH + KCN. HCHO was measured in the EMD assay, and p-aminophenol was measured in the AH assay.

aniline (Figure 15). This antiserum also strongly inhibits the NADPH-supported lipid peroxidase (Figure 16), indicating the same reductase is involved in this system as in the mixed function oxidase system. The antiserum did not, however, affect the NADH-supported fatty acid desaturase reaction. Of interest was the observation that the antiserum to FP_2 did not effect NADPH-supported stearyl CoA desaturase activity (Figure 17). Either NADPH-derived reducing equivalents enter the desaturase system via some route other than FP_2, or electrons from this enzyme pass to cytochrome b_5 via a non-inhibited site on the enzyme to cytochrome b_5; Oshino and Omura (13) have already demonstrated with antibodies to cytochrome b_5 that this hemoprotein is involved in NADPH-supported desaturase activity.

In conclusion, we have presented evidence which shows that three microsomal transfer systems, the mixed function oxidase, the

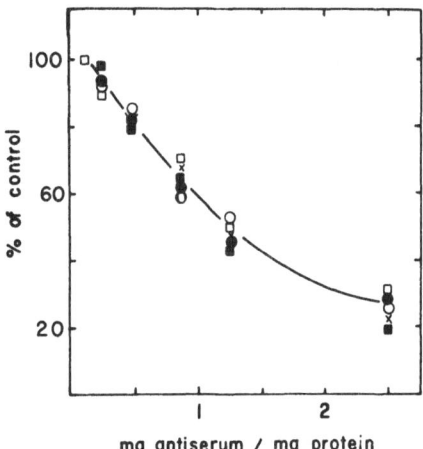

Fig. 15. Inhibition of aminopyrine demethylase (X), aniline
hydroxylation (●), ethylmorphine demethylase (0), NADPH-cytochrome
c reductase (■), and NADPH-cytochrome P-450 reductase (□) activities by preincubation of rat liver microsomes with rabbit antiserum to rat liver (trypsin isolated) cycochrome c reductase.
Control serum was used for control activity. Unpublished data of
Moldeus and Schenkman.

fatty acid desaturase and the lipid peroxidase, interact. This
interaction is at the level of electron sources as well as at components of the electron transfer system, and at substrate level.
For example, we have shown that substrates of one system can
strongly inhibit the activity of the other systems; e.g., mixed
function oxidase substrates inhibit fatty acid desaturase and
lipid peroxidase, stearyl CoA inhibits mixed function oxidase.
Changes in the level of components of the electron transfer system
alter the activities of the different systems differently. For
example, since fortification of microsomes with cytochrome b_5
enhances fatty acid desaturase activity, in which it is known to
function, as well as NADH-cytochrome c reductase activity, but has
no effect on lipid peroxidation where it is not known to function,
b_5 inhibition of the mixed function oxidase must indicate a physiological interaction between the desaturase and the mixed function
oxidase. Increasing the level of FP_1 elevates the NADPH-supported
mixed function oxidase and the NADH-supported fatty acid desaturase.
Since the three electron transfer systems appear to depend upon
the same input flavoproteins, changes in the redox state of these
and any subsequent electron carriers during operation of one pathway cannot necessarily be taken as evidence of involvement of the

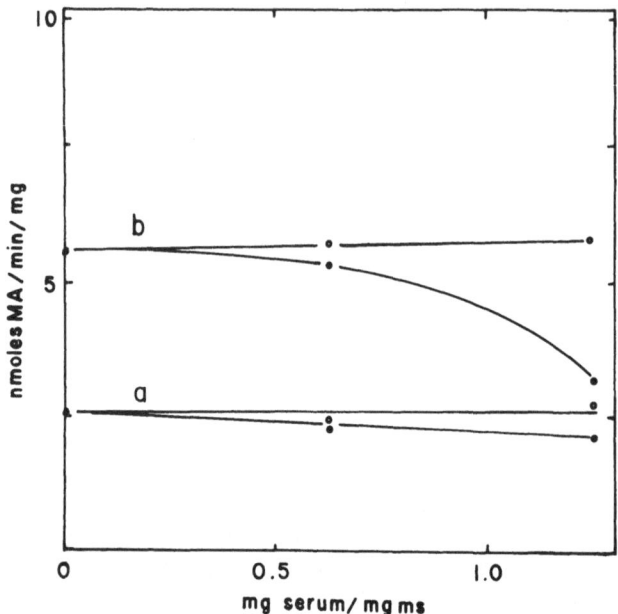

Fig. 16. Inhibition of NADPH supported but not NADH supported
lipid peroxidation by rabbit antiserum to rat liver NADPH-cytochrome c reductase (tryptic preparation). The assay was as
described in Figure 8 with microsomes preincubated with control
serum (0) or antiserum (●). a = 1 mM NADH; b = 1 mM NADPH.

carrier in that pathway. For example, cytochrome b_5 is clearly a
member of the fatty acid desaturase system (see Figure 3) and, as
determined by inhibition of both NADPH and NADH-supported desaturase reactions by b_5 antibody (13), accepts electrons from both
pyridine nucleotides (see Figure 2). However, during operation of
the mixed function oxidase, supported by NADPH, the steady state
reduced level of cytochrome b_5 is decreased (Figure 2 and
reference 4). This is probably due to the faster flow of electrons
to cytochrome P-450 in the presence of Type I substrates of the
mixed function oxidase (22), which would decrease the availability
of electrons to cytochrome b_5. The addition of NADH to the NADPH-
supported system accelerates electron flow to cytochrome b_5 and
simultaneously stimulates the mixed function oxidase activity.
Although we feel this is due to reversal of inhibition by stopping
the electron drain, the reported (20,23) enhancement of NADH
oxidation by the presence of NADPH would suggest stimulation of
some function for the former pyridine nucleotide by the latter one.

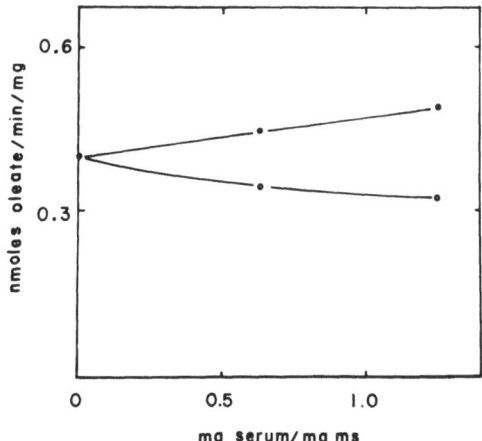

Fig. 17. Lack of inhibition of NADPH-supported fatty acid desaturase activity by rabbit antiserum to rat liver NADPH-cytochrome c reductase (tryptic preparation). The assay was as described in Figure 7 with microsomes pretreated with control serum (0) or antiserum (●).

In view of the inhibition of mixed function oxidase activity caused by increasing the cytochrome b_5 content of microsomes, it is, at present, not completely clear as to how the cytochrome b_5 containing desaturase system alters the activity of the mixed function oxidase system.

REFERENCES

1. Remmer, H., Schenkman, J. B., Estabrook, R. W., Sasame, H., Gillette, J., Narasimhulu, S., Cooper, D. Y. and Rosenthal, O. Mol. Pharmacol. 2, 187 (1966).
2. Conney, A. H., Brown, R. R., Miller, J. A. and Miller, E. C., Cancer Res. 17, 628 (1957).
3. Nilsson, A. and Johnson, B. C., Arch. Biochem. Biophys. 101, 494 (1964).
4. Hildebrandt, A. G. and Estabrook, R. W., Arch. Biochem. Biophys. 143, 66 (1971).
5. Bjorkhem, I. and Danielson, H., Eur. J. Biochem. 17, 450 (1970).
6. Holloway, P. W., Peluffo, R. O. and Wakil, S. J., Biochem. Biophys. Res. Commun. 12, 300 (1963).
7. Oshino, N. and Sato, R., J. Biochem. 69, 169 (1971).

8. May, H. E. and McCay, P. BP., J. Biol. Chem. 243, 2288 (1968).
9. Enamoto, K. and Sato, R., Biochem. Biophys. Res. Commun. 51, 1 (1973).
10. Strittmatter, P., Rogers, M. J. and Spatz, L., J. Biol. Chem. 247, 7188 (1972).
11. Jansson, I. and Schenkman, J. B., Mol. Pharmacol. 9, 840 (1973).
12. Shimakata, T., Mihara, V. and Sato, R., J. Biochem. 72, 1163 (1972).
13. Oshino, N. and Omura, T., Arch. Biochem. Biophys., 157, 395 (1973).
14. Spatz, L. and Strittmatter, P., J. Biol. Chem. 248, 793 (1973).
15. Oshino, N. and Sato, R., Arch. Biochem. Biophys. 149, 369 (1972).
16. Orrenius, S., Dallner, G. and Ernster, L., Biochem. Biophys. Res. Commun. 14, 329 (1964).
17. Schenkman, J. B., Ball, J. A. and Estabrook, R. W., Biochem. Pharmacol. 16, 1071 (1967).
18. Jacobson, M., Levin, W., Lu, A. Y. H., Conney, A. H. and Kuntzman, R., Drug Metab. Dispos. 1, 766 (1973).
19. Schenkman, J. B., Mol. Pharmacol. 8, 178 (1972).
20. Correia, M. A. and Mannering, G. J., Mol. Pharmacol. 9, 470 (1973).
21. Oshino, N., Imai, Y. and Sato, R., Biochim. Biophys. Acta. 128, 13 (1966).
22. Schenkman, J. B., Hoppe-Seyler's Z. Physiol. Chem. 349, 1624 (1968).
23. Cohen, B. S., and Estabrook, R. W., Arch. Biochem. Biophys. 143, 54 (1971).

ROLE OF CYTOCHROME b_5 IN THE NADH SYNERGISM OF NADPH-DEPENDENT

REACTIONS OF THE CYTOCHROME P-450 MONOOXYGENASE SYSTEM OF HEPATIC

MICROSOMES

G. J. Mannering
Department of Pharmacology, University of Minnesota
Medical School
Minneapolis, Minnesota 55455

INTRODUCTION

Two systems are known to transfer electrons in hepatic microsomes: 1) the NADPH-dependent cytochrome P-450 monooxygenase system, and 2) the NADH dependent cytochrome b_5 system. Their co-
existence in the same organelle suggested to Estabrook and associates
(Estabrook et al., 1971; Hildebrandt and Estabrook, 1971) that the
systems might interact during the oxidation of drug substrates in
much the same way electron transfer systems interact in mitochrondria.
As evidence for the interaction of the two systems, they showed that
the rate of NADH oxidation by liver microsomes was enhanced in the
presence of NADPH and drug substrate, and that the rate of oxidation
of NADH was related to the rate of oxidation of the substrate. They
further implicated cytochrome b_5 in cytochrome P-450 monooxygenase
reactions by showing that the steady state of reduced cytochrome b_5
in the presence of NADPH and NADH was decreased by the addition of
drug substrate. Other experiments eliminated some alternative possibilities that might explain these observations, e.g., the possibility that NADH was converted to NADPH, or that NADH was sparing
NADPH used in competing reactions occurring simultaneously in microsomes.

The role of NADPH as the donor of reducing equivalents for the
microsomal oxidation of drugs, other xenobiotics and steroids in
the liver is thoroughly documented (Gillette, 1966; Conney, 1967;
Mannering, 1968; Estabrook et al., 1970; Mannering, 1971). NADH
is a poor substitute for NADPH in these reactions; it supports reaction rates to only about to 10 or 15 percent of those observed
with NADPH. However, in the presence of both NADH and NADPH,

reaction rates may be 30 to 100% faster than when NADPH is the sole
donor of reducing equivalents (Conney et al., 1957; Cohen and
Estabrook, 1971; Gillette, 1971; Cinti et al., 1972; Correia and
Mannering, 1973a,b,c; Björkhem and Danielsson, 1973; Sitar and
Mannering, 1973; Netter and Illing, 1974). The phenomenon is known
as NADH synergism.

Figure 1 shows how the two electron transfer systems may act
in concert during the metabolism of drugs and other xenobiotics.
The first of the two electrons required for the oxidation of substrates is used to reduce oxidized cytochrome P-450-substrate complex (step 8a); it is derived almost exclusively from NADPH (steps
5→7→8a). The second electron involves the reduction of the oxygenated, reduced cytochrome P-450-substrate complex (step 8b). This
electron is accepted from cytochrome b_5, which receives a reducing
equivalent from NADPH when it is the sole donor (steps 5→6), or
from either NADPH or NADH (steps 1→2), when both nucleotides are
present. NADH is the preferred donor of the second electron; it
exerts a "synergistic" effect because it maintains a higher steady
state level of reduced cytochrome b_5 during substrate oxidation than
does NADPH. Whether the pathway of the second electron through
cytochrome b_5 is obligatory when NADPH is the only source of electrons, remains a question; considerable evidence suggests that
the second electron from NADPH travels by a route that circumvents
cytochrome b_5 (steps 5→14→8b)*.

Figure 1 also illustrates two other systems which react with
NADPH and NADH electron transfer systems, the lipid peroxidase system (step 5→13), and the fatty acyl-CoA desaturation system (steps
1→2→3→4, or steps 5→6→3→4). Steps 11, 12 and 15 do not occur or
do not contribute substantially to product formation by the monooxygenase system (Cohen and Estabrook, 1971; Hildebrandt and
Estabrook, 1971; Correia and Mannering, 1973b).

This concept was supported by studies in our laboratory
(Correia and Mannering, 1973a,b,c) in which the flow of electrons
from NADH to the monooxygenase system was manipulated by activating
or inhibiting the competing cytochrome b_5 fatty acyl CoA desaturation system, and by Sasame et al., 1973, and Mannering et al.,
1974 , who employed cytochrome b_5 antisera to inhibit NADH synergism. On the other hand, studies employing solubilized, partially
purified systems indicate that cytochrome b_5 does not participate
in reactions mediated by the monooxygenase system (Lu et al.,
1974, 1975). The observation that the addition of highly purified

*   We have been credited with insisting that cytochrome b_5 is an
obligatory component of the monooxygenase system when NADPH is the
only source of electrons (Lu et al., 1974; Lu et al., 1975); the
fact is that we left this question open to conjecture (Correia and
Mannering, 1973b, 1973c; Sasame et al., 1973).

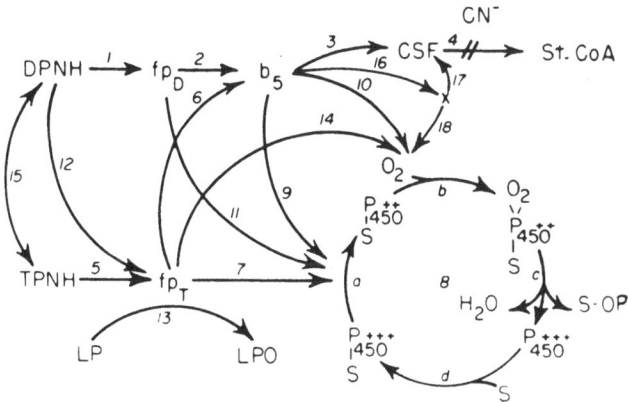

Figure 1. Possible ways in which NADPH-cytochrome P-450 and
NADH-cytochrome b₅ electron transfer systems might interact
(Correia and Mannering, 1973b). CSF, cyanide-sensitive factor,
St. CoA, stearyl-CoA, f_{pD}, DPNH-dependent flavoprotein, f_{pT}, TPNH-
dependent flavoprotein; LP, lipid; LPO, lipid peroxide.

cytochrome b₅ to microsomes inhibits drug metabolism has also been
offered as evidence against the involvement of cytochrome b₅ in
these reactions (Jansson and Schenkman, 1973; Schenkman, 1975).
Clarification of the role of cytochrome b₅ in cytochrome P-450
mediated reactions, whether obligatory or facilitative, is clearly
needed. An attempt will be made in this presentation to evaluate
the evidence for and against the participation of cytochrome b₅ in
monooxygenase reactions, particularly with reference to NADH
synergism.

EFFECT OF SUBSTRATE ON THE STEADY STATE OF REDUCTION
OF CYTOCHROME b₅

Of the experiments performed by Estabrook and collaborators
which attempted to establish the role of cytochrome b₅ in microsomal monooxygenase reactions, the most persuasive was the demonstration of an effect of substrate on the steady state of reduction
of cytochrome b₅ (Hildebrandt and Estabrook, 1971). When NADPH
was introduced into a closed cuvette containing a buffered suspension of microsomes and dissolved air, there was an immediate
increase in the level of reduction of cytochrome b₅ as determined
by the absorbance change at 557 nm relative to 567 nm. Cytochrome
b₅ was reduced further after the addition of NADH, thereby demonstrating the ability of NADH to maintain a higher steady state
level of reduced cytochrome b₅. Subsequent addition of ethylmorphine

caused a substantial reoxidation of cytochrome b_5. This was interpreted to mean that cytochrome b_5 was donating electrons for the N-demethylation of ethylmorphine at a rate that exceeded its capacity to accept electrons from NADH. An ethylmorphine-enhanced rate of oxygen utilization and product formation linked the oxidation of ethylmorphine by the monooxygenase system to the alteration of the steady state reduction of cytochrome b_5.

Recent studies conducted in our laboratory (Nelson and Mannering, 1974) suggest that NADH and cytochrome b_5 may be involved in drug metabolism by the isolated, continuously perfused rat liver, and by inference, by the in situ liver. The rate of hexobarbital metabolism by livers perfused for six hours was only about half that by livers perfused for one hour. Microsomes from the livers perfused for six hours were compared with those from non-perfused livers and from livers which had been perfused for only one hour. The following activities and components were assayed: ethylmorphine N-demethylase, aniline hydroxylase, NADPH-cytochrome c reductase, NADPH-cytochrome P-450 reductase, NADH-cytochrome b_5 reductase, cytochrome P-450, cytochrome b_5, Type I (ethylmorphine) and Type II (aniline) binding, and NADH synergism of ethylmorphine N-demethylation. No differences were seen except for NADH synergism of ethylmorphine N-demethylase, which was greatly reduced in microsomes from livers which had been perfused for six hours. The role of NADH synergism in the metabolism of hexobarbital by the perfused liver was investigated further by examining the effect of ethylmorphine on the steady state of reduced cytochrome b_5 in microsomes from perfused and non-perfused livers under conditions similar to those described by Hildebrandt and Estabrook (1971). As shown in Figure 2, prolonged perfusion lowered both the rate and degree of reoxidation of cytochrome b_5. While the total reducible cytochrome b_5 was affected little by prolonged perfusion, the initial rate of substrate-induced reoxidation of cytochrome b_5 was slowed about 60% (perfused and non-perfused, 5.7 ± S. E. 0.7 and 13.6 ± S. E. 2.2 ΔA/mg of microsomal protein/min, respectively) and the percentage of cytochrome b_5 reoxidized was reduced by about 40% (perfused and non-perfused, 23 ± S. E. 1 or 39 ± S. E. 2, respectively). Since NADH synergism of ethylmorphine N-demethylase activity and the effect of substrate on the steady state of reduced cytochrome b_5 were the only parameters of drug metabolism affected by prolonged perfusion, one might conclude from this study that the loss of hexobarbital metabolism that occurs during perfusion is due to a loss of NADH synergism and that NADH synergism may be mediated by cytochrome b_5.

Because cytochrome b_5 is reduced very rapidly by NADH-cytochrome b_5 reductase (Modurzadeh and Kamin, 1965), the nagging question persists as to how the slow-moving monooxygenation reaction can alter its steady state of reduction.

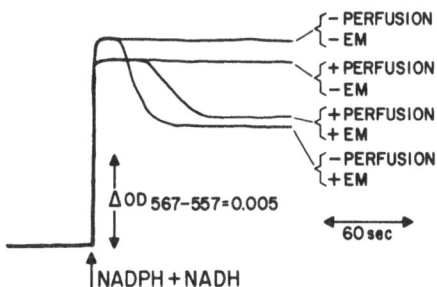

Figure 2. Effect of prolonged perfusion of isolated rat livers on the reoxidation of nucleotide-reduced microsomal cytochrome b_5 in the presence and absence of ethylmorphine (EM). Livers from male rats (200-250 g) were perfused continuously for 6 hr with a perfusate mixture containing 70% whole blood. Microsomes (6 mg of protein) from these livers or from non-perfused livers were suspended in 3 ml of a reaction mixture containing sodium isocitrate (7.0 mM), isocitrate dehydrogenase (0.4 units), $MgCl_2$ (7.5 mM), ethylmorphine (4 mM), and phosphate buffer (0.1 M, pH 7.4. Twenty μl of a mixture of sufficient NADPH and NADH to give final concentrations of 5 μM and 5 mM, respectively, were added rapidly using an Aminco plunger system. Reduced cytochrome b_5 was monitored by recording the absorbance changes at 557 nm relative to 567 nm.

Certainly the reductase systems should be able to furnish electrons to cytochrome b_5 at rates which would more than compensate for the meager loss of electrons to the monooxygenase system. The lowering of the steady state of reduction of cytochrome b_5 by substrate would be more understandable if it were to be demonstrated that substrate somehow restricted the input of electrons from NADH or NADPH to cytochrome b_5 rather than the output.

EFFECTS OF ACTIVATION AND INHIBITION OF THE CYTOCHROME b_5-DEPENDENT FATTY ACYL CoA DESATURATION SYSTEM

Cytochrome b_5 existed for many years as an electron transfer component without a purpose. A function for cytochrome b_5 was provided when Oshino et al., (1971) presented evidence for the participation of cytochrome b_5 as an electron carrier in the desaturation of stearyl-CoA to oleyl-CoA (Figure 1, steps 1→2→3→4). In this reaction, reducing equivalents from NADH or NADPH are transferred via cytochrome b_5 to a cyanide-sensitive factor (CSF), which is presumably the terminal component of the electron transport chain leading to desaturation (Oshino and Sato, 1971, 1972; Oshino, 1972). The discovery of the cytochrome b_5-dependent desaturation system suggested to us a means of investigating the role of cytochrome b_5 in the oxidation of drugs by the monooxygenase system, particularly with regard to NADH synergism (Correia and Mannering, 1973a, b,c). It seemed plausible that stearyl CoA, a natural substrate for the desaturation system, might be used to divert the flow of electrons from NADH that would normally flow to the monooxygenase system, and thus decrease or abolish NADH synergism. Moreover, cyanide should reverse any such effect of stearyl CoA by inhibiting CSF. Stearyl CoA did not produce results that allowed a conclusion regarding the role of cytochrome b_5 in NADH synergism because not only did it inhibit ethylmorphine N-demethylation when both NADH and NADPH were present, but also when NADPH was the only source of reducing equivalents (Figure 3). This was interpreted to mean that the desaturation system was sufficiently activated by stearyl CoA to attract electrons from NADPH as well as from NADH thereby depriving the monooxygenase system of an optimal pool of first electrons for the reduction of cytochrome P-450-substrate complex.

As anticipated, cyanide reversed the inhibitory effect of stearyl CoA on ethylmorphine N-demethylation (Figure 4), but it also produced an unexpected result; it enhanced NADH synergism in the absence of stearyl CoA (Figure 5). This was explained on the basis that CSF-mediated desaturation of membrane fatty acids occurs when microsomes are incubated with NADPH or NADH, and that this endogenous reaction normally inhibits ethylmorphine N-demethylation by channeling electrons away from the monooxygenase system via cytochrome b_5 (Correia and Mannering, 1973a,b,c). Cyanide enhances the activity of the monooxygenase system by inhibiting CSF, thereby diverting to the monooxygenase system those electrons from NADPH and NADH normally lost to the desaturation system. This concept does not exclude the possibility that CSF may be shunting electrons from the monooxygenase system by functioning in systems other than the desaturation system.

In concentrations which enhanced NADH synergism, cyanide was without appreciable effect on ethylmorphine N-demethylation when NADPH was the only source of electrons (Figure 5). This was interpreted to mean that in the absence of stearyl CoA, the drain of electrons from NADPH to the desaturation system via cytochrome b_5 was not great enough to limit the rate of turnover of the monooxygenase system.

When I visited Dr. Sato's laboratory last year, I learned that he was unable to demonstrate an appreciable enhancing effect of cyanide on the NADH synergism of the monooxygenase system. Moreover, I was unable to show a marked cyanide effect in Dr. Omura's laboratory under a variety of experimental conditions, including fasting of the rats, fasting and refeeding of the rats, and the feeding of a highly purified diet. Obviously, Japanese rats and

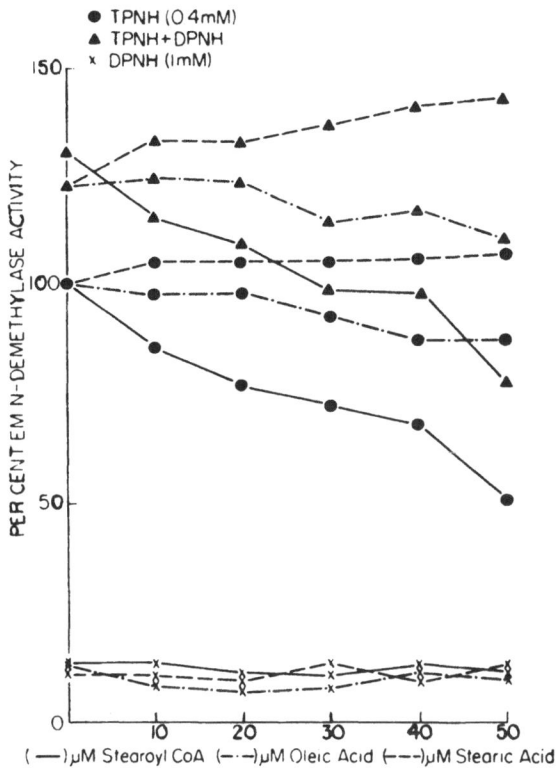

Figure 3. Effects of stearyl CoA, stearic acid, and oleic acid on ethylmorphine (EM) N-demethylation by hepatic microsomes from male rats (Correia and Mannering, 1973b). The incubation mixture contained the following materials in a volume of 5 ml: TPN^+ (0.4 mm), glucose-6-PO_4 (4 mM), glucose-6-PO_4 dehydrogenase (2 units), semi-carbazide HCl (7.5 mM), $MgCl_2$ (2 mM), 0.04 M phosphate buffer (pH 7.4), 1.15% KCl, 17.5-25 mg of microsomal protein, ethylmorphine (2 mM) and specified concentrations of stearyl CoA, stearic acid or oleic acid. Incubation time was 15 min. In systems containing both DPNH and TPNH, DPNH was added as a freshly prepared solution. in 1.15% KCl just prior to incubation. In mixtures containing DPNH only, the TPNH-generating system was omitted. Reaction rates were determined by measuring formaldehyde formed. Values (means of two experiments) are recorded as percentages of the ethylmorphine N-de-methylase activity observed when TPNH was the only source of elec-trons and stearyl CoA, oleic acid, and stearic acid were absent from the medium (100% = 0.43 μmole of formaldehyde/mg of protein/hr).

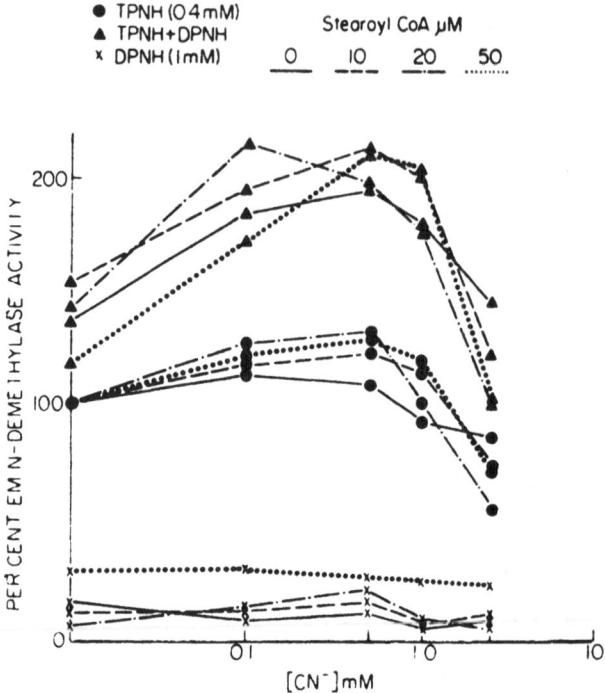

Figure 4. Reversal of inhibitory effect of stearyl CoA on
hepatic microsomal ethylmorphine (EM) N-demethylation by cyanide
(Correia and Mannering, 1973b). Stearyl CoA (0-50 µM) and various
concentrations of cyanide were included in the incubation mixture,
and the reaction was carried out as described in Figure 3. All
values (means of two experiments) are recorded as percentages of
the ethylmorphine N-demethylase activities observed at each of the
concentrations of stearyl CoA when TPNH was the only source of
electrons and cyanide was absent from the medium (100% at 0, 10,
20 and 50 µM stearyl CoA = 0.45, 0.41, 0.27, and 0.21 µmole of
formaldehyde/mg of protein/hr, respectively).

Minnesota rats are different. We continue to observe a marked
cyanide effect in our laboratory, and Dr. Holtzman informs me that
he also observes the effect. However, on one occasion, we observed
no enhancing effects of cyanide. It is of interest that in this
case we observed an NADH synergism in the absence of cyanide of
about 100% rather than the usual 30 to 50%. The 100% NADH synergism is about what we usually observe in the presence of cyanide.
This might suggest that the endogenous desaturation system was not
functioning in these microsomes, conceivably because it lacked an

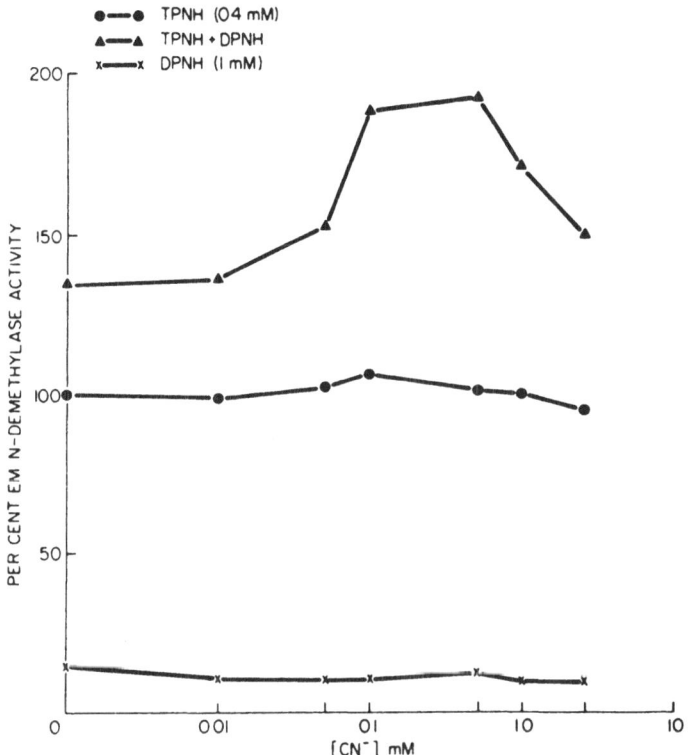

Figure 5. Effect of cyanide on ethylmorphine (EM) N-demethylase activity of hepatic microsomes from male rats in the presence of TPNH, DPNH, or both (Correia and Mannering, 1973b). Incubation conditions are described in Figure 3. Values (means of two experiments) are recorded as percentages of the ethylmorphine N-demethylase activity observed when TPNH was the only source of electrons and cyanide was absent from the medium (100% = 0.32 μmole of formaldehyde/mg of protein/hr).

available substrate. If this was the case, the monooxygenase system would have been expected to function maximally without the benefit of cyanide. The degree of availability of membrane fatty acids to the desaturation system might depend upon unrecognized differences in the environment of the animals or upon subtle differences in the manner in which the microsomes are prepared.

When these studies were expanded to include Type II substrates and Type I substrates other than ethylmorphine, it became apparent that both NADH synergism and its enhancement by cyanide occurred only with Type I substrates; no NADH synergism or cyanide enhance-

ment was observed with the Type II substrates, aniline and N-methyl-
aniline (Correia and Mannering, 1973c). As will be discussed later,
this observation contributed to our interpretation of the mechanism
of NADH synergism and explained why Imai and Omura (1974) did not
observe an inhibition of aniline hydroxylation with anti-cytochrome
b_5 immunoglobulin.

EFFECTS OF ANTIBODY AGAINST CYTOCHROME b_5

While strongly supporting a role of cytochrome b_5 in the cytochrome P-450 monooxygenase system, the foregoing evidence is indirect and subject to alternative interpretations. Immunochemical
methods have provided more direct and conclusive evidence for the
concept. Sasame and associates (1973; 1975) studied the effects of
antibody against cytochrome b_5 on NADPH-supported N-demethylation
of ethylmorphine, both in the presence and absence of NADH. The
antibody was provided in serum from rabbits which had been immunized
to trypsin-solubilized cytochrome b_5 isolated from rat liver. In
the presence of NADPH alone, the antibody had only a small inhibitory effect on either the N-demethylation of ethylmorphine or the
substrate-stimulated oxidation of NADPH, but in the presence of
both NADPH and NADH, antibody greatly decreased both the N-demethylation of ethylmorphine and the substrate-stimulated oxidation of
NADPH. Significantly, the rate of N-demethylation was reduced only
to the rate observed when NADPH was the sole source of electrons.
They concluded that the transfer of second electrons from NADH to
the monooxygenase system is mediated by cytochrome b_5, but that the
transfer of electrons from NADPH to the monooxygenase system is not.
These studies both agreed and disagreed with the original view of
Hildebrandt and Estabrook (1971) regarding the role of cytochrome
b_5 in monooxygenase reactions; they supported the role of cytochrome
b_5 in the transfer of second electrons from NADH, but did not support the contention that the donation of second electrons from NADPH
is also mediated via cytochrome b_5.

Using anti-cytochrome b_5 immunoglobulin (AIg), obtained from
rabbits treated with trypsin-solubilized, homogenous cytochrome b_5
isolated from rat liver, we were able to confirm the results of
Sasame and associates (Mannering et al., 1974). AIg inhibited NADH
synergism of ethylmorphine N-demethylation by rat hepatic microsomes,
but had little effect on the demethylase reaction when the incubation
medium contained NADPH as the only source of electrons (Figure 6).
We were in agreement with Sasame and associates that cytochrome b_5
is involved in the transfer of second electrons from NADH, but left
unanswered the question of whether or not cytochrome b_5 is involved
in the transfer of second electrons from NADPH. We believed this
point was not proved because the amount of microsomal cytochrome b_5
which may not have reacted with AIg may have been considerable,
and while the unreacted cytochrome b_5 may have been insufficient to
support the NADH-synergized rate of ethylmorphine N-demethylation,

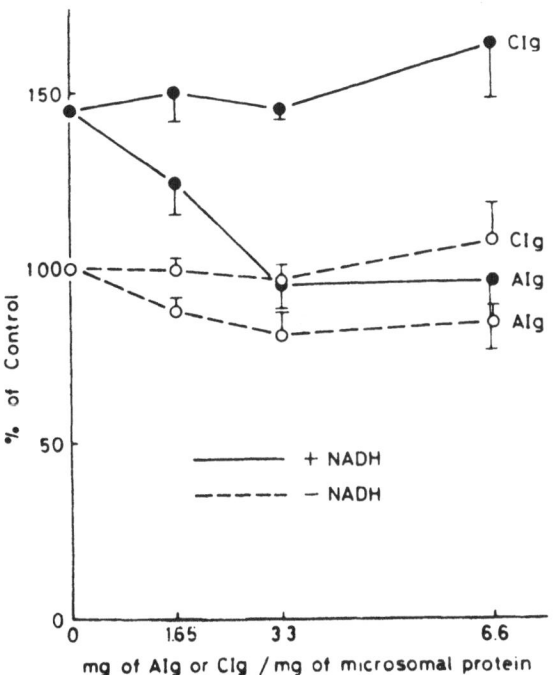

Figure 6. Effects of anti-cytochrome b₅ immunoglobulin (AIg) and control immunoglobulin (CIg) on ethylmorphine (EM) N-demethylation by hepatic microsomes (Mannering et al., 1974). A mixture (final vol: 1.0 ml) containing 0.5 mg of microsomal protein from phenobarbital-treated rats, specified amounts of AIg or CIg, isocitrate (4 mM), isocitrate dehydrogenase (0.4 units), NADP⁺ (0.4 mM), Mg⁺⁺ (2 mM), semicarbazide (7.5 mM), and Na-K PO₄ buffer (40 mM, pH 7.4) in 1.15% KC1 solution was preincubated in an open test tube for 5 min (37° C) with shaking (180 oscillations/min); the N-demethylation reaction was initiated by the addition of EM (2.0 mM) or EM (2.0 mM) + NADH (1.0 mM), and terminated 10 min later by the addition of ZnSO₄. Rates of EM N-demethylation were obtained by duplicate analyses of formaldehyde. Values are means (± S. E.) of of 3 experiments; combined microsomes from two livers were used in each experiment. Control = velocity of formaldehyde formation in the absence of AIg, CIg, and NADH (64.5 ± 2.4 nmoles/mg of microsomal protein/min). Microsomes contained 1.14 ± 0.27 , 0.49 ± 0.07 nmoles of cytochrome P-450 and b₅/mg of protein, respectively.

it may have been sufficient to support the maximal rate of ethylmorphine metabolism that is possible in the absence of NADH. However, this would have to mean that not all of the cytochrome b_5
was capable of reacting with AIg because doubling that amount of
AIg required to produce a maximum inhibitory effect on NADH synergism did not further increase the small inhibitory effect on ethylmorphine N-demethylation produced by AIg in the absence of NADH.

AIg almost completely inhibited cytochrome b_5-mediated cytochrome c reduction by NADH, but Cinti and Ozols (1975) state that
these studies "do not unequivocally establish cytochrome b_5 as the
donor for the transfer of the second electrons from NADH to the
mixed function oxidase since the effect of the cytochrome b_5 antibody on microsomal NADH cytochrome b_5 reductase was not reported."
In other words, the question is raised that our AIg preparation
may have complexed the reductase in some unexplained way rather
than, or in addition to, cytochrome b_5. If this were the case, it
would be difficult to explain why homogenous cytochrome b_5 completely
reversed the inhibitory effect of AIg on the NADH synergism of ethylmorphine N-demethylase (Mannering et al., 1974).

The case for the involvement of cytochrome b_5 in the transfer
of second electrons from NADPH to the monooxygenase system was re-
opened by recent immunochemical studies of Sasame et al., (1974b)
which employed lauric acid as a substrate for microsomes from the
kidney cortex and the liver of the rat. Antibody against cytochrome
b_5 inhibited both the NADH- and the NADPH-dependent reactions in
microsomal preparations from liver and kidney. The microsomal hydroxylation of lauric acid is not as NADPH-dependent as is the case
for the hydroxylation of most drug substrates; the rate of laurate
hydroxylation is as much as 40% as rapid with NADH as that observed
with NADPH when microsomes from either kidney cortex or liver are
used. NADH synergism of NADPH-supported laurate hydroxylation by
microsomes was not evident; in fact, with hepatic microsomes, somewhat less hydroxylation occurred when both nucleotides were present
than the sum of the hydroxylations observed with each of the nucleotides, suggesting perhaps, the presence of two monooxygenases of
unequal reactivity. The cytochrome P-450 monooxygenase system of
the kidney cortex does not metabolize most drug substrates. These
differences in the monooxygenase reactions involving fatty acid
hydroxylation and drug hydroxylation suggest that different monooxygenase systems may be responsible for the metabolism of these
two substrates, both in the kidney and in the liver. If this is
the case, one might explain the difference in the effects of antibody against cytochrome b_5 on NADPH-dependent laurate and ethylmorphine oxidation as being due to a difference in the abilities
of the two monooxygenase systems to accept second electrons from
NADPH without benefit of cytochrome b_5; the system functioning in
the hydroxylation of laurate would require cytochrome b_5 for its

acceptance of electrons from NADPH, the system functioning in the N-demethylation of ethylmorphine would not.

Sasame et al., (1974a) have studied the effects of antibody against cytochrome b_5 on the metabolism of a variety of substrates using microsomes from several tissues. Antibody caused a 20 to 30% inhibition of the NADPH-mediated metabolism of lauric acid, ethylmorphine, codeine, aminopyrine, hexobarbital, 3,4-benzpyrene and aniline by rat liver microsomes, but not the 0-dealkylation of p-nitroanisole and ethoxycoumarin, nor the N-hydroxylation of p-chloroacetanilide. Antibody inhibited the hydroxylation of 3,4-benzpyrene by microsomes from the lung, adrenal and liver, but not by kidney microsomes. The hydroxylation of testosterone by hepatic microsomes at the 6β position was inhibited by about 50%, but hydroxylation at the 2β, 7α or 16α positions was not affected. Whether or not cytochrome b_5 plays a role in monooxygenase reactions would appear to depend not only upon the substrate and the tissue, but also upon the nature of the biotransformation of a given substrate when that substrate can be metabolized in different ways.

ROLE OF CYTOCHROME b_5 IN RECONSTITUTED SYSTEMS

Using a reconstituted sytem consisting of NADPH, lipid and purified preparations of NADPH cytochrome c reductase, cytochrome P-450 or P-448 (P_1-450), which were demonstrated by spectral and biochemical procedures to be free of cytochrome b_5, Lu et al., (1974; 1975) showed that cytochrome b_5 played neither an essential nor a facilitative role in the hydroxylation of benzphetamine or 3,4-benzpyrene. However, o-, m- or p-hydroxylation of chlorobenzene was stimulated two-fold by the addition of cytochrome b_5. Cytochrome b_5 stimulated 16α hydroxylation of testosterone by the reconstituted system, but had little or no effect on 7α or 6β hydroxylation. Coon and associates (van der Hoeven et al., 1974; Coon et al., 1975) have also employed a reconstituted system to show that cytochrome b_5 is not required for the hydroxylation of benzphetamine, cyclohexane, aniline and laurate. The generalization can be made from these observations that when NADPH is the only source of electrons in reconstituted systems, cytochrome b_5 is not required in these monooxygenase reactions which have been studied, but depending upon the substrate, it may or may not be facilitative. If the cytochrome P-450 preparations employed in these studies were homogenous, it can be concluded that the substrate determines whether or not the cytochrome P-450-substrate complex will accept electrons from cytochrome b_5. If the cytochrome P-450 preparations contained more than one species of cytochrome P-450, each with its peculiar substrate specificity, it can be speculated that the species of cytochrome P-450 determines whether or not the cytochrome P-450-substrate complex will accept electrons from cytochrome b_5.

There are reasons to suspect that the roles of cytochrome b_5 in microsomes and reconstituted systems may not be identical:

1) The organization of the cytochrome P-450 and b_5 systems in the membrane may afford an opportunity for interaction not provided in reconstituted systems.

2) NADH synergism is observed with many substrates in microsomal systems, but it has not been observed with the same substrates in reconstituted systems. For example, NADPH-supported benzphetamine demethylation is synergized by NADH in microsomes, but not in a reconstituted system (Lu et al., 1974; 1975). If cytochrome b_5 is required for NADH synergism, as much evidence would indicate, it follows that if a cytochrome P-450-substrate complex cannot accept second electrons by this route, as is apparently the case with certain substrates in reconstituted systems, neither will it accept second electrons from NADPH via cytochrome b_5.

3) It is generally accepted that microsomes contain more than one cytochrome P-450. The likelihood is therefore great that one species of cytochrome P-450 may be concentrated preferentially during the solubilization and purification processes used to prepare the cytochrome P-450 fractions for the reconstituted systems. Evidence of this is seen when the specific activity of membrane bound cytochrome P-450 is compared with that of solubilized purified cytochrome P-450. Ethylmorphine N-demethylase activity is 30-50% higher than benzphetamine N-demethylase activity in hepatic microsomes from phenobarbital-treated rats. However, after partial purification of the cytochrome P-450, ethylmorphine N-demethylase activity is only 10-30% of that of benzphetamine N-demethylase (Lu, 1974). Turnover numbers (nmoles of substrate metabolized per nmole of cytochrome P-450 per min) of several substrates by rabbit microsomes were increased when cytochrome P-450 was solubilized and purified (van der Hoeven et al., 1974). Moreover, the increases observed with different substrates were not parallel; solubilization and purification caused 4.5-fold and 1.8-fold increases in benzphetamine and ethylmorphine demethylase activities, respectively, and the turnover number of laurate hydroxylation by soluble cytochrome P-450 was only about half that observed with microsomes. Treatment of microsomes with non-ionic detergents greatly reduces ethylmorphine N-demethylase activity, but has little effect on benzamphetamine demethylase activity; ammonium sulfate fractionation preferentially decreases ethylmorphine demethylase activity (Lu, 1974).

It is reasonable to expect that any given cytochrome P-450, selected through solubilization and purification, would not behave exactly like the mixture of cytochrome P-450's contained in microsomes.

4) Cytochrome b_5 stimulated the 16α hydroxylation of testosterone by the reconstituted system, but had little or no effect on 7α or 6β hydroxylation (Lu et al., 1974). Antibody against cytochrome b_5 inhibited the hydroxylation of testosterone by hepatic microsomes in the 6β position, but was without effect on hydroxylation in the 2β, 7α or 16α positions (Sasame et al., 1974a). These two observations are incompatible if the role of cytochrome b_5 in reconstituted systems is the same as it is in microsomes.

5) Cytochrome b_5 not only stimulates the metabolism of certain substrates (e.g., chlorobenzene and testosterone) by reconstituted systems, but, as will be discussed in the next section, it inhibits the metabolism of certain substrates (e.g., benzamphetamine and 3,4-benzpyrene). If inhibition and stimulation of metabolism by cytochrome b_5 occur simultaneously during the metabolism of a given substrate, the facilitative role of cytochrome b_5 might be masked by the inhibitory role. Immunochemical evidence employing antibody against b_5 indicates that cytochrome b_5 does not inhibit the monooxygenase system in microsomes (see next section). This may explain why a facilitative effect of cytochrome b_5 on the metabolism of a given substrate may be seen in microsomes, but not in reconstituted systems.

6) An unknown factor (e.g., "x" in Figure 1) may be required for the transfer of electrons from NADH to the monooxygenase system via cytochrome b_5. This hypothetical factor may be missing in reconstituted systems.

Recent studies by Lu and associates (Lu et al., 1974, 1975; West et al., 1974) employing a reconstituted cytochrome P-448-containing system, revealed an NADH- and cytochrome b_5-dependent system which hydroxylates 3,4-benzpyrene. The NADH system is about one-tenth as active as the NADPH system; it accounts for the small increase in benzpyrene hydroxylation observed when NADH is included with NADPH in the incubation medium. This raises the question as to how much of the stimulatory effect of NADH on the NADPH-dependent metabolism of other drug substrates is due to activation of an NADH-dependent system, and how much is due to NADH synergism of the NADPH-dependent system.

EFFECT OF THE ADDITION OF CYTOCHROME b_5 TO MICROSOMES AND RECONSTITUTED SYSTEMS

Purified cytochrome b_5 from detergent-solubilized hepatic microsomes can be incorporated into the microsomal membrane where it functions like endogenous cytochrome b_5 in the cytochrome b_5-dependent NADH-cytochrome c reductase and fatty acyl CoA desaturase systems (Strittmatter et al., 1972; Enomoto and Sato, 1973). Jansson and Schenkman (1973) showed a progressive inhibition of

NADPH-supported aminopyrine demethylation with increasing concentrations of incorporated cytochrome b_5 and concluded from this that cytochrome b_5 does not participate in cytochrome P-450 monooxygenase systems. They suggested that cytochrome b_5 inhibits the reactions by serving as an alternate electron acceptor diverting electrons from cytochrome P-450. The following comments are addressed to the interpretation of this study and to the validity of the procedure as a means of establishing the role of cytochrome b_5 in monooxygenase reactions:

1) Studies which employ exogenous cytochrome b_5 in an attempt to answer the question of whether or not endogenous cytochrome b_5 is involved in cytochrome P-450-dependent monooxygenase reactions in microsomal membranes are non sequiturs because the assumption must be made that added cytochrome b_5 can interact with the monooxygenase system in the same manner that endogenous cytochrome b_5 may interact with this system, when, in fact, this is largely the question the study is attempting to resolve. It cannot be assumed that because membrane bound exogenous cytochrome b_5 reacts with NADH-cytochrome c reductase and with the desaturation system, it also reacts with the monooxygenase system. In fact, there may be some reason to believe that it may not since cytochrome P-450 is thought to be more deeply imbedded in the membrane than the reductase and desaturase systems.

2) Not only can it not be assumed that added cytochrome b_5 will function in the same way as endogenous cytochrome b_5 in all microsomal reactions that may involve this hemoprotein, but it also cannot be assumed that it will not possess greater or lesser activity than endogenous cytochrome b_5 in those reactions or that it may possess properties other than those attributable to the endogenous hemoprotein. Added cytochrome b_5 might participate more actively than the endogenous hemoprotein in reactions not directly coupled to the monooxygenase system. For example, exogenous cytochrome b_5 might be more vulnerable than endogenous cytochrome b_5 to peroxidation and subsequent autoxidation, thereby creating an electron sink not seen normally in microsomes.

3) If added cytochrome b_5 inhibits aminopyrine metabolism by diverting electrons from NADPH, it might be expected that NADH would reverse this inhibition by sparing the loss of electrons from NADPH. This was not the case; the increment of inhibition at each level of added cytochrome b_5 was the same regardless of whether or not NADH was present. Moreover, the increment of NADH synergism of aminopyrine metabolism was also the same at every level of cytochrome b_5 addition. Clearly, the rate of acceptance of first electrons controlled the rate of aminopyrine metabolism regardless of whether or not NADH was added with NADPH. In short, the study does not provide a mechanism for NADH synergism.

4) If endogenous cytochrome b_5 is diverting electrons from NADPH in the manner credited to bound exogenous cytochrome b_5, as implied, the decrease in functional cytochrome b_5 that occurs when antibody against cytochrome b_5 is added to unsupplemented microsomes should increase the NADPH-supported metabolism of drug substrates by decreasing the diversion of electrons from the monooxygenase system. This does not occur; in fact, antibody to cytochrome b_5 decreases the N-demethylation of ethylmorphine slightly when NADPH is the only source of electrons (Sasame et al., 1973, 1975; Mannering et al., 1974).

5) When detergents are used to solubilize microsomal components, the possibility of contamination must be considered. The inhibitory effect of added cytochrome b_5 could be explained without complicated hypotheses if it were to be demonstrated that the added cytochrome b_5 was contaminated with the detergent used in its preparation. Unfortunately, rigorous methods do not exist for the detection of small amounts of unlabeled detergent. However, the question of whether the inhibitory effect of added cytochrome b_5 is due to contaminating detergent appears to have been answered affirmatively by Cinti and Ozols (1975). They repeated the study of Jansson and Schenkman using highly purified detergent-solubilized cytochrome b_5 and found that when bound to microsomes, it functioned in the membrane bound NADH cytochrome c reductase system. Ethylmorphine and aminopyrine metabolism were decreased with increasing amounts of bound exogenous cytochrome b_5. These studies were repeated using reconstituted cytochrome b_5 prepared from apocytochrome b_5 and heme. The bound reconstituted cytochrome b_5 was shown to be functional by its performance in the NADH-cytochrome c reductase and desaturase systems. In contrast with added native bound cytochrome b_5, added reconstituted cytochrome b_5 had no inhibitory effect on ethylmorphine or aminopyrine metabolism, NADPH-cytochrome c reductase, or NADPH-cytochrome P-450 reductase. Cinti and Ozols concluded that the most likely explanation for the inhibitory effect of added native cytochrome b_5, and the lack of inhibitory effect of added reconstituted cytochrome b_5, is that the native cytochrome b_5 contained contaminating amounts of tightly bound detergent, and that this detergent was removed when heme was extracted from the native cytochrome b_5 to form the apocytochrome b_5 used in the reconstitution of cytochrome b_5. They point out that as little as 0.005% of the detergent (Triton N-101) in the assay mixture will inhibit demethylase activity by 15-20%.

Hrycay and Estabrook (1974) extended the study of Jansson and Schenkman by including measurements of the effects of extra bound cytochrome b_5 on the activities of NADH-cytochrome P-450 reductase and NADH-peroxidase, systems that are both cytochrome b_5- and cytochrome P-450 dependent. NADPH- and NADH-benzphetamine demethylase, and NADPH- and NADH-oxidases were also studied. They observed concomitant decreases in NADPH-supported benzphetamine demethylation

and NADPH-cytochrome P-450 reductase activity with increasing
additions of cytochrome b_5. Conversely, NADH-cytochrome P-450 reductase and NADH-peroxidase activities were markedly enhanced with
increasing levels of bound exogenous cytochrome b_5, thus suggesting
that cytochrome b_5 functions as an intermediate electron carrier to
cytochrome P-450 in these two reactions. In short, Jansson and
Schenkman and Hrycay and Estabrook used the same experiment to derive opposite conclusions regarding the role of cytochrome b_5 in
cytochrome P-450 monooxygenase reactions.

Hrycay and Estabrook observed that in cytochrome b_5-fortified
microsomes the rate of the NADPH-dependent benzamphetamine demethylase
reaction was several times greater than the NADH-dependent reaction although rates of cytochrome P-450 reduction by NADPH and
NADH were similar. This could mean that the rate of reduction of
cytochrome P-450 by NADH under anaerobic conditions cannot be used
as a criterion of cytochrome b_5 participation in the aerobic benzamphetamine reaction, thus invalidating the conclusion derived
from these experiments that cytochrome b_5 participates in monooxygenase reactions. These studies would have been more meaningful
if NADPH- and NADH-cytochrome P-450 reductase activities had been
determined both in the presence and absence of benzamphetamine
since it is the increment of substrate stimulation of cytochrome
P-450 reductase that correlates with the rate of substrate metabolism, not the rate of P-450 reduction per se (Holtzman, 1970;
Holtzman and Carr, 1970; Holtzman and Rumack, 1971; Cohen and
Mannering, 1974). The discrepancy in benzphetamine demethylase and
cytochrome P-450 reductase activities might also be rationalized
if two cytochrome P-450's are involved in the reductase reactions,
one NADPH-dependent, the other NADH-dependent.

The observation that exogenous cytochrome b_5 caused no increase
in NADPH oxidase activity was used as an argument against the alternate electron acceptor hypothesis of Jansson and Schenkman.
However, an increase in NADPH utilization would not be required to
support the alternative acceptor concept; added cytochrome b_5 might
cause the redistribution of a given supply of electrons from NADPH
among alternative systems, with fewer electrons going to the monooxygenase system than formerly.

The urge to add excess cytochrome b_5 to soluble systems has
not been ignored. Lu and associates (1974; 1975) showed that cytochrome b_5 inhibited benzphetamine demethylation and 3,4-benzpyrene
hydroxylation by a reconstituted system consisting of cytochrome
P-450, NADPH-cytochrome c reductase, lipid and NADPH, and that the
inhibition was reversed by further addition of NADPH-cytochrome
c reductase or NADH plus NADH-cytochrome b_5 reductase. Cytochrome
b_5 was thought to produce its inhibitory effect in two ways: 1)
cytochrome b_5 complexes NADPH-cytochrome c reductase, thereby

decreasing its availability to the monooxygenase system. The addition of excess NADPH-cytochrome c reductase reverses cytochrome b$_5$ inhibition by replacing the reductase lost by the formation of the cytochrome b$_5$-reductase complex. NADH-cytochrome b$_5$ reductase reverses cytochrome b$_5$ inhibition by forming a complex with cytochrome b$_5$, thus preventing the formation of the NADPH cytochrome c reductase-cytochrome b$_5$ complex; 2) cytochrome b$_5$ acts as an electron drain and competes with cytochrome P-450 for reducing equivalents from NADPH-cytochrome c reductase. Added NADPH-cytochrome c reductase, or NADH plus NADH-cytochrome b$_5$ reductase, reverses cytochrome b$_5$ inhibition by providing the drain with an abundance of reducing equivalents, thereby sparing the monooxygenase a loss of electrons from NADPH.

These observations were used to explain NADH synergism of NADPH-supported monooxygenase reactions in microsomes as follows: the rate of the monooxygenase reaction normally observed with microsomes is an inhibited rate when NADPH is the only source of electrons. Inhibition is due to the competition between cytochrome b$_5$ and cytochrome P-450 for NADPH-cytochrome c reductase. NADH "synergism" is the sum of two effects: 1) NADH reverses cytochrome b$_5$ inhibition in the manner described previously for the reconstituted system; 2) a low rate of monooxygenase activity is supported by NADH via cytochrome b$_5$ and cytochrome b$_5$ reductase. In addition to the previously mentioned caution that must be taken in equating results obtained by reconstituted and microsomal systems, the following questions challenge the validity of this "cytochrome b$_5$ inhibitor" concept of NADH synergism:

1) Does cytochrome b$_5$ serve as an electron drain in microsomes to the same degree that it putatively serves as a drain in reconstituted systems? Where does the drain empty? Either the reconstituted system is impure enough to provide substrates for the electron drain via cytochrome b$_5$, or cytochrome b$_5$ is rendered autoxidizable. One may question whether either of these events occurs in microsomes to the degree that it may occur in reconstituted systems.

2) If cytochrome b$_5$ is causing a loss of NADPH electrons to the reconstituted monooxygenase system, why should the metabolism of some substrates be inhibited (e.g., benzamphetamine and 3,4-benzpyrene) and others be stimulated (e.g., chlorbenzene and testosterone)?

3) Why does antibody against cytochrome b$_5$ decrease rather than increase NADPH-supported monooxygenase activity? If endogenous cytochrome b$_5$ inhibits the monooxygenase system, then its inactivation by antibody against b$_5$ should relieve this inhibition, and an increase rather than the observed slight decrease in

monooxygenase activity (Sasame et al., 1973; 1975; Mannering et
al., 1974) should occur. This decrease would be expected if some
second electrons from NADPH are directed to the monooxygenase system through cytochrome b_5.

ROLE OF CYTOCHROME b_5 in UNCOUPLED REACTIONS

Ullrich and associates (Staudt et al., 1974; Lichtenberger,
1975) have described a role of NADH in uncoupled monooxygenase
reactions. The stoichiometry of the monooxygenase system was
studied using hepatic microsomes from phenobarbital-treated rats,
and cyclohexane, n-hexane and perfluoro-n-hexane as substrates.
With cyclohexane, the stoichiometry for cyclohexane: $NADPH:O_2$
was close to 1:1:1.   With the uncoupler, perfluoro-n-hexane,
no product was formed, and the stoichiometry for $NADPH:O_2$ was
2:1. The stoichiometry of n-hexane hydroxylations suggested that
partial uncoupling had occurred. NADH did not synergize NADPH-
supported cyclohexane hydroxylation, but caused more than additive
hydroxylation in the case of n-hexane. With perfluoro-n-hexane,
NADH caused almost a doubling of the oxygen used when NADPH was
the only source of electrons. An increased oxidation of NADH via
cytochrome b_5 was shown to occur during uncoupling. The reduction
of cytochrome b_5 by NADH occurs at a faster rate than by NADPH;
this explains the greater efficiency of NADH in providing the
electrons for the reduction of the uncoupled active oxygen. It
was concluded that active oxygen not used for product formation is
reduced to water by the NADH-cytochrome b_5 system rather than by
the NADPH-cytochrome c system, and that this sparing effect accounts for the synergistic action of NADH on monooxygenase reactions.

As pointed out by these authors, the question remains as to
the importance of the uncoupling process in monooxygenase reactions
when typical substrates are used. Recent studies in our laboratory
of the stoichiometry of some common substrates suggest that uncoupling may not be an important factor in most monooxygenase reactions (Jeffery and Mannering, 1974a). The hepatic microsome is
a complex membrane, containing a variety of enzymes, some of which
interact by sharing electron transfer components. Membrane components provide the substrates for these systems. Stoichiometry
for substrate, NADPH and oxygen utilization will vary from the required 1:1:1 relationship in accordance with the activities of
these other systems relative to that of the monooxygenase system.
Attempts to achieve stoichiometry by subtracting endogenous NADPH
utilization by microsomes in the absence of substrate from NADPH
utilization in the presence of substrate, have not proved satisfactory. In our study, NADPH was monitored both biochemically and
by observing the decline in spectral absorbance at 340 nm. Values

obtained with the two methods became increasingly divergent with
incubation time, with more of the apparent NADPH present when the
spectral procedure was used. The high values were shown to be due
to nicotinamidemono-nucleotide (NMNH) which also gives a maximum
absorption at 340 nm. NMNH is formed from NADPH through the action
of microsomal nucleotide pyrophosphatase. Moreover, substrate
(ethylmorphine) inhibited NMNH formation by competing for NADPH
for its metabolism. A 1:1 stoichiometry of NADPH utilization to
formaldehyde formation was obtained for ethylmorphine N-demethylation when calculations were corrected to include rates of NMNH
formation. Buening and Franklin (1974) obtained stoichiometry of
ethylmorphine:oxygen utilization of 1:1 after taking NMNH formation
into consideration when making calculations. EDTA inhibits nucleotide pyrophosphatase activity to the degree that the pyrophosphatase and the monooxygenase no longer compete for NADPH and 1:1
stoichiometry for NADPH:substrate utilization, monitored spectrophotometrically at 340 nm, was observed for ethylmorphine, aminopyrine, hexobarbital, codeine and benzphetamine simply by adding
EDTA to the incubation medium and including cyanide to inhibit the
fatty acyl CoA desaturation system (Jeffery and Mannering, 1974b).
Stoichiometry was achieved with ethylmorphine when several sources
of microsomes were employed, including male rats, mice and rabbits,
female rats, and rats treated with phenobarbital or 3-methylcholan-
threne. Using the same assay procedure, it was also possible to
show NADH synergism of ethylmorphine N-demethylation. Since NADH
synergism was observed in the absence of uncoupling, not all NADH
synergism can be explained by the uncoupling phenomenon. Normal
hexane may be the exception rather than the rule.

ROLE OF TYPE I BINDING IN NADH SYNERGISM

A study of the effect of NADH on rates of metabolism of a
variety of substrates by hepatic microsomes from the rat revealed
that NADH synergism occurred only when Type I substrates were employed (Correia and Mannering, 1973c). NADH synergism was observed with the Type I substrates, ethylmorphine, codeine, nor-
codeine, benzphetamine and aminopyrine, but not with the Type II
substrates, aniline and p-chloro-N-methylaniline. Netter (1974)
has also observed that, in contrast to Type I compounds, Type II
compounds do not increase the utilization of NADPH. Type I binding
does not insure that the metabolism of a substrate will be synergized by NADH; for example, the hydroxylation of cyclohexane,
which in very low concentration exhibits a Type I spectrum with
rabbit liver microsomes, is not synergized by NADH (Staudt et al.,
1974).

The role of Type I binding in NADH synergism was investigated
further by employing microsomes which varied in their abilities to

produce a Type I binding spectrum with ethylmorphine. This was
accomplished by selecting microsomes from different animal sources
or by subjecting microsomes from a given source to treatments known
to diminish Type I binding. The cytochrome P-450 in microsomes
from female rats and young male rats shows less Type I binding than
that from adult male rats (Schenkman et al., 1967). The cytochrome
P-450 that results after the administration of 3-methylcholanthrene
(cytochrome P_1-450 or P-448) is deficient in an available Type I
binding site (Shoeman et al., 1969; Mannering, 1971). Microsomes
lose their ability to produce a Type I binding spectrum with ethylmorphine during storage (Shoeman et al., 1969), or after they have
been treated with phospholipase C (Chaplin and Mannering, 1970).
The degree of NADH synergism of NADPH-dependent ethylmorphine N-
demethylation and NADH utilization was shown to be correlated with
the abilities of these microsomes to produce a Type I binding spectrum (Figures 7 and 8).

Cyclohexane is an exception to the generalization that reactions
employing Type I substrates are synergized by NADH (Staudt et al.,
1974). Cyclohexane is a Type I binding substrate with a low binding constant and it is hydroxylated with a high maximum velocity.
However, NADPH-supported cyclohexane hydroxylation is not synergized by NADH, and NADH utilization by microsomes is not increased
in the presence of cyclohexane.

CONCEPT OF AN NADH-INDUCED SHIFT IN RATE LIMITATION OF FIRST AND SECOND ELECTRONS

The rate of entry of the first electron is thought by many
investigators to be rate-limiting in the overall monooxygenase reaction. This view is based partly on the observations that 1)
Type I substrates stimulate NADPH-cytochrome P-450 reductase
activity (Gigon et al., 1968; 1969); 2) the degree of stimulation
of the rate of NADPH-cytochrome P-450 reductase by a given Type I
substrate correlates with the rate of metabolism of that substrate
(Holtzman, 1970; Holtzman and Carr, 1970; Holtzman and Rumack,
1971; Cohen and Mannering, 1974); 3) binding constants for Type I
spectra of substrates relate to corresponding Michaelis constants,
although not as ideally as one would prefer (Remmer et al., 1966;
Imai and Sato, 1966; Mannering, 1971). Our observation that NADH
synergism is observed with Type I substrates, but not with Type II
substrates (Correia and Mannering, 1973c), suggested that when
NADPH is the only source of electrons, the rate of entry of the
second electron is rate-limiting and that the rate of entry of
the first electron becomes rate-limiting only when both NADPH and
NADH are present. In accordance with this concept, the introduction of Type I substrates creates an increased demand for first
and second electrons. NADPH is able to satisfy the demand for
first electrons, but not for second electrons; i.e., the second
electron is rate-limiting. NADH is a more efficient donor of

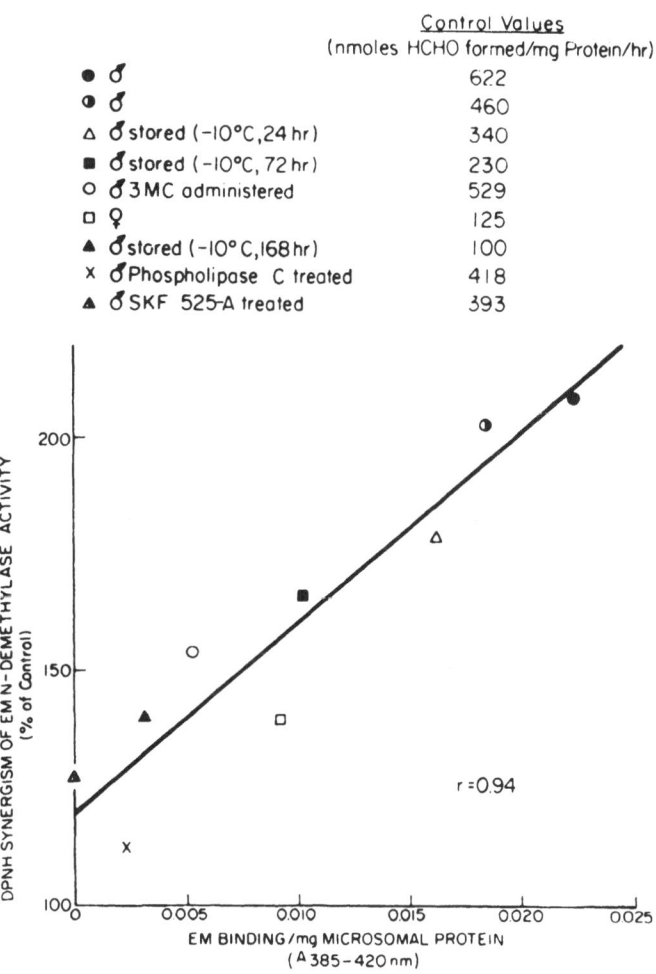

Figure 7. Relationship of DPNH synergism of TPNH-supported ethylmorphine (EM) N-demethylase activity to Type I (ethylmorphine) binding (Correia and Mannering, 1973c). Microsomes with varying abilities to bind ethylmorphine (EM) were selected as indicated. DPNH synergism was measured using the procedure given in Figure 5. Values are those obtained when the medium contained either 0.1 or 0.5 mM cyanide, whichever concentration gave the maximum DPNH synergism. 3MC, 3-methylcholanthrene.

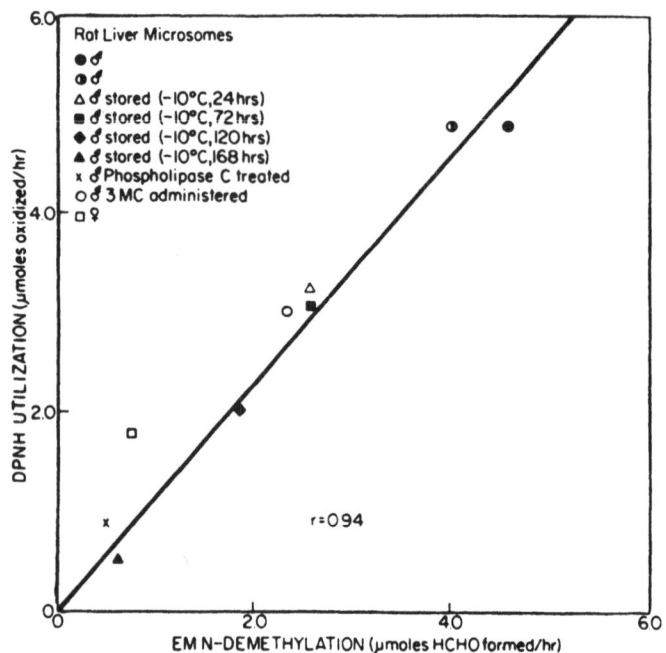

Figure 8. Relationship of DPNH utilization to formaldehyde formed during maximal DPNH synergism of TPNH-dependent ethylmorphine (EM) N-demethylase activity in the presence of cyanide (Correia and Mannering, 1973c). Values for DPNH utilization and formaldehyde formation were obtained from the experiment described in Figure 7. 3MC, 3-methylcholanthrene.

second electrons than NADPH either because NADH can provide second electrons to the monooxygenase system via cytochrome b_5, whereas NADPH cannot, or because NADPH is a less effective donor of electrons via cytochrome b_5. After addition of NADH to the NADPH-supported reaction, second electrons are now more available to the system than first electrons; i.e., first electrons become rate-limiting. However, the rate-limitation is now functioning at a higher level of electron flux than when NADPH was the sole electron donor. The result is "NADH synergism."

NADH synergism is not seen with Type II substrates because the first electron is rate-limiting regardless of whether or not NADH is present with NADPH. Type II substrates do not activate oxidized cytochrome P-450 (Gigon et al., 1968; Schenkman, 1968), therefore the metabolism of Type II substrates is much slower than that of Type I substrates, usually about an order of magnitude lower. The demand for first electrons is low enough that second electrons from NADPH are able to keep pace with the input of first electrons. This being the case, no benefit is derived from the addition of NADH although NADH is potentially a more efficient source of electrons; NADH is not offered the opportunity to synergize the reaction.

CONCLUDING REMARKS

Studies which have attempted to assess the role of the cytochrome b_5 electron transfer system in monooxygenase reactions involving P-450 hemoproteins have been addressed primarily to three questions: 1) Is cytochrome b_5 an obligatory component of NADPH-supported reactions? 2) Is cytochrome b_5 a facilitative component of NADPH-supported reactions? 3) Is cytochrome b_5 involved in NADH synergism of NADPH-supported reactions? Five approaches to these questions have been taken: 1) kinetic studies of the effects of substrate on the utilization of NADH and the steady state of reduction of cytochrome b_5; 2) activation and inhibition of the microsomal alkyl-CoA desaturation system to redistribute electrons toward or away from the monooxygenase system; 3) immunochemical studies employing antibody to cytochrome b_5; 4) use of reconstituted systems devoid of cytochrome b_5; 5) addition of homogenous cytochrome b_5 to microsomes and reconstituted systems.

Is cytochrome b_5 an obligatory component of NADPH-supported reactions? Studies with reconstituted systems show it is not. These systems appear to be sufficiently free of cytochrome b₅ that the obligatory nature of cytochrome b₅ can be ruled out, at least for the substrates that have been studied. The possibility is not excluded that other reconstituted systems, containing other species of cytochrome P-450 than those which have been employed, may require cytochrome b_5 as an obligatory component. Immunochemical studies support the view that cytochrome b_5 is not required in microsomal reactions. The observation that substrate lowers the steady state of reduction of cytochrome b_5 in microsomes supports a facilitative, but not an obligatory role of cytochrome b_5 in monooxygenase reactions. Studies which have employed microsomes fortified with homogenous cytochrome b_5 neither support nor oppose the view that cytochrome b_5 is a required component of monooxygenase reactions. Studies that manipulated the flow of electrons from NADPH to the monooxygenase system by stimulating or inhibiting the fatty acyl CoA desaturation system neither supported nor excluded an obligatory role of cytochrome b_5. An obligatory role of cytochrome b_5 in reactions supported by NADH in the absence of NADPH has been observed in reconstituted systems. Whether NADH- and NADPH-dependent monooxygenase reactions employ the same or a different species of cytochrome P-450 is not known. The probability is good that NADH-dependent reactions in microsomes also require cytochrome b_5.

Is cytochrome b_5 a facilitative component of NADPH-supported
reactions? Cytochrome b_5 facilitates the NADPH-supported metabolism
of some substrates by reconstituted systems, but not of other substrates. It remains to be determined whether this is due to preferential substrate activation of a single P-450 hemoprotein, or to the
selectivity of different P-450 hemoproteins. A slight facilitative
role of cytochrome b_5 in the NADPH-supported reactions in microsomes is indicated by the 10-30% inhibition of the metabolism of
certain substrates by antibody against cytochrome b_5. When homogenous cytochrome b_5 is added to a reconstituted system, some reactions are stimulated, others are inhibited. These results are
subject to a variety of interpretations, which may or may not apply
to microsomes. Inhibition is also observed when homogenous cytochrome b_5 is bound to microsomes. Again interpretation is difficult because it is not known whether incorporated cytochrome b_5
behaves in all respects like endogenous cytochrome b_5 in the microsomes; immunochemical evidence suggests that it does not. The question of whether an inhibitory amount of detergent was introduced into
the reaction with the added cytochrome b_5 must also be considered.
Studies employing cytochrome b_5-fortified microsomes have also
been used to support a role of cytochrome b_5 in monooxygenase
reactions.

Is cytochrome b_5 involved in NADH synergism of NADPH-supported
reactions? We need to be less speculative in responding to this
question than we were to the two previous questions. In microsomal
suspensions containing NADPH, substrate causes an increase in NADH
utilization commensurate with the increase in product formed, oxygen uptake, replacement of second electrons from NADPH by second
electrons from NADH, and the reoxidation of cytochrome b_5. Antibody against cytochrome b_5 blocks NADH synergism. Stimulation or
inhibition of fatty acyl-CoA desaturase, a cytochrome b_5-dependent
system, affects the rate of substrate formation or NADH utilization
in a manner compatible with the involvement of cytochrome b_5 in
the monooxygenase reaction. Cytochrome b_5 can also be involved in
uncoupled reactions without direct interaction with the monooxygenase system by using electrons from NADH to spare electrons from
NADPH which would have been lost to the monooxygenase system via
the uncoupling mechanism. However, uncoupling does not occur with
many typical substrates known to be synergized by NADH, and may in
fact, be a rare phenomenon. NADH synergism has not been observed
with reconstituted systems. This suggests that the organization
of the components of the monooxygenase system provided by the membrane may determine to a large degree the interactability of cytochrome b_5, or, alternatively, that the components of the reconstituted system are not identical in all respects with their native
counterparts.

REFERENCES

Björkhem, I. and Danielsson, H. 1973. Heterogeneity of hepatic mixed function oxidases. Biochem. Biophys. Res. Comm. 51: 766-774.

Buening, M. and Franklin, M. 1974. Limitations in the use of the 340 nm absorbance maximum of NADPH for the determination of oxidation rates and stoichiometry during rat hepatic microsomal metabolism. Mol. Pharmacol. in press.

Chaplin, M. D. and Mannering, G. J. 1970. Role of phospholipids in the hepatic microsomal drug metabolizing system. Mol. Pharmacol. 6: 631-640.

Cinti, D. L. and Ozols, J. 1975. The role of cytochrome b$_5$ in mixed function oxidations: effect of microsomal binding of the hemoprotein on hepatic N-demethylations. This publication.

Cinti, D. L., Moldeus, P. and Schenkman, J. B. 1972. The role of the mitochondria in rat liver mixed function oxidation reactions. Biochem. Biophys. Res. Comm. 47: 1028-1035.

Cohen, B. S. and Estabrook, R. W. 1971. Microsomal electron transport reactions. II. The use of reduced triphosphopyridine nucleotide for the oxidative N-demethylation of aminopyrine and other drug substrates. Arch. Biochem. Biophys. 143: 46-53.

Cohen, G. M. and Mannering, G. J. 1974. Sex-dependent differences in drug metabolism in the rat. III. Temporal changes in Type I binding and NADPH-cytochrome P-450 reductase during sexual maturation. Drug Metab. Disp. 2: 285-292.

Conney, A. H. 1967. Pharmacological implications of microsomal enzyme induction. Pharmacol. Rev. 19: 317-366.

Conney, A. H., Brown, R. R., Miller, J. A. and Miller, E. C. 1957. The metabolism of methylated aminoazo dyes. VI. Intracellular distribution and properties of the demethylase system. Cancer Res. 17: 628-633.

Coon, M. J., van der Hoeven, T. A., Haugen, D. A., Guengerich, F. P., Vermilion, J. L. and Ballou, D. P. 1975. Biochemical characterization of highly purified cytochrome P-450 and other components of the mixed function oxidase system of liver microsomal membranes. This volume, p. 26.

Correia, M. A. and Mannering, G. J. 1973a. DPNH synergism of TPNH-dependent mixed function oxidase reactions. Drug Metab. Disp. 1: 139-149.

Correia, M. A. and Mannering, G. J. 1973b. Reduced diphosphopyridine nucleotide synergism of the reduced triphosphopyridine. I. Effects of activation and inhibition of the fatty acyl coenzyme A desaturation. Mol. Pharmacol. 9: 455-469.

Correia, M. A. and Mannering, G. J. 1973c. Reduced diphosphopyridine nucleotide synergism of the reduced triphosphopyridine. II. Role of Type I drug-binding site of cytochrome P-450. Mol. Pharmacol. 9: 470-485.

Enomoto, K. and Sato, R. 1973. Incorporation in vitro of purified
    cytochrome b_5 into liver microsomal membranes. Biochem.
    Biophys. Res. Comm. 51: 1-7.

Estabrook, R. W., Shigematsu, A. and Schenkman, J. B. 1970. The
    contribution of the microsomal electron transport pathway to
    the oxidative metabolism of liver. Adv. Enzyme Reg. 8:
    121-130.

Estabrook, R. W., Franklin, M., Baron, J., Shigematsu, A. and
    Hildebrandt, A. 1971. Properties of the membrane-bound
    electron transfer complex of the hepatic endoplasmic reticu-
    lum associated with drug metabolism. IN E. Mihich, Ed.,
    Drugs and Cell Regulation, 227-257.

Gigon, P. L., Gram, T. E. and Gillette, J. R. 1968. Effect of
    drug substrates on the reduction of hepatic microsomal cyto-
    chrome P-450 by NADPH. Biochem. Biophys. Res. Comm. 31:
    558-562.

Gigon, P. L., Gram, T. E. and Gillette, J. R. 1969. Studies on
    the rate of reduction of hepatic microsomal cytochrome P-450
    by reduced nicotinamide adenine dinucleotide phosphate:
    effect of drug substrates. Mol. Pharmacol. 5: 109-122.

Gillette, J. R. 1966. Biochemistry of drug oxidation and reduc-
    tion by enzymes in hepatic endoplasmic reticulum. Advan.
    Pharmacol. 4: 219-261.

Gillette, J. R. 1971. Factors affecting drug metabolism. Ann.
    N. Y. Acad. Sci. 179: 43-66.

Hildebrandt, A. and Estabrook, R. W. 1971. Evidence for the
    participation of cytochrome b_5 in hepatic microsomal mixed
    function oxidation reactions. Arch. Biochem. Biophys. 143:
    66-79.

Holtzman, J. L. 1970. Effect of 4,4-dideuteration of reduced
    nicotinamide adenine dinucleotide phosphate on the mixed
    function oxidases of hepatic microsomes. Biochemistry 9:
    995-1001.

Holtzman, J. L. and Carr, M. L. 1970. Inhibition of hepatic micro-
    somal mixed function oxidases by D_2O. Life Sci. 9: 1033-1038.

Holtzman, J. L. and Rumack, B. H. 1971. The kinetics of cyto-
    chrome P-450 reductase stimulation by ethylmorphine. Life
    Sci. 10: 669-677.

Hrycay, E. G. and Estabrook, R. W. 1974. The effect of extra bound
    cytochrome b_5 on cytochrome P-450-dependent enzyme activities
    in liver microsomes. Biochem. Biophys. Res. Comm. 60: 771-
    778.

Imai, Y. and Omura, T. 1974. Unpublished results cited by Mannering,
    Kuwahara and Omura (Biochem. Biophys. Res. Comm. 57:476-481).

Imai, Y. and Sato, R. 1966. Substrate interaction with hydroxylase
    system in liver microsomes. Biochem. Biophys. Res. Comm. 22:
    620-626.

Jansson, I. and Schenkman, J. B. 1973. Evidence against partici-
    pation of cytochrome b_5 in the hepatic microsomal mixed-

function oxidase reaction. Mol. Pharmacol. 9: 840-845.

Jeffery, E. and Mannering, G. J. 1974a. Discrepancy in the measurement of TPNH oxidized during N-demethylation due to the presence of nucleotide pyrophosphatase. Mol. Pharmacol. in press.

Jeffery, E. and Mannering, G. J. 1974b. Unpublished results.

Lichtenberger, F. 1975. Discussion in this publication.

Lu, A. Y. H. 1974. Private communication.

Lu, A. Y. H., West, S. B., Vore, M., Ryan, D. and Levin, W. 1974. Role of cytochrome b₅ in hydroxylation by a reconstituted cytochrome P-450-containing system. J. Biol. Chem. 249, in press.

Lu, A. Y. H., Levin, W., West, S. B., Vore, M., Ryan, D. Kuntzman, R. and Conney, A. H. 1975. Role of cytochrome b₅ in NADPH- and NADH-dependent hydroxylation by the reconstituted cytochrome P-450- or P-448-containing system. This publication.

Mannering, G. J. 1968. Significance of stimulation and inhibition of drug metabolism in pharmacological testing. IN A. Burger, Ed. Selected Pharmacological Testing Methods, 51-119.

Mannering, G. J. 1971a. Role of substrate binding to P-450 hemoprotein in drug metabolism. IN E. Mihich, Ed., Drugs and Cell Regulation, 197-225.

Mannering, G. J. 1971b. Properties of cytochrome P-450 as affected by environmental factors: qualitative changes due to administration of polycyclic hydrocarbons. Metabolism 20: 228-245.

Mannering, G. J. 1971c. Microsomal enzyme systems which catalyze drug metabolism. IN B. N. LaDu, H. G. Mandel and E. L. Way, Eds. Fundamentals of Drug Metabolism and Drug Disposition, 206-252.

Mannering, G. J., Kuwahara, S. and Omura, T. 1974. Immunochemical evidence for the participation of cytochrome b₅ in the NADH synergism of the NADH-dependent mono-oxidase system of hepatic microsomes. Biochem. Biophys. Res. Comm. 57: 476-481.

Modurzadeh, J. and Kamin, H. 1965. Reduction of microsomal cytochromes by pyridine nucleotides. Biochem. Biophys. Acta. 99: 205-226.

Netter, K. J. and Illing, H. P. A. 1974. Kinetic experiments on the synergistic effect of NADH on microsomal drug oxidation. Xenobiotica 4: 549-561.

Nelson, D. and Mannering, G. J. 1974. Unpublished results.

Oshino, N. 1972. Dynamic behavior during dietary induction of the terminal enzyme (cyanide-sensitive factor) of the stearyl CoA desaturation system of rat liver microsomes. Arch. Biochem. Biophys. 149: 378-387.

Oshino, N., Imai, Y. and Sato, R. 1971. A function of cytochrome b₅ in fatty acid desaturation by rat liver microsomes. J. Biochem. 69: 155-168.

Oshino, N. and Sato, R. 1971. Stimulation by phenols of the reoxidation of microsomal bound cytochrome b₅ and its implication to fatty acid desaturation. J. Biochem. 69: 169-180.

Oshino, N. and Sato, R. 1972. Dietary control of the microsomal stearyl CoA desaturation enzyme system in rat liver. Arch. Biochem. Biophys. 149: 369-377.

Sasame, H. A., Mitchell, J. R., Thorgeirsson, S. and Gillette, J. R. 1973. Relationship between NADH and NADPH oxidation during drug metabolism. Drug Metab. Disp. 1: 150-155.

Sasame, H. A., Thorgeirsson, S. S., Mitchell, J. R. and Gillette, J. R. 1974a. The possible involvement of cytochrome b_5 in the oxidation of lauric acid by microsomes from kidney cortex and liver of rats. Life Sci. 14: 35-46.

Sasame, H. A. Thorgeirsson, S. S., Menard, R. H., Hinson, J. A., Mitchell, J. R., and Gillette, J. R. 1974b. A role of cytochrome b_5 in both NADH and NADPH-mediated cytochrome P-450 enzymatic reaction in mammalian tissues. Fed. Proc. 33: 1437.

Sasame, H. A., Thorgeirsson, S. S., Mitchell, J. R. and Gillette, J. R. 1975. The role of cytochrome b_5 in cytochrome P-450 enzymes. This publication.

Schenkman, J. B. 1968. Effect of substrates on hepatic microsomal cytochrome P-450. Hoppe-Seyler's Z. Physiol. Chem. 349: 1624-1628.

Schenkman, J. B. and Jansson, I. 1975. Interaction between microsomal electron transfer pathways. This publication.

Schenkman, J. B., Frey, I., Remmer, H. and Estabrook, R. W. 1967. Sex differences in drug metabolism by rat liver microsomes. Mol. Pharmacol. 3: 516-525.

Shoeman, D. W., Chaplin, M. D. and Mannering, G. J. 1969. Induction of drug metabolism. III. Further evidence for the formation of a new P-450 hemoprotein after treatment of rats with 3-methylcholanthrene. Mol. Pharmacol. 5: 412-419.

Sitar, D. S. and Mannering, G. J. 1973. Determination of apparent kinetic constants of the microsomal hydroxylation of amobarbital, hexobarbital and pentobarbital. Drug Metab. Disp. 1: 663-668.

Strittmatter, P., Rogers, M. J. and Spatz, L. 1972. The binding of cytochrome b_5 to liver microsomes. J. Biol. Chem. 247: 7188-7194.

Staudt, H., Lichtenberger, F. and Ullrich, V. 1974. The role of NADH in uncoupled microsomal monoxygenations. Eur. J. Biochem. 46: 99-106.

van der Hoeven, T. A., Haugen, D. A., and Coon, M. J. 1974. Cytochrome P-450 purified to apparent homogeneity from phenobarbital-induced rabbit liver microsomes: catalytic activity and other properties. Biochem. Biophys. Res. Comm. 60: 569-575.

West, S. B., Levin, W., Ryan, D., Vore, M. and Lu, A. Y. H. 1974. Liver microsomal electron transport systems. II. The involvement of cytochrome b_5 in the NADH-dependent hydroxylation of 3,4-benzpyrene by a reconstituted cytochrome P-448-containing system. Biochem. Biophys. Res. Comm. 58: 516-522.

THE ROLE OF CYTOCHROME b_5 IN CYTOCHROME P-450 ENZYMES

H. A. Sasame, S. S. Thorgeirsson, J. R. Mitchell and
J. R. Gillette, Laboratory of Chemical Pharmacology,
National Heart and Lung Institute, National Institutes
of Health
Bethesda, Maryland 20014

The possible involvement of cytochrome b_5 in NADPH-mediated
cytochrome P-450 reactions in liver microsomes was first suggested
by Estabrook and co-workers (1,2). Two years ago we presented
immunochemical evidence that the synergistic effect of NADH on
NADPH-mediated ethylmorphine metabolism by rat liver microsomes is
indeed mediated through cytochrome b_5 by showing that an antibody
against rat liver cytochrome b_5 blocks the increase in NADH-oxidation and ethylmorphine N-demethylation in the presence of NADPH,
ethylmorphine and liver microsomes (3). We recently presented evidence that a sheep antibody against cytochrome b_5 (4,5) also inhibits the omega hydroxylation of lauric acid by rat liver and
kidney cortex microsomes in the presence of either NADH or NADPH
(Table I). In order to differentiate the inhibitory effect of
sheep antibody against cytochrome b_5 on the omega hydroxylation
of lauric acid from some nonspecific inhibitor present in the
antibody preparation, we preincubated the antibody with varying
amounts of purified, trypsin-solubilized cytochrome b_5 from rat
liver and then examined the reversibility of the inhibitory effect.
As shown in Figure 1, the inhibition of the omega hydroxylation of
lauric acid by rat kidney cortex microsomes was decreased as the
antibody sites were titrated with purified cytochrome b_5. Moreover,
NADH-cytochrome $\underline{c}$ reductase activity increases in parallel to the
omega hydroxylase activity. Similar results were obtained with
liver microsomal systems (Figure 2). Thus, the inhibition of omega
hydroxylation must be due to the anti-cytochrome b_5 antibody and
not to a nonspecific inhibitor in the preparation.

If cytochrome b_5 participates in the NADPH-mediated omega
hydroxylation of lauric acid by donating the second electron

TABLE I

Effect of the Antibody against Cytochrome b_5 on the Hydroxylation
of Lauric Acid by Microsomes from Rat Liver and Kidney Cortex

Tissue and Cofactor	Preimmune γ-globulin	Anti-cytochrome b_5 γ-globulin	Percent Inhibition
Liver			
NADPH	4.87 ± .33	3.30 ± .17	32
NADH	1.00 ± .05	0.51 ± .05	49
NADPH + NADH	5.14 ± .10	3.22 ± .05	37
Kidney Cortex			
NADPH	3.62 ± .12	1.64 ± .10	55
NADH	0.35 ± .04	0.18 ± .04	50
NADPH + NADH	4.48 ± .20	2.03 ± .22	55

Each incubation mixture contained 30 mg of preimmune or immune γ-globulin, 1.5 mg of microsomes isolated from either liver or kidney cortex, 0.15 μmoles ^{14}C-lauric acid, 15 μmoles EDTA, 0.45 mg rotenone, and either 3 μmoles of NADH or 1.0 μmole NADP, 30 μmole glucose-6-phosphate and 2 units of glucose-6-phosphate dehydrogenase, in a total volume of 1.5 ml of 0.05 M potassium phosphate buffer, pH 7.4. Incubations were carried out in air at 37°C for either 10 min in the experiments with NADPH and NADPH plus NADH, or for 20 min in the experiments with NADH alone.

The anti-cytochrome b_5 γ-globulin inhibited the NADH-cytochrome c reductase in liver and kidney cortex by 63% and 32% respectively, but did not inhibit NADPH-cytochrome c reductase in these tissues.

Values are means ± standard deviation of 3 determinations.

required by a mixed-function oxidase, however, one should expect to see some change in total electron flux through microsomes in the presence of antibody against cytochrome b_5, that is, a change in total NADPH-oxidation. Since the omega hydroxylation of lauric acid

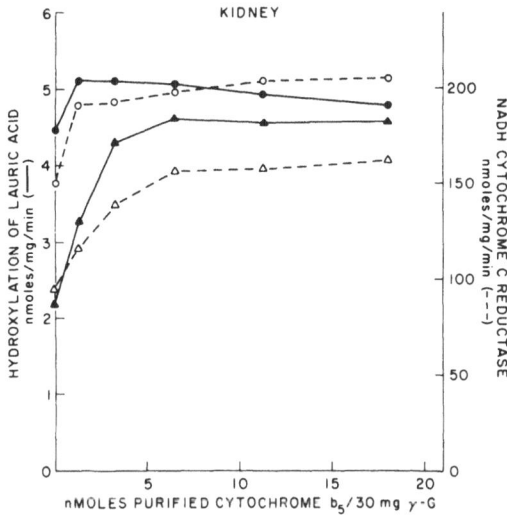

Figure 1. Effect of purified cytochrome b_5 on the ability of anti-cytochrome b_5 antibody to inhibit lauric acid hydroxylation by kidney cortex microsomes of rats. Preimmune and immune γ-globulin fractions were preincubated with purified cytochrome b_5 for 25 min at room temperature and then added to rat kidney cortex microsomes and preincubated for another 10 min at room temperature. To each of the mixtures consisting of 30 mg of preimmune or immune γ-globulin, 1.5 mg of microsomes, and various amounts of cytochrome b_5, were added 0.15 μmoles of ^{14}C-lauric acid, 1 μmole of NADP, 30 μmoles of glucose-6-phosphate, 2 units of glucose-6-phosphate dehydrogenase, 15 μmoles of $MgCl_2$ and 1.5 μmoles of EDTA in a total volume of 1.5 ml of 0.05 M potassium phosphate buffer, pH 7.4

Incubations were carried out in air for 10 min at 37°C. Experiments with the preimmune γ-globulin fraction are shown by the plot ●————● for the hydroxylation of lauric acid and by the plot o- - - - -o for the reduction of cytochrome c. Experiments with the immune γ-globulin fraction are shown by the plot ▲————▲ for the hydroxylation of lauric acid and the plot Δ- - - -Δ for the reduction of cytochrome c.

in liver microsomes represents only 14% of the total NADPH-oxidation, one might not expect to see a significant change in total electron flux in the liver system, but in kidney cortex microsomes, the omega hydroxylation accounts for 40% of the total NADPH-oxidation and yet the antibody did not change the rate of NADPH-oxidation (Table II).

We also studied the effects of anti-cytochrome b_5 on the

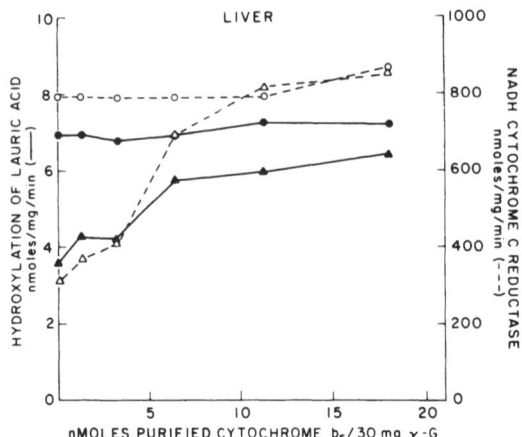

Figure 2. Effect of purified cytochrome b_5 on the ability of the anti-cytochrome b_5 antibody to inhibit lauric acid hydroxylation by rat liver microsomes. Pretreatment of the anti-cytochrome b_5 antibody with purified cytochrome b_5 and the incubation conditions were the same as those described in Figure 1.

Experiments with the preimmune γ-globulin fraction are shown by the plot ●———● for the hydroxylation of lauric acid and the plot o- - - -o for the reduction of cytochrome c. Experiments with the immune γ-globulin fraction are shown by the plot ▲———▲ for the hydroxylation of lauric acid and the plot Δ - - -Δ for the reduction of cytochrome c.

metabolism of various substrates in the presence of NADH or NADPH. As one might expect, the sheep antibody against cytochrome b_5 selectively blocked not only endogenous reduction of cytochrome P-450 by NADH but also that in the presence of ethylmorphine. But the antibody did not alter the rate of cytochrome P-450 reduction by NADPH in either the presence or absence of ethylmorphine (Table III). Moreover, when NADH was the sole electron donor the antibody blocked not only the metabolism of various substrates, but also the NADH-oxidation (Table IV). Thus, cytochrome b_5 mediates the NADH-dependent reduction of cytochrome P-450 and NADH-dependent oxidation of various substrates. However, when NADPH was the sole source of electron donor, the antibody against cytochrome b_5 failed to inhibit the NADPH-oxidation (Table V).

The addition of NADH to systems containing NADPH and rat liver microsomes increases the metabolism of a number of different substrates (Table VI). Assays of the oxidation of NADPH and NADH, however, revealed that the increase in substrate metabolism

TABLE II

Effect of Antibody against Cytochrome b$_5$ on NADPH-oxidation and Hydroxylation of Lauric Acid by Rat Liver and Kidney Cortex Microsomes

Tissue	NADPH-oxidation (nmoles/mg protein/min)		Lauric Acid Hydroxylation (nmoles/mg protein/min)	
	Preimmune γ-globulin	Anti-cytochrome b$_5$ γ-globulin	Preimmune γ-globulin	Anti-cytochrome b$_5$ γ-globulin
Liver	23.30 ± .50	21.70 ± 1.00	3.30 ± .12	1.56 ± .02
Kidney Cortex	6.03 ± .50	6.49 ± .40	2.33 ± .13	.88 ± .02

Each incubation mixture contained 40 mg of preimmune or immune γ-globulin, 2.0 mg of microsomes isolated from either liver or kidney cortex, 0.2 μmoles of 14C-lauric acid, 20 μmoles of EDTA, 1.5 μmoles of NADPH in a total volume of 2.0 ml of 0.05 M potassium phosphate buffer, pH 7.4. The mixtures were incubated for 10 min at 37°.

Values are means ± standard deviation of 3 determinations.

TABLE III

Effect of Anti-cytochrome b_5 against NADPH and NADH-dependent P-450 Reductase in Rat Liver Microsomes

| | Reduction rate nmoles/mg/min | | | |
| | NADPH | | NADH | |
	without EM	with EM	without EM	with EM
Sheep γ-globulin	22.2 ± 1.8	27.7 ± .2	2.3 ± .4	6.36 ± .6
Anti-cytochrome b_5	23.3 ± 1.3	27.5 ± .2	1.0 ± .04	1.75 ± .04

Thirty milligrams of preimmune or immune γ-globulin from sheep were preincubated with 1.5 mg of rat liver microsomes for 10 min at room temperature. Subsequently, the preincubated microsomes were placed in an anaerobic cell containing 150 μmoles of potassium phosphate buffer, pH 7.4, in a final volume of 3.0 ml with or without 10 μmoles of ethylmorphine. The cytochrome P-450 reduction rate was measured according to the method described by Sasame and Gillette (6).

paralleled an increase in NADH-oxidation rather than an increase in NADPH-oxidation. Moreover, the increases in the metabolism of the substrates and the NADH-oxidation were blocked by the antibody against cytochrome b_5. These findings thus indicate that when NADH is added to a NADPH system, the electrons transferred from cytochrome b_5 to cytochrome P-450 are exclusively donated by NADH and not by NADPH.

As demonstrated with the omega hydroxylation of lauric acid, however, the antibody against cytochrome b_5 inhibited the metabolism of a number of substrates even when NADPH was the sole electron donor (Table VI). Moreover, preincubation of the antibody with purified cytochrome b_5 prevented this inhibition (Figure 3). By contrast, the addition of exogenous cytochrome b_5 to control γ-globulin resulted in a slight decrease in ethylmorphine metabolism, presumably due to its uncoupling effect on the electron flux in liver microsomes. Thus, the inhibitory effect of antibody on ethylmorphine is not due to the presence of endogenous inhibitors in the antibody fraction, but is due to the antibody itself.

In conclusion, the following mechanisms of cytochrome b_5 in cytochrome P-450 systems can be proposed:

TABLE IV

Effect of Anti-cytochrome b$_5$ on the Metabolism of Substrate by Rat Liver Microsomes in the Presence of NADH Alone

Substrate	Total NADH-oxidation nmoles/mg/min		Percent Inhibition	Metabolite(s) nmoles/mg/min		Percent Inhibition
	Control	Anti-cyto-chrome b$_5$		Control	Anti-cyto-chrome b$_5$	
Ethylmorphine	10.4 ± .4	6.3 ± .3	40	3.9 ± .4	2.1 ± .3	46
Codeine	11.0 ± .5	6.0 ± .2	45	6.0 ± .5	3.6 ± .2	40
Benzphetamine	7.1 ± .3	4.8 ± .2	32	3.3 ± .2	1.51± .1	55
Aminopyrine	6.6 ± .4	4.3 ± .4	35	3.9 ± .2	2.1 ± .1	46
Hexobarbital	5.0 ± .3	3.4 ± .2	32	4.3 ± .3	3.0 ± .4	30

Each incubation mixture contained 1.5 mg of rat liver microsomes which had been preincubated with 30 mg of either preimmune or immune γ-globulin fractions, isolated from sheep serum, for 10 min at room temperature, 6 μmoles of a substrate (1.5 μmoles of hexobarbital), 1.5 μmoles of MgCl$_2$, 15 μmoles of nicotinamide, 0.45 mg of rotenone, 1.5 μmoles of EDTA and 1.5 μmoles of NADH, in a total volume of 1.5 ml of 0.05 M potassium phosphate buffer, pH 7.4. The mixtures were incubated for 9 min at 37°C in air. The oxidation of NADH was determined by a method utilizing alcohol dehydrogenase (7).
Values are means ± standard deviation of 3 determinations.

TABLE V

Effect of Anti-cytochrome b_5 on the Oxidation of NADPH and NADH of Rat Liver Microsomes in the Presence of Various Substrates

Substrate	Total NADPH-oxidation nmoles/mg/min				Total NADH-oxidation nmoles/mg/min	
	NADPH		NADPH + NADH		NADH	
	Control	Anti-cyto-chrome b_5	Control	Anti-cyto-chrome b_5	Control	Anti-cyto-chrome b_5
Ethylmorphine	59.2 ± 3.3	54.0 ± 0.2	51.8 ± 4.3	52.1 ± 4.3	32.7 ± 1.8	18.2 ± 0.2
Codeine	54.1 ± 2.1	57.7 ± 1.0	54.6 ± 0.7	52.4 ± 0.4	33.0 ± 0.2	18.1 ± 0.1
Morphine	37.3 ± 0.2	37.1 ± 0.1	30.6 ± 0.3	31.1 ± 0.2	24.9 ± 0.9	23.3 ± 1.0
Benzphetamine	60.0 ± 0.2	56.7 ± 1.2	57.1 ± 2.8	59.3 ± 2.7	28.0 ± 0.6	17.0 ± 0.4
Aminopyrine	50.1 ± 0.2	46.9 ± 2.7	45.3 ± 0.7	41.0 ± 1.0	26.5 ± 1.6	17.3 ± 0.6
Hexobarbital	53.5 ± 3.3	48.7 ± 0.4	47.0 ± 0.9	47.0 ± 3.4	24.7 ± 0.2	18.2 ± 0.7

The pretreatment of liver microsomes with the γ-globulin fractions and the incubation conditions were the same as those described in Table IV except that NADPH (1.5 μmoles) was added as indicated. The oxidation of NADPH was determined by a method utilizing glucose-6-phosphate dehydrogenase (10). The oxidation of NADH was determined by a method utilizing alcohol dehydrogenase (10). Values are means ± standard deviation of 3 determinations.

TABLE VI

Effect of Anti-cytochrome b$_5$ on the Formation of Various Substrate Metabolites by Rat Liver Microsomes

Substrate	Metabolite(s) nmoles/mg/min			
	NADPH		NADPH + NADH	
	Control	Anti-cytochrome b$_5$	Control	Anti-cytochrome b$_5$
Ethylmorphine	38.0 ± 2.6	28.2 ± 1.2	61.9 ± 0.2	31.8 ± 3.7
Codeine	31.3 ± 0.9	22.3 ± 1.7	45.8 ± 0.7	26.2 ± 0.1
Morphine	0	0	0	0
Benzphetamine	23.3 ± 0.7	23.2 ± 1.3	28.9 ± 0.4	23.8 ± 0.2
Aminopyrine	29.2 ± 0.7	23.0 ± 1.6	36.3 ± 0.3	22.6 ± 1.5
Hexobarbital	14.0 ± 1.9	12.3 ± 1.2	21.9 ± 0.1	17.7 ± 0.7

The pretreatment of liver microsomes with the γ-globulin fractions and the incubation conditions were the same as those described in Table IV except that some incubation mixtures contained both 1.5 μmoles of NADPH and 1.5 μmoles of NADH.

Values are means ± standard deviation of 3 determinations.

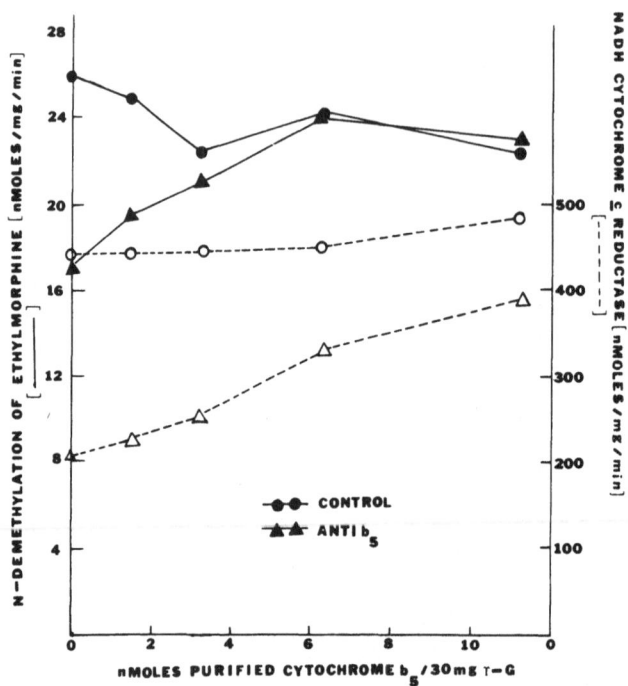

Figure 3. Effect of purified cytochrome b_5 on the ability of the anti-cytochrome b_5 antibody to inhibit ethylmorphine N-demethylation by rat liver microsomes. The pretreatment of the anti-cytochrome b_5 antibody with purified cytochrome b_5 and the incubation conditions were the same as those described in Figure 1, except that 6 μmoles of ethylmorphine were added instead of lauric acid in the incubation mixture. Experiments with the preimmune γ-globulin fraction are shown by the plot ●——● for the N-demethylation of ethylmorphine and by the plot o- - -o for the reduction of cytochrome c. Experiments with the immune γ-globulin fraction are shown by the plot ▲——▲ for the N-demethylation of ethylmorphine and by the plot Δ- - -Δ for the reduction of cytochrome c.

a) Cytochrome b$_5$ mediates NADH drug metabolism by reducing cytochrome P-450 because the anti-cytochrome b$_5$ antibody blocks drug metabolism by NADH.

b) Other investigators have shown that cytochrome b$_5$ may decrease drug metabolism through channeling electrons from NADPH-cytochrome c reductase. NADH reverses the inhibition by keeping cytochrome b$_5$ reduced. This mechanism, however, is not important in this series of experiments, because 1) the anti-cytochrome b$_5$ should stimulate rather than inhibit drug metabolism and 2) the mechanism does not account for the finding that NADPH stimulates NADH-oxidation in the presence of substrates.

c) Since the anti-cytochrome b$_5$ antibody inhibits the drug metabolism not only when both NADPH and NADH are present but also when NADPH is present alone, it seems likely that the second electron is mediated preferentially by cytochrome b$_5$. Since the antibody frequently inhibits NADH-oxidation more than it inhibits drug metabolism, however, the second electron may be alternatively mediated by NADPH-cytochrome c reductase when the antibody is present.

d) Since the anti-cytochrome b$_5$ antibody inhibits drug metabolism but not NADPH-oxidation even when NADPH alone is present, the antibody may lead to uncoupling of the relationship between NADPH-oxidation and drug metabolism.

REFERENCES

1   Hildebrandt, A. G. and Estabrook, R. W. Arch. Biochem. Biophys. 143: 66 (1971).
2   Cohen, B. S. and Estabrook, R. W. Arch. Biochem. Biophys. 143: 54 (1971).
3   Sasame, H., Mitchell, J. R., Thorgeirsson, S. S. and Gillette, J. R. Drug Metabolism and Disposition 1:150 (1973).
4   Sasame, H., Thorgeirsson, S. S., Mitchell, J. R. and Gillette, J. R. Pharmacologist 15: 170 (1973).
5   Sasame, H., Thorgeirsson, S. S., Mitchell, J. R. and Gillette, J. R. Life Sciences 14: 35 (1974).
6   Sasame, H. A. and Gillette, J. R. Mol. Pharmacol. 5: 123 (1969).
7   Kornberg, A. J. Biol. Chem. 182: 779 (1950).
8   Nash, T. Biochem. J. 55: 416 (1953).
9   Cooper, J. R. and Brodie, B. B. J. Pharmacol. Exptl. Ther. 114: 409 (1955).
10 Horecker, B. L. and Kornberg, A. in: Methods in Enzymology, Vol. III: 879 Academic Press, New York. (1957).

ROLE OF CYTOCHROME b_5 IN NADPH- AND NADH-DEPENDENT HYDROXYLATION

BY THE RECONSTITUTED CYTOCHROME P-450- OR P-448-CONTAINING SYSTEM

Anthony Y. H. Lu, Wayne Levin, Susan B. West, Mary Vore,
Dene Ryan, Ronald Kuntzman and A. H. Conney
Department of Biochemistry and Drug Metabolism
Hoffman-LaRoche Inc.
Nutley, New Jersey 07110

INTRODUCTION

The possible interaction between the NADPH-dependent hydroxylation system and the NADH-dependent desaturation system is an important aspect of the liver microsomal electron transport system. From the discussions we have had in this meeting, and recent studies (Cohen and Estabrook, 1971a, 1971b, 1971c; Correia and Mannering, 1973a, 1973b; Hildebrandt and Estabrook, 1971; Modirzadeh and Kamin, 1965; Sasame et al., 1973a, 1973b), it is obvious that these two systems do interact and that cytochrome b_5 plays an important role in this interaction.

In this paper, we shall summarize our recent studies on the role of cytochrome b_5 in both NADPH- and NADH-dependent hydroxylation reactions as catalyzed by our reconstituted system. The results indicate that cytochrome b_5 is not an obligatory component in the NADPH-dependent benzphetamine or 3,4-benzpyrene hydroxylation systems, but appears to be involved in the NADPH-dependent hydroxylation of testosterone at the 16α position and the hydroxylation of chlorobenzene. In addition, cytochrome b_5 is an obligatory component of the NADH-mediated hydroxylation of several substrates.

PREPARATION OF VARIOUS COMPONENTS

The reconstituted, NADPH-dependent hydroxylation system consists of cytochrome P-450 (from phenobarbital*-treated rats) or P-448 (from 3-methylcholanthrene-treated rats), NADPH-cytochrome c reductase and lipid (Lu et al., 1969; Lu et al., 1972). In order to test the suggested role of cytochrome b_5 in NADPH-dependent hydroxylation, it is necessary to remove cytochrome b_5 completely from cytochrome P-450, P-448 and the NADPH-dependent reductase. The lipid factor presents a lesser problem because it can be boiled before use (to destroy b_5 but not lipid), or it can be replaced by synthetic phosphatidylcholine (Strobel et al., 1970). Earlier preparations of cytochrome P-450, P-448 and NADPH-cytochrome c reductase always contained some cytochrome b_5 (Lu et al., 1972); Levin et al., 1972), but the improved preparations of these en- zymes have been shown to contain no detectable cytochrome b_5 (Levin et al., 1974; Lu et al., 1974).

Two methods were used to establish the absence of cytochrome b_5 in the cytochrome P-450, P-448 and NADPH-cytochrome c reductase preparations. The first method is based on the reduction of cyto- chrome b_5 by NADH and NADH-cytochrome b_5 reductase as measured by the difference between the reduced versus the oxidized spectrum of cytochrome b_5, and the second method involves the reduction of cytochrome c by NADH, which is mediated by cytochrome b_5 reductase. For example, when an earlier cytochrome P-450 preparation--termed Step III P-450 (Lu and Levin, 1972; Levin et al., 1972)--was re- duced by NADH and cytochrome b_5 reductase, the reduced minus oxi- dized spectrum showed the typical trough at 410 nm and peak at 424 nm characteristic of cytochrome b_5. In contrast, the improved preparation--termed Step IV P-450 (Levin et al., 1974)--contained no detectable cytochrome b_5, as judged by the lack of a trough at 410 nm or a peak at 424 nm (Figure 1). Both crude and purified cytochrome b_5 reductases could be used in these experiments, but deoxycholate was required for the complete reduction of cytochrome b_5 when purified cytochrome b_5 reductase was used (Lu et al., 1974). The absence of cytochrome b_5 in Step IV P-450 was further confirmed by the lack of NADH-dependent reduction of cytochrome c in a re- action mixture containing Step IV cytochrome P-450 and cytochrome b_5 reductase; cytochrome c could only be reduced by NADH when exogeneous cytochrome b_5 was added to the reaction mixture (Lu et al., 1974). Using these two methods, we could also demonstrate the absence of cytochrome b_5 in our improved cytochrome P-448 and NADPH-cytochrome c reductase preparations.

Cytochrome b_5 and NADH-cytochrome b_5 reductase were purified from the liver microsomes of PB-treated rats according to the method of Spatz and Strittmatter (1971 and 1973). The lipid fraction was prepared by a previously described method (Lu et al., 1969). All the enzymes used in the present study were solubilized and purified in the presence of detergents.

NADPH-DEPENDENT BENZPHETAMINE N-DEMETHYLATION

A comparison of the rate of benzphetamine N-demethylation obtained from liver microsomal suspensions (from PB-treated rats)

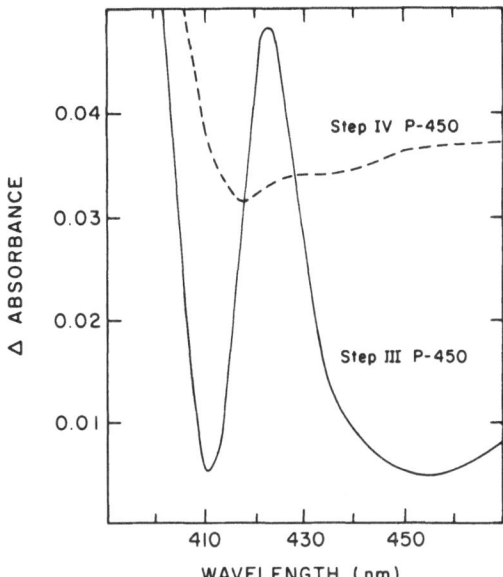

Fig. 1. Reduction of cytochrome b₅ by NADH. The reaction
mixture contained Step III cytochrome P-450 (13.5 nmoles, 2.6 mg
protein, solid line ———) or Step IV cytochrome P-450 (13.5
nmoles, 1.3 mg protein, dotted line - - -), purified NADH-
cytochrome b₅ reductase (10 units, 0.16 mg protein), deoxycholate
(0.05 ml of a 5% solution) and 0.02M potassium phosphate buffer
(pH 7.4) in a final volume of 2.5 ml. One ml of the mixture was
transferred to each of the two cuvettes, and a baseline of equal
light absorbance was established. NADH (10 μl, 55 mg/ml) was
added to the sample cuvette, and the difference spectrum was
recorded.

and the reconstituted system is shown in Table I. It is clear
that, in the absence of cytochrome b₅, benzphetamine could still
be readily metabolized by cytochrome P-450, NADPH-cytochrome c
reductase and lipid in the presence of NADPH and molecular oxygen.
It should be noted that in the reconstituted system, the experimental conditions can be adjusted so that cytochrome P-450 is the
rate-limiting component, whereas this may not be the case with intact microsomes. Nevertheless, the specific activity, or the turnover number for benzphetamine in the reconstituted system, is
better than that in liver microsomes, indicating that cytochrome
b₅ is not an obligatory component in the NADPH-dependent metabolism
of benzphetamine.

When cytochrome b₅ was added to the reaction mixture containing cytochrome P-450, NADPH-cytochrome c reductase and lipid,

TABLE I

Benzphetamine N-Demethylation by Liver Microsomes and Reconstituted
System from PB-Treated Rats

	Formaldehyde formed	
Conditions	nmoles/min/mg protein	nmoles/min/nmole P-450
Microsomes	20- 40	10-14
Reconstituted system	80-140	30-50

The values in this table represent the range of activity observed in our

laboratory from numerous experiments and many different preparations over

the past several years. Male, Long-Evans rats (50-60 g) pretreated with

PB were used in these studies. In the reconstituted system, protein

concentration was the sum of the amounts of cytochrome P-450 and NADPH-

cytochrome c reductase used in the incubation mixture. Benzphetamine

N-demethylation was assayed by measuring the formation of formaldehyde

(Lu et al., 1974).

NADPH-dependent benzphetamine N-demethylation was inhibited
(Figure 2). At a ratio of cytochrome b_5 to P-450 normally found
in microsomes (0.25 in microsomes from PB-treated rats, 0.5 in
microsomes from untreated rats), the reaction rate was inhibited by
about 30-40%. This inhibition is probably not due to the presence
of detergent in the cytochrome b_5 preparation for the following
reasons: a) Boiled cytochrome b_5 had no effect on the rate of
benzphetamine N-demethylation. b) At the ratio of cytochrome b_5
to P-450 which inhibited benzphetamine metabolism, the hydroxylation of 3,4-benzpyrene (Figure 3) and of testosterone at the 6β
position (Figure 6) were not affected, whereas both the hydroxylation of testosterone at the 16α position (Figure 6) and of chlorobenzene were stimulated. Finally, c) as will be shown (Table III),
the inhibition of cytochrome b_5 could be reversed under certain
experimental conditions.

Iyanagi and Mason (1973) have reported that, using NADPH as an
electron donor, detergent-solubilized NADPH-cytochrome c reductase
will reduce cytochrome b_5. We have confirmed this observation with

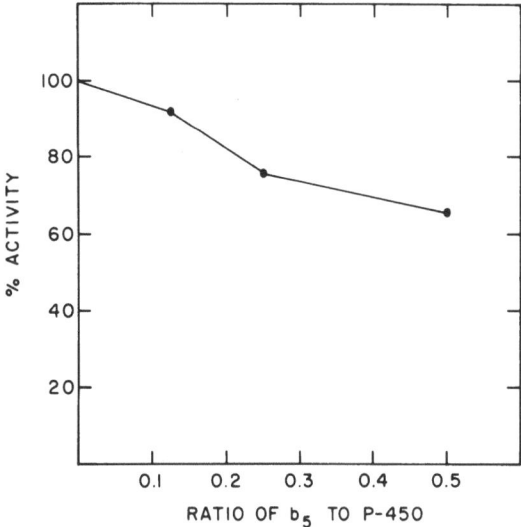

Fig. 2. Effect of cytochrome b_5 on NADPH-dependent benzpheta-mine N-demethylation. The reaction mixture contained cytochrome P-450 (0.27 nmole; 0.028 mg protein), NADPH-cytochrome c reductase (250 units; 0.07 mg protein) lipid (0.12 mg) and variable amounts of cytochrome b_5 in a final volume of 1.5 ml. NADPH was the sole electron donor in this experiment. The 100% activity (i.e., in the absence of b_5) was 136 nmoles HCHO formed/10 min. One unit of NADPH-cytochrome c reductase = 1.0 nmole cytochrome c reduced per min assay by the method of Phillips and Langdon (1962). For-maldehyde formation was determined as previously described (Lu et al., 1974).

our preparation of NADPH-cytochrome c reductase. It is therefore possible that cytochrome b_5 inhibits benzphetamine oxidation by competing with cytochrome P-450 for NADPH-cytochrome c reductase. If this hypothesis is correct, then at a fixed ratio of cytochrome b_5 to P-450 one should be able to reverse this inhibition by in-creasing the concentration of NADPH-cytochrome c reductase. Table II (Experiment A) shows that this was indeed the case. Fur-thermore, the inhibition caused by cytochrome b_5 could also be reversed by adding a high concentration of NADH-cytochrome b_5 reductase (Experiment B), presumably the reversal of inhibition being due to cytochrome b_5 complexing more efficiently with NADH-dependent cytochrome b_5 reductase than with NADPH-cytochrome c reductase. It should be noted that the amount of cytochrome b_5 reductase and cytochrome c reductase needed to reverse the inhibi-tory effect of cytochrome b_5 at a b_5 to P-450 ratio of 0.25 -- the ratio present in PB microsomes--is considerably higher than the amount of both reductases found in microsomes from PB-treated rats.

TABLE II

NADPH-Dependent Benzphetamine N-Demethylation: Effect of NADPH-
Cytochrome c Reductase and NADH-Cytochrome b_5 Reductase
Concentrations on Cytochrome b_5 Inhibition

		Activity		
		(A)	(B)	%
Expt.	Conditions	no b_5	with b_5	(B/A)
		nmoles	HCHO/10 min	
A.	Cytochrome P-450 + lipid + cytochrome c			
	reductase (214 units)	69.8	44.6	64
	Cytochrome P-450 + lipid + cytochrome c			
	reductase (428 units)	103.1	89.6	87
B.	Cytochrome P-450 + lipid + cytochrome c			
	reductase (214 units)	84.2	60.3	72
	Cytochrome P-450 + lipid + cytochrome c			
	reductase (214 units) + b_5 reductase			
	(4 units)	80.1	77.4	97

The reaction mixture contained cytochrome P-450 (0.27 nmole; 0.028 mg
protein), lipid (0. 12 mg), NADPH-cytochrome c reductase and NADH-
cytochrome b_5 reductase as indicated. The ratio of cytochrome b_5 to
P-450 was 0.25. NADPH was the sole electron donor in this experiment.
One unit of NADH-cytochrome b_5 reductase = 1.0 µmole ferricyanide reduced per min assayed by the method of Mihara and Sato (1972).

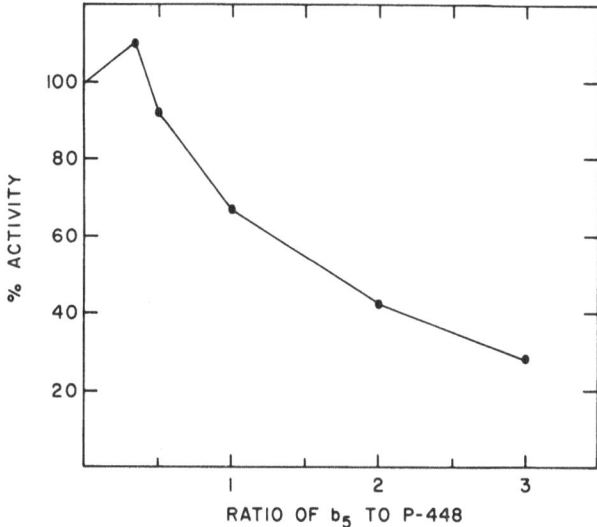

Fig. 3. Effect of cytochrome b_5 on NADPH-dependent 3,4-benz-pyrene hydroxylation. The reaction mixture contained cytochrome P-448 (0.1 nmole; 0.009 mg protein), NADPH-cytochrome c reductase (54 units; 0.01 mg protein), lipid (0.1 mg) and variable amounts of cytochrome b_5 in a final volume of 1.0 ml. NADPH was the sole electron donor in this experiment. The 100% activity (i.e., in the absence of b_5) was 1.24 nmole 8-hydroxy-3,4 benzpyrene formed per 5 min.

Thus, in microsomes, one would expect to see inhibition of benz-phetamine N-demethylation by cytochrome b_5.

The competition between cytochrome b_5 and cytochrome P-450 for NADPH-cytochrome c reductase probably results in the following two effects: a) Cytochrome b_5 limits the availability of the re-ductase itself and thus decreases the concentration of NADPH-cytochrome c reductase available to cytochrome P-450, and b) Cytochrome b_5 acts as an electron drain and competes with cyto-chrome P-450 for reducing equivalents from NADPH-cytochrome c reductase. Both of these factors are probably important.

The effect of cytochrome b_5 and cytochrome b_5 reductase on benzphetamine N-demethylation by the reconstituted hydroxylation system was also examined in the presence of either NADPH or NADH alone, or of both. As shown in Table III, the NADPH-dependent reaction was inhibited by cytochrome b_5 at a ratio of b_5 to P-450 of 0.5 (Expt. A-2). Again, the addition of cytochrome b_5 re-ductase (at a concentration which resembles that found in micro-somes containing 0.27 nmole of cytochrome P-450) partially re-versed the inhibition (Expt. A-3). When NADH was the sole

TABLE III

Effect of Cytochrome b_5 and b_5 Reductase on Benzphetamine N-Demethylation in the Presence of NADPH, or NADH, or Both

		% Activity		
		with	with	with NADPH
Expt.	Conditions	NADPH	NADH	and NADH
A-1	Cytochrome P-450 + cytochrome c reductase			
	+ lipid	100	0	103
A-2	Cytochrome P-450 + cytochrome c reductase			
	+ lipid + cytochrome b_5	66	3	82
A-3	Cytochrome P-450 + cytochrome c reductase			
	+ lipid + cytochrome b_5 + b_5 reductase	77	8	104
B	Microsomes from PB-treated rats	100	3	122

The reaction mixture in Experiment A-3 contained cytochrome P-450 (0.27 nmole; 0.028 mg protein), NADPH-cytochrome c reductase (250 units; 0.07 mg protein), lipid (0.12 mg), cytochrome b_5 (0.14 nmole; 0.006 mg. protein), and b_5 reductase (1.3 units; 0.02 mg protein) in a final volume of 1.5 ml. The ratio of cytochrome b_5 to P-450 was 0.5. In Experiment B, microsomes from PB-treated rats (3.56 nmoles P-450; 1.5 mg protein in a final volume of 4.5 ml) were used. The concentrations of NADPH and NADH were 0.33 mM and 1 mM, respectively. The 1-0% activities were 78.6 nmoles HCHO formed/10 min. for Experiment A and 258 nmoles HCHO formed/10 min. for Experiment B.

electron donor, no activity was detected with the three components alone (Expt. A-1); however, low but detectable NADH-dependent activity was consistently observed in the presence of cytochrome b_5, b_5 reductase and the three components (Expt. A-3). When both NADPH and NADH were present, there was no stimulation by NADH with the three components (Expt. A-1), but the inhibition caused by cytochrome b_5 could be completely reversed by cytochrome b_5 reductase (Expt. A-3), presumably because NADH and cytochrome b_5

reductase can: a) keep cytochrome b_5 reduced, and b) prevent cytochrome b_5 from interacting with NADPH-cytochrome c reductase.

The synergistic effect of NADH on the NADPH-dependent N-demethylation of benzphetamine in microsomal suspensions (Expt. B; Correia and Mannering, 1973b) may be explained in the following way. Since cytochrome b_5 at a ratio of b_5 to P-450 normally found in microsomes inhibits the NADPH-dependent reaction, the rate normally obtained with liver microsomes is probably an inhibited rate when NADPH is used as the sole electron donor. This inhibition is due to the competition between cytochrome b_5 and P-450 for NADPH-cytochrome c reductase. The addition of NADH could have two effects: a) the reversal of the inhibition caused by cytochrome b_5 and b) a low rate of N-demethylation supported by NADH via cytochrome b_5 and cytochrome b_5 reductase. Thus, in the presence of cytochrome b_5, the addition of NADH causes a synergistic effect on NADPH-dependent benzphetamine oxidation (Expt. A-2 and A-3). In contrast, no NADH synergistic effect is observed when cytochrome b_5 is absent (Expt. A-1). Microsomal metabolism (Expt. B) would be analogous to Expt. A-3; i.e., the sum of the rates with NADPH alone and NADH alone is less than the rate with both NADPH and NADH together.

NADPH-DEPENDENT 3,4-BENZPYRENE HYDROXYLATION

A comparison of the rate of 3,4-benzpyrene hydroxylation obtained in intact microsomes (from 3-MC-treated rats) and the reconstituted system is shown in Table IV. Again, the activity obtained with the reconstituted system which was free of cytochrome b_5 was at least as good as that in liver microsomes, as judged by specific activity and turnover number. These results indicate that cytochrome b_5 is not an obligatory component of the NADPH-dependent 3,4-benzpyrene hydroxylation system.

When cytochrome b_5 was added to the reaction mixture containing cytochrome P-448, NADPH-cytochrome c reductase and lipid, the NADPH-dependent hydroxylation of 3,4-benzpyrene was not affected until the ratio of cytochrome b_5 to P-448 was greater than 0.5 (Figure 3). However, since in microsomes from 3-MC-treated rats the ratio of cytochrome b_5 to P-448 is approximately 0.33, cytochrome b_5 probably does not inhibit the microsomal hydroxylation of 3,4-benzpyrene. The lack of an effect of cytochrome b_5 on 3,4-benzpyrene hydroxylation at the ratio of cytochrome b_5 to P-448 normally found in microsomes is not a unique characteristic of cytochrome P-448, since at the same ratio of cytochrome b_5 to P-450 (obtained from PB-treated rats) the cytochrome P-450-mediated 3,4-benzpyrene hydroxylation was also not inhibited by cytochrome b_5.

If high concentrations of cytochrome b_5 were added to the

TABLE IV

3,4-Benzpyrene Hydroxylation by Liver Microsomes and Reconstituted
System from 3-MC-Treated Rats

	8-Hydroxy-3,4-benzpyrene formed	
Conditions	nmoles/min/mg protein	nmoles/min/nmole P-448
Microsomes	2.5-3.5	1.5-2.5
Reconstituted system	6-13	2.5-5.0

The values in this table represent the range of activity observed in our
laboratory from numerous experiments and many different preparations.
Male, Long-Evans rats (50-60 g) pretreated with 3-MC were used in these
studies. In the reconstituted system, protein concentration was the sum
of the amounts of cytochrome P-448 and NADPH-cytochrome c reductase used
in the incubation mixture. 3,4-Benzpyrene hydroxylation was assayed
as previously described (Lu et al., 1972).

reconstituted system, the NADPH-mediated hydroxylation of 3,4-
benzpyrene was inhibited (Figure 3). This inhibition could be reversed by increasing the concentration of NADPH-cytochrome c reductase or NADH-cytochrome b_5 reductase in the reaction mixture
while keeping the ratio of cytochrome b_5 to P-448 constant. These
results are consistent with the idea that the mechanism of cytochrome b_5 inhibition is competition between cytochrome b_5 and
cytochrome P-448 for NADPH-cytochrome c reductase.

Table V (Experiment A) shows that at a ratio of cytochrome b_5
to P-448 normally found in microsomes from 3-MC-treated rats
(i.e., 0.33), the addition of cytochrome b_5 and b_5 reductase to
the reconstituted system containing the three components had little
effect on the NADPH-dependent reaction. When NADH was the sole
electron donor, significant activity was observed only in the
presence of all five components. In the presence of both NADH and
NADPH and all five components, hydroxylation activity was

TABLE V

Effect of Cytochrome b_5 and b_5 Reductase on 3,4-Benzpyrene Hydroxylation in the Presence of NADPH, or NADH, or NADPH and NADH

		% Activity		
		with	with	with NADPH
Expt.	Conditions	NADPH	NADH	and NADH
A	Cytochrome P-448 + cytochrome c reductase + lipid	100	5	101
	Cytochrome P-448 + cytochrome c reductase + lipid + cytochrome b_5	107	-	-
	Cytochrome P-448 + cytochrome c reductase + lipid + cytochrome b_5 + b_5 reductase	109	24	128
B	Microsomes from 3-MC-treated rats	100	19	126

The reaction mixture of Experiment A contained cytochrome P-448 (0.1 nmole; 0.009 mg protein), NADPH-cytochrome c reductase (31 units; 0.007 mg protein), lipid (0.1 mg), cytochrome b_5 (0.033 nmole; 0.0016 mg protein) and b_5 reductase (1 unit; 0.015 mg protein) as indicated in a final volume of 1.0 ml. The ratio of cytochrome b_5 to P-448 was 0.33. In Experiment B, microsomes from 3-MC-treated rats (0.056 nmole cytochrome P-448; 0.040 mg protein) were used. The concentrations of NADPH and NADH were 0.4 mM and 1 mM, respectively. The 100% activities were 1.27 nmoles/5 min for Experiment A and 0.46 nmole/5 min for Experiment B.

approximately equal to the sum of the rates with NADPH and NADH alone. Similarly, the rate of 3,4-benzpyrene hydroxylation with 3-MC microsomes measured in the presence of both NADPH and NADH was about equal to the sum of the rates measured with either NADPH or NADH alone (Table V, Experiment B). Thus, the NADH dependent pathway is responsible for the effect of NADH on 3,4-benzpyrene hydroxylation.

NADH-DEPENDENT 3,4-BENZPYRENE HYDROXYLATION

Although cytochrome b_5 is not an obligatory component of the NADPH-dependent 3,4-benzpyrene hydroxylation system, the results shown in Table V indicate that cytochrome b_5 is involved in the NADH-dependent pathway. Further studies have established that the NADH-mediated hydroxylation of 3,4-benzpyrene requires cytochrome b_5, cytochrome b_5 reductase, lipid and cytochrome P-448 (Table VI). Activity was abolished, or greatly decreased, when any one of these components was omitted from the reaction mixture. Because of the contamination of the cytochrome P-448 preparation with a small amount of cytochrome b_5 reductase, a complete dependence on cytochrome b_5 reductase was not observed. Figure 4 shows that the NADH-mediated hydroxylation pathway was dependent on the concentration of cytochrome b_5 when the other components were present in excess. Maximal activity occurred at a ratio of cytochrome b_5 to P-448 of 1. In addition, this experiment also serves as further evidence that our cytochrome P-448 preparation contains no cytochrome b_5.

The cyanide-sensitive factor--an essential component of the NADH-dependent fatty acid desaturase pathway in liver microsomes--does not appear to be involved in the NADH-dependent hydroxylation of 3,4-benzpyrene. Figure 5 shows that cyanide at concentrations known to inhibit the desaturase system (Oshino et al., 1966) did not inhibit this NADH-mediated pathway in either 3-MC microsomes or in the reconstituted system.

OTHER NADPH-AND NADH-DEPENDENT REACTIONS

In view of the different effects of cytochrome b_5 on the NADPH-dependent metabolism of benzphetamine and 3,4-benzpyrene (i.e., at the ratio of b_5 to P-450 or P-448 normally found in microsomes, cytochrome b_5 inhibits the former but has no effect on the latter), the metabolism of several other substrates was also studied in the absence and in the presence of cytochrome b_5 with the reconstituted system. When cytochrome b_5 was added to the reconstituted system containing cytochrome P-450, NADPH-cytochrome c reductase and lipid, different effects were observed on the NADPH-dependent hydroxylation of testosterone at three different positions. As shown in Figure 6, hydroxylation at the 16α-position was stimulated by cytochrome b_5; maximal stimulation occurred at the ratio of cytochrome b_5 to P-450 normally found in microsomes. In contrast, hydroxylation at the 6β-position was not affected, whereas 7α-hydroxylation was only very slightly stimulated.

TABLE VI

Required Components for NADH-Dependent, 3,4-Benzpyrene Hydroxylation by the Reconstituted System

Conditions	% Activity
Complete (lipid + P-448 + b_5 + b_5 reductase)	100
$-b_5$	1
-P-448	0
$-b_5$ reductase	21
-lipid	1

The reaction mixture, in a final volume of 1.0 ml, contained the following microsomal components as indicated: cytochrome P-448 (0.31 nmole, 0.034 mg), cytochrome b_5 (0.29 nmole, 0.017 mg), NADH-cytochrome b_5 reductase (2 units, 0.030 mg), and lipid (0.1 mg).   One-hundred percent activity was 0.61 nmole per 5 min. NADH concentration was 0.2 µmole.

In collaboration with Drs. D. Jerina and H. Selander of the National Institutes of Health, we have found that the NADPH- and cytochrome P-450-mediated hydroxylation of chlorobenzene at the ortho, meta and para positions was stimulated approximately two-fold by cytochrome b_5; maximal stimulation occurred at a ratio of cytochrome b_5 to P-450 normally found in microsomes (i.e., 0.25 and 0.5). This stimulation was observed in the absence of NADH and NADH-cytochrome b_5 reductase. On the other hand, the cytochrome P-448-mediated NADPH-dependent hydroxylation of chlorobenzene was not affected by cytochrome b_5 under exactly the same assay conditions. This result is quite different from the effect of cytochrome b_5 on 3,4-benzpyrene hydroxylation. With 3,4-benzpyrene, the effect of cytochrome b_5 is the same whether the reaction is mediated by cytochrome P-450 or cytochrome P-448.

In collaboration with Dr. E. Hrycay (University of Texas Southwestern Medical School in Dallas), we have also studied the effect of cytochrome b_5 on the NADPH- and NADH-peroxidase activities using the reconstituted system. Our results indicate that cumene hydroperoxide-dependent NADH peroxidation required cytochrome b_5, cytochrome b_5 reductase, lipid, cytochrome P-450 or

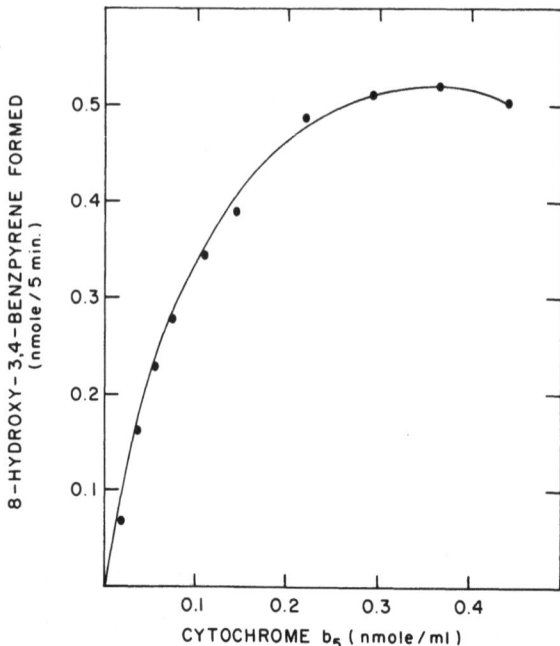

Fig. 4. Effect of cytochrome b_5 concentrations on NADH-dependent 3,4-benzpyrene hydroxylation. Assay conditions were similar to those described in Table VI.

P-448. Thus, like the NADH-dependent hydroxylation of 3,4-benz-pyrene, cytochrome b_5 is an obligatory component of the NADH-peroxidase system. In contrast, cytochrome b_5 did not appear to be required in cumene hydroperoxide-dependent NADPH oxidation. The NADPH-peroxidase system required cytochrome P-450 or P-448, NADPH-cytochrome c reductase and lipid.

CONCLUDING REMARKS

Recent studies from several laboratories have suggested that cytochrome b_5 serves as an essential electron carrier in liver microsomal NADPH-dependent hydroxylation (Hildebrandt and Estabrook, 1971; Correia and Mannering, 1973a, 1973b). The results presented in this paper indicate that the involvement of cytochrome b_5 in the liver microsomal hydroxylation system is very complex. It appears that one cannot draw a generalized conclusion on the involvement of cytochrome b_5 unless one specifies

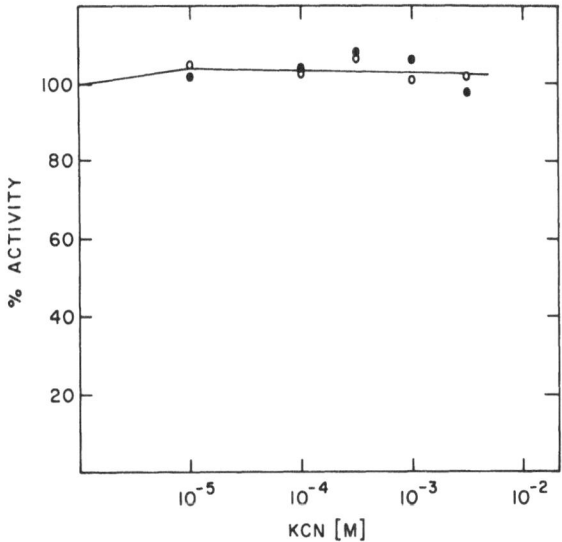

Fig. 5. Effect of cyanide on NADH-dependent 3,4-benzpyrene hydroxylation in the microsomal (solid circle) and reconstituted (open circle) systems. Microsomes from 3-MC-treated rats (0.05 mg microsomal protein) had a control activity (100%) of 0.1 nmole per 5 min. The 100% activity of the reconstituted system was 0.54 nmole per 5 min.

the particular substrate and reaction studied. The role of cytochrome b_5 may also depend on the tissue, sex, age, strain, species and the various pretreatments of animals.

The results of our studies with the reconstituted hydroxylation system are generally in good agreement in many respects with the results of studies by Sasame and coworkers (1973, 1973b, 1974a, 1974b) on the role of cytochrome b_5 in NADPH- and NADH-dependent hydroxylation in intact microsomes. Based upon our studies, several comments can be made:

a) It is clear that cytochrome b_5 is an obligatory component of the NADH-dependent hydroxylation pathway. This conclusion is supported by the findings of Sasame et al., (1973a) that the reduction of cytochrome P-450 by NADH in liver microsomes is blocked by an antibody against cytochrome b_5. Several recent studies also indicate a role for cytochrome b_5 in this pathway (Mannering et al., 1974; Hrycay, 1974). The results of our studies indicate that 3,4-benzpyrene can undergo hydroxylation by an NADH-dependent pathway that requires both cytochrome b_5 and cytochrome P-448.

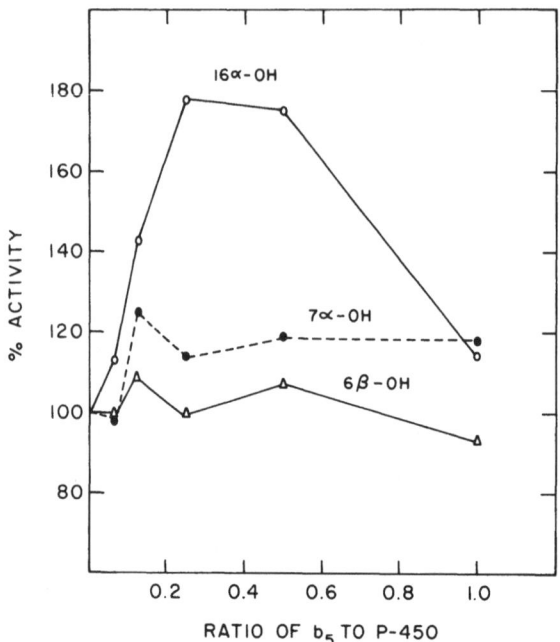

Fig. 6.   Effect of cytochrome b_5 on NADPH-dependent hydroxylation of testosterone at the 6β, 7α and 16α positions by the reconstituted system.   The reaction mixture contained cytochrome P-450 (0.25 nmole; 0.028 mg protein), NADPH-cytochrome c reductase (250 units; 0.05 mg protein), lipid (0.1 mg) and variable amounts of cytochrome b_5 in a final volume of 1.0 ml.   NADPH was the sole electron donor in this experiment.   The 100% activities (i.e., in the absence of cytochrome b_5, for 6β-, 7α- and 16α-hydroxylation were 0.35, 0.83 and 1.04 nmoles per 7.5 min, respectively.   Testosterone hydroxylation was assayed as previously described (Lu et al., 1972).

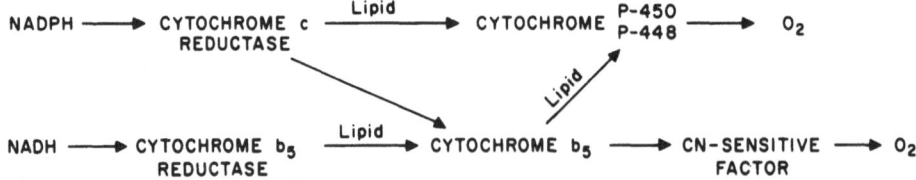

Fig. 7.   Proposed scheme for liver microsomal electron transport systems.

b) The second electron from NADPH is probably not transferred to cytochrome P-450 via b_5, since NADPH-dependent hydroxylation of benzphetamine and 3,4-benzpyrene can occur in the absence of cytochrome b_5. Based on immunochemical studies, Mannering et al., (1974) have reached a similar conclusion.

c) The addition of cytochrome b_5 to the reconstituted, NADPH-mediated hydroxylation system containing cytochrome P-450 or P-448, NADPH-cytochrome c reductase and lipid at a ratio of cytochrome b_5 to P-450 or P-448 normally found in microsomes can have three different effects, depending on the substrate studied: 1) it inhibits benzphetamine N-demethylation; 2) it has little or no effect on the hydroxylation of 3,4-benzpyrene or the hydroxylation of testosterone at the 6β and 7α positions; and 3) it stimulates the ortho-, meta-, and para-hydroxylation of chlorobenzene and the hydroxylation of testosterone at the 16α position.

d) Cytochrome P-450 or P-448, NADPH-cytochrome c reductase and lipid are obligatory components of the NADPH-dependent hydroxylation system; omission of any one of these components results in a total, or substantial, loss of activity. In those NADPH-dependent reactions in which cytochrome b_5 is involved, cytochrome b_5 may be considered a stimulatory but not an obligatory component, since, unlike the other three components, the omission of cytochrome b_5 only results in partial loss of activity. The mechanism of the stimulatory effect of cytochrome b_5 on the hydroxylation of chlorobenzene and on the 16α-hydroxylation of testosterone is unknown.

(e) Cytochrome b_5 plays a central role in the synergistic effect of NADH on NADPH-dependent hydroxylation reactions. The two microsomal electron transport systems are suggested to interact with each other as shown in Figure 7.

FOOTNOTE

*   The abbreviations used are: PB, phenobarbital; 3-MC, 3-methylcholanthrene.

ACKNOWLEDGMENTS

We thank Mrs. MaryAnn Augustin for her assistance in preparing the manuscript.

REFERENCES

Cohen, B. S., and Estabrook, R. W., 1971a. Microsomal electron transport reactions. I. Interaction of reduced triphosphopyridine nucleotide during the oxidative demethylation of aminopyrine and cytochrome b_5 reduction. Arch. Biochem. Biophys. 143: 37-45.

Cohen, B. S., and Estabrook, R. W., 1971b. Microsomal electron transport reactions. II. The use of reduced triphospho-pyridine nucleotide for the oxidative N-demethylation of aminopyrine and other drug substrates. Arch. Biochem. Biophys. 143: 46-53.

Cohen, B. S., and Estabrook, R. W., 1971c. Microsomal electron transport reactions. III. Cooperative interactions between reduced disphosphopyridine nucleotide and reduced triphos-phopyridine nucleotide linked reactions. Arch. Biochem. Biophys. 143: 54-65.

Correia, M. A. and Mannering, G. J., 1973a. Reduced disphospho-pyridine nucleotide-dependent mixed-function oxidase system of hepatic microsomes. I. Effects of activation and in-hibition of the fatty acyl coenzyme A desaturation system. Mol. Pharmacol. 9: 455-469.

Correia, M. A., and Mannering, G. J., 1973b. Reduced disphospho-pyridine nucleotide synergism of the reduced triphospho-pyridine nucleotide-dependent mixed-function oxidase system of hepatic microsomes. II. Role of the Type I drug-binding site of cytochrome P-450. Mol. Pharmacol. 9: 470-485.

Hildebrandt, A., and Estabrook, R. W. 1971. Evidence for the participation of cytochrome b_5 in hepatic microsomal mixed-function oxidation reactions. Arch. Biochem. Biophys. 143: 66-79.

Hrycay, E. G. 1974. Cytochrome b_5 as an electron carrier in hepatic microsomal NADH-supported cytochrome P-450-dependent activities. Fed. Proc. 33: 1288.

Iyanagi, T., and Mason, H. W. 1973. Some properties of hepatic reduced nicotinamide adenine dinucleotide phosphate-cyto-chrome c reductase. Biochemistry 12: 2297-2308.

Levin, W., Lu, A. Y. H., Ryan, D., West, S., Kuntzman, R. and Conney, A. H. 1972. Partial purification and properties of cytochrome P-450 and P-448 from rat liver microsomes. Arch. Biochem. Biophys. 153: 543-553.

Levin, W., Ryan, D., West, S., and Lu, A. Y. H. 1974. Preparation of partially purified, lipid-depleted cytochrome P-450 and reduced nicotinamide adenine dinucleotide phosphate-cytochrome c reductase from rat liver microsomes. J. Biol. Chem. 249: 1747-1754.

Lu, A. Y. H., Junk, K. W., and Coon, M. J. 1969. Resolution of the cytochrome P-450 containing ω-hydroxylation system of liver microsomes into its three components. J. Biol. Chem. 244: 3714-3721.

Lu, A. Y. H., Kuntzman, R., West, S., Jacobson, M. and Conney, A. H. 1972. Reconstituted liver microsomal enzyme system that hydroxylates drugs, other foreign compounds, and endogenous substrates. II. Role of the cytochrome P-450 and P-448 fractions in drug and steroid hydroxylation. J. Biol. Chem. 247: 1727-1734.

Lu, A. Y. H., West, S. B., Vore, M., Ryan, D., and Levin, W. 1974. Role of cytochrome b$_5$ in hydroxylation by a reconstituted cytochrome P-450-containing system. J. Biol. Chem. 249: (in press).

Lu, A. Y. H., and Levin, W. 1972. Partial purification of cytochromes P-450 and P-448 from rat liver microsomes. Biochem. Biophys. Res. Commun. 46: 1334-1339.

Mannering, G. J.,Kuwahara, S. I. and Omura, T. 1974. Immunochemical evidence for the participation of cytochrome b$_5$ in the NADH synergism of the NADPH-dependent mono-oxidase system of hepatic microsomes. Biochem. Biophys. Res. Commun. 57: 476-481.

Mihara, K., and Sato, R. 1972. Partial purification of NADH-cytochrome b$_5$ reductase from rabbit liver microsomes with detergents and its properties. J. Biochem. 71: 725-735.

Modirzadeh, J., and Kamin, H. 1965. Reduction of microsomal cytochromes by pyridine nucleotides. Biochim. Biophys. Acta 99: 205-226.

Oshino, N., Imai, Y., and Sato, R. 1966. Electron-transfer mechanism associated with fatty acid desaturation catalyzed by liver microsomes. Biochim. Biophys. Acta 128: 13-28.

Phillips, A. H. and Langdon, R. G. 1962. Hepatic triphosphopyridine nucleotide-cytochrome c reductase: isolation, characterization and kinetic studies. J. Biol. Chem. 237: 2652-2660.

Sasame, H. A., Thorgeirsson, S. S., Mitchell, J. R. and Gillette, J. R. 1973a. NADPH and NADH electron transport systems in rat liver microsomes. Pharmacologist 15: 170.

Sasame, H. A., Mitchell, J. R., Thorgeirsson, S. S. and Gillette, J. R. 1973b. Relationship between NADH and NADPH oxidation during drug metabolism. Drug Metabolism and Disposition 1: 150-155.

Sasame, H. A., Thorgeirsson, S. S., Mitchell, J. R. and Gillette, J. R. 1974a. The possible involvement of cytochrome b$_5$ in the oxidation of lauric acid by microsomes from kidney cortex and liver of rats. Life Sciences 14: 35-46.

Sasame, H. S., Thorgeirsson, S. S., Menard, R. H., Hinson, J. A., Mitchell, J. R. and Gillette, J. R. 1974b. A role of cytochrome b$_5$ in both NADH and NADPH-mediated cytochrome P-450 enzymatic reactions in mammalian tissues. Fed. Proc. 133: 1437.

Spatz, L., and Strittmatter, P. 1971. A form of cytochrome b$_5$ that contains an additional hydrophobic sequence of 40 amino acid residues. Proc. Nat. Acad. Sci. USA 68: 1042-1046.

Spatz, L., and Strittmatter, P. 1973. A form of reduced nicotinamide adenine dinucleotide-cytochrome b$_5$ reductase containing both the catalytic site and an additional hydrophobic membrane-binding segment. <u>J. Biol. Chem</u>. 248: 793-799.

Strobel, H. W., Lu, A. Y. H., Heidema, J. and Coon. M. J. 1970. Phosphatidylcholine requirement in the enzymatic reduction of hemoprotein P-450 and in fatty acid, hydrocarbon and drug hydroxylation. <u>J. Biol. Chem</u>. 245: 4851-4854.

THE ROLE OF CYTOCHROME b_5 IN MIXED FUNCTION OXIDATIONS: EFFECT OF MICROSOMAL BINDING OF THE HEMOPROTEIN ON HEPATIC N-DEMETHYLATIONS

Dominick L. Cinti and Juris Ozols

Departments of Pharmacology and Biochemistry
University of Connecticut
Health Center
Farmington, Connecticut 06032

SUMMARY

Incubation of rat cytochrome b_5 (D-b_5) with rat liver microsomes resulted in specific binding of the hemoprotein. The bound hemoprotein was rapidly reduced by NADH. The NADH cytochrome c reductase activity in these preparations increased in proportion to the amount of cytochrome. In contrast to D-b_5, which inhibited N-demethylation and the NADH synergism, the binding of cytochrome b_5 preparations, reconstituted from heme and apocytochrome b_5 had no effect on either the NADPH-dependent N-demethylation of aminopyrine or ethylmorphine or the NADH synergism observed with rat liver microsomes. In addition, manganese protoporphyrin-apocytochrome complex, when bound to microsomes in amounts equivalent to D-b_5, showed no effect on N-demethylation activity. These results suggest that homogeneous cytochrome b_5 contains contaminating amounts of tightly bound detergent which presumably is removed during the extraction of the heme from the apocytochrome.

INTRODUCTION

Cytochrome b_5 is one of two hemoproteins found in the endoplasmic reticulum of liver and other mammalian tissues, and, until recently, its physiological role has remained obscure. Recently, Oshino et al., (1971) reported that rat liver microsomal cytochrome b_5 participated as an electron carrier in the fatty acid desaturation reaction, transferring reducing equivalents

from either NADH or NADPH to a cyanide-sensitive factor, presumably
the terminal component of the electron-transport chain (Oshino and
Sato, 1971, 1972; Oshino, 1972). Confirmation of the involvement
of cytochrome b_5 in fatty acid desaturation came from the laboratories of Strittmatter (Strittmatter et al., 1972) and Omura
(Oshino and Omura, 1973) using liver microsomes containing
exogenously bound cytochrome b_5 and the antibody prepared against
the hemoprotein, respectively. Cytochrome b_5 and its NADH-
dependent reductase have also been implicated in catalyzing the
reduction of methemoglobin in red blood cells (Hultquist and
Passon, 1971); moreover the hemoprotein has been reported to be
involved in microsomal lipid peroxidation (Bidlack, Okita and
Hochstein, 1973).

Another area of paramount interest and still unsettled is the
role of cytochrome b_5 in hepatic microsomal mixed function oxidation reactions. The overall cytochrome P-450 catalyzed mixed
function oxidation reaction requires two electrons for the
coupled reduction of oxygen and hydroxylation of substrate
(Estabrook et al., 1969). Cytochrome P-450 has been shown to be
a one-electron carrier by anaerobic titration of hepatic microsomes with NADPH (Ullrich et al., 1969); however, a recent report
by Ballou et al., (1974) indicates that partially purified liver
microsomal cytochrome P-450 accepts two electrons from dithionite
under anaerobic conditions. On the basis of the well-documented
synergistic effect of NADH on NADPH-dependent drug hydroxylation
reactions (Conney et al., 1957; Nilsson and Johnson, 1963; Cohen
and Estabrook, 1971, 1971a) and on the evidence that partial
reoxidation of reduced cytochrome b_5 occurs in the presence of
excess NADH, NADPH and ethylmorphine, Hildebrandt and Estabrook
(1971) postulated that the second reducing equivalent is transferred from NADH to the oxygenated cytochrome P-450-substrate
ternary complex via cytochrome b_5. Recent reports have provided
evidence in both supporting and opposing the hypothesis. For
example, Correia and Mannering (1973) observed a significant increase in the NADH synergism of ethylmorphine N-demethylation in
the presence of cyanide which was used to inhibit the desaturase
pathway. More recently, Mannering et al., (1974) reported that
an antibody to trypsin-solubilized cytochrome b_5 inhibited the
NADH-stimulated ethylmorphine N-demethylation, but not the
NADPH-dependent demethylation, suggesting that the second electron
can arise from NADH via cytochrome b_5. These results strongly
suggest that the second electron reaches the cytochrome P-450
system via cytochrome b_5. However, Ichikawa and Loehr (1972)
reported the reduction of cytochrome P-450 by NADH in submicro-
somal particles which contained NADH ferricyanide (cytochrome b_5)
reductase but were devoid of cytochrome b_5. Jansson and
Schenkman (1973) in a recent communication, reported an inhibition
of aminopyrine dealkylation following addition of detergent-

solubilized cytochrome b_5 to liver microsomes, and the addition of
NADH to this NADPH-dependent dealkylation did not reverse the
inhibition. Thus, the precise role of native cytochrome b_5 in
microsomal hydroxylation reactions has not been established.

We undertook the isolation from rat liver microsomes of
homogeneous cytochrome b_5 with the membranous segment intact,
since spectrally and biologically active cytochrome b_5 preparations often contain contaminating amounts of tightly bound
membranous proteins, lipids and detergents (Ozols, 1974), which
may introduce ambiguities in determining the precise role of
cytochrome b_5 in the above reactions. Moreover, we have investigated the binding of this hemoprotein and its manganese derivative
to the microsomes, as well as characterized its effect on NADPH-
dependent microsomal N-demethylations.

METHODS

Normal male Sprague-Dawley rats (225-275g) were fed and
watered ad libitum until sacrifice. Animals were sacrificed,
livers removed and perfused as previously described (Cinti and
Schenkman, 1972). The microsomal fraction used for isolation of
cytochrome b_5 was obtained by centrifugation (25,000 x g) of the
post-mitochondrial supernatant after addition of $CaCl_2$ (Cinti et
al., 1972). Microsomes used for assaying drug metabolizing enzymes were prepared by the Ca^{2+}-dependent sedimentation procedure
or by the conventional differential ultracentrifugation method as
previously described (Schenkman and Cinti, 1972). Either method
of isolation gave microsomal preparations with identical drug
metabolizing activities.

The solubilization of microsomes and the subsequent isolation
of cytochrome b_5 were accomplished as described by Ozols (1974).
The binding of homogeneous cytochrome b_5 to rat liver microsomes
was performed as described by Cinti and Ozols (1974).

Manganese derivatives of porphyrins were obtained as
described by Ozols and Strittmatter (1964). A manganese mesoporphyrin preparation with properties identical to the synthetic
compound, and bovine cytochrome b_5 obtained by trypsin treatment
of microsomes were generously supplied by Dr. Strittmatter's
laboratory. Mn-apocytochrome complex was prepared by the procedure of Rogers and Strittmatter (1974) as modified by Cinti
and Ozols (1974). The binding of Mn cytochrome b_5 to rat liver
microsomes employed the same procedure for binding homogeneous
cytochrome b_5 to microsomes. Both bound Mn cytochrome b_5 and
apocytochrome b_5 were measured spectrally by the method recently
described by Rogers and Strittmatter (1974).

RESULTS

Isolation of Cytochrome b$_5$ from Rat Liver Microsomes. The
cytochrome b$_5$ preparations migrated as a single protein band when
subjected to acrylamide gel electrophoresis in sodium dodecyl
sulfate. The absorption spectra of oxidized rat liver cytochrome
b$_5$ in 10 mM Tris-acetate buffer (pH 8.1) consisted of a 413 nm
absorption maximum with a shift to 423 nm upon reduction with
dithionite including absorption bands in the α and β region at 526
nm and 557 nm, respectively. No changes in absorbance occurred
when either oxidized or reduced hemoprotein preparations were
treated with carbon monoxide indicating the complete absence of
cytochrome P-450 and its denatured form (P-420). The cytochrome
b$_5$ was not reduced by NADH (1mM), indicating that the preparation
was free of the flavoprotein b$_5$ reductase. In addition, there was
no demonstrable NADH or NADPH cytochrome c reductase activity.

Binding of Cytochrome b$_5$ to Rat Liver Microsomes. As seen
in Figure 1, incubating rat liver microsomes for 20 min with
increasing concentrations of detergent-solubilized homogeneous
cytochrome b$_5$ (D-b$_5$) resulted in a slow but significant binding
at 20°, plateauing in the presence of 80 μM D-b$_5$. Incubations at
37°C resulted in a greater than two-fold increase in bound D-b$_5$
(7 nmoles/mg protein vs 2.8 nmoles/mg protein) in the presence of
80 μM D-b$_5$; saturation of binding was not observed at 37° even in
the presence of 100 μM D-b$_5$. The amount of D-b$_5$ bound to the microsomal membrane at 37° was 13-fold higher than the control level.

When microsomes were incubated with 60 μM D-b$_5$ at various
time intervals at 37°C, a rapid binding of cytochrome b$_5$ occurred
within 15 min and reached a maximum by 22 min (Fig. 2); 0.26 nmoles
of D-b$_5$ were bound per min per mg microsomal protein. At 20°C,
the rate of binding was markedly less, increasing at a rate of 0.1
nmoles of D-b$_5$ per min. That the D-b$_5$ microsomal interaction was
not a result of non-specific protein adsorption to the microsomes
was suggested by the failure of washing with salt solutions
(0.1 - 0.8 M NaCl) and sonication of the preparations to remove
the bound cytochrome. In addition, tryspin-solubilized cytochrome
b$_5$, in which the membranous segment of the molecule is removed
during the proteolytic isolation (Strittmatter et al., 1972; Ito
and Sato, 1968), did not bind to rat liver microsomes.

Enzymatic Properties of Bound Detergent Cytochrome b$_5$.
Homogeneous preparations of cytochrome b$_5$ were not reduced by the
addition of NADH, whereas upon binding of the hemoprotein to
microsomes, complete reduction by NADH was obtained. These results
imply that the bound cytochrome b$_5$ has attained a native orien-
tation in the microsomal membrane. Furthermore, with increasing
amounts of bound D-b$_5$, there was a proportional increase in NADH

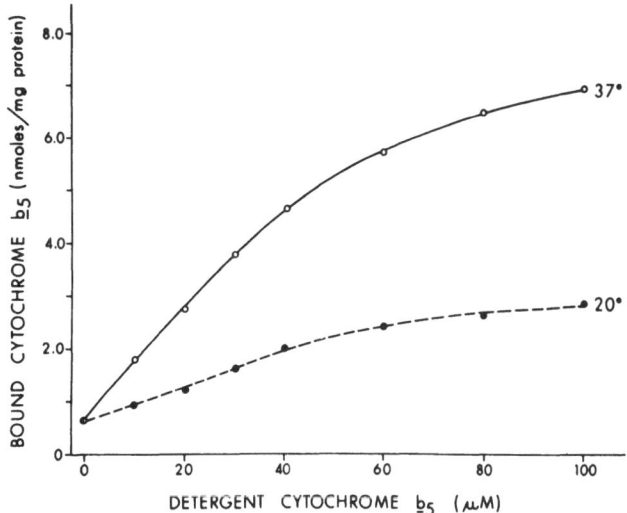

Fig. 1. The effect of temperature on the extent of cytochrome b₅ binding to rat liver microsomes. Microsomes were incubated with various amounts of D-b₅ in a 0.1 M Tris-Cl buffer pH 8.0 for 20 minutes. The concentration of endogenous cytochrome b₅ was 0.62 nmoles/mg of microsomal protein.

cytochrome c reductase activity (Table 1), suggesting that the D-b₅ molecules are functionally active. In this reaction, reducing equivalents from NADH are transferred by the flavoprotein reductase to cytochrome b_5, which in turn donates electrons to cytochrome c (Strittmatter and Velick, 1956). Although fatty acyl CoA desaturase activity was not measured in these experiments, Strittmatter et al.,(1972) reported that bound cytochrome b₅ is an effective electron donor to the cyanide-sensitive factor during conversion of stearyl CoA to oleyl CoA, providing further evidence that exogenously bound cytochrome b₅ is functionally active.

Effect of Bound Cytochrome b_5 on N-demethylation Activities. The N-demethylation of ethylmorphine by normal rat liver microsomes prior to incubation with D-b₅ was approximately 7.0 nmoles/min/mg microsomal protein (Fig. 3). In the presence of both pyridine nucleotides, the synergistic effect of NADH was observed, resulting in a 60% stimulation of ethylmorphine demethylase activity. However, when microsomes were preincubated with 60 μM D-b₅ for various time periods (from 5 min to 60 min) and subsequently used for determining demethylase activity, an unexpected decrease in the demethylation of ethylmorphine occurred with increasing amounts of bound D-b_5 (Fig. 3); the stimulatory effect of NADH was also

TABLE I

Enzymatic Properties of Bound Cytochrome b_5

INCUBATION TIME min	BOUND CYTOCHROME b_5 nmoles/mg microsomal protein	NADH-CYTOCHROME c REDUCTASE nmoles/min/mg microsomal protein
0	0.55	430
2	1.05	820
4	1.51	1290
6	2.17	1730
8	2.55	1850
10	3.02	2360

60 μM D-b_5 was incubated with 7 ml of rat liver microsomes (10 mg protein/ml). At the end of each incubation period, 1 ml of microsomes was removed, centrifuged for 30 minutes at 104,000 x g and resuspended in 1 ml of 0.1 M Tris-Cl buffer (pH 7.5).

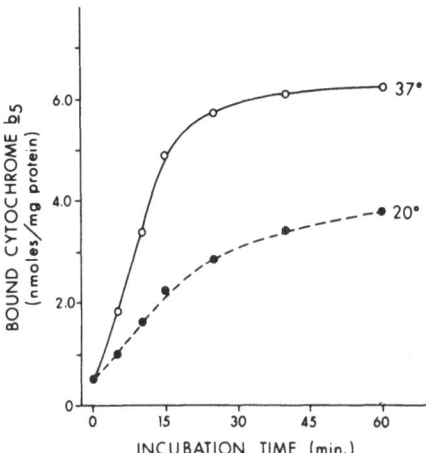

Fig. 2.   Time course of and temperature effect on cytochrome b_5 binding to rat liver microsomes. Microsomes (10 mg/ml) were incubated with 60 µM cytochrome b_5 for the indicated time in a 0.1 M Tris-Cl buffer pH 8.0. The content of endogenous cytochrome b_5 in the microsomal preparation used was 0.55 nmoles/mg protein.

abolished by exogenously bound D-b_5. Identical results were obtained when aminopyrine replaced ethylmorphine. Furthermore, both NADPH cytochrome c reductase and NADPH cytochrome P-450 reductase activities were markedly inhibited (45% and 65%, respectively) when the microsomal cytochrome b_5 concentration was almost doubled (Table 2, Prep. Ia). An explanation of the data could be found in one of several possibilities: 1) that bound D-b_5 is acting as an electron sink channelling reducing equivalents away from the mixed function oxidase system, 2) that bound D-b_5 by its very presence in large amounts (relative to endogenous cytochrome b_5) on the membrane is disrupting the normal sequential flow of electrons through the mixed function oxidase pathway, 3) that incorporation of cytochrome b_5 into the microsomes results in an increase in the membrane protein-lipid ratio -- such alteration of the membrane components could affect the architecture or fluidity of the membrane, which in turn may have an adverse effect on the mixed function oxidase system, 4) that our cytochrome b_5 preparation, although homogeneous, contains trace amounts of tightly bound detergent and 5) the inhibitory effect observed could be attributed to any combination of these possibilities.

Binding of Manganese Derivative of Cytochrome b_5 to Rat Liver Microsomes and The Effect on N-Demethylation. In an attempt to determine the mechanism by which exogenously bound D-b_5 inhibited drug

TABLE II

Effect of Bound Native Cytochrome b_5, Reconstituted Cytochrome b_5, and Manganese Derivative on Microsomal NADPH Cytochromes c and P-450 Reductase Activities

Treatment	Cytochrome b_5 nmoles/mg protein	NADPH Cytochrome c Reductase	NADPH Cytochrome P-450 Reductase
		nmoles/min/mg protein	
Microsomes	0.55	65	6.4
Microsomes + Prep. I_a	0.96	38	2.3
+ Prep. I_b	2.40	21	1.7
Microsomes + Prep. II_a	1.13	59	6.9
+ Prep. II_b	3.08	70	6.0
Microsomes + Prep. III_a	0.82*	62	5.9
+ Prep. III_b	2.15*	66	6.7

Prep. I consists of microsomes containing native cytochrome b_5 in which N-demethylations were inhibited; Prep. II consists of microsomes which had been incubated with the reconstituted cytochrome b_5; Prep. III is the Mn b_5 derivative bound to microsomes.
* Represents the amount of Mn b_5 bound to rat liver microsomes (nmoles/mg protein).

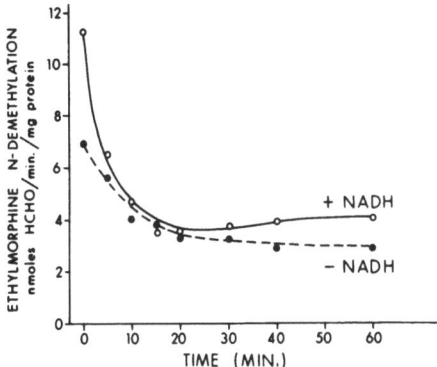

Fig. 3. Effect of increasing amounts of bound D-b₅ on ethylmorphine N-demethylation in the presence and absence of NADH.
Microsomes (10 mg/ml) were pre-incubated with 60 μM D-b₅ for
0, 5, 10, 15, 20, 30, 40 and 60 minutes. At the end of each time
period, the microsomes were diluted with cold Tris-Cl buffer
(pH 8.0) and centrifuged at 104,000 x g for 30 minutes; the microsomes were then suspended to the original incubating volume with
0.25 M sucrose containing 0.1 M Tris-HCl (pH 7.4). Ethylmorphine
demethylation reaction was initiated with the pre-incubated microsomes and run for 8 minutes. Microsomes pre-incubated without
D-b₅ for the same time periods were used as controls. Equimolar
concentrations of NADPH and NADH were used (0.5 mM); similar results were obtained with 1.0 mM of each pyridine nucleotide. The
microsomal content of cytochrome b₅ for the above pre-incubation
times was 0.51, 1.04, 1.65, 2.33, 2.64, 3.01, 3.35 and 3.74 nmoles/
mg protein.

metabolism, manganese analogue of cytochrome b₅ was prepared and
its binding to microsomes examined. The reason for using the
manganese derivative of cytochrome b₅ was two-fold. Firstly, to
delineate the role of heme in the above reaction and secondly, to
determine whether the physical presence of the hemoprotein was
disrupting sequential flow of electrons.

The absolute absorption spectrum of the manganese protoporphyrin complex shown in Fig. 4 (Curve A) consisted of 2 major
peaks at 369 nm and 461 nm with shoulders at 390 nm and 415 nm.
The spectrum of the metalloporphyrin upon reduction with dithionite
(Curve B) resulted in a decrease in the absorption maxima and the
formation of a new peak at 421 nm. When the oxidized derivative
was complexed with apocytochrome b₅, a 4 nm spectral shift to 465
nm and a loss of both shoulders was observed.

The rate and extent of binding of the metalloporphyrin-

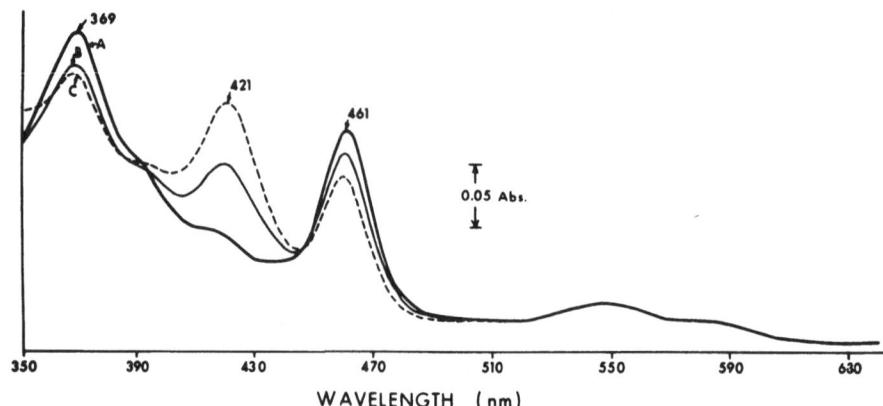

Fig. 4. Absorption spectra of manganese protoporphyrin in 20 mM Tris HCl buffer (pH 7.5) containing 50% ethanol. The concentration of metalloporphyrin was 5 μM. Curve A, oxidized form; Curve B, dithionite-reduced form; Curve C, CO complex of the reduced form. Substitution of Mn mesoporphyrin for the protoporphyrin gave identical spectra.

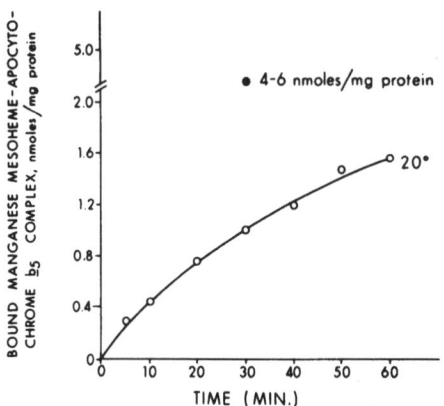

Fig. 5. Time course of binding of manganese porphyrin-apocytochrome b_5 complex to rat liver microsomes at 20°C. The metalloporphyrin-apocytochrome complex (60 μM) was incubated with microsomes (10 mg/ml) as described in Methods. The single solid circle represents the amount of the Mn derivative bound to microsomes following incubation at 37° for 30 min.

apocytochrome b_5 (Mn b_5) complex to rat liver microsomes is shown in Figure 5. About 1 nmole of Mn b_5 was bound to microsomes when 30 nmoles were incubated at 20° for 30 min; as much as 4-6 nmoles/ mg microsomal protein were bound when the temperature was raised to 37° and the concentration increased to 50-60 nmoles Mn b_5 (Fig. 5).

When microsomal preparations containing various amounts of bound Mn b_5 were tested for N-demethylation of either ethylmorphine or aminopyrine, no inhibition of drug metabolism was observed (Fig. 6). Moreover, neither NADPH cytochrome c reductase nor NADPH cytochrome P-450 reductase activity was inhibited in microsomes containing 0.82 nmoles and 2.15 nmoles of Mn b_5/mg microsomal protein (Table 2. Prep. III$_a$ and III$_b$), strongly suggesting that the physical presence of extra bound cytochrome b_5 is not the reason for the observed inhibition of demethylation activity. Results obtained using the Mn protoporphyrin derivative were identical to those obtained with Mn mesoporphyrin derivative.

Since the Mn derivatives of cytochrome b_5 were prepared by removing the iron protoporphyrin from the cytochrome and then reconstituting the apocytochrome with the metalloporphyrin, control experiments using reconstituted cytochrome b_5 prepared from heme and apocytochrome were also performed. Contrary to the native bound cytochrome b_5, the microsomal preparations containing reconstituted cytochrome b_5 did not inhibit ethylmorphine or aminopyrine N-demethylation (Fig. 6) in the absence or presence of NADH even at concentrations of 3 nmoles b_5/mg microsomal protein; as shown in Figure 3 native cytochrome b_5 inhibited demethylation approximately 50% at this concentration. Furthermore, no inhibition of either NADPH cytochrome c reductase or NADPH cytochrome P-450 reductase activity was observed in microsomal preparation containing reconstituted cytochrome b_5 (Table 2, Prep. II).

Strittmatter (1960) had previously reported that microsomal cytochrome b_5 can be completely resolved into an apoprotein which will recombine with 1 mole of heme to yield the original cytochrome b_5 absorption spectrum and a hemoprotein having the same physical and enzymatic characteristics as the native molecule. As observed with native D-b_5, the NADH cytochrome c reductase activity increased in proportion to the amount of reconstituted cytochrome b_5 bound to microsomes (Fig. 7). The NADH cytochrome b_5 reductase activity, determined with ferricyanide as the electron acceptor, did not change with increasing amounts of cytochrome b_5 as expected, since the rate-limiting step in the reduction of microsomal cytochrome b_5 is the transfer of electrons from the flavoprotein reductase to the hemoprotein, Strittmatter et al., (1972). The reconstituted cytochrome b_5 was active in the desaturation of stearyl CoA, as previously observed by Strittmatter et al., (1972).

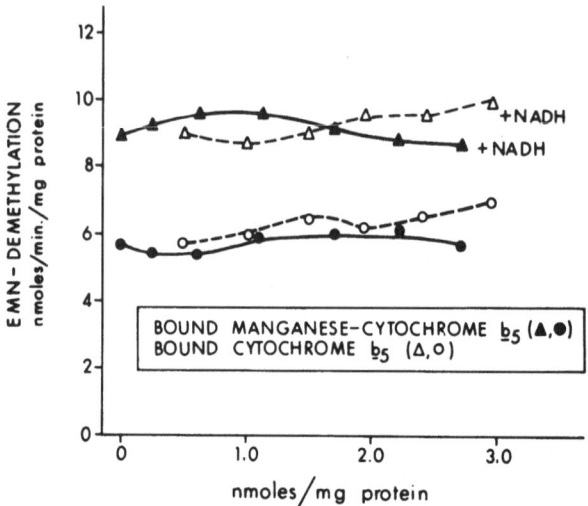

Fig. 6.   Effect of increasing amounts of bound Mn b$_5$ (▲,●)
and reconstituted cytochrome b$_5$ (Δ, 0) on demethylation activity
in the presence and absence of NADH.   The binding of cytochrome
b$_5$ and the manganese derivative to microsomes and the determination
of demethylase activity were carried out as in Figure 3.

DISCUSSION

Cohen and Estabrook (1971, 1971a) and Hildebrandt and
Estabrook (1971) put forward the concept that cytochtome b$_5$ participates in the transfer of the second electron to mixed function
oxidase system during the oxidation of drugs, other xenobiotics and
steroids.   This postulated mechanism is supported by the elegant
studies of Correia and Mannering (1973) who showed that stearyl
CoA decreased the rate of metabolism of ethylmorphine by shunting
electrons from cytochrome b$_5$ to the microsomal fatty acyl CoA desaturase system and that low concentrations of cyanide increased
the rate of drug metabolism by diverting electrons away from the
desaturase system.   More recently, Mannering et al., (1974) reported that anti-cytochrome b$_5$ immunoglobulin inhibited the NADH
synergism of NADPH-dependent ethylmorphine N-demethylation implicating cytochrome b$_5$ participation in the transfer of reducing
equivalents from NADH to the mixed function oxidase system.

Our data are not necessarily in disagreement with the afore-
mentioned evidence.   The inhibition of drug metabolism (N-demethylation of aminopyrine and ethylmorphine) observed with increasing
amounts of bound native cytochrome b$_5$ presumably was not due to

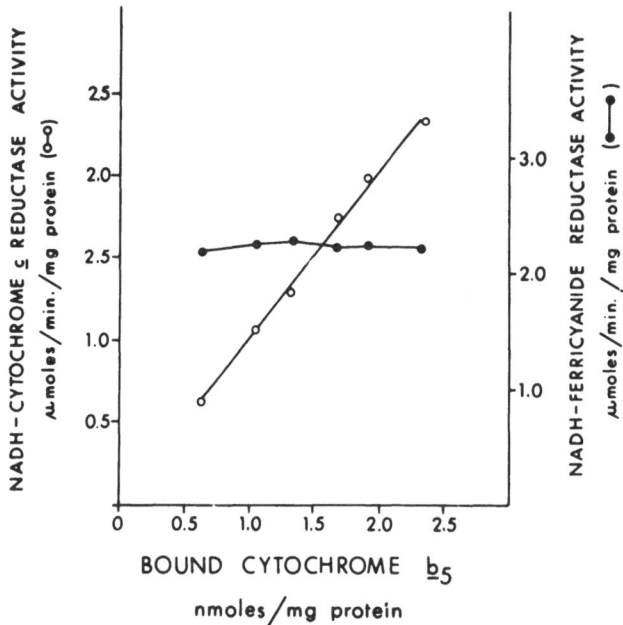

Fig. 7. Effect of reconstituted bound cytochrome b_5 on NADH-cytochrome c reductase and NADH cytochrome b_5 reductase activities. Heme-free apocytochrome was prepared and reconstituted with heme as described by Cinti and Ozols. The preparation was then incubated with 10 mg/ml microsomes in the usual manner. Microsomal preparations used in this experiment contained 0.62 nmoles of endogenous cytochrome b_5 and the endogenous NADH cytochrome c and b_5 reductase activities were 0.6 and 2.2 μmoles/min/ mg protein, respectively.

disruption of electron flow through the mixed function oxidase pathway by the physical presence of D-b_5, since the binding of the Mn porphyrin-apocytochrome complex to microsomes failed to show inhibition of N-demethylation. As much as 6 nmoles of the Mn derivative per mg protein were bound to rat liver microsomes without affecting either NADPH cytochrome c reductase or NADPH cytochrome P-450 reductase activities. The Mn derivative was the analogue of choice since it combined stoichiometrically with apocytochrome b_5 and was not reduced by the cytochrome b5 reductase system, despite its ability to accept electrons from dithionite.

Contrary to the native cytochrome b_5, binding of reconstituted cytochrome b_5 to liver microsomes did not inhibit drug metabolism. The inhibition observed with native cytochrome b_5 could be attributed to 1) the presence of Triton bound to cytochrome b_5 since as little as 0.005% Triton N-101 present in an assay mixture inhibited demethylase activity by 15-20% (Cinti, unpublished observation) and since this detergent was used in the isolation of cytochrome b_5. Jansson and Schenkman (1973) in a recent communication observed an inhibition of aminopyrine demethylase activity following binding of cytochrome b_5 to rat liver microsomes, although they indicate that their cytochrome b_5 preparation was free of detergent. 2) The reconstitution procedure may induce conformational changes not present in the native b_5 molecule; however, this seems unlikely since the reconstituted cytochrome b_5 functions as effectively as native cytochrome b_5 in NADH-dependent reduction of cytochrome c and in stearyl CoA desaturation.

Although Ichikawa and Loehr (1972) reported that electrons from NADH could be transferred to cytochrome P-450 in the absence of cytochrome b_5 this does not exclude the possibility that in the presence of b_5 the preferential route is from NADH to b_5 via the flavoprotein and ultimately to P-450. Also immunological studies reported by Mannering et al., (1974) do not unequivocally establish cytochrome b_5 as the donor for the transfer of the second electron from NADH to the mixed function oxidase since the effect of the cytochrome b_5 antibody on microsomal NADH cytochrome b_5 reductase activity was not reported. Also, the data of Hildebrandt and Estabrook (1971) do not necessarily demonstrate the involvement of cytochrome b_5 in drug oxidations, since changes in the steady-state levels of reduced cytochrome b_5 could be explained by a competition between cytochrome b_5 and the cytochrome P-450 ternary complex for the reducing equivalents from NADH. Hence, when the mixed function oxidase pathway is functioning optimally, i.e., when NADPH, oxygen and substrate are present, reducing equivalents in the form of NADH, which normally flow to cytochrome b_5 and ultimately to the desaturase system, are now diverted to the P-450-dependent pathway; the result is a change in the redox state of cytochrome b_5.

Finally, the fact that increased amounts of bound cytochrome b_5 have no effect on N-demethylase activity does not necessarily mean lack of involvement of cytochrome b_5 in drug metabolism. If sufficient quantities of cytochrome b_5 are already present on the membrane so that the hemoprotein is not rate-limiting, then increasing the amount of cytochrome b_5 bound to the membrane would not stimulate drug oxidations.

ACKNOWLEDGMENTS

The authors wish to express their thanks to Dr. P. Strittmatter and Dr. M. Rogers for their valuable advice.

This study was supported in part by National Institute of Health Grant AM 16678 and American Cancer Society Grant NP-134.

REFERENCES

Ballou, D. P., Veeger, C., van der Hoeven, T. A. and Coon, M. M. 1974. Properties of partially purified liver microsomal cytochrome P-450: acceptance of two electrons during anaerobic titration. FEBS Letters 38:337-340.

Bidlack, W. R., Okita, R. T. and Hochstein, P. 1973. The role of NADPH cytochrome b_5 reductase in microsomal lipid peroxidation. Biochem. Biophys. Res. Commun. 53:459-465.

Cinti, D. L. and Ozols, J. 1974. Binding of homogenous cytochrome b_5 to rat liver microsomes: effect on N-demethylation reactions. Submitted to Mol. Pharmacol.

Cinti, D. L. and Schenkman, J. B. 1972. Hepatic organelle interaction I. Spectral investigation during drug biotransformation. Mol. Pharmacol. 8:327-338.

Cinti, D. L., Moldeus, P. and Schenkman, J. B. 1972. Kinetic parameters of drug-metabolizing enzymes in Ca^{2+}-sedimented microsomes from rat liver. Biochem. Pharmacol. 21:3249-3256.

Cohen, B. S. and Estabrook, R. W. 1971. Microsomal electron transport reactions. I. Interaction of reduced triphosphopyridine nucleotide during the oxidative demethylation of aminopyrine and cytochrome b_5 reduction. Arch. Biochem. Biophys. 143:37-45.

Cohen, B. S. and Estabrook, R. W. 1971a. Microsomal electron transport reactions. III. Cooperative interactions between reduced nucleotide linked reactions. Arch. Biochem. Biophys. 143:54-65.

Conney, A., Brown, R. R., Miller, J. A. and Miller, E. C. 1957. The metabolism of methylated aminoazo dyes. VI. Intracellular distribution and properties of the demethylase system. Cancer Res. 16: 628-633.

Correia, M. A. and Mannering, G. J. 1973. Reduced diphosphopyridine nucleotide synergism of the reduced triphosphopyridine nucleotide-dependent mixed function oxidase system of hepatic microsomes. I. Effects of activation and inhibition of the fatty acyl coenzyme A desaturation system. Mol. Pharmacol. 9:455-469.

Estabrook, R. W., Hildebrandt, A., Remmer, H., Schenkman, J. B., Rosenthal, O. and Cooper, D. Y. 1968. The role of cytochrome P-450 in microsomal mixed function oxidation reactions, pp. 142-177. In B. Hess and H. Staudinger (eds.) Biochemie des Sauerstoffs. Springer, Berlin.

Hildebrandt, A. and Estabrook, R. W. 1971. Evidence for the participation of cytochrome b_5 in hepatic microsomal mixed-function oxidation reaction. Arch. Biochem. Biophys. 143:66-79.

Hultquist, D. E. and Passon, P. G. 1971. Catalysis of methemo-
    globin reduction by erythrocyte cytochrome b_5 and cytochrome
    b_5 reductase. Nature (New Biology) 229:252-254.
Ichikawa, Y. and Loehr, J. S. 1972. NADH-dependent cytochrome
    P-450 oxidase system in submicrosomal particles. Biochem.
    Biophys. Res. Commun. 46:1187-1193.
Ito, A. and Sato, R. 1968. Purification by means of detergent
    and properties of cytochrome b_5 from liver microsomes. J.
    Biol. Chem. 243:4922-4923.
Jansson, I. and Schenkman, J. B. 1973. Evidence against partici-
    pation of cytochrome b_5 in the hepatic microsomal mixed
    function oxidase system. Mol. Pharmacol. 9:840-845.
Mannering, G. J., Kuwahara, S. and Omura, T. 1974. Immuno-
    chemical evidence for the participation of cytochrome b_5
    in the NADH synergism of the NADPH-dependent monooxidase
    system of hepatic microsomes. Biochem. Biophys. Res.
    Commun. 57:476-481.
Nilsson, A. and Johnson, B. C. 1963. Cofactor requirements of the
    0-demethylating liver microsomal enzyme system. Arch.
    Biochem. Biophys. 104:494-498.
Oshino, N. 1972. Dynamic behavior during dietary induction of
    the terminal enzyme (cyanide-sensitive factor) of the stearyl
    CoA desaturation system of rat liver microsomes. Arch.
    Biochem. Biophys. 149:378-387.
Oshino, N. and Omura, T. 1973. Immunochemical evidence for the
    participation of cytochrome b_5 in microsomal stearyl CoA
    desaturation reaction. Arch. Biochem. Biophys. 157:395-404.
Oshino, N. and Sato, R. 1971. Stimulation by phenols of the re-
    oxidation of microsomal bound cytochrome b_5 and its impli-
    cation to fatty acid desaturation. J. Biochem. 69:169-180.
Oshino, N. and Sato, R. 1972. Dietary control of the microsomal
    stearyl CoA desaturation enzyme system in rat liver. Arch.
    Biochem. Biophys. 149:369-377.
Oshino, N., Imai, Y. and Sato, R. 1971. A function of cytochrome
    b_5 in fatty acid desaturation by rat liver microsomes.
    J. Biochem. 69:155-168.
Ozols, J. 1974. Cytochrome b_5 from microsomal membranes of
    equine, bovine, and porcine livers. Isolation and properties
    of preparations containing the membranous segment. Biochem.
    13:426-434.
Ozols, J. and Strittmatter, P. 1964. Interaction of porphyrins and
    metalloporphyrins with apocytochrome b_5. J. Biol. Chem.
    239:1018-1023.
Rogers, M. and Strittmatter, P. 1974. Evidence for random dis-
    tribution and translational movement of cytochrome b_5 in
    endoplasmic reticulum. J. Biol. Chem. 249:895-900.
Schenkman, J. B. and Cinti, D. L. 1972. Hepatic mixed function
    oxidase activity in rapidly prepared microsomes. Life Sci.
    11:247-257.

Strittmatter, P. 1960. The nature of the heme binding in micro-
    somal cytochrome b$_5$. J. Biol. Chem. 235:2492-2497.
Strittmatter, P. and Velick, S. 1956. A microsomal cytochrome
    reductase specific for diphosphopyridine nucleotide. J.
    Biol. Chem. 221:277-286.
Strittmatter, P., Rogers, M. J. and Spatz, L. 1972. The binding
    of cytochrome b$_5$ to liver microsomes. J. Biol. Chem.
    247:7188-7194.
Ullrich, V., Cohen, B., Cooper, D. Y. and Estabrook, R. W. 1969.
    Reactions of cytochrome P-450, pp. 649-654. In K. Okunuki,
    M. D. Kamen and I. Sekuzu (eds.). Structure and function of
    Cytochromes. Osaka, Japan.

COMPARISON OF METHODS TO STUDY ENZYME INDUCTION IN MAN

Ivar Roots, Klaus Saalfrank and Alfred G. Hildebrandt
Department of Clinical Pharmacology, Free University
of Berlin, D-1000 Berlin 45
Hindenburgdamm 30, Germany

SUMMARY

A combination of several in vivo parameters has been applied
in male healthy volunteers to test the suitability of these
parameters to indicate enzyme induction in man: Urinary excretion
of D-glucaric acid and 6β-hydroxycortisol, activity of serum γ-glut
amyltranspeptidase, and pharmacokinetics of aminopyrine respond
significantly to phenobarbital treatment.

Glucaric acid excretion is enhanced about 7-fold. Its response to induction overcomes the large individual and interindividual variations which exist in the untreated state for glucaric acid excretion and the other parameters applied, as well.

Total body clearance of aminopyrine as obtained after an oral
test dose increases more than twofold from 251 to 551 ml/min upon
phenobarbital treatment. This arises from increases in both the
elimination constant and the apparent volume of distribution, as
well. Urinary excretion of aminoantipyrine during 24 hr is about
doubled, whereas the elimination of acetyl-aminoantipyrine is not
much affected. 6β-hydroxycortisol excretion in urine and activity
of serum γ-glutamyltranspeptidase are significantly augmented to
about 150% of control values.

Half life times of phenobarbital measured after termination
of treatment are in normal range, suggesting no self-induction of
phenobarbital metabolism.

Because of the complexity of drug metabolizing enzymes only
a combination of different parameters reliably indicates alterations in this enzyme system by inducing agents.

INTRODUCTION

The diversity of substrates which are metabolized via cytochrome P-450 in the endoplasmic reticulum of the hepatocyte and
many obvious but as yet unexplained changes in the pattern of
drug metabolism which are due to age, sex differences, species and
treatment with various inducing agents have led to an increased
interest in investigating drug metabolism. Many attempts have
been made to modify drug metabolism by treating animals with inducing agents to study the properties of cytochrome P-450 as well
as the importance of such treatments for drug metabolism in microsomes. As a consequence of such investigations the existence of
a family of cytochrome P-450 pigments with significant differences
in their chemical and physical properties is assumed (1) and discussed with respect to their role in regulating rate limiting
steps in drug metabolism. However, the more diverse the results
become which derive from in vitro studies with microsomes from
animals, the more difficult it appears to transfer from such data
to the human, i.e., the patient. This is the concern of a clinical pharmacologist insofar as he is engaged in investigating the
control mechanism of drug metabolism in man, and is considering
alterations of drug metabolizing enzyme activity in a sense of induction or inhibition.

To promote our understanding of the role of induction in man,
multiple in vivo parameters have been developed and numerously
applied (2-6). However, it has been recognized that various in
vivo tests obviously do not correlate with each other (7-9). As
a consequence of the various reactions and alterations of cytochrome P-450 occurring in animals, this is exactly the situation
one would expect in man as well. However, an examination of induction in man appears difficult. Because of the complexity of
drug metabolism one simple test hardly can be expected to yield
sufficient information. On the other hand, a combination of different in vivo tests should not affect the patient severely.

As phenobarbital is considered a prototype of drugs which
cause one type of induction (1,10,11), we have asked ourselves
whether and to what extent its administration might affect in vivo
parameters of drug metabolizing enzymes, and thus to allow us to
recognize one of possibly several types of induction in man. The
tests applied comprise the determination of endogenous substances
like 6β-hydroxycortisol and glucaric acid, the activity of enzymes
in serum like γ-glutamyltranspeptidase and, on the other hand,

the pharmacokinetics of a prototype drug of drug metabolism such as aminopyrine.

In addition to the phenobarbital-treated group, these tests were also applied to a placebo group. To test the sensitivity and applicability of these parameters a third group received clemastine, the inductive or inhibitory effects of which were not known before.

METHODS

Parts of the procedures have been recently described (12). The investigation was carried out in 23 healthy male volunteers, 20 - 32 years old. To obtain a standardized environment, 18 of these subjects were kept in a hotel, which was equipped to serve as a "metabolic ward." Additionally, 5 volunteers, who received higher doses of phenobarbital, were surveilled in a regular hospital.

The examination of the main group of 18 healthy subjects was conducted as a double blind study with a time schedule as follows: a 3 days control period, to obtain basic data, was followed by a treatment period of 14 days, and 3 subsequent days as washout period. The medication consisted of 100 mg phenobarbital 3 times daily in 7 subjects, 2 mg clemastine[*]3 times daily in 5 subjects, and 6 subjects received a placebo medication. Physical examinations and standard laboratory tests were repeatedly obtained. 24 hour urine samples were collected 3 times during control period, 5 times during the treatment period and 2 further collections were carried out during washout period. Aliquots of these urines were frozen and later tested for D-glucaric acid (glucaric acid), 6β-hydroxycortisol (6β-OH-cortisol) and 17-hydroxycorticosteroid excretion. The methods applied have already been described (12). Briefly, glucaric acid was determined according to Marsh (13) with several modifications. Calibration curves were obtained by adding calcium D-saccharate (Merck, Darmstadt) to urine. 6β-OH-cortisol determination was based on the procedures of Franz et al., (14) and Trasher 35 al., (15) which were modified (12). 17-hydroxycorticosteroids were determined according to Silber and Porter (16). The ratio of 24 hr 6β-hydroxycortisol to 17-hydroxycorticosteroid output was computed to compensate for variations in cortisol production.

At day 3 of the control period a test dose of 600 mg of aminopyrine was given to the volunteers. 5 blood samples were taken within 7 hours. Concentrations of aminopyrine in serum were determined in these samples by a gas chromatographic technique (12) to obtain pharmacokinetic data such as half life time, elimination constant, volume of distribution and total body clearance. Urine was collected in portions so that the appearance of 4-aminopyrine

TABLE I

Parameter	n	mean value	coefficient of variation [%]
6β-OH-cortisol 17-OH-corticosteroids	18	0.060	31.0
glucaric acid [μmoles/day]	23	32.7	48.5
elimination constant k_2 of Aminopyrine [h^{-1}] dose: 600 mg	19	0.377	29.8
relative volume of distribution [ml/g] of Aminopyrine	19	0.594	20.7
serum γ-glutamyltrans- peptidase [mU/ml]	18	20.8	47.2

Mean value and coefficient of variation of several in vivo tests of drug metabolism obtained from male volunteers (20-32 years old) under control conditions. Coefficient of variation represents the standard deviation expressed as per cent of the mean.

and acetyl-4-aminoantipyrine could be determined six, twelve and 24 hours after dosage. The method described by Brodie et al., (17) was applied. Activity of serum γ-glutamyl-transpeptidase was measured, using a commercial kit (Boehringer, Manheim).

These procedures were repeated at day 2 of the washout period to obtain posttreatment data. Serum levels of phenobarbital were determined (18) at the last 4 days of the trial to obtain half life times. The second group of 5 subjects received 500 mg of phenobarbital daily for 5 days. Glucaric acid excretion was determined during a two day control period and during the treatment period. The aminopyrine test was carried out by applying an oral dose of 1.2 g of aminopyrine one day before and one day after phenobarbital administration.

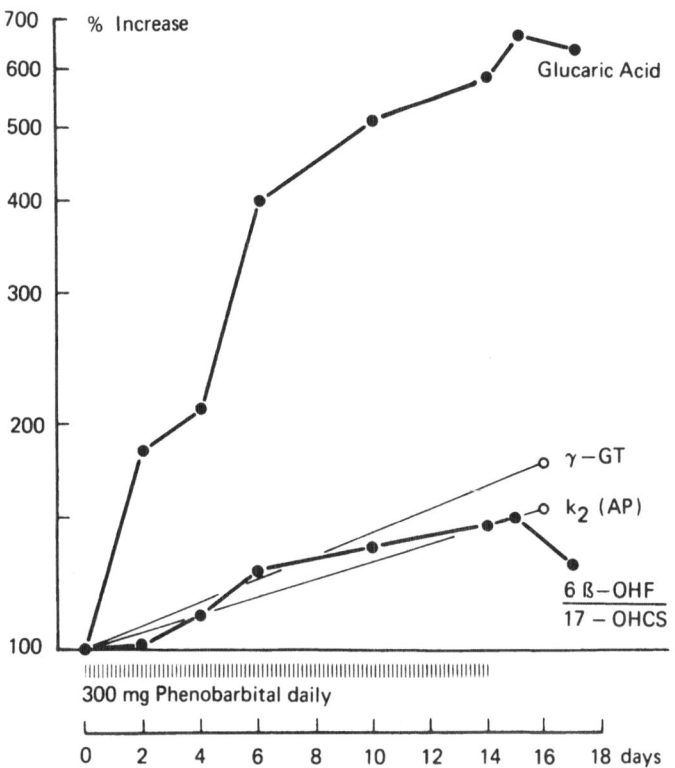

Figure 1.   Relative increase of <u>in vivo</u> parameters of drug metabolizing enzyme activity upon treatment with 300 mg phenobarbital daily in 7 male healthy subjects aged 20 - 32.

Values from control period are taken as 100% (= day 0). For glucaric acid excretion and ratio 6β-hydroxycortisol/17-hydroxycorticosteroids (6β-OHF/17-OHCS) values from all 3 days of control period have been averaged. Ordinate is scaled logarithmically. γ-GT = serum γ-glutamyltranspeptidase activity; k_2 (AP) = elimination constant of aminopyrine (test dose = 600 mg per os).

RESULTS AND DISCUSSION

The mean values of all four tests obtained from the volunteers during control period are listed in Table I. The variation coefficient ranges between 20 and 50% indicating considerable interindividual variations of all parameters, though the collective of subjects was rather homogenous. In addition, the individual variations from one day to another have to be taken into account.

Because of these variations under control conditions, induction
of drug metabolism in a patient can be best recognized if pretreatment values have been obtained. Furthermore, the tests applied
should be highly sensitive.

The question has been asked above to what extent these parameters
might increase after a relatively "strong" induction with 300 mg of
phenobarbital daily for two weeks. Figure 1 shows that all tests
respond positively to the treatment with phenobarbital. The increases of all parameters are statistically significant. However,
the increases of about 50-60% in serum γ-glutamyltranspeptidase
activity, in the elimination constant of aminopyrine as well as in
the ratio of 6β-OH-cortisol/17-OH-corticosteroids are - though note-
worthy - not sufficient to clearly overcome the variations mentioned
above. These tests, indeed, only allow reliable interpretations of
their results, if a control value before treatment is available.

In contrast, glucaric acid excretion increases nearly 7-fold
during treatment, with an early and significant rise already within
two days. At this time the ratio of 6β-OH-cortisol to 17-OH-cortisol
has not yet changed significantly. Therefore, determination of the
amount of glucaric acid excretion might be of special value for
studying the pattern of phenobarbital induction.

In Figure 2 the mean value of glucaric acid excretion as obtained from all 23 subjects is shown with several multiples of the
standard deviation. The dotted columns indicate the increases of
each individual upon 14 days of treatment with phenobarbital. As
one can see, all subjects are outside mean + 4 standard deviations;
most are even far outside mean + 6 S.D. This shows that a strong
induction by drugs acting like phenobarbital could be recognized
even without knowledge of the individual control value, when the
value exceeds the arbitrary limit of 80.4 μmoles/day (mean + 3 S.D.).
The horizontal bars in the columns represent the individual values
at day 6 of phenobarbital treatment. In all but one subject, this
value is already outside mean + 3 S.D. Similarly, Sotaniemi et al.,
(4) estimated from a single determination of glucaric acid excretion
in a patient, with unknown medical history, whether he was in normal
range or received inducers in small doses or even inducers in large
doses.

It is interesting to note that the posttreatment values of
five of the seven volunteers are relatively similar, ranging between about 150 to 200 μmoles/day, though the respective control
values differ considerably as can be seen from the abscissa of
Figure 3. In this diagram increases of glucaric acid excretion upon
14 days treatment with phenobarbital are expressed as percentage

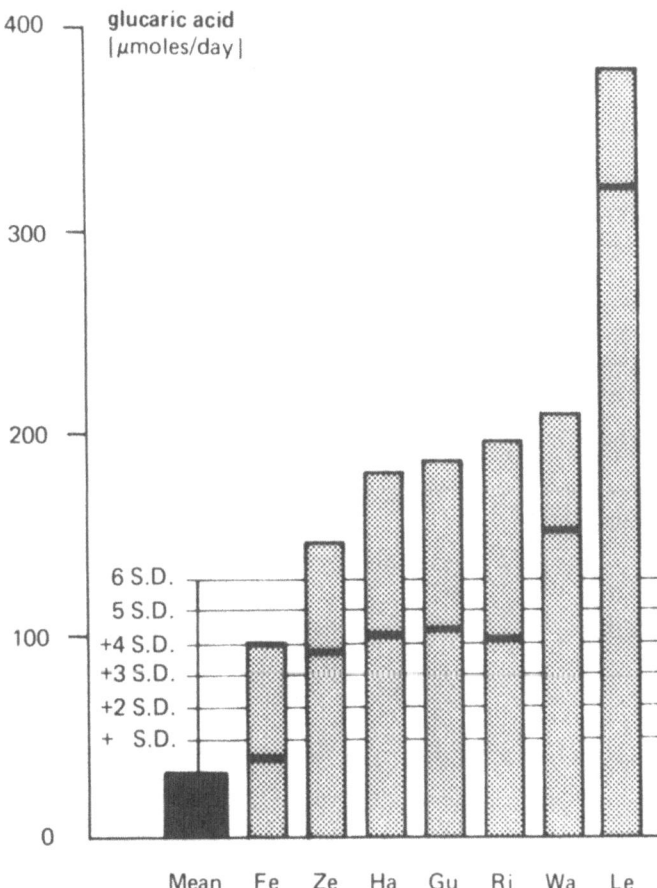

Figure 2.   Single values of glucaric acid excretion in 7 male volunteers after 14 days of phenobarbital treatment (300 mg daily) as compared to the mean value obtained under control conditions from all 23 male volunteers examined.

Mean value (32.7 μmoles/day) plus several multiples of its standard deviation (15.9) are given to indicate that post-treatment values lie outside of limits of confidence.   Horizontal bars in the dotted columns represent glucaric acid output after 6 days of phenobarbital treatment.

of individual control values.   There is a trend in that low control values yield a high percentage increase upon treatment with phenobarbital, whereas those values which are already high are not stimulated to the same extent.   It may be tentatively

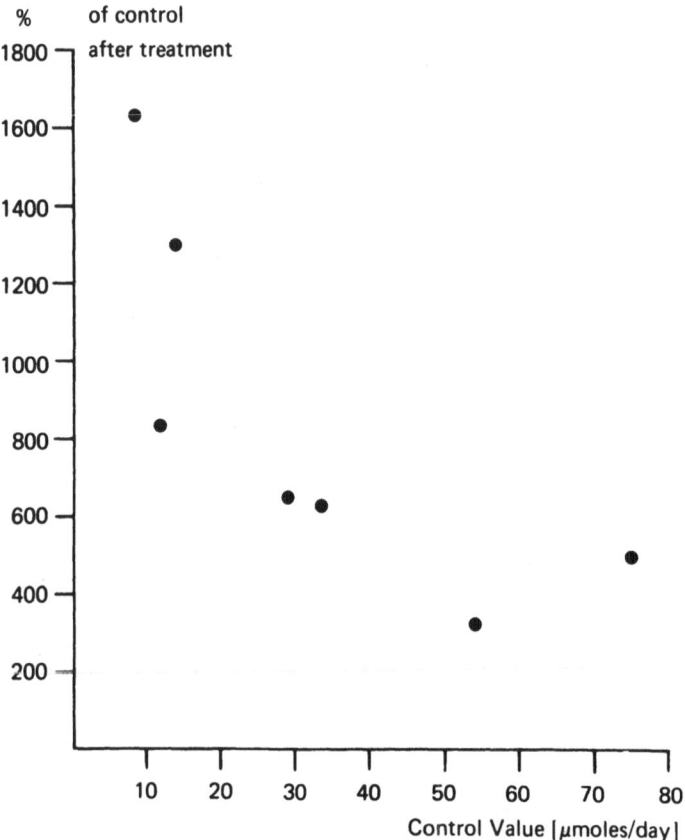

Figure 3. Relationship between individual control values and increases of glucaric acid excretion after 14 days treatment with 300 mg phenobarbital.

Control values are obtained by averaging results from 3 consecutive days before treatment (male volunteers aged 20 - 32).

concluded that the high interindividual variation of control values possibly reflects genetically determined differences as well as a certain extent of induction in some individuals caused by environmental, nutritional or other factors.

In Table II results are given of glucaric acid excretion in

TABLE II

dose of Phenobarbital	mean of control period	Phenobarbital treatment					
		day 1	day 2	day 3	day 4	day 5	day 6
300 mg/day (n = 7)	32.4	-	60.8	-	70.2	-	128
500 mg/day (n = 5)	33.7	47.0	51.6	71.3	96.4	109	141

Daily urinary excretion of glucaric acid (μmoles/day) before and during treatment with 300 and 500 mg of phenobarbital daily in male healthy volunteers.

those subjects receiving 500 mg of phenobarbital daily for 5 days. Values are not significantly different from those obtained after treatment with 300 mg of phenobarbital. From that it follows that in man, as well as in animal, duration of treatment plays an important role in enzyme induction. The dose of 300 mg seems to result in a maximal level of induction in humans. However, a dose-effect-relationship may very well be obtained with lower doses.

Though glucaric acid excretion in many respects appears to be an ideal parameter to test enzyme induction in man, all the other tests should not be neglected. They paradigmatically show in how many ways phenobarbital influences the steady state of metabolism and excretion of drugs and endogenous compounds. The application of a combination of several tests allows one to estimate the diverse effects of a single inducing agent on such functions. It would be of interest to know whether enzyme inducers other than phenobarbital stimulate these parameters to a different extent than phenobarbital does.

This aspect is not only important in patients on chronic drug treatment, but also during "phase II" of evaluation of new drugs in humans, as has been shown with clemastine (12). The latter stimulated none of the parameters; therefore, it was concluded that clemastine fails to induce drug metabolizing enzymes, at least in the dosage given and the time span examined. Such an inference would obviously be less trustworthy, if only one single parameter, for instance daily urinary excretion of 6β -hydroxycortisol,would have been tested.

TABLE III

	Control	Phenobarbital
t 1/2 [h]	2.05 (0.158)	1.32 (0.122)
relative volume of distribution [ml/g]	0.630 (0.134)	0.911 (0.134)
Cl_{tot} [ml/min]	251 (21.2)	551 (222)
k_2 [h^{-1}]	0.346 (0.0237)	0.548 (0.0530)

Pharmacokinetic data obtained after an oral test dose of 600 mg of aminopyrine, before and after 14 days treatment with 300 mg phenobarbital daily.   Numbers in parentheses represent standard error of the mean.   n = 6; Cl_{tot} = total clearance, k_2 = elimination constant.

       One would like to apply to a patient diverse test drugs which are all metabolized in different ways (e.g. aromatic hydroxylation, N-, O-, C-demethylation, aliphatic hydroxylation, etc.), thus trying to examine different rate limiting steps in drug metabolism and elimination. However, aside from ethical reasons, such a procedure would suffer from the fact that meaningful correlations of pharmacokinetic data of two different drugs in untreated subjects are rare. Aminopyrine, the test drug applied here, is in Germany a frequently used analgesic, antipyretic and antirheumatic agent. Its disappearance from plasma is mainly controlled by its metabolism, i.e., N-demethylation to form aminoantipyrine. This metabolite and its acetylated derivative are the primary products, the appearance of which has been determined in urine.

TABLE IV

subject	before treatment	after treatment
Y. B.	3.2	1.4
H. B.	3.4	1.3
G. H.	6.5	1.2
M. J.	3.7	2.3
H. Ba.	1.7	1.1
Mean	3.70	1.46
S.D.	1.74	0.49
S.E.M.	0.78	0.22

Half life time [hr]of aminopyrine after an oral test dose of
1.2 g before and after 5 days treatment with 500 mg phenobarbital
in male healthy subjects, aged 22 - 26.

The application of aminopyrine allows the study of various
effects arising from phenobarbital treatment in man. Phenobarbital administration increases the elimination constant k_2 by
about 50% as shown in Table III. Additionally, the apparent volume of distribution is enlarged so that total clearance values
are more than doubled (Table III). This enhanced disposition of
aminopyrine from the body is partly accomplished by the kidney.
About one third of the dose given is found in urine within 24 hr
(Figure 4). Since only in 4 of the 7 volunteers was a complete
excretion performed, the data in Figure 4 represent rather a
trend, since the intraindividual variations in the excretion pattern of the metabolites are considerable. Aminoantipyrine excretion increases by about 100% at all periods, whereas the
amount of the acetylated product eliminated remains almost constant after phenobarbital treatment. Since aminoantipyrine contributes only to a minor degree to the total urinary excretion
of aminopyrine (17, 9), the sum of the amounts of aminoantipyrine

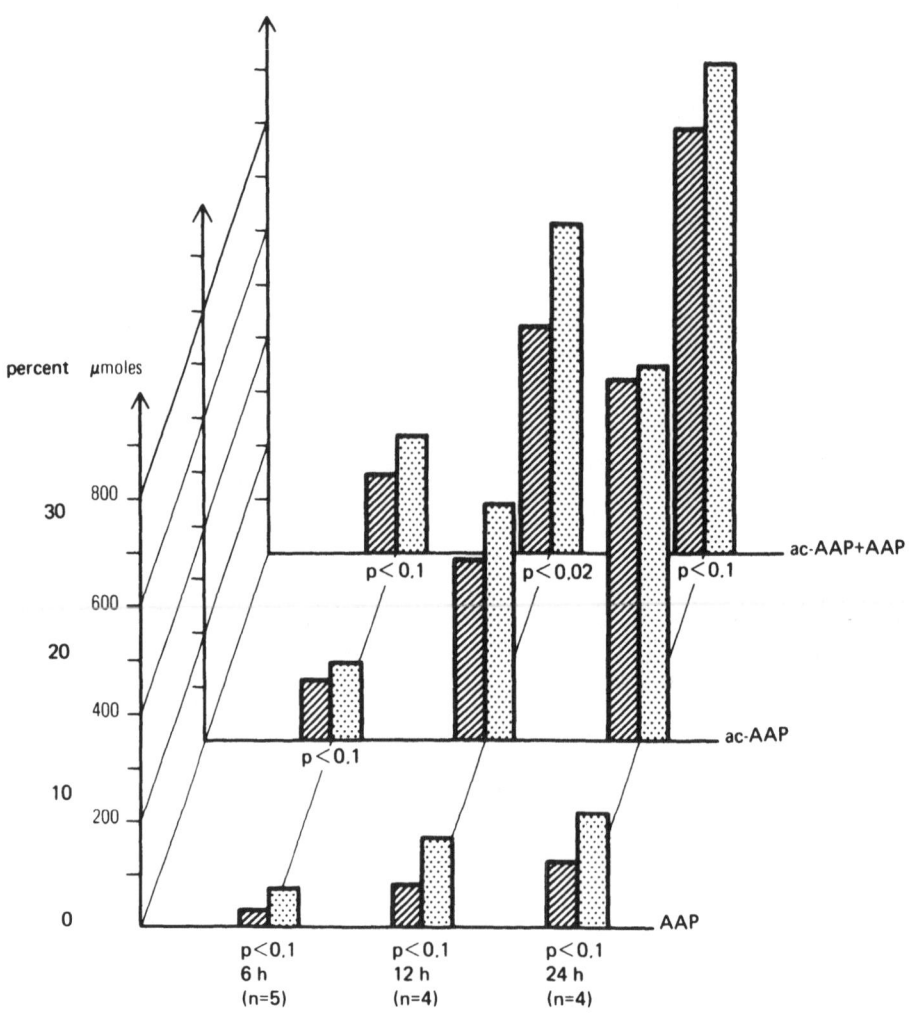

Figure 4. Urinary excretion of aminoantipyrine (AAP) and
acetyl-aminoantipyrine (ac-AAP) after an oral dose of 600 mg
(= 2600 μmoles = 100%) of aminopyrine.

Dashed columns represent cumulative values obtained after
6, 12 or 24 hr under control conditions. Excretion data from a
second aminopyrine test after 14 days treatment with 300 mg phenobarbital daily are depicted in the dotted columns.

and acetyl-aminoantipyrine formed in urine does not reflect the high increase in total body clearance. With the reservation that these data are based on only 4 subjects and that the urinary output of other metabolites is not significantly affected, it should follow that the biliary excretion is also enhanced considerably by phenobarbital treatment. An alternative possibility would be that, in contrast to control conditions (19), high urinary excretion rates persist longer than 24 hours upon phenobarbital induction.

Another aspect of aminopyrine pharmacokinetics arises from this study: under control conditions, half lives are considerably longer, when the relatively high dose of 1.2 g is applied (Table IV). After application of 500 mg of phenobarbital daily for 5 consecutive days, however, the mean half life time drops from 3.7 hr to 1.46 hr. This corresponds to an increase of about 150% for the elimination constant k_2, which is much more than can be gained after a test dose of only 600 mg (Figure 1). Interestingly, the half life times after phenobarbital treatment are almost the same with both test doses (Tables III and IV), indicating a maximal rate of aminopyrine metabolism in all subjects. This suggests that saturation phenomena play a role in the rate of liver uptake and metabolism of aminopyrine. Additionally, it follows that if aminopyrine is used as a test drug, higher doses than 0.6 g are advisable.

In summarizing these results it becomes evident that this "aminopyrine test" represents already a complex test combination, including various parameters which are effected by phenobarbital treatment.

Phenobarbital itself, though applied as an enzyme inducer, may also be regarded as a test drug. Serum levels have not been followed in the beginning, but values obtained from day 1 of the posttreatment period (Figure 5) represent steady state levels after 14 days of treatment. From the decline of the phenobarbital concentration after termination of treatment a mean half life time of 3.7 days (range: 1.7 - 5.0 days) can be computed. This number being within the range of normal values (20) does not indicate a significant degree of self induction of phenobarbital metabolism [mainly aromatic hydroxylation (21)], as far as can be judged by comparison with values from the literature. This example shows that not all drugs are influenced as markedly as aminopyrine, even if they are, at least partly, hydroxylated in the liver.

The specificity of γ-glutamyltranspeptidase as an indicator of enzyme induction is doubtful, though it responds to treatment with phenobarbital and other enzyme inducers, e.g., diphenylhydantoin, as shown here and by others (5,22,23). However, elevated serum levels sometimes are found also in subjects who should be regarded as healthy as judged from all other laboratory parameters. Since γ-glutamyltranspeptidase activity is increased in many diseased states, it does not seem to be a parameter as reliable as the others mentioned above. However, in a controlled study, especially in healthy subjects where pretreatment values are available, measurements of γ-glutamyltranspeptidase should be included.

6β-hydroxycortisol excretion has been applied frequently as an indicator of enzyme induction in man (3,9,14,24). As it represents an endogenous substance arising from cortisol metabolism via cytochrome P-450, 6β-hydroxycortisol should be an ideal in vivo parameter to test the activity of this mixed function oxidase. The interpretation of the meaning of 6β-hydroxycortisol excretion rates in a single patient is difficult due to high individual and intraindividual variations. Reliable conclusions can only be drawn by following excretion rates over a longer period instead of simply determining endpoint values. However, in a controlled trial with more subjects involved, determination of 6β-hydroxy-cortisol excretion rates should be considered an essential part of the combination of tests to be applied in order to study and detect enzyme induction of various types.

Because of the diversity of the parameters tested, a correlation among them cannot primarily be expected. The lack of correlation under control conditions has been reported by many authors (6,7,9,25). However, some parameters do correlate (7,8,25,26,27) which is also shown in this study. There is a positive correlation (P < 0.05) between the elimination constant k_2 of amino-pyrine and glucaric acid excretion. Perhaps this reflects a latent induction, yielding an increase of these two parameters in some individuals. Furthermore, Cl_{tot} significantly (P < 0.01) correlates with the apparent volume of distribution V and with the elimination constant k_2 (P < 0.05). Total clearance is computed by the equation

$$Cl_{tot} = V \cdot k_2.$$

Since Cl_{tot} significantly correlates with both V and k_2, it follows that these two variables are of influence in determining an individual's total body clearance of aminopyrine. Variations of the apparent volume of distribution V from one subject to another are not automatically compensated by changes of elimination constants in the opposite direction. Thus, the apparent volume of distribution should be determined as well, since k_2-values or half life times only partly reflect pharmacokinetics of aminopyrine.

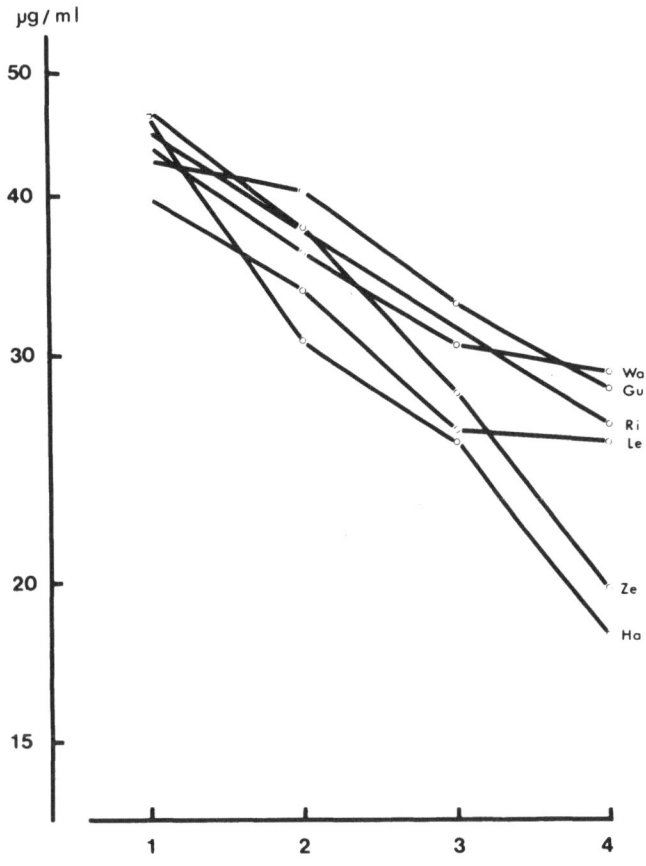

Figure 5. Decline of serum levels of phenobarbital in 6 male volunteers after the end of 14 days treatment with 300 mg phenobarbital daily. Days refer to washout period.

By applying this combination of tests in a patient it is possible to detect enzyme induction. But from the knowledge of this effect it is by now not possible to make precise predictions how enzyme induction influences pharmacokinetics and metabolism of any drug. Much systematic research has to be done until a catalogue exists from which information can be obtained on how a drug behaves in a patient with a certain disease and to what extent and in which direction the drug metabolizing enzyme system of this patient has been stimulated.

FOOTNOTES

*   (+)-2- [(2-parachloro-α-methyl-α-phenylbenzyl) oxyethyl] -1-
methyl-pyrrolidine, used as antihistaminic.

ACKNOWLEDGEMENTS

The authors would like to thank Dr. W. H. Aellig, Sandoz
Ltd., Basle, for his constant interest and advice. The skilled
technical aid of Mrs. E. Berg, Mrs. B. Schwaneberg, and Mrs. I.
Tinnes is acknowledged.

This study has been supported by Deutsche Forschungsgemeinschaft (Schwerpunktprogram: Biologische Grundlagen der
Arzneimittel- und Fremdstoffwirkungen [Hi 156/2]).

REFERENCES

1   Conney, A. H., Lu, A. Y. H., Levin, W., Somogyi, A., West,
S., Jacobson, M., Ryan, D., Kuntzman, R. Effect of
enzyme inducers on substrate specificity of the cytochrome P-450's. Drug Metab. and Dispos. 1, 199 (1973).
2   Aarts, E. M. Evidence for the function of D-glucaric acid
as an indicator for drug induced enhanced metabolism
through the glucuronic acid pathway in man. Biochem.
Pharmacol. 14, 359 (1965).
3   Burstein, S., Kimball, H. L., Klaiber, E. L., Gut, M. J.
Metabolism of 2α- and 6β-hydroxycortisol in man:
determination of production rates of 6β-hydroxycortisol
with and without phenobarbital administration. Clin.
Endocr. 27, 491 (1967).
4   Sotaniemi, E. A., Medzihradsky, F. and Eliasson, G.:
Glucaric acid as an indicator of use of enzyme-inducing
drugs. Clin. Pharmacol. Therap. 15, 417 (1974)
5   Davidson, D. C., McIntosh, W. B. and Ford, J. A. Assessment
of plasma glutamyl transpeptidase activity and urinary
D-glucaric acid excretion as indices of enzyme induction.
Clin. Science and Molec. Med. 47, 279 (1974)
6   Cunningham, J. L. and Price Evans, D. A. Urinary D-glucaric
acid excretion and acetanilide pharmacokinetics before
and during diphenylhydantoin administration. Europ. J.
Clin. Pharmacol. 7, 387 (1974)
7   Sjöqvist, F. and von Bahr, C. Interindividual differences
in drug oxidation: clinical importance. Drug Metab.
and Dispos. 1, 469 (1973).
8   Davies, D. S., Thorgeirsson, S. S., Breckenridge, A. and
Orme, M. Interindividual differences in rates of drug
oxidation in man. Drug Metabol. and Dispos. 1, 411
(1973)

9   Smith, S. E. and Rawlins, M. D. Prediction of drug oxidation
    rates in man: lack of correlation with serum gamma-
    glutamyl-transpeptidase and urinary excretion of D-glucaric
    acid and 6β-hydroxycortisol. Eur. J. Clin. Pharmac. 7,
    71 (1974).

10 Conney, A. H. Pharmacological implications of microsomal en-
    zyme induction. Pharmac. Rev. 19, 317 (1967).

11 Remmer, H., Merker, H. J. Drug induced changes in the liver
    endoplasmic reticulum: association with drug-metabolizing
    enzymes. Science, N. Y. 142, 1657 (1963)

12 Hildebrandt, A. G., Roots, I., Speck, M., Saalfrank, K. and
    Kewitz, H. Evaluation of in vivo parameters of drug
    metabolizing enzyme activity in man following clemastine,
    phenobarbital or placebo administration. Europ. J. Clin.
    Pharmacol. (in press)

13 Marsh, C. A. Metabolism of D-glucuronolactone in mammalian
    systems. Identification of D-glucaric acid as a normal
    constituent of urine. Biochem. J. 86, 77 (1963).

14 Frantz, A. G., Katz, F. H., Jailer, J. W. 6β-OH-cortisol and
    other polar corticosteroids measurement and significance
    in human urine. J. Clin. Endocr. 21, 1290 (1961).

15 Trasher, K., Werk, E. E., Choi, J. Y., Sholiton, L. J., Meyer,
    W., Olinger, Ch. The measurement, excretion and source
    or urinary 6β-hydroxycortisol in humans. Steroids 14,
    455 (1969).

16 Silber, R. H., Porter, C. C. The determination of 17,21-di-
    hydroxy-20-ketosteroids in urine and plasma. J. Biol.
    Chem. 210, 923 (1954)

17 Brodie, B. B., Axelrod, J. The fate of aminopyrine (pyrami-
    don) in man and methods of estimation of aminopyrine
    and its metabolites in biological materials. J.
    Pharmacol. Exptl. Therap. 99, 171 (1950).

18 Kipnich, S. Differenzspektroskopische Bestimmung von
    Barbituraten. Beckman-report No. 1+2, p. 45 (1974).

19 Gradnik, R. and Fleischmann, L. Quantitative Ausscheidung
    verschiedener Aminopyrin-Stoffwechselprodukte beim
    Menschen. Pharmaceut. Acta. Helv. 48, 181 (1973).

20 Kampffmeyer, H. G. Elimination of phenacetin and phenazone
    by man before and after treatment with phenobarbital.
    Europ. J. Clin. Pharmacol. 3, 113 (1971).

21 Parke, V. D. The Biochemistry of Foreign Compounds, p. 186,
    Pergamon Press Ltd., 1968.

22 Rosalki, S. B., Parlow, D., Rau, D. Plasma gamma-glutamyl-
    transpeptidase elevation in patients receiving enzyme
    inducing drugs. Lancet 2, 376 (1971).

23 Whitfield, J. B., Moss, D. W., Neale, G., Orme, M. and
    Breckenridge, A. Changes in plasma γ-glutamyl-trans-
    peptidase activity associated with alterations in drug
    metabolism in man. Brit. Med. J. 1, 316 (1973).

24   Kuntzman, R., Jacobson, M., Levin, W. and Conney, A. H.
        Stimulatory effect of N-phenylbarbital (phetharbital)
        on cortisol hydroxylation in man.   Biochem. Pharmacol.
        17, 565 (1968).
25   Kadar, D., Inaba, R., Endrenyi, L., Johnson, G. E. and Kalow,
        W.   Comparative drug elimination capacity in man -
        glutethimide, amobarbital, antipyrine and sulfinpyrazone.
        Clin. Pharmacol. and Therap. 14, 552 (1973).
26   Cunningham, J. L., Bullen, M. F. and Price Evans, D. A.   The
        pharmacokinetics of acetanilide and of diphenylhydantoin
        sodium.   Europ. J. Clin. Pharmacol. 7, 461 (1974).
27   Hammer, W., Mårtens, S., Sjöqvist, F.   A comparative study of
        the metabolism of desmethylimipramine, nortriptyline, and
        oxyphenylbutazone in man.   Clin. Pharmacol. Therap. 10,
        44 (1969).

GENERAL DISCUSSION

Session IV

HOLTZMAN: I would like to comment on the studies concerning the participation of cytochrome b_5 in the hydroxylation reactions. As Drs. Mannering and Sasame have indicated, they feel that this cytochrome only participates in the NADH stimulation of the hydroxylation. Similarly I believe that there is some kinetic evidence supporting their immunological evidence that cytochrome b_5 is not involved in the hydroxylations when NADPH is the sole source of reducing equivalents. A couple of years ago we published a study in the Archives of Biochemistry and Biophysics (150:227, 1972) in which we demonstrated that the activation energy of ethylmorphine N-demethylase was significantly higher than the activation energy of the cytochrome P-450 reductase in the presence or absence of ethylmorphine. This suggested to us that there are two steps of nearly equal rate which control the rate of the overall demethylase reaction. We tried to determine whether the other step could be the reduction of cytochrome b_5, but it is so rapid that we could not measure it. Hence, there would always seem to be enough of the reduced cytochrome around so that at least its reduction could not be rate limiting and this could not be the cryptic second slow step we were searching for.

LU: The inhibition of NADPH-dependent benzphetamine N-demethylation by cytochrome b_5 could be due to two things: 1) Cytochrome b_5 reacts with NADPH-cytochrome c reductase limiting the availability of the reductase itself and decreasing the concentration of NADPH-cytochrome c reductase available for cytochrome P-450; and 2) cytochrome b_5 acts as an electron drain and competes with cytochrome P-450 for reducing equivalents from NADPH-cytochrome c reductase.

HOLTZMAN: Yet, Estabrook's laboratory has reported that in the presence of a substrate, the steady state concentration of reduced cytochrome b_5 decreased by 30-40%. This is not hard evidence that it participates, since in air cytochrome b_5 is readily autooxidized. It is possible that in the presence of ethylmorphine or aminopyrine there is increased competition for the electrons necessary to keep the b_5 in the reduced state. I feel, therefore, that when NADPH is the sole source of electrons, the addition of cytochrome b_5 would at best do nothing.

MANNERING: In our studies the effects of activation or inhibition of the desaturation system on NADH synergism of the monooxidase system using stearyl CoA or cyanide (Correia and Mannering, Mol. Pharmacol., 9, 455, 1973), we concluded that the desaturation system competed with the monoxidase system by sharing electrons from cytochrome b_5 which were derived from NADH. This raised the

question as to how two systems with very low turnovers could
create a limitation of electrons from cytochrome b_5 when cytochrome
b_5 is so rapidly reduced by NADH-cytochrome b_5 reductase. We
"solved" this problem by inserting an "x" component between cytochrome b_5 and the desaturation and monooxidase systems, with "x"
being shared by the two systems. I should also like to comment on
benzamphetamine, the substrate of choice for solubilized systems.
The turnover number of benzamphetamine, based on cytochrome P-450
content, when solubilized systems are used, is about twice that observed when microsomes are employed. The turnover number of ethylmorphine observed with solubilized systems is only about one-third
that seen with microsomes. This suggests that the solubilized system may not be typical of the microsomal systems. This being the
case, it may not be wise to equate the bound and soluble systems
when cytochrome b_5 and NADH synergism are being considered.

LU: For most substrates we have studied, the turnover numbers
obtained with the resolved system are either the same or greater
than those obtained in microsomes. Ethylmorphine is one of the
few substrates which has a lower turnover number in the resolved
system. We have reason to believe that different P-450 species
are responsible for benzphetamine N-demethylation and ethylmorphine
N-demethylation, and that the species responsible for ethylmorphine
N-demethylation may be destroyed during isolation.

KAMIN: My question is directed at Dr. Lichtenberger.* If
it is not appropriate now it can wait. What interests me particularly are two factors I am curious about - one is in the uncoupling
phenomenon as it relates to NADH and cytochrome b_5. In a formal
sense it is not entirely clear that P-450 is involved. Some adduct
could make b_5 itself autoxidizable. Do you have any evidence
whether P-450 is indeed involved in the uptake of oxygen from NADH
in the presence of your uncoupler? The other question is whether
water, rather than peroxide, could be the direct product. The
system probably contains catalase. I wonder whether you have any
independent evidence that water is the direct product of oxygen
reduction.

LICHTENBERGER: The stoichiometry evidently indicates: 2
NADPH to 1 oxygen.

KAMIN: That would be the same with peroxide and catalase.

HOLTZMAN: The disproportionation of peroxide going to O_2
and water - you cannot tell the difference.

---

*   No manuscript received. Content of presentation taken from
Staudt, H., Lichtenberger, F. and Ullrich, V. (1974). Eur. J.
Biochem. 46: 99-106.

ROSENTHAL: Whether it goes directly? You see, there is acti-vated P-450(Fe^{2+})·O_2, actually an oxygen complex between, and this is either reduced to water or involved in the hydroxylation. The question is whether it goes <u>directly</u> from P-450(Fe^{2+})·O_2 to water or via peroxide.

HILDEBRANDT: We have recently shown that hydrogen peroxide is produced during hydroxylation reaction. Therefore, one can directly measure peroxide formation enzymatically if one adds azide to the system to inhibit catalase. In addition, in the presence of methanol, peroxide, and active catalase, formaldehyde is formed. So, one of the uncoupling products could be hydrogen peroxide and, of course, if this is the uncoupling product, one should be able to measure the degree of uncoupling in your system by deter-mining the rate of hydrogen peroxide formation.

LICHTENBERGER: I believe if I measure oxygen consumption and I measure NADPH consumption, I clearly assume water is produced by the system. There are two molecules of NADPH needed per molecule of oxygen.

SCHLEYER: That's the overall result. It doesn't say anything about the mechanism which is what I think Dr. Kamin wanted to know.

LICHTENBERGER: About the mechanism, I don't know anything.

SCHLEYER: It was perfectly all right. Dr. Kamin wanted to say: Could you rule out that it doesn't go via catalase which might be present in the microsomes and hydrogen peroxide?

ROSENTHAL: The stoichiometry would be the same.

SCHLEYER: The other question of Dr. Kamin was: What about the role of P-450; is it actually involved? Does P-450 directly interact or could you explain your result by a modification of your b_5? Could a modification of b_5 be the "uncoupling agent?"

LICHTENBERGER: It could be a modification of the b_5.

NARASIMHULU: Dr. Lichtenberger's paper was particularly interesting to me because I was the first one to demonstrate an uncoupling of oxygen activation from hydroxylation in a P-450-dependent hydroxylase system. This was found when studying the stimulation of NADPH oxidation by the non-hydroxylatable Type I substrate Δ^4-androstenedione in bovine adrenocortical microsomes (Arch. Biochem. Biophys. 147, 384, 1971). When I observed that 2 moles of NADPH were oxidized per mole of O_2 consumed, I first thought that catalase, which was present in the microsomes, might have converted an intermediary peroxide to water and oxygen,

putting half of the oxygen back into the system, resulting in a
2:1 stoichiometry of the overall reaction. In the presence of
mM azide there was indeed some NADPH background oxidation that
produced peroxide. The stoichiometry of the Type I substrate-
dependent NADPH and O_2 consumption, however, remained 2:1 in the
presence of azide. This is consistent with the formation of water
without involvement of catalase.

ORRENIUS: A comment to Dr. Schenkman and Dr. Mannering which
relates to the use of stearyl CoA as a competitor with the P-450
system. In this system it is always assumed that stearyl CoA acts
as a substrate for the desaturase system. Now we know that stearyl
CoA belongs to those fatty acids that are poorly hydroxylated by
the microsomes and it has been recently shown that the CoA derivatives of those fatty acids may possibly serve as substrates for
the P-450 system. In fact, we have shown with palmityl CoA that
this is bound with a higher affinity to P-450 than palmitate itself. So I was wondering whether it is a possibility that the inhibition by stearyl CoA might be due to an effect on P-450 directly?
Now Dr. Mannering's evidence that cyanide relieves this inhibition
would be against this thought unless one takes the work of Gaylor
into consideration and assumes that the fatty acid hydroxylating
P-450 may be more sensitive to cyanide than the demethylase P-450.
Then it would explain his finding why cyanide produces these effects.

MANNERING: With respect to Gaylor's observations, it should
be understood that the concentration of cyanide required to give a
spectral shift with cytochrome P-450 is considerably higher than is
necessary to inhibit the cyanide sensitive factor of the desaturation system. If, in our experiments, stearyl CoA inhibited the
monooxidase system directly, as has been suggested by Dr. Orrenius,
cyanide should not have affected the inhibition. In our studies,
cyanide completely reversed the inhibitory effect of stearyl
CoA on ethylmorphine N-demethylation.

SCHENKMAN: Would you comment on the concentrations of cyanide
shown on your slide?

MANNERING: We always use several concentrations of cyanide
to ensure that we obtain a maximum effect. The optimal concentration of cyanide varies from one batch of microsomes to another.

SCHENKMAN: Well, the activity was the same with stearyl CoA
and cyanide as with the stearyl CoA alone so I think I tend to stay
with Dr. Orrenius' suggestion that the inhibition of ethylmorphine
demethylation by stearyl CoA which we are seeing here is not so
much inhibition due to competition for electrons in the presence
of stearyl CoA, as just competition in the P-450 system. Our
data would tend to agree with the suggestion of Dr. Orrenius. At

concentrations of cyanide that inhibit the stearyl CoA desaturase activity by 50%, we still see inhibition of ethylmorphine N-demethylase. Thus it is very possible that this latter inhibition is due to the binding of stearyl CoA to P-450.

MANNERING: What concentration of cyanide did you use?

SCHENKMAN: 200 μM. This concentration was shown by Dr. Correia and you to reverse the inhibition by stearyl CoA. We varied the cyanide concentration and measured stearyl CoA desaturase activity. That's about the concentration at which you get 50% inhibition.

MANNERING: At that concentration you may observe some inhibition of the monooxygenase system.

SCHENKMAN: No, even in the absence of stearyl CoA, cyanide binding to P-450 has a K_s of about 2.4 mM, hence doesn't explain the inhibition we are seeing. With regard to the inhibition of drug metabolism by cytochrome b_5 which has been bound to microsomes, I don't think it is due to detergents. First of all, we are using very high purity cytochrome b_5 of comparable concentration shown by Enamoto and Sato, and containing 42 nmole/mg protein. The concentration Strittmatter obtained with rabbit cytochrome b_5 is slightly higher. The absorbance of detergents in the ultra violet region is quite high. By taking the ratio of the 280 band to the 413 nm peaks of cytochrome b_5 as an indication of purity, our results would indicate we don't have detergents present, at least not appreciably. Then again, there is another point. In collaboration with Dr. Hutterer, we studied the effects of detergents on the microsomal system. The detergent deoxycholate is a Type I binding compound, stimulating reductase activity. If anything, we would expect to see an increase in TPNH-cytochrome c reductase activity with an increase in b_5 concentration. We don't see this. Then, too, as I showed, the stearyl CoA desaturase activity is enhanced by increased microsomal concentration of b_5. I feel detergent presence is not the cause of the inhibition of drug metabolism by added b_5. I think your first suggestion of an electron sink is probably the more likely one.

LEVIN: I would like to go back to Dr. Mannering's comment concerning the possibility of detergent in the cytochrome b_5 preparation resulting in an inhibition of hydroxylase activity. The inhibition of benzphetamine demethylation by cytochrome b_5 can be _reversed_ by the addition of cytochrome b_5 reductase and NADH or _by using_ much higher amounts of NADPH-cytochrome c reductase as Dr. Lu stated. If the inhibition were due to detergent in the cytochrome b_5 preparation, how could the addition of these other enzymes and NADH reverse detergent inhibition? These data argue very strongly for the inhibition of activity being due to cyto-chrome b_5 per se. You also stated that the resolved systems may behave quite differently from microsomes. It is interesting that Dr. Schenkman also gets inhibition of metabolism by the addition of purified b_5 to microsomes. Thus, both microsomes and the partially purified system are acting in a similar manner upon addition of cytochrome b_5.

I agree that Dr. Cinti has somewhat different results, but the fact remains that in the resolved system, the b_5 inhibition cannot be explained by detergent contamination of the cytochrome b_5 for the reasons that I just gave.

MANNERING: As I remember, Dr. Lu got a very small inhibition by addition of cytochrome b_5, about 30-40%.

LEVIN: Dr. Lu got only a 30-40% inhibition by addition of cytochrome b_5 to the resolved system because he added much less cytochrome b_5 than Dr. Schenkman did. The inhibition by cyto-chrome b_5 is a concentration dependent relationship. When Dr. Lu added much more cytochrome b_5 to the system (as Dr. Schenkman did), the inhibition was much greater than 30-40%.

SASAME: From Dr. Lu's presentation and from our data, I would like to direct a question to Dr. Cinti. Have you checked DPNH dependent P-450 reductase in microsomes after you incor-porated cytochrome b_5? From our data, using antibody against cytochrome b_5, it is evident that DPNH-dependent P-450 reductase is inhibited by the antibody. Likewise from Dr. Lu's data on DPNH-dependent 3,4-benzpyrene hydroxylation by the reconstituted system it appears that cytochrome b_5 is absolutely essential. These observations contradict those of the Japanese workers who removed cytochrome b_5 from microsomes but still found DPNH being able to reduce P-450. I would like to know if Dr. Cinti did check DPNH-dependent P-450 reductase after incorporation of cytochrome b_5.

CINTI: We did not check for NADH-P-450 reductase activity after binding cytochrome b_5 to rat liver microsomes.

LU: I would like to make a comment about the work of Ichikawa and Loehr (Biochem. Biophys. Res. Comm. 46, 1187, 1972). They treated rabbit liver microsomes with a protease to remove cytochrome b_5, and found that cytochrome P-450 in the treated microsomes can still be reduced by NADH, suggesting that cytochrome b_5 is not involved in NADH-dependent reduction of cytochrome P-450. Treatment of microsomes with a protease is a very risky thing. Studies from Sato's laboratory and Strittmatter's laboratory have clearly shown that protease-solubilized cytochrome b_5 and cytochrome b_5 reductase have been altered. Another interesting observation was the work of Shimakata et al., J. Biochem, 72, 1163, 1972, who have shown that when combined with the cyanide-sensitive factor, protease-solubilized cytochrome b_5 reductase catalyzes the desaturation even in the absence of cytochrome b_5. In contrast, cytochrome b_5 is absolutely required for desaturation in the presence of the cyanide-sensitive factor and detergent-solubilized cytochrome b_5 reductase. Therefore, the treatment of microsomes with protease may change the interactions between microsomal electron carriers.

CINTI: I don't know what is gained by adding protease in order to solubilize b_5. Certainly, b_5 concentration will be decreased in the microsomes by proteolytic digestion but unfortunately so will other microsomal enzymes be affected - NADH and NADPH cytochrome c reductase and NADH b_5 reductase activities are also diminished.

LU: I just want to mention the risk one takes in using protease since there are many precedents in the literature describing alterations in the properties of protease-solubilized enzymes. I also want to ask if anyone has shown that even though the cytochrome P-450 is still spectrally intact, trypsin-solubilized microsomes are still active in hydroxylation.

SCHENKMAN: We have used the Nishibayashi protease and Sato's Emulgen 911 preparations, but they are not active.

LU: That's right. We really should be careful about using protease, unless we can show that the catalytic activity has not been altered.

LOTLIKAR: We have used protease treated preparations, solubilized with non-ionic detergent and we have been able to reconstitute carcinogen activation systems and they are very active. So we have enzymatic activity.

SCHENKMAN: I have a question for Dr. Sasame. I was not able to understand some of your slides in which you were showing the b_5 additions: were you adding cytochrome b_5 to the medium, or adding antibody to b_5?

SASAME: No, I am sorry, we were preincubating cytochrome b_5 with the antibodies against cytochrome b_5 for 5 min.

LEVIN: Dr. Lotlikar, if you use protease solubilized microsomes, is the enzymatic activity you obtain only for N-hydroxylation?

LOTLIKAR: No.   It's also for ring-hydroxylation.

LEVIN: Do these preparations metabolize benzo[α]pyrene? it is important to define what carcinogen(s) are metabolized by the protease treated microsomes.

LOTLIKAR: Not for benzpyrene.   Carcinogen activation and ring-hydroxylation of 2-acetylaminofluorene.

HILDEBRANDT: I would like to address a question to Dr. Lu. Dr. Lu, I enjoyed your beautiful presentation.   Do you have evidence for a synergistic effect with respect to NADH and NADPH? If you put your reconstituted preparation into cuvettes and look for the steady state kinetics of cytochrome b_5, do you get similar redox changes as we observed with whole microsomes which exceeded about 40%?

LU: No, we have not looked into this.   This is certainly one of the things we want to look into,

HILDEBRANDT: You should do so, because I think this experiment, and not so much the synergistic effect, is the very point with respect to b_5 involvement.

CONTRIBUTORS

Bäckström, Dan
Department of Biophysics, University of Stockholm,
Stockholm, Sweden

Ballou, David P
Department of Biological Chemistry, Medical School, University of
Michigan, Ann Arbor, Michigan

Baron, Jeffrey
The Toxicology Center, Department of Pharmacology, University
of Iowa, Iowa City, Iowa

Blumberg, William E.
Bell Laboratories, Inc., Murray Hill, New Jersey

Bock, Karl W.
Institut für Toxikologie der Universität Tübingen,
Tübingen, Germany

Bresnick, Edward
Department of Cell and Molecular Biology, Medical College of
Georgia, Augusta, Georgia

Cinti, Dominick L.
Department of Pharmacology, University of Connecticut Health
Center, Farmington, Connecticut

Conney, Allan H.
Department of Biochemistry and Drug Metabolism, Hoffman-La
Roche, Inc., Nutley, New Jersey

Coon, Minor J.
Department of Biological Chemistry, Medical School, University
of Michigan, Ann Arbor, Michigan.

Cooper, David Y.
Harrison Department of Surgical Research, School of Medicine,
University of Pennsylvania, Philadelphia, Pennsylvania

Czygan, Peter
Ludolf-Krehl-Klinik, Heidelberg, Germany

Dean, William L.
Department of Biological Chemistry, Medical School, University
of Michigan, Ann Arbor, Michigan.

511

Dressler, Kenneth
The Stratton Laboratory for the Study of Liver Disease, Mount
Sinai School of Medicine of the City University of New York,
New York, New York

Dus, Karl
Present address: Department of Biochemistry, St. Louis University Medical School, St. Louis, Missouri

Ehrenberg, Anders
Department of Biophysics, University of Stockholm, Stockholm,
Sweden

Felton, James S.
Section on Developmental Pharmacology, Laboratory of Biomedical
Sciences, National Institute of Child Health and Human Develop-
ment, National Institutes of Health, Bethesda, Maryland

Garro, Anthony J.
The Stratton Laboratory for the Study of Liver Disease, Mount
Sinai School of Medicine of the City University of New York,
New York, New York

Gillette, James R.
Laboratory of Chemical Pharmacology, National Heart and Lung
Institute, National Institutes of Health, Bethesda, Maryland

Graves, T.
Department of Population Sciences and Department of Obstetrics
and Gynecology, Harvard University, Boston, Massachusetts

Greim, Helmut
Institut für Toxikologie, Tübingen, Germany.

Grundin, Robert
Department of Forensic Medicine, Karolinska Institutet,
Stockholm, Sweden

Guengerich, F. Peter
Department of Biological Chemistry, Medical School, University
of Michigan, Ann Arbor, Michigan.

Haugen, David A.
Department of Biological Chemistry, Medical School, University
of Michigan, Ann Arbor, Michigan

Hildebrandt, Alfred G.
Department of Clinical Pharmacology, Free University of Berlin,
Berlin, West Germany.

Holtzman, Jordan L.
Clinical Pharmacology Section, Veterans Administration Hospital
and Departments of Pharmacology and Medicine, University of
Minnesota, Minneapolis, Minnesota.

Hopkinson, John
Department of Cell and Molecular Biology, Medical College of
Georgia, Augusta, Georgia

Hutterer, Ferenc
The Stratton Laboratory for the Study of Liver Disease, Mount
Sinai School of Medicine of the City University of New York,
New York, New York

Jansson, Ingela
Department of Pharmacology, Yale University School of Medicine,
New Haven, Connecticut

Krauss, Peter
Institut für Toxikologie der Universität Tübingen, Tübingen,
Germany.

Kuntzman, Ronald G.
Department of Biochemistry and Drug Metabolism, Hoffman-La Roche,
Inc., Nutley, New Jersey

Levin, Wayne
Department of Biochemistry and Drug Metabolism, Hoffman-La Roche,
Inc., Nutley, New Jersey

Litchfield, William J.
Department of Biochemistry, University of Illinois, Urbana,
Illinois

Lu, Anthony Y. H.
Department of Biochemistry and Drug Metabolism, Hoffman-La Roche,
Inc., Nutley, New Jersey

Mannering, Gilbert J.
Department of Pharmacology, University of Minnesota Medical
School, Minneapolis, Minnesota

Marazzo, Donald P.
Johnson Research Foundation, University of Pennsylvania,
Philadelphia, Pennsylvania

McIntosh, E. Noel
Department of Population Sciences and Department of Obstetrics
and Gynecology, Harvard University, Boston, Massachusetts

Miguel, Anne G.
Department of Biochemistry, University of Illinois, Urbana,
Illinois

Mitchell, J. R.
Laboratory of Chemical Pharmacology, National Heart and Lung
Institute, National Institutes of Health, Bethesda, Maryland

Mock, Donald M.
Department of Biochemistry, University of Texas Health Science
Center, Southwestern Medical School, Dallas, Texas.

Moldéus, Peter
Department of Forensic Medicine, Karolinska Institutet, Stockholm,
Sweden

Narasimhulu, Shakunthala
Harrison Department of Surgical Research, School of Medicine,
University of Pennsylvania, Philadelphia, Pennsylvania

Nebert, Daniel W.
Section on Developmental Pharmacology, Laboratory of Biomedical
Sciences, National Institute of Child Health and Human Development,
National Institutes of Health, Bethesda, Maryland

Nehls, Peter
Institut für Toxikologie der Universität Tübingen, Tübingen,
Germany

Orrenius, Sten
Department of Forensic Medicine, Karolinska Institutet, Stockholm,
Sweden

Ozols, Juris
Department of Biochemistry, University of Connecticut Health
Center, Farmington, Connecticut

Peisach, E.
Institute for Developmental Studies, New York University, New
York, New York

Peisach, Jack
Departments of Pharmacology and Molecular Biology, Albert
Einstein College of Medicine of Yeshiva University, Bronx, New
York

Peterson, Julian, A.
Department of Biochemistry, University of Texas Health Science
Center, Southwestern Medical School, Dallas, Texas

Popper, Hans
The Stratton Laboratory for the Study of Liver Disease, Mount
Sinai School of Medicine of the City University of New York,
New York, New York

Prichard, P. M.
Department of Cell and Molecular Biology, Medical College of
Georgia, Augusta, Georgia

Remmer, Herbert
Institut für Toxikologie der Universität Tübingen, Tübingen,
Germany

Rexer, Bernhard
Institut für Toxikologie der Universität Tübingen, Tübingen,
Germany

Roots, Ivar
Department of Clinical Pharmacology, Free University of Berlin,
Berlin, Germany

Rosenthal, Otto
Harrison Department of Surgical Research, School of Medicine,
University of Pennsylvania, Philadelphia, Pennsylvania

Ryan, Dene
Department of Biochemistry and Drug Metabolism, Hoffman-La Roche,
Inc., Nutley, New Jersey

Saalfrank, Klaus
Department of Clinical Pharmacology, Free University of Berlin,
Berlin, West Germany

Salhanick, Hilton A.
Department of Population Sciences and Department of Obstetrics
and Gynecology, Harvard University, Boston, Massachusetts

Sasame, Henry A.
Laboratory of Chemical Pharmacology, National Heart and Lung
Institute, National Institutes of Health, Bethesda, Maryland

Schaffner, Fenton
The Stratton Laboratory for the Study of Liver Disease, Mount
Sinai School of Medicine of the City University of New York,
New York, New York

Schenkman, John B.
Department of Pharmacology, Yale University School of Medicine,
New Haven, Connecticut

Schleyer, Heinz
Harrison Department of Surgical Research and Johnson Research
Foundation, University of Pennsylvania, Philadelphia, Pennsylvania

Scholz, Roland
Institut für Physiologische Chemie und Physikalische Biochemie,
Universität München, Munich, Germany

Schulman, Martin P.
Department of Pharmacology, School of Basic Medical Sciences,
University of Illinois School of Medicine, Chicago, Illinois

Snyder, Robert
Department of Pharmacology, Thomas Jefferson University,
Philadelphia, Pennsylvania

Stern, Gerald O.
Departments of Pharmacology and Molecular Biology, Albert
Einstein College of Medicine of Yeshiva University, Bronx, New
York

Tephly, Thomas R.
The Toxicology Center, Department of Pharmacology, University of
Iowa, Iowa City, Iowa

Thomas, James H.
Harrison Department of Surgical Research, School of Medicine,
University of Pennsylvania, Philadelphia, Pennsylvania

Thorgeirsson, S. S.
Laboratory of Chemical Pharmacology, National Heart and Lung
Institute, National Institutes of Health, Bethesda, Maryland

Thurman, Ronald G.
Johnson Research Foundation, University of Pennsylvania,
Philadelphia, Pennsylvania

Uzgiris, V. I.
Department of Population Sciences and Department of Obstetrics
and Gynecology, Harvard University, Boston, Massachusetts

Vadi, Helena
Department of Forensic Medicine, Karolinska Institutet, Stockholm,
Sweden

van der Hoeven, Theodore A.
Department of Biological Chemistry, Medical School, University
of Michigan, Ann Arbor, Michigan

Vars, Harry M.
Harrison Department of Surgical Research, School of Medicine,
University of Pennsylvania, Philadelphia, Pennsylvania

Vatsis, Kostas
Department of Pharmacology, School of Basic Medical Sciences,
University of Illinois School of Medicine, Chicago, Illinois

Vermilion, Janice L.
Department of Biological Chemistry, Medical School, University
of Michigan, Ann Arbor, Michigan

von Bahr, Christer
Department of Medicine, Huddinge University Hospital, Stockholm,
Sweden

Vore, Mary
Department of Biochemistry and Drug Metabolism, Hoffman-La Roche,
Inc., Nutley, New Jersey

Wagner, T. R.
The Toxicology Center, Department of Pharmacology, University of
Iowa, Iowa City, Iowa

West, Susan B.
Department of Biochemistry and Drug Metabolism, Hoffman-La Roche,
Inc., Nutley, New Jersey

Witmer, Charlotte
Department of Pharmacology, Thomas Jefferson University,
Philadelphia, Pennsylvania

DISCUSSION PARTICIPANTS

Jerina, Donald M.
Section Oxidation Mechanisms, National Institute Arthritis,
Metabolism and Digestive Disorders, National Institutes of
Health, Bethesda, Maryland

Kamin, Henry
Department of Biochemistry, Duke University, Durham, North
Carolina

Lichtenberger, Fritz
Fachbereich Theoretische Medizin der Universität des Saarlandes,
Homburg (Saar)

Lotlikar, Prebhakar D.
Fels Research Institute, Temple University School of Medicine,
Philadelphia, Pennsylvania

Masters, Bettie Sue
Department of Biochemistry, Southwestern Medical School,
Dallas, Texas

AUTHOR INDEX

If an author's name is cited in text and references of an article, only the text page is listed. For names cited only in the references the page number of the latter is given enclosed in parentheses. Names cited in the discussions are not included. No name is cited more than once per page.

Aarts, E.M.
 (500)
Alonso, C.
 (226),(227)
Alvares, A.P.
 2, 8, 104, (146), 190, 191,
 197, 198, 201
Ames, B.N.
 104, 105, 109, (145)
Amruthavolli, E.
 121
Appleby, C.A.
 (212)
Arcos, J.C.
 92
Arcos, M.
 92
Aspiras, L.
 (149)
Aust, S.D.
 (146)
Autor, A.P.
 26, 37
Axelrod, J.
 (501)
Bacchin, P.G.
 (125), (126)
Baird, W.M.
 (148)
Ball, J.A.
 (404)
Ballou, D.P.
 26, (431), 468
Barakat, H.A.
 370
Barnes, J.M.
 104
Barnett, C.A.
 (126)
Baron, J.
 56, 60, 61, 62, 63, 65, 68,
 (70), 121, (286), (323),
 (324), 343, 345, (354), (432)
Barth, Chr.
 (367)
Bartlett, G.R.
 118, 273
Baudhuin, P.
 373

Bayer, E.
 (212)
Beaufray, H.
 (379)
Beetlestone, J.
 (186), 175
Beinert, H.
 (227)
Benedict, W.F.
 (147)
Berry, M.N.
 (269)
Berzofsky, J.A.
 192
Bickers, D.R.
 (113)
Bidlack, W.R.
 481
Billiar, R.B.
 67
Birks, J.B.
 92
Björkem, J.
 242, 245
Björkhem, I.
 (403), 406
Blackburn, W.R.
 (354)
Blake, D.A.
 (126)
Bligh, E.G.
 5, 31
Blumberg, W.E.
 190, 192, 193, 197, (201),
 (202), (212), (227), (269)
Bock, K.W.
 (341), 364
Bodin, N.O.
 259
Bondesen, S.
 (269)
Borg, K.O.
 (268)
Boutry, M.
 (366)
Boyd, G.S.
 55, 62, 65, 67, (70)
Boyer, R.F.
 (43)

Brauser, B.
  (366)
Brawerman, G.M.
  (164)
Breckenridge, A.
  (500)
Brentzel, H.J.
  (367)
Bresnick, E.
  158, (164)
Brierley, G.
  (286)
Brodie, B.B.
  (148), (445), (501)
Bromfeld, E.
  (187)
Brookes, P.
  (148)
Brown, H.D.
  (381)
Brown, R.R.
  (114), (403), (431), (481)
Brunengraber, H.
  (366)
Bücher, Th.
  (366)
Buenig, M.
  425
Bullen, M.F.
  (502)
Burleigh, B.D.
  (43)
Burns, J.J.
  (164)
Burstein, S.
  (226), (500)
Buterbaugh, G.G.
  (126)
Buu-Hoi, Ng. Pk.
  (100)
Bye, A.
  (113)
Cammer, W.
  254
Campbell, T.C.
  (114)
Carafoli, E.
  (353)

Carr, M.L.
  241, 422, 426
Carter, E.A.
  369, 370
Cartwright, G.E.
  (353)
Chang, R.
  (113)
Chaplin, M.D.
  12, 271, 426, (434)
Chattopadhyay, S.K.
  (381)
Cheng, S.C.
  (226)
Choi, J.Y.
  (501)
Cinti, D.L.
  124, (186), (227), 406, 469,
  479 (482)
Clemens, M.J.
  (164)
Co, N.
  (226)
Cohen, B.S.
  (404), 406, (445), 447, 468,
  478
Cohen, G.M.
  422, 426
Comai, K.
  2, 197
Conney, A.H.
  1, (22), (23), (24), (44),
  (45), (100), 104, (146),
  (147), (164), 175, (181),
  (186), 190, (202), (403),
  (404), 405, 406, 433, (464),
  468, (500), (501), (502)
Considine, N.
  (146), (147), (202)
Coon, M.J.
  1, 11, 25, 26, 28, 29, 30,
  33, 39, (43), (44), (45),
  48, (147), (186), (285),
  (308), 417, 418, (434),
  (464), (466), (481)
Cooper, D.Y.
  63, (71), 89, 95, 96 (101),
  (113), (149), (226), (227),
  (323), (324), (403), 468

Cooper, J.R.
  (445)
Copp, D.H.
  (353)
Cori, C.F.
  (227)
Correia, M.A.
  (404), 406, 407, 410, 412,
  413, 414, 425, 426, 427, 428,
  447, 455, 460, 468, 478
Cotzias, G.C.
  346
Creaver, P.J.
  (147), (381)
Crystal, R.G.
  (164)
Cunningham, J.L.
  (500), (502)
Cuthbertson, E.M.
  (353)
Czygan, P.
  105, 106, 107, (114), (125),
  (149)
Dallner, G.
  (404)
Daly, J.W.
  17, (22), (23), (145), (147)
Dambach, G.
  124
Danielson, H.
  (403), 406
Dansette, P.M.
  3
Davidson, D.C.
  (500)
Davies, D.S.
  375 (500)
Davis, D.C.
  (146)
Dean, W.L.
  (308)
Debrunner, P.
  80, (323)
DeCarli, L.M.
  369, 370
Decker, K.
  (367)
Deckert, F.W.
  377

DeDuve, C.
  123, (379)
Delwiche, C.V.
  3
Denk, H.
  (125)
Deter, R.L.
  123
Dewaide, J.H.
  251
DiAugustine, R.P.
  12
Dikon, W.R.
  (226)
Dipple, A.
  (148)
Drake, J.W.
  (145)
Dresbach, L.
  (286)
Dressler, K.
  117
Druyan, R.
  (353)
Duppel, W.
  33, (43)
Durston, W.E.
  (113), (145)
Dus, K.
  48, 52, 287, 293, 295, 296,
  299, 303, (309)
Dyer, W.T.
  5, 31
Eliasson, G.
  (500)
Eling, T.E.
  12
Emptage, M
  (308)
Endrenyi, L.
  (502)
Engel, L.L.
  219, 311
Enomoto, K.
  124, 390, 419
Erbes, D.L.
  (53), (308)
Ernster, L.
  89, 379, (404)

Estabrook, R.W.
  12, (70), (71), (101), (186),
  (196), 240, 245, (268), (269),
  271, 311, 312, 321, (324),
  (403), (404), 405, 406, 407,
  408, 414, (434), 435,
  447, 460, 468, 478, 480,
  (483)
Exton, J.H.
  117
Fahmy, M.J.
  (148)
Fahmy, O.G.
  (148)
Falk, H.L.
  104
Falk, J.E.
  345
Farber, E.
  (114)
Farr, A.L.
  (23), (44), (126), (164),
  (353)
Feher, G.
  192
Felton, J.S.
  (146)
Finster, M.
  (113)
Fleischer, S.
  276, (286)
Fleischmann, L.
  (501)
Flesher, J.W.
  (148)
Folch, J.
  118, 273
Ford, J.A.

Foust, G.P.
  (43)
Fouts, J.R.
  81, 190
Franklin, M.
  425, (432)
Franklin, R.M.
  (145)
Frantz, A.G.
  484

Freommer, U.
  (147)
Frey, I.
  (434)
Frey, R.J.A.
  (114)
Fried, J.
  (148)
Friedmann, N.
  124
Fröhlich, W.
  (367)
Frohling, W.
  (341)
Fujita, P.
  1, 26, 95, 191, 198
Gang, H.
  (381)
Ganguli, B.N.
  (324)
Garro, A.J.
  (113), (114), (149)
Gaylor, J.L.
  2, 3, 191, 197, (202)
Gelboin, H.V.
  152, (164), 192
George, P.
  175
Getz, G.S.
  (353)
Gibson, Q.H.
  (70), (380)
Gielen, J.E.
  (146), (147), (202)
Gigon, P.L.
  241, 321, 375, (379), 426, 428
Gillette, J.R.
  117, (146), (147), (148),
  (201), (248), (323), 364,
  (379), (380), (403), 405, 406,
  (432), (434), 440, (465)
Girke, W.
  (148)
Glascoe, P.K.
  243
Gonasun, L.
  (187)
Goujon, F.M.
  (146), 191, 197, 198, 200

Gradnick, R.
  (501)
Graf, H.
  (164)
Gram, T.E.
  (248), (323), (380), (432)
Granick, S.
  343
Green, F.E.
  (354)
Green, I.
  (148)
Greenberg, D.M.
  346
Greene, F.E.
  (379)
Greenwood, F.C.
  (308)
Greim, H.
  (113), (114), (125), (149),
  (202)
Griffin, B.W.
  312, 313
Griffin, P.
  210
Griffith, J.S.
  (186)
Grollman, A.P.
  (164)
Gros, F.
  (164)
Grover, P.L.
  (149)
Grundin, R.
  253, 254, 263, (268), (269)
Grunnet, N.
  (269), 369
Guengerich, F.P.
  (431)
Gunsalus, I.C.
  33, (53), 80, (227), 287, 301,
  (308), (309), 311, 312 (324)
Gut, M.J.
  (226), (500)
Hackenschmidt, J.
  (367)
Hagihara, B.
  (248)
Hallinant, T.
  (286)

Hamilton, J.G.
  (24)
Hammer, W.
  (502)
Hamrick, M.
  (147)
Haniu, M.
  (309)
Hänninen, O.
  267
Hansen, W.
  358
Hardesty, B.
  (164)
Harding, B.W.
  126
Harrison, J.E.
  (53), (308)
Hasegawa, E.
  343, (354)
Hasumura, Y.
  (381)
Haugen, D.A.
  (45), (147), (308), (431),
  (434)
Hayaishi, O.
  (366)
Hayes, J.R.
  (114)
Haynes, R.
  (227)
Heidelberger, C.
  (146), (149)
Heidema, J.
  (23), (43), (45), (186),
  (466)
Henderson, P. Th.
  251 (341)
Henseleit, K.
  365
Hesse, S.
  (367), (382)
Heywood, S.M.
  (164)
Hibbeln, P.
  343

Hildebrandt, A.G.
  43, 175, 197, 201, (323),
  (403), 405, 406, 407, 408,
  414, (432), (445), 447, 460,
  468, 478, 480, (501)
Hill, H.A.O.
  177, (212)
Hinson, J.A.
  (434), (465)
Hochberg, R.B.
  (226)
Hochstein, P.
  379, (481)
Holloway, P.W.
  (403)
Holtzman, J.L.
  241, 247, 251, 422, 426
Hopkinson, J.
  (164)
Horecker, B.L.
  (445)
Horie, S.
  175, 213
Hrycay, E.G.
  461
Hsia, Y.
  (114)
Hsie, A.W.
  (125)
Hubbard, N.
  343
Huberman, E.
  (149)
Hultin, T.
  (164)
Hultquist, D.E.
  468
Hume, R.
  (70)
Hutterer, F.
  (113), (125), 118, (126),
  (149)
Hutton, J.J.
  (146)
Hwang, M.T.
  (164)
Iball, J.
  (147)
Ichii, S.
  219

Ichikawa, Y.
  468, 480
Iizuka, T.
  (248)
Ikeda, M.
  245
Illing, H.P.A.
  433
Imai, Y.
  8, (45), 184, 197, 203, 205,
  206, 207, 210, 211, (404),
  (433), (465), (482)
Inaba, R.
  (502)
Inscoe, J.K.
  (201)
Isaacson, E.L.
  (70), (286)
Ishii, H.
  (381)
Ishimura, Y.
  (324)
Israel, Y.
  369
Isselbacher, K.J.
  369, 370
Ito, A.
  470
Ito, N.
  (114)
Iyanagi, T.
  33, 450
Jacobson, M.
  (45), (113), (147), (186),
  (202), (380), (404), (464),
  (502)
Jacques, P.
  (379)
Jailer, J.W.
  (501)
Jain, S.C.
  (145)
Jansen, P.L.M.
  (341)
Jansson, I.
  (404), 434, 468, 480
Jefcoate, C.R.
  63, 191, 201
Jeffery, E.
  425

Jergill, B.
  124
Jerina, D.M.
  17, (22), (23), (145)
Johnson, B.C.
  (403), 468
Johnson, G.E.
  (502)
Jollow, D.J.
  (147), (148)
Joly, J-G.
  (381)
Jones, M.S.
  343
Jones, O.T.G.
  343
Jost, J.P.
  124
Jungas, R.L.
  (366)
Junk, K.W.
  (44), (285), (464)
Kacew, S.
  120
Kadar, D.
  (502)
Kalant, H.
  (380)
Kalow, W.
  (502)
Kalststein, A.
  (366)
Kamin, H.
  30, 63, (380), 408, 447
Kampffmeyer, H.G.
  (501)
Kappas, A.
  (113)
Katagiri, M.
  (53), (308), (309), 312, 313
Kato, R.
  117, 364, 375
Katz, F.H.
  (501)
Khanna, J.M.
  369, 370, (380)
Kimball, H.L.
  (500)
Kimura, H.
  124

Kimura, T.
  55
King, E.J.
  31
Kinoshita, T.
  175
Kipnich, S.
  (501)
Klaiber, E.L.
  (500)
Klingenberg, M.
  (227), 264
Klouwen, H.
  (286)
Kniffen, T.
  (354)
Kobayashi, S.
  (226)
Kohl, K.B.
  (381)
Kon, H.
  191, 197, (202)
Konishi, Y.
  (114)
Kornberg, A.
  (445)
Kouri, R.E.
  (146), (147), (148)
Kraschnitz, R.M.
  33, (43), (187), (227)
Krauss, P.
  (341)
Krebs, H.A.
  251, 264, (269), 365
Kumaki, K.
  (146)
Kung, W.
  104
Kuntzman, R.
  (21), (22), (23), (24), (44),
  (45), (104), (146), (147),
  (186), (201), (202), (212),
  (380), (404), (433), (464),
  (502)
Kuroki, T.
  (149)
Kusunose, E.
  40
Kusunose, M.
  (44)

Kuwahara, S.I.
  (433), 465, 468, 478, 480
Labbe, R.F.
  343, (344)
Laemmli, U.K.
  29
Lanceos, K.D.
  158, (164)
Landon, J.
  295
Langdon, R.G.
  3, 450
Lardy, H.A.
  (366)
Lawley, P.D.
  (148)
Laycock, D.G.
  (164)
Lebeault, S.M.
  33
Lee, F.D.
  (113), (145)
Lee, R.G.
  350
Lees, M.
  (125), (285)
Lehninger, A.L.
  346
Leibman, K.C.
  12, 271, (323)
Leloir, L.F.
  (367)
Levin, S.S.
  (70), (101), (227)
Levin, W.
  1, 2, 3, 5, 14, (21), (22),
  (23), (24), 26, (146), (147),
  193, 197, 197, (201), (202),
  (212), 379, (433), (434),
  448, (465), (500), (502)
Levine, W.G.
  194
Ley, H.G.
  (382)
Lichtenberger, F.
  424, (434)
Lieber, C.S.
  369, 370, (381)
Lieberman, S.
  217, (226)

Liersch, M.
  (367)
Lin, G.
  370, (380)
Lin, R.
  (380)
Lin, S.W.
  (164)
Lipscomb, J.D.
  80, 301, 303, (308), (309),
  (323)
Litchfield, W.J.
  (53), 291, 294, 295, (308)
Little, B.
  67
Livanou, T.
  (308)
Lode, E.
  (44)
Loehr, J.S.
  468, 480
Long, F.A.
  243
LoSpalluto, J.
  (70)
Lowenstein, J.M.
  (366)
Lowenstein, L.
  (366)
Lowry, O.H.
  3, 27, 118, (164), 192, 345
Lu, A.Y.H.
  1, 2, 3, 11, 14, 16, (22),
  (23), (24), 25, 30, 33,
  (146), 175, 192, 193, 197,
  198, (202), (212), 271,
  (380), 406, 417, 418, 419,
  (434), 447, 448, 462, (466),
  500
Luken, J.N.
  (353)
Luttrell, B.
  (226)
McDonald, J.C.
  (114)
McDonald, P.D.
  (226)
McIntosh, E.N.
  (226), (227)

McIntosh, W.B.
  (500)
McKay, P.B.
  379, (404)
McKay, R.
  343
McKenna, W.R.
  (367), 369, 375
McKibbin, J.B.
  (126)
McLean, A.E.M.
  104
McLean, E.K.
  104
Magee, P.N.
  104
Magour, S.
  104
Malling, H.V.
  104
Mannering, G.J.
  1, 2, 8, 12, (44), (45), 81,
  (101), 104, 197, (202), 204,
  207, 208, 211, 271, (286),
  (404), 408, 421, 422, 424,
  447, 455, 460, 461, 463, 468,
  478, 480, (481)
Marazzo, D.
  (367)
Margolis, S.
  (269)
Marquardt, H.
  (149)
Marsh, C.A.
  487
Marshall, V.
  80, 104
Märtens, S.
  (502)
Marugami, M.
  104
Marver, H.S.
  (381)
Mason, H.S.
  33, 95, 193, (202), 210,
  (366)
Mason, H.W.
  450
Mason, J.I.
  55, 66, 67

Masters, B.S.S.
  56, 60, (70), 281, 375,
  (381)
May, H.E.
  379, (404)
Maynard, L.S.
  346
Mazel, P.
  104
Medzihradsky, F.
  500
Meeks, J.R.
  (323)
Meijer, A.J.
  (269)
Menard, R.H.
  (434), (465)
Merker, H.J.
  (366), (501)
Merrick, W.C.
  (164)
Meyer, W.
  (501)
Mezey, E.
  370
Mgbodlie, M.U.K.
  104
Miguel, A.G.
  (53), (308)
Mihara, V.
  (404), (465)
Miller, C.
  (382)
Miller, E.C.
  104, (145), (403), (431),
  (481)
Miller, J.A.
  104, (145), (403), (431),
  (481)
Miller, R.L.
  (164)
Misra, P.
  (381)
Mitani, F.
  (226), 213
Mitchell, J.R.
  (147), (148), (434), (445),
  (465)
Mitoma, C.
  271

Modurzadeh, J.
    408, 447
Moffitt, A.E.
    347
Moldeus, P.
    (186), (227), 252, 257, 259,
    261, 262, 264, (268), (431),
    (481)
Morton, R.K.
    272
Moss, D.W.
    (501)
Mosse, H.
    (164)
Mostellen, R.D.
    (164)
Munck, E.
    80, (323)
Munro-Faure, A.D.
    (113)
Murad, F.
    (126)
Murakami, K.
    193
Murphy, S.D. •
    347
Myer, U.A.
    (381)
Nacht, S.
    (353)
Narasimhulu, S.
    (285), (286), (403)
Nash, T.
    377, (445)
Neale, G.
    (501)
Nebert, D.W.
    33, 40, (146), 175, 191, 192,
    197, 198, 200, 201, (202)
Nehls, P.
    (341)
Nelson, E.B.
    370, 408
Netter, K.J.
    (323), (367), 408
Neville, A.M.
    219
Newberne, P.M.
    104
Newton, N.A.
    (354)

Nierel, J.G.
    104
Nilsson, A.
    (403), 468
Niwa, A.
    (146)
Oesch, F.
    3, (147), (148)
Ohlsson, R.
    124
Ohno, H.
    55
Okita, R.
    (481)
Olinger, Ch.
    (501)
Olsen, N.S.
    (227)
Omata, S.
    (266)
Omura, T.
    3, 27, 31, 55, 60, 62, 68,
    82, 87, (147), (212), 312,
    (367), 377, (404), 414,
    (433), (465), 468, (482)
Onoda, K.-I.
    (380)
Oppelt, W.W.
    (126)
Orme, M.
    (500), (501)
Orme-Johnson, W.H.
    (227)
Orrenius, S.
    89, (147), (186), (202),
    (228), (268), (269), (404)
Oshino, N.
    393, 400, (404), 409, 458,
    467, 468
Ottolenghi, A.
    (381)
Ozols, J.
    (431), 469, 479
Pain, V.M.
    (164)
Papa, S.
    (269)
Papenberg, J.
    369

Park, C.R.
(125)
Parke, D.V.
(147), (501)
Parker, F.
(285)
Parlow, D.
(501)
Passon, P.G.
468
Peck, A.W.
(113)
Peisach, J.
190, 192, 193, 194, (201),
(202), 210, (212), (227), 254
Peluffo, R.O.
(403)
Pennington. S.N.
374 (379)
Peraino, C.
104
Peterson, J.A.
210, 311, 312, 313
Peterson, N.F.
(285)
Phillips, A.H.
3, 450
Picciano, D.J.
(164)
Piper, W.
(354)
Pogell, B.M.
(367)
Poland, A.
40, (113), (146)
Polis, B.D.
(286)
Popper, H.
(113), 104, (149)
Poppers, P.J.
(113)
Porra, R.J.
344, 349
Porter, C.C.
487
Posner, H.S.
(201)
Potter, J.J.
(381)

Potter, W.Z.
(148)
Poyer, J.L.
(380)
Price Evans, D.A.
(500), (502)
Prichard, P.M.
(164)
Purvis, J.L.
(70)
Quagliariello, E.
265
Quistorff, B.
251, (380)
Rabinowitz, M.
(353)
Rahman-Li, Y.
(379)
Raiha, N.C.R.
(126)
Raisfeld, I.H.
121
Rall, T.W.
(126)
Ramasarma, T.
121
Ramseyer, J.
(226)
Randall, R.J.
(23), (44), (126), (164),
(202), (353)
Ratrie, III, H.
(147), (148)
Rau, D.
(501)
Ravel, M.
(164)
Rawlins, M.D.
(501)
Ray, G.S.
(227)
Raygatt, P.R.
(227)
Reed, W.D.
(381)
Reese, W.N.
(381)
Reich, E.
(145)

Remmer, H.
  (101), (186), (202), (269),
  (323), (341), (366), 377,
  (380), 403, (434), (481),
  (501)
Rexer, B.
  (341)
Rickenberg, H.V.
  124, (125)
Rickert, D.E.
  190
Rizack, M.A.
  123
Roach, M.K.
  370
Robinson, G.R.
  (146), (202)
Robison, G.A.
  (125)
Röder, A.
  (186), (212)
Rogers, A.E.
  104, 124
Rogers, M.J.
  (404), 419, 469, (483)
Roots, I.
  (501)
Rosalki, S.B.
  (501)
Rosebrough, N.J.
  (23), (44), (126), (164),
  (196), (202), (353)
Rosenthal, O.
  (70), (71), 89, 96, (101),
  (149), (226), (227), (323),
  (324), (403)
Ross, B.P.
  (269)
Ross, W.E.
  117
Rossi, C.S.
  (353)
Rothman, V.
  (269)
Rubin, E.
  370
Rudick, J.
  (113)
Rumack, B.H.
  422, 426

Ryan, D.
  (22), (44), (146), 202,
  (212), (433), (434), (500)
Ryan, K.J.
  311, (464), (465)
Saalfrank, K.
  (501)
Salhanick, H.A.
  (226), (227)
Salser, W.
  (164)
Sanders, E.
  (70), (71), (82), (101),
  (324)
Sasame, H.A.
  (146), (379), (403), 406,
  414, 417, 419, 421, 424,
  440, 447, 461
Satake, H.
  (45)
Sato, R.
  3, 8, 26, 27, 31, (82), 87,
  (101), 124, 147, 184, 192,
  (202), 203, 205, 206, 207,
  211, (212), (367), 377, 390,
  393, (403), 404, 409, 419,
  426, (465), 467, 468, 470,
  (482)
Schacter, B.A.
  379
Schaffner, F.
  (113), (114), (125), (126),
  (149)
Schafritz, D.A.
  (164)
Schauder, G.
  (113)
Schenkman, J.
  82, 125, 177, 178, (186), 190,
  197, (227), 252, (268),
  (323), (403), (404), 407, 419,
  421, 426, 428, (431), (432),
  468, 469, 480, (481)
Schilling, G.R.
  (21), (201)
Schleyer, H.
  (70), (101), (226), (227)
Schlotte, W.
  (367)

Schoenthal, R.
  104
Scholz, R.
  265, 358, (366), (367), 370,
  (382)
Schultze, H.U.
  (286)
Schumm, D.E.
  (148)
Schuppel, R.
  (367)
Schwabe, U.
  (366)
Schwartz, A.L.
  120
Schwartz, F.
  367
Schwartz, H.P.
  (286)
Schweet, R.
  (164)
Seglen, P.O.
  251
Seidel, G.
  (367)
Selkirk, J.K.
  (149)
Sellinger, O.F.
  (379)
Shargel, L.
  104
Sharrock, M.
  80
Shatkin, A.J.
  (145)
Shigematsu, A.
  (432)
Shimakata, T.
  (404)
Shoeman, D.W.
  38, (44), (202), 426
Sholiton, L.J.
  (501)
Shrago, E.
  (366)
Siekevitz, P.
  (146), 197
Sies, H.
  (366)

Silber, R.H.
  487
Silverstone, H.
  104
Simpson, E.R.
  55, 62, 65
Simrell, C.
  (126)
Sims, P.
  (148), (149)
Singer, R.W.
  194
Singhal, R.
  120
Sitar, D.S.
  406
Sjögvist, F.
  (500), (502)
Skanberg, I.
  (268)
Skews, T.
  (125)
Sladek, N.E.
  2, 8, 81, 204, 207, 208, 211
Slauterback, B.D.
  (286)
Sloan-Stanley, G.H.
  (285)
Smith, A.L.
  222
Smith, C.
  (353)
Smith, P.C.
  (53), (308)
Smith, S.E.
  (501)
Snyder, R.
  176, (341)
Sobell, H.B.
  (145)
Soehring, K.
  (367)
Soffer, E.
  (286)
Sokoloff, L.
  (164)
Somogyi, A.
  (500)

Sotaniemi, E.A.
  490
Spatz, L.
  27, 240, (404), 419, 448,
  (483)
Speck, M.
  (501)
Staffelt, E.
  (114)
Stanley, G.H.S.
  (125)
Staudinger, Hj.
  (286)
Staudt, H.
  424, 425, 426
Stern, J.O.
  5, 191, 193, 197, (202), 211,
  (212), (269)
Stewart, M.L.
  (164)
Stripp, B.
  (147)
Strittmatter, P.
  27, 124, 240, 390, (404),
  406, 448, 468, 469, 470,
  471, 477
Strobel, H.W.
  1, 11, 25, 39, (45), (186),
  448
Sulimorici, S.
  55
Sutherland, E.W.
  (125), 214
Sydnor, K.L.
  (148)
Szarkowska, L.
  (227)
Tager, J.M.
  (269)
Tagg, J.
  271
Takanaka, A.
  (380)
Takao, H.
  (248)
Tanaka, M.
  287, 288
Tannenbaum, A.
  104

Tapscott, E.
  (379)
Tatum, E.L.
  (145)
Taylor, W.E.
  (70)
Tephly, T.R.
  121, 343, 345 (353)
Tepperman, H.M.
  (367)
Tepperman, J.
  (367)
Teshka, R.
  370
Thieden, H.I.
  369 (380)
Thomas, C.
  (113)
Thomas, E.
  (126)
Thomas, P.E.
  (146)
Thompson, J.A.
  247
Thor, H.
  (202)
Thorgeirsson, S.S.
  (148), (434), (445), (465),
  (500)
Thurman, R.G.
  265, 358, (366), (367), 369,
  370, 375
Trams, E.R.
  (201)
Trasher, K.
  487
Trussler, P.
  (354)
Tsai, R.
  (227), (309)
Tyler, D.D.
  240, 245
Tyler, W.E.
  (286)
Tyson, C.A.
  287, (323)
Udenfriend, S.
  (22)

Ullman, H.
   (367)
Ullrich, V.
   (147), 253, (434), 468
Umbreit, J.N.
   286
Urata, C.
   343
Uzgiris, V.I.
   (226), 227
Uzuki, F.
   (187)
Vadi, H.
   (268), (269)
Van der Hoeven, T.A.
   1, 26, 29, 37, (43), 48,
   (147), (226), (308), (431),
   (434), (481)
Vatsis, K.P.
   369, 370, 371, 376
Veech, R.L.
   (366)
Veeger, C.
   (43), (481)
Velick, S.
   240, 471
Vermilion, J.L.
   (431)
Vitols, K.S.
   (354)
Von Bahr, C.
   (227), 252, 256, (268),
   (269), (500)
Von der Decken, A.
   (164)
Vore, M.
   12, (433), (434), (465)
Wakil, S.
   (366), (403)
Walkden, V.
   (286)
Warburg, O.
   110
Wartburg, J.P.
   (381)
Waterman, M.R.
   45
Wattenberg, L.W.
   (148)

Wattiaux, R.
   (379)
Webb, C.
   (354)
Weiner, M.
   117, 120
Welch, M.
   (226)
Welch, R.
   (113)
Wells, A.
   (187)
Welton, A.R.
   (146)
Werk, E.E.
   (501)
West, S.B.
   16, (44), (146), (186),
   (202), (212), (433), (464),
   (465), (500)
White, J.G.
   (45)
Whitehouse, M.W.
   (227)
Whitfield, J.B.
   (501)
Whitmire, C.E.
   (147), (148)
Whysner, J.A.
   (226)
Wicks, W.D.
   124
Wilk, M.
   (148)
Williams, C.H.
   30, (43), (70), (380)
Williams, R.J.P.
   (186), (212)
Wilson, B.J.
   (227)
Wilson, R.D.
   (164)
Winer, B.J.
   193
Witkop, B.
   (22)
Witmer, C.
   (341)

Wolff, J.A.
  (113)
Wortham, J.S.
  (164)
Wyard, S.J.
  180
Yagi, H.
  (22)
Yamasaki, E.
  (113), (145)
Yasunobu, K.T.
  (309)
Yonetani, T.
  (227)

Young, F.W.
  (366)
Yu, C.A.
  (53), (227), 287, 301, (308)
Zaltzman-Nirenberg, P.
  (22)
Zampaglione, N.
  (147)
Zange, M.
  (202)
Zemaitas, M.A.
  347

SUBJECT INDEX

To simplify use of the index, all entries related to the heme protein P-450 appear as
P-450.
Words such as cytochrome, heme protein, hemoprotein, etc., are omitted. The listing in the index has been further simplified by not distinguishing between different forms of the heme protein, such as P-450, P-448, P_1-450 or P-450 $_{LM}$; all appear under the general heading P-450.

Oxidized and reduced pyridine nucleotides are listed as NAD and NADP, and NADH and NADPH, respectively. Other abbreviations used include PB (phenobarbital), MC (3-methylcholanthrene), ISP (iron sulfur protein), AHH (aryl hydrocarbon hydroxylase), DMN (dimethylnitrosamine), and MNNG (N-methyl-N'-nitro-N-nitroso-guanidine).

Absorption spectra, optical
  correlation with EPR
  spectra   175 ff
  modification by
  ligands   213 ff
  of P-450 (Fe^{2+})·CO,
  rabbit liver preparations
  after induction by PB
  or MC   182-183
  of partially purified
  P-450 and P-448   5-6, 9
  of partially purified
  P-450 preparations from
  rabbit liver   175 ff
Actinomycin D,
  interaction with DNA
  base pairs   128-129
Activation
  of adrenocortical
  microsomal hydroxylation
  by detergents   271 ff
  of carcinogens   103 ff
                             170 (disc.)
  of MC as carcinogen   127 ff
  of polycyclic hydro-
  carbons to mutagens   136
Adrenal mitochondria   58, 63
  (see also Mitochondria)
  interaction with anti-
  bodies against adrenal
  ISP   58
Adrenodoxin (see also
  Iron sulfur protein
  (ISP), Antibodies to
  (ISP)
  amino acid composition
  of -   289
  ISP isolated from
  adrenal glands   79 (disc.)
Aflatoxin
  as mutagen   128 ff
Alkyl derivatives
  of SH groups from
  P-450   301-307
Alprenolol$_{cam}^m$ [1-(2-allyl-
  phenoxy)-3-isopropyl-
  aminopropanol]
  binding to P-450 in
  isolated rat liver cells
                             251 ff

effect on EPR spectra of
  P-450 in isolated rat
  liver cells   258-259
  metabolism in iso-
  lated rat liver
  cells   258 ff
Amino acid composition
  of adrenodoxin and
  putidaredoxin   287 ff
  of P-450$_{cam}$ 49
Amino acid sequence,
  proposed for adreno-
  doxin and putidare-
  doxin   290
Aminoglutethimide [2-
  ethyl-2(p-amino-
  phenyl)-glutarimide],
  effects on bovine
  corpus luteum
  preparations   216, 226
Aminopyrine
  as substrate   40
  clearance (total
  body) of   485 ff
                             494-495
  interaction with
  gluconeogenesis and
  lipid synthesis in per-
  fused rat liver   355 ff
  N-demethylation of -,
  inhibition by pretreat-
  ment of rat liver microsomes with detergent-
  treated cytochrome b$_5$ 391
Anaerobic redox titration
  of P-450 preparations   33-38
                             73-74 (disc.)
  suggested model   75-78 (disc.)
Aniline
  as substrate   41
  binding to partially
  purified P-450 and
  P-448   16-17
  effect of D$_2$0 on aniline binding
  spectra   243-244
  effect of D$_2$0 on
  hydroxylation of -   241

hydroxylation of - by
rat liver microsomes,
time course of
induction   89-91, 94
photochemical action
spectra (rat liver
microsomes)   89-91, 96
                             166 ff (disc.)
                             172 (disc.)
Antibodies
  (see also Immuno-
  chemistry)
  effect on NADPH-cyto-
  chrome c reductase   57, 59-62
  inhibition of laurate
  hydroxylation in rat
  liver and kidney
  cortex   436
  interaction with ISP(s)
  from other species and
  organs   56, 58
  to adrenal ISP   56 ff
  to cytochrome b_5 414 ff
                             435 ff
  to P-450$_{cam}$ 48-51, 292 ff
Arene oxide
  (see also epoxide)
  as hydroxylation product
  of MC, role in transport
  and carcinogenesis   127 ff
                             172-173 (disc.)
  as intermediate in
  metabolism of poly-
  cyclic aromatic hydro-
  carbons   130 ff
  criteria for possible
  involvement in car-
  ginogenesis   168-169 (disc.)
  formation of -, via aryl
  hydrocarbon hydroxylase
  (AHH)   16-17, 19, 81 ff
  possible transport of -
                        168-169 (disc.)
Aryl hydrocarbons,
  induction of P-450
  by a series of -   97-98
Aryl hydrocarbon hydroxylase
  (AHH),
  (aryl hydroxylase, benzpyrene-
  hydroxylase)   81 ff, 127 ff

induction by substrates and
relation to
carcinogenesis   139-140
                             168-169 (disc.)

Benz[a]anthracene, activation to
  frameshift mutagen   127 ff
Benz[a]pyrene
  (see Benzo[a]pyrene)
Benzene, spectral effect
  of interaction with
  P-450   175 ff
                             230 (disc.)
Benzo[a]pyrene
  activation to frameshift
  mutagen   128-130
  after MC treatment   235-236
  as inducer   97-98, 127 ff
  comparison of hydroxy-
  lation activities in
  liver microsomes and
  reconstituted systems
  (MC-treated rats)   456
  effect of cytochrome b_5
  on NADPH dependent hy-
  droxylation of -
                             453, 455-457
  hydroxylase activity in
  rat and rabbit liver
  microsomes   229-230
  hydroxylase activity in
  rat liver microsomes
  after induction by MC 189 ff
  hydroxylation of - 16-17, 20
  hydroxylation of -, and
  P-450 content after
  MC treatment   235-236
  NADH-dependent hy-
  droxylation of -   457-458
3,4-Benzpyrene
  (see Benzo[a]pyrene)
Benzo[a]pyrene hydroxylase
  (see aryl hydrocarbon
  hydroxylase)
Benzphetamine
  effects of NADPH-cytochrome
  c reductase, and NADH-cytochrome b_5 reductase on
  N-demethylation of -   452

inhibition of hydroxyla-
tion by antibodies to
P-450 (liver microsomes)
and P-450$_{cam}$ 51
N-demethylation of -   14-15
                              18-19
                          32, 40-41
                            448 ff
spectral changes on
addition to partially
purified P-450
preparations   16-17
synergistic effects in
in N-demethylation of -   455
1-benzyl-phenylimidazole,
effects on bovine
corpus luteum mito-
chondria   216
Binding spectra
  (see Difference spectra,
  Type I and Type II spectral
  changes)

Camphor 5-exo-methylene
hydroxylase of P. putida
  (see also P-450$_{cam}$)   287 ff
  metabolic control of -   311 ff
Cancer
  (see carcinogens)
Carbon monoxide
  (see also individual
  reactions)
  inhibition of camphor
  hydroxylation by -   311-312
                         315-316
  inhibition of mixed
  function oxidase
  systems by -   81 ff
  inhibition of mutagenic
  action of DMN   103 ff
  inhibition of NADPH-
  oxidation in rat liver
  microsomes by -   247
Carcinogens
  (see also Mutagens,
  Mutagenicity)
  activation of -   103 ff
                              127 ff
                        170 (disc.)

carcinogenic activity
and role of P-450
                     167-169 (disc.)
carcinogenic index   143
DMN as secondary -, correlation of mutagenic
activity with microsomal
enzyme activity   105-107
inactivation of -   103 ff
mechanisms of initia-
tion of cancer by -   134 ff
MC as -, genetic dif-
ferences   142
MNNG as primary -   105-107
relationship to DNA
repair   169 (disc.)
Catalase
activity in liver slices
and liver microsomes of
several strains of
mice   369 ff
and its possible link to
ethanol oxidation in
liver microsomes 384-385 (disc.)
Cations
effects of divalent -, on
ferrochelatase activity
in rat liver mitochon-
dria   347
Chloramphenicol
as substrate for UDP-
glucuronosyltransferase
activity   335 ff
Chlorobenzene
effects of cytochrome b$_5$
on hydroxylation of -   459
Cholesterol
inhibition of side chain
cleavage reaction of -,
by antibodies to adrenal
ISP   62
side chain cleavage reaction
of -,   55 ff
side chain cleavage re-
action of -, in mitochon-
dria from bovine corpus
luteum   213 ff

Clemastine
(2- (2-p-chloro-α-methyl-α-
phenylbenzyl) oxyethyl - 1-
methyl-pyrrolidine)
effects on induction   487 ff
Cobalt
inhibition of ferro-
chelatase activity in
rat liver mitochondria
by Co^{2+} ions   346-347
Codeine demethylation
(in rat
liver microsomes)
effects of MC on -   89, 95
induction by PB   89, 95
photochemical action
spectra
89-91
96, 165-167 (disc.)
Competitive binding,
studies on P-450$_{cam}$ 298
Components,
of mixed function oxi-
dase systems   14-15
25-26
Conjugation
with glucuronic acid,
of hydroxylated sub-
strates in isolated
brain cells   267
Copper
(see also ferrochelatase)
activation of ferrochela-
tase in rat liver mito-
chondria by Cu^{2+} ions   347
Corpus luteum (bovine),
P-450 in mitochondria
from -   312 ff
Cross reactions, immuno-
logical
(see also Antibodies)
Anti-putidaredoxin,
with adrenodoxin   292 ff
microsomal P-450 with
Anti-P-450$_{cam}$ 47-50
Cyanide ions
effects on NADH-dependent
3,4-benzpyrene
hydroxylation -   461

effects on stearyl-CoA-
desaturase   401 ff
506-507 (disc.)
Cyanogenbromide (BrCN)
modification reagent for
P-450   48
modification reagent for
P-450$_{cam}$ 297
"Cyanoketone"
[2α-cyano-17β-hydroxy-
4,4,17α-trimethyl-androst
5-en-3-one],
bovine corpus luteum
mitochondria   216, 220-223
Cyclic AMP
(cyclic adenosine-3',5'-mono-
phosphate, N^6, O^2 -dibutyryl
derivative;   cAMP)
effect on induction of
P-450 by PB   117 ff
165 (disc.)
Cyclohexane
as substrate of mixed
function oxidation   41
Cytochrome b$_5$
as electron sink in microsomal electron transfer
pathways   387 ff
binding to rat liver
microsomes   471
comparison of effects
of native with recon-
stituted -   477-480
effects of -, on benz-
phetamine N-demethyla-
tion   467 ff
effects of PB pretreatment
and c-AMP on b$_5$ level   123-124
effects on anti-cytochrome
b$_5$ inhibition of ethyl-
morphine-demethyla-
tion   444
incubation of liver
microsomes with cyto-
chrome b$_5$ (detergent
treated)   392-393
in reconstituted
systems   417-424
isolation of - from

rat liver
microsomes 470
participation of -, in
hydroxylation reactions
503-504 (disc.)
preparation of Mn con-
taining derivative 469
proteolytic digestion,
detergent -treatment
of - 239-240
reduction by NADH 449
role of - in NADH
synergism of P-450
monooxygenase systems
of liver microsomes 405 ff
role in NADPH-and NADH-
dependent hydroxylation
(reconstituted P-448
and P-450 systems) 447 ff
role of -, in P-450
enzyme systems 435 ff
properties of Mn-deriva-
tive of - 476
Cytochrome b$_5$ reductase
(see NADH-cytochrome b$_5$
reductase)
Cytochrome c
as an artificial
electron acceptor 30, 32
Cytochrome c reductase
(see NADH- and NADPH-
cytochrome c reductase)
Cytochrome P-420
(see P-420)
Cytochrome P-450, P-448, P$_1$-450
(see P-450)

Delta-amino-levulinic acid
(δ-ALA)
effect of c-AMP on
121, 165 (disc.)
Detergents
(Na) cholate 26
(Na) deoxycholate 31
effects on hydroxylation
activities in reconsti-
tuted systems 74 (disc.)
Emulgen 911 2, 18
Renex 690 26

Sodium dodecyl sulfate
(SDS) 28-30
Triton N-101, X-114 (and
other members of the
Triton series) 271 ff, 472
Tween 80 345
Diet,
effects of dietary pro-
teins on microsomal
P-450 and mutagenicity
of MNNG and DMN 107
Difference spectra (optical)
interpretation of significance in bovine corpus
luteum mitochondria 214-215
modifications by ligands
and reagents in P-450 from
corpus luteum
mitochondria 213 ff
Difference spectra (optical) with
isocyanides 8-13
analysis of titration
results at 430 and 455
nm in rat and rabbit
livers 213 ff
Difference spectra (optical)
with carbon monoxide
$[\text{P-450}(Fe^{2+})\cdot CO]$
8-10, 86-88
99, 79 (disc.), 175 ff
of isolated rat liver
cells 254-255
Difference spectra (optical)
with substrates 12, 16-17
with aniline 89-90
with barbiturates or
benzene 175 ff
with hexobarbital 88-90
with hexobarbital, in
relation to time course
of induction 93
7,12-dimethyl benz[a]anthracene
(DMBA)
activation to frameshift
mutagen 127 ff
as inducer 98, 127 ff
Dimethylnitrosamine (DMN)
activation to frame
shift mutagen 127 ff

activation to a
mutagen   103 ff, 109
oxidative demethyla-
tion of -   106-108, 110
photochemical
action spectra of
activation to a
mutagen   110
DNA
   base pair arrangement
   of -   129
   repair and
   carcinogenesis   169
Drug metabolism
   in isolated rat liver
   cells   251 ff
   lack of parallelism
   with ethanol oxidation
   in hepatic microsomes
   of mice   369 ff
   time course of induc-
   tion with PB and MC
                           89, 94-95

Electron acceptor
   2,6-dichlorophenol-
   indophenol (DCPIP) as -,
   in reductive titration
   of microsomal P-450
                  38, 75-78 (disc.)
Electron donor
   2,6-dichlorophenol-
   indophenol (DCPIP)
   as -, to P-450 in
   microsomal electron
   transport
          38, 73-78 (disc.),393 ff
Electron microscopy
   of endoplasmic reticulum
   after cyclic AMP
   (dibutyryl-derivative)   119
   treatment of control and
   lipid-depleted micro-
   somes   275-276
Electron paramagnetic resonance
   (EPR) spectroscopy
   Fe^{3+}(S=1/2) in redox
   studies   73 (disc.), 99
   of ISP (adrenal and P.
   putida), effects of

antibodies on   79 (disc.)
of isolated rat
liver cells
              254-255, 258-259
of P-450$_{cam}$ 307
of rabbit liver micro-
somes and partially
purified P-450   175 ff
of rat liver
microsomes   189 ff
proposed model of
redox titration
events and EPR spectra
              75-77 (disc.)
Electron transfer
   systems
   in camphor hydroxyla-
   tion by P. putida   311 ff
   of microsomal mixed
   function oxidations   81 ff
   pathways of electron
   transfer in liver
   microsomes   387 ff, 405 ff
Endoplasmic reticulum,
   hypertrophy of smooth -,
   caused by c-AMP (di-
   butyryl derivative)
   treatment   117 ff
Enzyme induction
   methods for study
   in man   485 ff
   time course of - 89, 94-95
Epoxide
   (see arene oxide)
Epoxide hydrase
              5, 18, 21, 132
              168-169 (disc.)
Ethanol
   dissociation of oxi-
   dation of -, from
   P-450 catalyzed drug
   metabolism   369 ff
              384-385 (disc.)
Ethionine (D,L-
   ethionine), effects
   on benzpyrene hydroxy-
   lase activity and P-450
   synthesis in rat liver
   microsomes   189 ff

Ethylisocyanide
   (ethylisonitrile)
   difference spectra,
   (rat liver micro-
   somes)   189 ff
   pK values associated
   with spectral changes
   at 430 and 455 nm (rat
   and rabbit liver)   13,203 ff
Ethylmorphine
   as substrate   42
   effect of D_2O on
   binding spectra in
   rat liver
   microsomes   242-245
   effects of D_2O on N-
   demethylation of -   242
Extraction
   of phospholipids from
   adrenocortical
   microsomes   272

Ferrochelatase
   possible role of Cu^{2+}
   in regulation of heme
   biosynthesis   343 ff
Flavin adenine dinucleo-
   tide (FAD)
   in NADPH-cytochrome c
   reductase   80 (disc.)
   in NADPH-P-450
   reductase   33
Flavin mononucleotide
   (FMN)
   in NADPH-cytochrome c
   reductase   80 (disc.)
   in NADPH-P-450
   reductase   33
Frameshift mutation   128-131

Genetic factors
   in strains of mice   127 ff
Gluconic acid
   urinary excretion
   of   485 ff, 493
Gluconeogenesis,
   and aminopyrine
   metabolism in per-
   fused rat liver   355 ff

Glucuronyl transferase
   (see UDP-glucuronosyl-
   transferase)

Half time ($t_{1/2}$),
   of drugs in vivo   485 ff
Heme biosynthesis
   in relation to P-450
   (scheme)   344
Heme peptide   48-52
   small - from $P-450_{cam}$
                               299-301
Heme protein P-450
   (see P-450)
Heme synthesis
   and cyclic AMP   121
   regulation of -, through
   ferrochelatase   343 ff
Hemoglobin
   as possible contaminant
   in liver microsomes
                              27, 167
   assay for synthesis
   of -   153
Hepatocytes
   aryl hydrocarbon hy-
   droxylase activity in
   reticulo-endothelial
   cells   328-330 (disc.)
n-Hexane as substrate
                                   41
Hexobarbital
   as substrate   41
   effects of NADPH on
   Type I spectral changes
   induced by -   388
   effects on EPR spectra
                              181-182
   effects on "off balance"
   absolute spectra of
   rabbit liver
   preparations   175 ff
   metabolism in perfused
   liver   383 (disc.)
High spin (S=5/2) form
   of $P-450(Fe^{3+})$
   effect of acetone   233 (disc.)
   effects of cholate
                        234-235 (disc.)

hypothetical ligands
in -
                        327, 328 (disc.)
in partially purified
preparations from rat
and rabbit liver
                        229-231 (disc.)
in rabbit liver
preparations   177 ff
in rat liver
microsomes   189 ff
species differences
(rat, rabbit,
mice)   231 (disc.)
Hydrophobicity
of the active site of
P-450   332 (disc.)
of heme environment -   52
Hydroxylation of substrates
and relation to glucu-
ronosyl transferase   335 ff
6β-hydroxycortisol
urinary excretion of -,
in man   485 ff
17-hydroxycorticosteroids
excretion of -, in man 487
20α-hydroxy-cholesterol,
interaction with bovine
corpus luteum mito-
chondria   214-215
Hypertrophy of endoplasmic
reticulum, effects of
cyclic AMP (dibutyryl
derivative)   117 ff

Immunochemistry
antibody for cytochrome
b_5, effects on P-450
related reactions in
liver and kidney
microsomes   435 ff
P-450$_{cam}$ 47 ff
effects of anti-
cytochrome b_5 on
ethylmorphine
demethylation   414 ff
of adrenal ISP   55 ff
of ovarian, testicular
and placental mito-
chondria   63-69

of putidaredoxin   289 ff
studies on adrenal
mitochondria   56-63
Inactivation of
carcinogens   103 ff
Incumbrance area
and relation to
carcinogenesis   92
Induction
difference in metabolic
profile after induction
with PB or MC   141-142
of hydroxylation
activities and UDP
glucuronosyl trans-
ferase activities in
rat liver microsomes   335 ff
of monooxygenase
activity   137
Induction of P-450
and mutagenicity   106
by PB, MC and
Arochlor 1254 (mice)   103 ff
effects of cyclic AMP
on -, by PB   117 ff
time course of -   81-87
by phenobarbital (PB),
rabbit   28, 47 ff
                        175 ff
by phenobarbital (PB),
rat   13, 81 ff, 117 ff
by 3-methylcholanthrene
(MC) rabbit   175 ff
by MC, rat   13, 81 ff
Inhibition of camphor
hydroxylation by
anti-P-450$_{cam}$ 292-295
Inhibitor studies
on bovine corpus
luteum mitochondria   213 ff
Initiation factors
exchange of -, from
rabbit and rat liver,
and protein synthesis   156
of protein synthesis   151 ff
preparation of -   152-153
Intercalation with DNA 128 ff
Interconversion of spin states
(see also spin states)
rabbit liver   175 ff

Intermediary metabolism
and mixed function
oxidation in liver 355 ff
^{125}Iodine labelling
of P-450$_{cam}$
47, 50, 294
297-298
Iron sulfur protein (ISP)
of adrenal and ovarian
mitochondria 63
of P. putida
(putidaredoxin) 287 ff
reactivity of several
different ISP(s)
78, 79 (disc.)
Isocyanide (isonitrile)
(see also difference
spectra) 8-13
Isoelectric focusing techniques for studies
of P-450$_{cam}$ - putidaredoxin complexes 292-294
Isolated rat liver cells
drug metabolism in 251 ff
isolation and prop-
erties of - 251 ff, 253
Isotope effects
with D_2O 239, 332 (disc.)

Kidney cortex microsomes
(rat)
effects of anti-cyto-
chrome b_5 on - 436-437

Laurate
as substrate, 41
ω-hydroxylation of -, in
rat liver and kidney
cortex microsomes; in-
hibition by antibody
to cytochrome b_5 435 ff
Ligand field strength 214
Ligands
modification of ab-
solute spectra of
P-450 by 213 ff
of P-450(Fe^{2+})·CO complex
327 (disc.)

Lipid depletion of adrenocortical microcomal
preparations 271 ff
Lipid peroxidation
in mouse liver micro-
somes; effects on
P-450 content and en-
zyme activity 378
in rat liver microsomes;
effects of cytochrome b_5
on 396 ff
pathway of electron
transfer in liver
microsomes 393 ff
Lipid requirement
in adrenocortical
microsomes 271 ff
in reconstituted
systems 14, 25
Liver microsomes
human, surgical
biopsy 105, 111-112
mouse strains 103 ff, 127 ff,
369 ff
rabbit 26 ff, 47 ff
175 ff
rat 1 ff, 81 ff, 117 ff

Low spin (S=1/2) form of
P-450(Fe^{3+}) 73-74
rabbit liver 178-182
rat liver 189 ff
Metabolic control
in liver endoplasmic
reticulum 355 ff
of P-450$_{cam}$ 311 ff
Metabolites
differences in profile
of - of hydroxylation
reactions 141 ff
3-Methylcholanthrene (MC)
activation as carcinogen
in vivo 127 ff, 142-144
aryl hydrocarbon hy-
droxylase induced
by -, in reticuloendo-
thelial cells of liver
328-329 (disc.)

as inducer   127 ff
as inducer (rabbit
liver)   175 ff
as mutagen   128
binding of -, difference
spectra   16
binding to rat liver
microsomes   193-195
effects of induction
by -, in rat liver
microsomes   82 ff
effects of induction
by -, on isocyanide
difference spectra   13
effects of -, on initia-
tion factor of protein
synthesis and trans-
lation   151 ff
possibility of covalent
binding to rabbit liver
microsomes   181
spin state of P-450
(Fe^{3+}) after induction
with -   189 ff
Methyl-methane-sulfonate
as mutagenic agent
                                    170 (disc.)
N-Methyl-N'-nitro-N-nitroso-
guanidine (MNNG)
as primary carcinogen   103 ff
inactivation as a
mutagen by mouse
liver microsomes   103 ff
inactivation by
microsomes from human
liver   112
Metopirone
(see also metyrapone)
as inhibitor of drug
metabolism in isolated
rat liver cells   260 ff
inhibition of UDP glucu-
ronosyl transferase
activity in rat liver
microsomes   338
optical difference
spectra   91, 93, 99
Microsomes
from bovine adrenal
cortex   271 ff

from human liver,
activation of DMN by -   111
from human liver, in-
activation of MNNG by -   112
from mouse liver
                                    127 ff, 369 ff
from rabbit liver   175 ff
from rat kidney cortex   435 ff
from rat liver   435 ff
Mitochondria
from adrenal cortex
                                    56 ff, 63-68
from bovine corpus
luteum   63-68
from placenta   63-68
from testis   63-68
Mixed function oxidase systems
and intermediary metabolism
in liver   355 ff
and role of hepatic
cytochrome b_5 in   405 ff
Mode of interaction
of $P-450_{cam.}$ with
putidaredoxin   284 ff
Model systems
for $P-450 \cdot CO$
complexes   325-327 (disc.)
Molecular weight
of liver microsomal
P-450   48
of NADPH-P-450 reductase
preparation   33
of P-450 and P-448
preparations   28-30
Mononuclear heme
in model compounds for
$P-450(Fe^{2+}) \cdot CO$
                                    325-327 (disc.)
Morphology
of microsomes of adrenal
cortex and phospholipid
composition   272-276
Mutagen(s)
DMN as -, after metabolic
activation   105 ff
MC as -, and relation
to P_1-450 137 ff
metabolic activation
of -   103 ff

microbiological assay
of -   105, 128 ff, 134
MNNG as -   105

NADH
and cytochrome b_5
dependent system
which hydroxylates 3,4-
benzpyrene   419
NADH-cytochrome b_5 reductase
in microsomes   408
preparation of   448
effects of cytochrome
b_5 and its Mn derivative
on -,   474
effects of -, on benz-
phetamine-N-demethyla-
tion   454
NADH-oxidation
effects of anti-cyto-
chrome b_5 on -   441

NADH-P-450-reductase
effect of anti-cyto-
chrome b_5 on -   440
NADH synergism
of NADPH-dependent
P-450-monooxygenase
systems in liver
microsomes   405 ff
NADPH
effect on Type I
spectral change with
hexobarbital   388
NADPH-cytochrome c reductase
effects of cytochrome
b_5 on -   474
effects of induction
on -   171 (disc.)
effects of PB and cyclic
AMP treatment on -   118 ff
FAD and FMN as pros-
thetic group of -   80
inhibition of -, by anti-
bodies to adrenal ISP   56-62
of placental mito-
chondria   66-69
of rat adrenal and
ovarian mitochondria,

interaction with anti-
bodies against adrenal
ISP   64-68
requirement in recon-
stituted systems   14 ff
solubilization and
purification of -   30-33
NADPH generation
interaction of gluconeo-
genesis with mixed func-
tion oxidation   355 ff
scheme of extramito-
chondrial -   357
NADPH oxidation
effects of anti-cyto-
chrome b_5 on -,   439
effects of anti-P-450$_{cam}$,
on substrate dependent
- of   51
NADPH-P-450 reductase
(see also NADPH-cytochrome c
reductase)
effects of anti-cyto-
chrome b_5 on -   440
effects of cytochrome b_5
on -   474
effects of D_2O on -   242
pretreatment of animals,
and -   321-322
solubilization and puri-
fication of -   14-15, 30-33
NADH and NADPH peroxidase
activities,
effects of cytochrome
b_5 on -   459
Naphthalene
hydroxylation of -;
metabolites   17-19, 21
N-ethyl-maleimide (NEM)
as titrant for SH groups
in P-450$_{cam}$ 301
Non-heme bound iron
(see iron sulfur proteins)
Non-ionic detergents
(see detergents)

Octane   41
Optical absorption spectra
(see Absorption spectra,
optical; Difference spectra)

Organic solvents
  effects of - on binding
  spectra of aniline to
  rat liver microsomes   246
  effects of isooctane
  treatment on rat liver
  microsomes   11
Overlap integral
  of hydrocarbons, and
  relation to
  carcinogenicity   92
Oxidation-reduction
  potential
  of P-450   95-100
Oxidation state
  of components of mixed
  function oxidase
  systems   73-78
Oxygen
  and the mechanism of
  hydroxylation reactions   99
  effects on reducibility
  of P-450   86, 96-100
                          171-172 (disc.)
Oxygen uptake
  in P-450$_{cam}$ system   317
  in perfused rat liver,
  stimulation by
  aminopyrine   363

P-420
  as impurity in partially
  purified P-450   30
  reaction with ethyliso-
  cyanide   204
P-450
  anaerobic redox titration
  of -   33-38
  autoxidation of -   34-35
  basis for differentia-
  tion between multiple
  forms of -
            8-9, 86-88, 141-142
  EPR spectra of rabbit
  liver microsomes and
  solubilized
  preparations   180-182
  from human liver   112
  from rabbit liver,

  partially purified   175 ff
  from rat liver
  microsomes   81 ff
  involvement in
  mutagenesis   106-107
  levels of - after in-
  duction with PB
  or 3-MC   81-86
  mice,hepatic microsomes,
  content and enzyme
  activities in   377-378
  mitochondrial - from
  endocrine glands   213 ff
  - of kidney cortex   416
  P$_1$-450 from liver
  microsomes   127 ff
  partially purified -
  from rabbit liver
  microsomes   25 ff
  properties of partially
  purified preparations   5-13
                             28-31
  reconstituted systems
  with P-450 and P-448   447 ff
  separation of multiple
  forms (P-448 and
  P-450)   3-5
P-450$_{cam}$ (from P. putida)
                              47 ff
                             287 ff
  metabolic control of -   311 ff
P-450(Fe^{2+})
  interaction of - with
  metopirone or SKF-525A   91-99
P-450(Fe^{2+})·CO
  (see difference spectra)
P-450(Fe^{3+})
  interaction of -, with
  metopirone or SKF-525A   91-99
Pathways of electron transfer
  in liver microsomes   387 ff
                             405 ff
Perfusion studies
  in rat liver   355 ff
                             233 (disc.)
pH effect
  on binding of ethyl
  isocyanide to liver
  microsomes   203 ff

on ethylmorphine binding
spectra 247
on NADPH oxidation and
ethylmorphine metabolism
in rat liver microsomes 247
Phenobarbital (PB)
effects of induction on
isocyanide difference
spectra 13
effects on mutagenicity
of DMN and MNNG 105 ff
induction effects in
presence of cyclic AMP
(dibutyryl derivative) 117 ff
induction effects in
rabbit liver 175 ff
induction effects in
rat liver microsomes 82 ff
metabolism of - (half
times) 485 ff
Philadelphia, as Home-
town for P-450 v
Phosphatidyl-choline
in adrenocortical
microsomes 277
incorporation of
^{14}C into - 7
restoration of
benzo[a]pyrene hydroxy-
lase activity by - 75 (disc.)
Phospholipids 11-14
25, 31, 39
composition in
adrenal microsomes 273
effects of cyclic AMP
(dibutyryl derivative)
on - 118 ff
extraction of -, from
adrenocortical
microsomes 271 ff
possible role of -, in
substrate binding and
enzyme activities
331 (disc.)
Photochemical action spectra
abnormal bands in 420 nm
region
165-167 (disc.)
effects of induction

by PB and MC (rat
liver microsomes) 81 ff
89-96
photochemical activa-
tion of oxidative de-
methylation of DMN 110
Polychlorinated biphenyls
as inducing agents 104
Polycyclic hydrocarbons
(see also aryl
hydrocarbons) 91-92
97-98
Polyethylene glycol
use in purification
of P-450 28
Pregnenolone
as feedback inhibitors
of side chain cleavage
of cholesterol 219
synthesis in mitochondria
from corpus luteum
213 ff
Preparation
(see also individual
enzymes)
of isolated rat liver
cells 251-253
of NADPH-cytochrome c-
reductase 30
of partially purified
P-450 3 ff, 26-30
Protein synthesis
and induction of
P-450 236 (disc.)
Pseudomonos putida
(see P-450$_{cam}$;
putidaredoxin)
Putidaredoxin
(see also iron sulfur
proteins, ISP)
formation of a func-
tional trimer by
crosslinking 291
Radioimmunassay
^{150}I-labelled P-450$_{cam}$
47, 50-51, 295

Reaction mechanism
  and redox behavior   95-100
  of camphor-hydroxy-
  lation in P. putida   312-315
  of hydroxylation
  reactions   38-42
Reconstituted systems
  properties of   14-22, 35-42
                 447 ff
Reconstitution
  of mixed function
  oxidase systems   15 ff
                 25 ff
  role of cytochrome
  b_5 in reconstituted
  systems   417 ff
                 447 ff
Redox potential
  (see oxidation-
  reduction potential)
Reducibility
  of cytochrome b_5 by
  NADH and NADPH   387 ff
  of P-450 by NADPH and
  $Na_2S_2O_4$ 84-86
Reduction of P-450
  -and lipid depletion,
  in adrenocortical
  microsomes   275 ff
  effect of aminoglute-
  thimide on rate of -   223
Reduction state
  of NADPH, effects on -
  in isolated rat liver
  cells   264
Resolution
  attempt at -, of
  steroid C_{21} hydroxylase (without deter-
  gent)   278 ff
Responsiveness in strains
  of mice   127 ff
Reverse Type I spectral
  changes (RI)   177 ff
  loss of -, with hexo-
  barbital, after
  butanol extraction   233
  with 20α-hydroxychol-
  esterol in bovine corpus

luteum mitochondria   214-215
  with barbiturates and
  benzene in rabbit liver
  microsomes and
  partially purified P-450
  preparations   177 ff
              230 (disc.)
  with hexobarbital in MC
  induced microsomes,
  relation to high spin
  form of P-450
           229-230 (disc.)
  with hexobarbital in
  MC - induced
  microsomes   89-90
RNA synthesis
  and induction of P-450   236
Rotenone
  effect of -, on al-
  prenolol metabolism   263

Salmonella typhimurium
  use of -, for detection
  of mutagens   128 ff
  use of strain TA 1535
  in bacterial assay for
  mutageniticty   105 ff
Sex specificity
  of microsomal hydroxylation and glucuronosyl
  transferase activity in
  rat liver microsomes   335 ff
SKF-525A [2-diethylaminoethyl-
  2,2-diphenyl-valerate·HCl]
  as inhibitor of drug
  metabolism in isolated
  rat liver cells   260 ff
  as inhibitor of micro-
  somal UDP glucuronosyl-
  transferase activity   340
  optical difference
  spectra   91-93, 99
Sodium dithionite ($Na_2S_2O_4$)
  as electron donor
           33-39, 75-78
Sodium dodecylsulfate (SDS)
  polyacrylamide gel
  electrophoresis, use
  for molecular weight
  determination   28-30, 48

Solubilization
  of microsomal rat liver and
  UDP glucuronosyl transferase
  activity   335 ff
  of NADPH-P-450-
  reductase   30-33
  of P-450   2-3, 26
Spectral changes
  (see difference spectra)
  correlation between - and
  enzyme activity   216
  with hexobarbital
                    11-14, 88-89, 90
Spin state
  in rat liver microsomes
  after MC induction   189 ff
  lack of correlation
  of -,with enzyme
  activity   189 ff
  of P-450(Fe^{3+}), in
  partially purified
  preparations from
  rabbit liver
  microsomes   176 ff
Steady state
  analysis of reaction
  cycle of camphor
  hydroxylation
  (preparation from P.
  putida)   316 ff
  effects of substrate
  on -, of reduction of
  cytochrome b_5 387 ff
Stearyl-CoA-desaturase (NADH-
  supported)
  activation and in-
  hibition of -, in
  liver microsomes   409 ff
  -and role of cytochrome
  b_5 in liver microsomal
  electron transport
                    393 ff, 405 ff
  relation to microsomal
  P-450 systems   506 (disc.)
Steroid 11β-hydroxylase
  inhibition of deoxy-
  corticosterone
  metabolism in
  adrenocortical

mitochondria by anti-
  bodies to adrenal
  ISP   60-63
Steroid C_{21}-hydroxylase
  in bovine adrenocortical
  microsomes   271 ff
Steroid hydroxylation
  partial restoration by
  phospholipids in
  lipid depleted adrenal
  microsomes   274
  reactivation by non-
  ionic detergents   274
Steroidogenesis   55 ff
  inhibition by
  "cyanoketone"   222
Stoichiometry
  of mixed function oxidase
  reactions (coupled and
  uncoupled   424
Structure
  of P-450$_{cam}$ 287 ff
  of putidaredoxin   288
Substrates
  for mixed function oxidase
  systems
  (see individual listings)
Substrate binding
  (see also difference spectra)
  to isolated rat liver
  cells   252 ff
  to lipid-depleted adrenocortical microsomes   278
  to liver microsomes   89, 191
  to partially purified
  P-450 preparations
                        11-17, 191
Substrate hydroxylation
  effects of anti-cyto-
  chrome b_5 on - (rat
  liver microsomes)   443
  reaction mechanism of -   40
Substrate specificity
  (see also individual
  substrates)   35-38
Sulfhydryl groups
  titration of -, in P-450$_{cam}$
            301-305, 325-327 (disc.)

Superoxide
  (and superoxide anion)   39
Synergistic effects
  and role of cytochrome
  b_5 in mixed function
  reactions of liver
  microsomes   405 ff
  of NADH and NADPH in
  mixed function
  oxidase systems   387
  of NADH on NADPH-de-
  pendent benzphetamine
  N-demethylation   455
  of NADH on NADPH-mediated
  ethylmorphine
  metabolism   435 ff

Testosterone
  NADPH-dependent hy-
  droxylation of -, (at
  6β, 7α, 16α positions),
  effects of cytochrome
  b_5 462
Time course of induction
  (see induction of
  P-450)
Titration
  reductive -, of P-450 with
  $Na_2S_2O_4$
              33-39, 73-78 (disc.)
Translation
  increased, after MC ad-
  ministration   151 ff
1,1,1-trichloropropene-2-3-
oxide (TCPO)
  as epoxide hydrase
  inhibitor   170
Trypan blue test
  for viability of
  isolated
  liver cells 329-330 (disc.)
Tumorigenesis
  (see also carcino-
  gens), by MC   127 ff
Turnover number
  for substrate hy-
  droxylation in
  liver microsomes   41
Type I spectral change

(see also difference
spectra)
and relation to NADH
synergism   425-436
effects of PB induction
on -   256
effect of pH on -, with
ethylmorphine   248
with cholesterol in
bovine corpus luteum
mitochondria   214-215
with hexobarbital   89-90
with hexobarbital in
perfused liver   233 (disc.)
Type II spectral change
(see also difference
spectra)
in bovine corpus luteum
mitochondria   214-216
with aniline   12, 89

UDP-glucuronosyl transferase
(E.C.2.4.1.17)
  and microsomal
  hydroxylation
              375 ff, 383 (disc.)
"Uncoupled" monooxygenase
  reactions   424

Water
  incorporation from ^3H_2O
  into lipids (rat
  liver)   360
  interaction with
  microsomal P-450   239 ff